Geology and Geophysics of an Arc-Continent Collision, Taiwan

Edited by

Timothy B. Byrne
Department of Geology and Geophysics, U-2045,
University of Connecticut
354 Mansfield Road
Storrs, Connecticut 06269-2045
USA

and

Char-Shine Liu
Institute of Oceanography
National Taiwan University
P.O. Box 23-13
Taipei, Taiwan 106
Republic of China

THE
GEOLOGICAL
SOCIETY
OF AMERICA

Special Paper 358

3300 Penrose Place, P.O. Box 9140 • Boulder, Colorado 80301-9140, USA

2002

Published by The Geological Society of America, Inc.
3300 Penrose Place, P.O. Box 9140, Boulder, Colorado 80301
www.geosociety.org

Printed in U.S.A.

GSA Books Science Editor Abhijit Basu
Cover design by Margo Good

Library of Congress Cataloging-in-Publication Data

Geology and geophysics of an arc-continent collision, Taiwain / edited by Timothy Byrne and Char-Shine Liu.
p. cm. — (Special paper ; 358)
Includes bibliographical references.
ISBN 0-8137-2358-2
1. Geology, Structural—Taiwan. 2. Subduction zones—Taiwan. 3. Plate tectonics—Taiwan. 4. Island arcs—Taiwan. I. Byrne, Timothy Briggs. II. Liu, Char-Shine, 1952- III. Special papers (Geological Society of America) ; 358.

QE634.T28 G46 2002
555.124'9—dc21 2002020337

Cover: Left: Weathered and fractured sandstone of probable Miocene age exposed along the southeastern coast of the Coastal Range (the Juchiangchiao Recreation Area). In this area, this sandstone unit is overturned and considered to be a large exotic block embedded in the Lichi mélange. Photo by Jacques Malavieille. Upper right: Track and soccer field behind the Kuang-Fu Middle School that was deformed and uplifted 2 m by the Chi-Chi earthquake ($M_L = 7.3$) on September 21, 1999. Lower right: Precipitous cliffs along the Ilan to Hualien highway where the Central Range trends east and probably overrides the Philippine Sea Plate. Photos by Timothy Byrne.

10 9 8 7 6 5 4 3 2 1

Contents

Geological Society of America
Special Paper 358
2002

Preface: Introduction to the geology and geophysics of Taiwan

Timothy B. Byrne
Department of Geology and Geophysics, University of Connecticut, 354 Mansfield Road, Storrs, Connecticut 06269-2045, USA
Char-Shine Liu
Institute of Oceanography, National Taiwan University, P.O. Box 23-13, Taipei, Taiwan 106, Republic of China

INTRODUCTION

Convergent plate boundaries represent sites of the world's greatest earthquakes and tsunamis and are often marked by substantial topographic or bathymetric anomalies. The island of Taiwan straddles the boundary between the converging Eurasian (mainland China) and Philippine Sea plates and is the site of thousands of earthquakes each year (Wang and Shin, 1998). Nearly a dozen major ($M > 7.0$) earthquakes have occurred this century alone (Huang et al., 1999), taking the lives of thousands of Taiwan's citizens and causing billions of dollars of damage. Clearly, understanding these systems has societal relevance. In addition, the seismogenic zone lies at the heart of the subduction process in which both oceanic and continental lithospheric plates are carried into the mantle. Studies of active convergent plate boundaries, like the boundary exposed in Taiwan, are therefore vital to understanding both the earthquake process as well as the growth of orogenic belts around the world.

The Taiwan arc-continent collision arguably represents one of the best-studied active orogenic belts in the world and is an ideal locality to investigate orogenic processes. The collision zone is reasonably accessible, has a relatively well understood tectonic setting, and is so active that the zone has some of the highest uplift and erosion rates on record. In addition, in the 26 years since publication of the most recent geologic map of Taiwan (Ho, 1974, 1975), the island of Taiwan has been extensively studied by a substantial number of national and international scientists, and the orogenic belt is often referred to as a classic example of an arc-continent collision or accretionary prism (e.g., Davis and Reynolds, 1996; Suppe, 1985; Twiss and Moores, 1992). C.S. Ho's geologic compilation (Ho, 1974) and explanatory text (Ho, 1975) came on the heels of the plate-tectonic revolution in the earth sciences and provided a modern framework for reinterpreting the geologic history of Taiwan and understanding arc-continent collision processes. For example, Suppe and his students relied extensively on this new compilation in reinterpreting and remapping the fold-and-thrust belt of western Taiwan (Suppe, 1976, 1980a, 1980b), in developing the concepts of fault-bend and fault-propagation folding (Suppe, 1983; Suppe and Medwedeff, 1984), and in applying the concept of a critically tapered accretionary wedge (Suppe, 1981) first suggested by Chapple (1978). The structural geology community now commonly refers to these concepts and models of orogenic processes that were developed primarily from observations of rocks exposed in Taiwan. These concepts and models continue to be evaluated and analyzed, both in Taiwan and elsewhere around the world.

Researchers have continued to build on the early fundamental contributions, including the papers in this volume, which address significant problems in orogenesis. A common theme of the papers is the authors' desire to document the three-dimensional architecture of the orogen and to understand its spatial and temporal evolution by using a variety of techniques. As a result, the papers provide new insights on the evolution of the orogenic belt and lay the groundwork for future studies. This preface serves to highlight previous research that has laid a framework for the papers that follow as well as controversial aspects of the geology and geophysics of Taiwan.

REGIONAL TECTONIC AND GEOLOGIC SETTING

The island of Taiwan is forming as the Luzon arc of the Philippine Sea plate obliquely collides with the Eurasian plate along the passive margin of southeast China. On the basis of data in Seno et al. (1993) and Yu et al. (1997), collision appears to have been initiated at 8–6 Ma and, as a result of the obliquity, is propagating southwestward at ~5.5 cm/yr to

Byrne, T.B, and Liu, C.-S., 2002, Preface: Introduction to the geology and geophysics of Taiwan, *in* Byrne, T.B., and Liu, C.-S., eds., Geology and Geophysics of an Arc-Continent Collision, Taiwan: Boulder, Colorado, Geological Society of America Special Paper 358, p. v–viii.

eventually close the South China Sea (Fig. 1). As a result of the collision, both the arc (on the western edge of the Philippine Sea plate) and the Chinese continental margin have been deformed and shortened. Paleomagnetic data also indicate that the arc rocks have progressively rotated clockwise as a result of the collision (Lee et al., 1991); rotations of the opposite sense may also be present in the continental-margin rocks, but this possibility remains untested. The southward propagation of the collision also results in time-space equivalence parallel to the structural grain of the island; the oldest parts of the collision-related rocks are exposed in the north and the youngest in the south. This time-space equivalence combined with the high uplift and erosion rates provides a unique opportunity to study the history and architecture of an actively evolving mountain belt. Much of the previous research in Taiwan has taken advantage of this opportunity.

The collision has also resulted in growth of an accretionary prism, which forms the western three-fourths of the island and is composed of uplifted and exhumed slices of the accreted Chinese passive margin. Like many accretionary prisms, the metamorphic grade and structural complexity of the exposed rock increase across the prism, from unmetamorphosed and mildly deformed rocks near the toe of the prism to highly metamorphosed and deformed rocks in the thick parts of the prism. In Taiwan, geologic studies have documented three geologic provinces that record this progressive change: (1) a fold-and-thrust belt in the west, (2) a slate belt in the middle, and (3) a belt of pre-Tertiary metamorphic basement in the east. Although the accretion and exhumation of these belts of rocks are consistent with a single, critically tapered wedge (e.g., Davis et al., 1983; Suppe, 1981), more recent interpretations have also been proposed that involve flow of the lower crust (Wu et al., 1997) or double (back-to-back) wedges with opposing vergence (Koons, 1990; Willet et al., 1993).

Recent detailed mapping of the offshore regions north and south of the island of Taiwan (Fig. 1) suggests, with some modification, continuation of many of the structural and geologic patterns recognized on land (Hsu et al., 1998; Huang et al., 1992; Liu et al., 1997, 1998). In the region south of Taiwan, subduction of oceanic rather than continental lithosphere results in a more subdued, but wider, accretionary prism (Reed et al., 1992) and an active, rather than inactive, magmatic arc. The transition from subduction to collision in southern Taiwan and the degree to which continental crust is involved are important areas of current research and are discussed in several papers in this volume. North and northeast of Taiwan, the collision zone appears to have subsided beneath sea level owing, at least in part, to subduction of the Philippine Sea plate along the Ryukyu Trench and to backarc spreading in the Okinawa Trough. The transition from subduction of the Philippine Sea plate in the Ryukyu Islands to its collision in Taiwan is also an important area of current research (e.g., Kao et al., 1998; Lallemand et al., 1997, 1999).

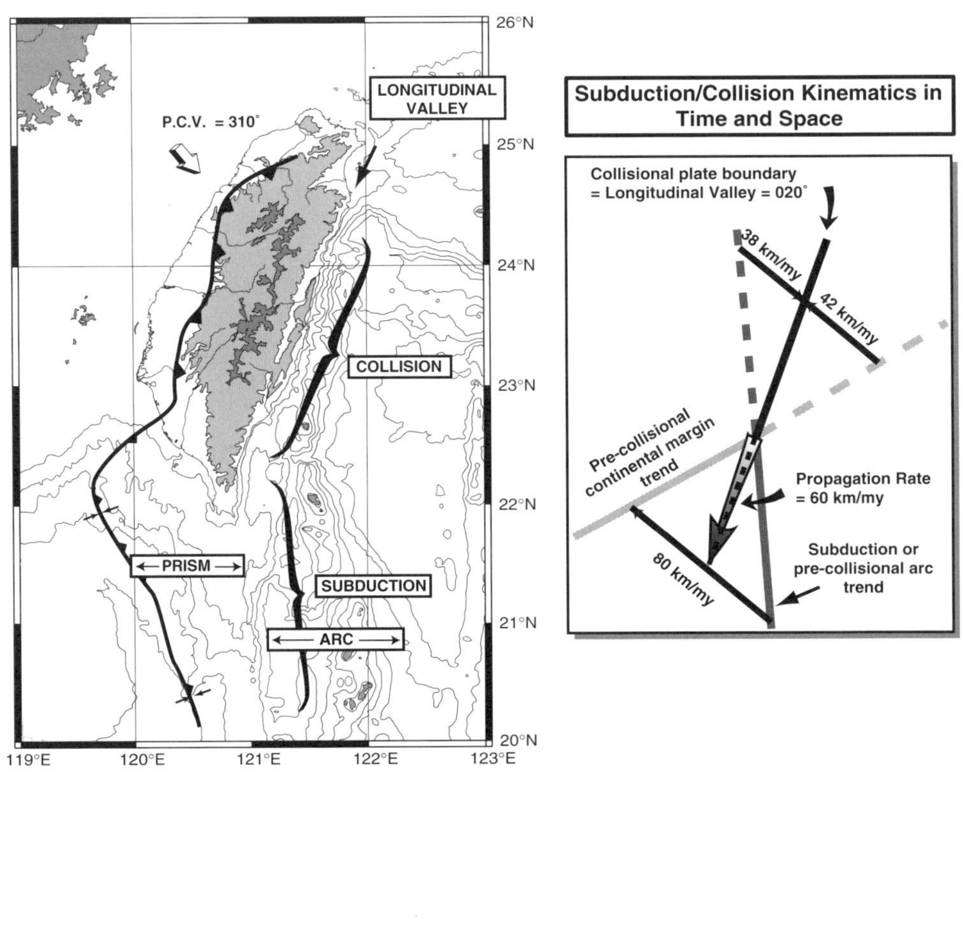

Figure 1. Bathymetry and topography of the area around Taiwan and kinematic model for the Taiwan arc-continent collision. The kinematic model indicates a propagation rate of 60 km/m.y. and nearly equal shortening rates (~40 km/m.y.) in the arc and accreted continental margin.

ORGANIZATION OF THE VOLUME

The papers that follow are organized around five research themes and are followed by Malavieille et al.'s synthesis of new observations and an interpretation of the tectonic evolution of the subduction-collision zone in southern Taiwan.

1. In the first three papers, Chou and Yu, Lee et al., and Mouthereau et al. discuss the role of extensional structures before and during thrusting in the fold-and-thrust belt.
2. Hsieh and Knuepfer consider the relationships between tectonically driven uplift and river terraces. Hickman et al. deal with fold-and-thrust belt development through the use of GPS (Global Positioning System) data.
3. Fisher et al. and Pulver et al. offer interpretations of deformation patterns and strain compatibility in the exposed, internal parts of the orogen.
4. Song and Ma, in one paper, and Wang and Hung, in a second, analyze the geothermal structure and the role of synorogenic extension in the internal parts of the orogen.
5. Finally, the papers by Hu et al., Cheng et al., Tang et al., and Malavieille et al. offer integrations of geologic and geophysical data sets with finite-element models to understand the three-dimensional evolution of the orogen, including the subduction-collision transition.

DIRECTIONS OF FUTURE RESEARCH

Within the research themes covered in this volume, a number of the papers address more specific research topics that will probably be the focus of more detailed studies in the future. In no particular order, possible areas of future research include the following.

Origin of the longitudinal valley in eastern Taiwan

This valley marks the "suture" between rocks of different affinities (Philippine Sea plate to the east and China passive margin to the west), yet it is currently at or near sea level for its entire length. Although the valley's eastern boundary against the Coastal Range is a generally well-mapped, top-to-the-west thrust fault, the western boundary with the Central Range is much less well understood and may also be a fault between the valley and the pre-Tertiary metamorphic rocks. Future research should focus on understanding the kinematic history of this apparent fault and specifically address whether the fault is dominated by normal-, thrust-, or strike-slip motion.

Tectonic denudation and exhumation

The recent recognition of the importance of extension and normal faulting through both field- and GPS-based geologic studies in central Taiwan raises the possibility that tectonic denudation and exhumation may have played a more important role than previously thought. The role of tectonic denudation could be addressed through a combination of structural, geodetic, geochronologic, and modeling studies.

Involvement of continental crust in the collision

The degree that continental crust is involved in the collision has also recently been questioned. Both field studies and geophysical data suggest a component of lateral flow, but the degree of flow and the relationship of the flow to the collision are unknown. Some authors have also argued that the crust is passively subducted into the mantle or imbricated and vertically or laterally extruded. All of these possibilities need to be investigated further.

Strain rates

Understanding the relationship between short-term (e.g., years or decades) and long-term (e.g., hundreds of thousands to millions of years) horizontal and vertical strain rates has been highlighted recently by the abundance of high-quality GPS data. Although some of the GPS results are consistent with geologic (and therefore generally long-term) structures, other data are inconsistent and suggest significant changes over time. Available surface uplift data (e.g., geodetic, GPS, and river terraces) also suggest a heterogeneous pattern of deformation. Future studies need to continue to integrate long- and short-term data sets with geologic structures and patterns so that the most complete deformation picture can be constructed.

THE MOTIVATION FOR GATHERING THESE PAPERS INTO A VOLUME

The success of both the 1997 Tectonics of East Asia (TEA) Conference in Chungli, Taiwan, and the associated field trip has provided the motivation for producing this volume. Since 1988, a series of international meetings have been held on a nearly biannual basis: Princeton, New Jersey, and Paris, France (1988); Taipei, Taiwan (TAICRUST Workshop, 1991); Taipei, Taiwan (ACT Symposium, 1995); Chungli, Taiwan (TEA Conference, 1997); and Montpellier, France (SEASIA Conference, 1999). Many of these conferences resulted in the publication of proceedings volumes. In keeping with this tradition, this volume contains a collection of edited, refereed papers resulting, in part, from the TEA Conference. The compilation brings together relevant components of the geology and geophysics of Taiwan with the overall goal of better understanding the evolution of arc-continent collisions and subduction processes in general.

ACKNOWLEDGMENTS

This volume is a contribution of the ongoing collaboration among U.S. and Taiwan scientists in studying the arc-continent collision in Taiwan. One of us (Byrne) and Louis Teng of the National Taiwan University have coordinated the cooperative research projects for the past several years. As one of the organizers of the TEA Conference, which stimulated interest in developing this Geological Society of America Special Paper, Byrne thanks Louis Teng for his help, encouragement, and leadership in organizing the conference. Both of us thank the nearly 120 participants who made the conference a success. Many of the participants also attended a seven-day preconference field trip led ably and graciously by Chu-Hao Hsu of the Central Geological Survey. Although this collection of papers has evolved since the conference, it still reflects the wide range of topics presented at the conference and discussed on the seven-day field trip that preceded the conference.

Financial backing for the conference and field trip came from the International Division of the National Science Foundation (Grant INT-9724732), which supported travel and logistical costs for 15 American scientists, and from the National Science Council in the Republic of China. The Earth Sciences Division of the National Science Foundation has also supported research conducted by Byrne during the past several years (through grants EAR-9706801 and EAR-9903248). Substantial support was also provided by three academic institutions: National Taiwan University, National Central University, and the University of Connecticut. A number of agencies and institutes in Taiwan also contributed to the success of the conference. These include the Ministry of Education, the Central Geological Survey, the Geological Society in Taipei, the Chinese Geophysical Society, National Cheng-Kong University, National Sun Yat-Sen University, National Taiwan Normal University, and the Institute of Earth Sciences in the Academia Sinica.

We sincerely thank Abhijit Basu (GSA Books Editor) for his extreme patience in the early development of this volume. We also thank Ray Joesten and Lanbo Liu for critically reading this introduction. A number of friends and colleagues graciously helped edit and review the manuscripts at various stages and their dedication and services are sincerely appreciated.

REFERENCES CITED

Chapple, W.M., 1978, Mechanics of thin-skinned fold-and-thrust belts: Geological Society of American Bulletin, v. 89, p. 1189–1198.

Davis, D., Suppe, J., and Dahlen, F.A., 1983, Mechanics of fold-and-thrust belts and accretionary wedges: Journal of Geophysical Research, v. 88, p. 1153–1172.

Davis, G.H., and Reynolds, S.J., 1996, Structural Geology: New York, John Wiley and Sons, 776 p.

Ho, C.S., 1974, Geologic map of Taiwan: Central Geological Survey, Ministry of Economic Affairs, scale 1:250 000, 1 sheet.

Ho, C.S., 1975, An introduction to the geology of Taiwan and explanatory text of the geologic map of Taiwan: Taipei, Taiwan, Republic of China, Central Geological Survey, Ministry of Economic Affairs, 153 p.

Hsu, S.K., Liu, C.-S., Shyu, C.T., Liu, S.Y., Sibuet, J.C., Lallemand, S., Wang, C., and Reed, D., 1998, New gravity and magnetic anomaly maps in the Taiwan-Luzon region and their preliminary interpretation: TAO, Terrestrial, Atmospheric and Oceanic Sciences, v. 9, no. 3, p. 509–532.

Huang, C.-Y., Lin, C.-W., Chen, W.-S., Chen, Y.-G., Yu, S.-B., Chia, I.-P., Lu, M.-D., Hou, C.-S., and Wang, Y.-S., 1999, Seismic geology of the Chi-Chi Earthquake, *in* International workshop on Chi-Chi, Taiwan, Earthquake of September 21, 1999: Taipei, Taiwan, National Science Council Report, p. 3-1 to 3-8.

Huang, C.Y., Shyu, C.T., Lin, S.B., Lee, T.Q., and Sheu, D.D., 1992, Marine geology in the arc-continent collision zone off southeastern Taiwan: Implications for the late Neogene evolution of the Coastal Range: Marine Geology, v. 107, p. 183–212.

Kao, H., Shen, S.-S.J., and Ma, K.-F., 1998, Transition from oblique subduction to collision: Earthquakes in the southernmost Ryukyu arc-Taiwan region: Journal of Geophysical Research, v. 103, no. 4, p. 7211–7229.

Koons, P.O., 1990, The two-sided orogen: Collision and erosion from the sandbox to the Southern Alps, New Zealand: Geology, v. 18, p. 679–682.

Lallemand, S., Liu, C.-S., Domingues, S., Schnurle, P., Malavieille, J., and the ACT Scientific Crew, 1999, Trench-parallel stretching and folding of forearc basins and lateral migration of the accretionary wedge in the southern Ryukyus: A case of strain partitioning caused by oblique convergence: Tectonics, v. 18, p. 231–247.

Lallemand, S.E., Liu, C.-S., and Font, Y., 1997, A tear fault boundary between the Taiwan orogen and the Ryukyu subduction zone: Tectonophysics, v. 274, p. 171–190.

Lee, T.-Q., Kissel, C., Barrier, E., Laj, C., and Chi, W.-R., 1991, Paleomagnetic evidence for a diachronic clockwise rotation of the Coastal Range, eastern Taiwan: Earth and Planetary Science Letters, v. 104, p. 245–257.

Liu, C.-S., Huang, I.L., and Teng, L.S., 1997, Structural features off southwestern Taiwan: Marine Geology, v. 137, p. 305–319.

Liu, C.-S., Liu, S.Y., Lallemend, S.E., Lundberg, N., and Reed, D., 1998, Digital elevation model offshore Taiwan and its tectonic implications: TAO, Terrestrial Atmospheric and Oceanic Sciences, v. 9, p. 705–738.

Reed, D.L., Lundberg, N., Liu, C.S., and Kuo, B.Y., 1992, Structural relations along the margins of the offshore Taiwan accretionary wedge: Acta Geologica Taiwanica, v. 30, p. 105–122.

Seno, T., Stein, S., and Gripp, A.E., 1993, A model for the motion of the Philippine Sea plate consistent with NUVEL-1 and geologic data: Journal Geophysical Research, v. 98, p. 17941–17948.

Suppe, J., 1976, Decollement folding in southwestern Taiwan: Petroleum Geology of Taiwan, v. 13, p. 25–35.

Suppe, J., 1980a, Imbricate structure of western foothills belt, south-central Taiwan: Petroleum Geology of Taiwan, v. 17, p. 1–16.

Suppe, J., 1980b, A retrodeformable cross section of northern Taiwan: Proceedings of the Geological Society of China, v. 23, p. 46–55.

Suppe, J., 1981, Mechanics of mountain building and metamorphism in Taiwan: Memoir of the Geological Society of China, v. 4, p. 67–89.

Suppe, J., 1983, Geometry and kinematics of fault-bend folding: American Journal of Science, v. 283, p. 684–721.

Suppe, J., 1985, Principles of structural geology: Englewood Cliffs, New Jersey, Prentice-Hall, 537 p.

Suppe, J., and Medwedeff, D.A., 1984, Fault-propagation folding: Geological Society of America Abstracts with Programs, v. 16, p. 670.

Twiss, R.J., and Moores, E.M., 1992, Structural geology: New York, W.H. Freeman and Company, 532 p.

Wang, C.-Y., and Shin, T.-C., 1998, Illustrating 100 years of Taiwan seismicity: TAO, Terrestrial, Atmospheric and Oceanic Sciences, v. 9, no. 4, p. 589–614.

Willet, S., Beaumont, C., and Fullsack, P., 1993, Mechanical model for the tectonics of doubly vergent compressional orogens: Geology, v. 21, no. 4, p. 371–374.

Wu, F., Rau, R.-J., and Salzberg, D., 1997, Taiwan orogeny: Thin-skinned or lithospheric collision?: Tectonophysics, v. 274, p. 191–220.

Yu, S.-B., Cheng, H.-Y., and Kuo, L.-C., 1997, Velocity field of GPS stations in the Taiwan area: Tectonophysics, v. 274, p. 41–59.

Manuscript Accepted by the Society April 12, 2001

Printed in the U.S.A.

Geological Society of America
Special Paper 358
2002

Structural expressions of flexural extension in the arc-continent collisional foredeep of western Taiwan

Ying-Wei Chou
Chinese Petroleum Corporation, Taipei, Taiwan, Republic of China
Ho-Shing Yu
Institute of Oceanography, National Taiwan University, Taipei, Taiwan, Republic of China

ABSTRACT

The western Taiwan foredeep developed in the Pliocene–Quaternary during the formation of the Taiwan orogen. The sedimentary sequences within the foredeep are offset by a number of normal faults dated as Pliocene–Quaternary. The presence and significance of these young normal faults have been neglected, or they have mistakenly been interpreted as rifting features. Industrial multichannel seismic profiles and borehole information have been used in this study to examine these normal faults in the context of the model of flexural extension and to document their structural expressions and spatial and temporal variations. Unlike the early Tertiary normal faults of synrift origin, data from Pliocene–Quaternary normal faults in the Taiwan Strait indicate a close relationship to the adjacent fold-and-thrust belt. The results of these observations and interpretations of seismic reflection profiles support the idea that the crust underlying the western Taiwan foredeep has been flexurally downwarped and deformed by lithospheric bending.

INTRODUCTION

The presence and significance of the Pliocene–Quaternary normal faults in the western Taiwan foredeep have been neglected during previous structural analysis in the Taiwan Strait (Bradley and Kidd, 1991). Normal faults cutting the Cenozoic strata in the Taiwan Strait have long been interpreted as features of rifting or rejuvenation of rifting (Sun, 1982; Wang, 1987). In addition, petroleum studies paid little attention to these young normal faults because they are considered to be of little importance in trapping hydrocarbons.

The purpose of this paper is to investigate the structural expressions of the Pliocene–Quaternary normal faults of the western Taiwan foredeep, in the context of flexural extension (Bradley and Kidd, 1991), by using industrial seismic reflection profiles from the Chinese Petroleum Corporation in Taiwan (CPC). This new set of observations contributes to the understanding of synorogenic normal faults and their consistency with the flexural model.

Tectonic setting of Taiwan

In the western Pacific, the island of Taiwan is situated at the junction between the Luzon and Ryukyu arcs (Fig. 1A). The Chinese passive margin formed as a result of rifting during early Cenozoic time, and the collision of the Luzon arc with the passive margin, beginning in the late Miocene, resulted in the formation of the Taiwan orogen (Biq, 1972; Chi et al., 1981; Suppe, 1981; Ho, 1986). The arc-continent collision is actively propagating to the south (Suppe, 1987).

The island of Taiwan can be divided into five major north-trending geologic provinces. From west to east, these are the Coastal Plain, the Western Foothills, the Hsüehshan Range, the Central Range, and the Coastal Range (Fig. 1A, inset). The

Chou, Y.-W., and Yu, H.-S., 2002, Structural expressions of flexural extension in the arc-continent collisional foredeep of western Taiwan, *in* Byrne, T.B., and Liu, C.-S., eds., Geology and Geophysics of an Arc-Continent Collision, Taiwan: Boulder, Colorado, Geological Society of America Special Paper 358, p. 1–12.

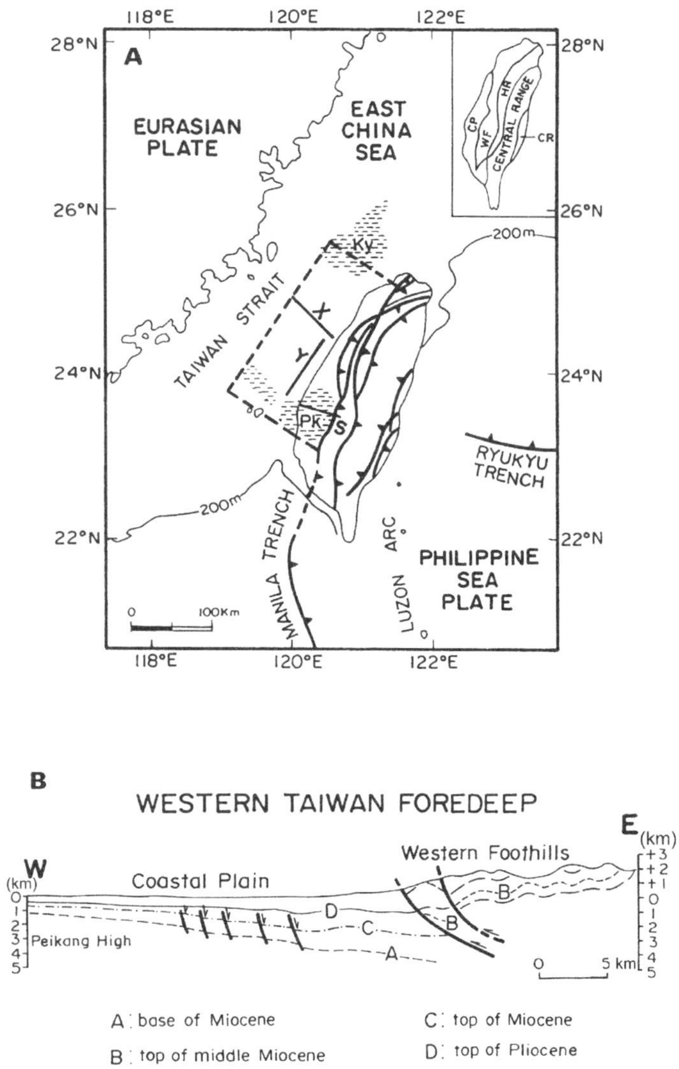

Figure 1. (A) Map of tectonic setting of Taiwan showing the Taiwan orogen and the western Taiwan foredeep to the west. The study area (outlined in dashes), two main basement highs (Ky—Kuanyin High, Pk—Peikang High), and locations of regional representative cross sections (X, Y, and S) are also shown. The inset shows the geologic provinces of Taiwan: CP—Coastal Plain, WF—Western Foothills, HR—Hsüehshan Range, CR—Coastal Range. Modified after Ho (1986). (B) Cross section S across the western Taiwan foredeep extending from the Western Foothills to the Coastal Plain overlying the Peikang basement high. Movement on the normal faults in the Coastal Plain occurred mainly during the Pliocene and offset the passive-margin sequences. Modified after Chou and Yang (1986).

Hsüehshan Range and Central Range together occupy the central parts of Taiwan Island and form a mountain belt; they are the principal sediment-source terranes. The Coastal Range in eastern Taiwan is an accreted island-arc body, resulting from the collision between the northern Luzon arc and the Chinese passive margin (Ho, 1986).

Accompanying the uplift of the Taiwan orogen, flexural subsidence of the continental-margin crust and rapid sedimentation produced both a forebulge and a foredeep, which flank the Taiwan orogen (Covey, 1984). Chou (1999) determined that the flexural forebulge was located in the middle of the Taiwan Strait and that the foredeep was located between the forebulge and Taiwan Island. This foredeep contains a wedge of Pliocene-Pleistocene clastic sedimentary deposits that thins westward from a maximum thickness of 5 km. The sediments were shed from the evolving Taiwan orogen; they form a series of sedimentary sequences that are present in the eastern parts of Taiwan Strait as well as the Coastal Plain and Western Foothills. In the area of the Peikang basement high (Fig. 1), the foredeep sedimentary deposits are relatively thin, suggesting less flexure in this area (Chou et al., 1994). Brusset et al. (1999) divided the western Taiwan foreland basin into an overfilled northern foreland-basin system and an underfilled southern foreland-basin system.

Western Taiwan foredeep

The western Taiwan foredeep runs approximately parallel to the Taiwan orogenic front from north to the south. The cross section of the western Taiwan foredeep shows the progressive deepening of the contact between rifted-margin sequences and Pliocene–Quaternary foreland-basin fill toward the Taiwan orogen (Fig. 1B). The Moho discontinuity, as determined from modeling of onshore and offshore wide-angle seismic data and Pn-wave arrivals from regional earthquakes, dips east beneath the Taiwan Strait, the Coastal Plain, and the Western Foothills (Huang et al., 1998; Shih et al., 1998). Both the onshore and offshore areas of western Taiwan are also associated with a negative gravity anomaly corresponding to the mass deficit of a flexural basin (Karner and Watts, 1983; Hsu et al., 1998).

The western Taiwan foredeep exhibits a two-stage tectonic subsidence history from Oligocene to the present. Slower passive-margin subsidence, at ca. 3 Ma was followed by an increase in the subsidence rate. This subsidence rate has generally increased toward the orogenic belt (Lee et al., 1993; Chou et al., 1994), which is consistent with the foreland-basin model (e.g., Flemings and Jordan, 1989; Sinclair et al., 1991). The Pliocene–Quaternary sedimentary deposits were derived principally from the Taiwan orogen to the east, whereas the Oligocene-Miocene sedimentary deposits, which are dominated by quartzofeldspathic particles, were shed from the Chinese continent to the west (e.g., Chou, 1973). In the lower Pliocene sequences, the first appearance of abundant slate fragments and reworked nannofossils from the young Taiwan orogen (Lee, 1963; Chang and Chi, 1983) marks the initial uplift of the Central Range in eastern Taiwan.

The depositional environments of the western Taiwan foredeep exhibit distinct deepening during a period of eustatic fall, suggesting rapid tectonic subsidence in the early Pliocene (Teng, 1987). The foredeep then shallowed with time as the orogen approached, as is evident by a coarsening-upward sequence whose environment of deposition progressed from deep water to subaerial (Covey, 1984, 1986).

Previous studies

Although comprehensive studies of the stratigraphy and sedimentology of upper Cenozoic strata in western Taiwan exist (Chou, 1973, 1980), the nature of the foreland basin has not been fully understood. Covey (1984), through study of 17 exploratory wells in the Coastal Plain region and 11 outcrops in the Western Foothills, recognized that the upper Pliocene and Quaternary sedimentary deposits in western Taiwan were accumulated in a rapidly subsiding basin along the western cratonic flank of the Hsüehshan Range and Central Range. Covey (1984), the first to recognize this eastward-thickening basin, named it the western Taiwan foredeep (e.g., Fig. 1B) and suggested that the asymmetric shape resulted from flexure of the Eurasian continental lithosphere under the orogenic load. However, Covey's work did not include the offshore area (i.e., the Taiwan Strait).

Teng (1987) interpreted the stratigraphic record of upper Cenozoic sedimentary rocks in western Taiwan in terms of foreland-basin sedimentation. Later, Shiao and Teng (1991) performed two-dimensional flexural modeling of the western Taiwan foredeep by using the depth variations of the basal Pliocene unconformity ascertained from seismic data. The effective elastic thickness of the continental lithosphere beneath western Taiwan was determined to range from 18 to 22 km. These previous studies were focused on the lithofacies and the mechanism of basin formation. However, little attention was given to structures involving the Pliocene–Quaternary strata in the western Taiwan foredeep.

Yu (1993) compared the structural styles of the offshore region, southwest of Taiwan, to the Taiwan Strait shelf to the west. Sediments derived from the Central Range accumulated in the region southwest of Taiwan and were deformed into compressional folds and thrusts during the late Pliocene and Quaternary. Yu (1993) proposed the term "southwest foreland basin" to designate the synorogenic offshore basin in southwest Taiwan. He limited the areal extent of the active foredeep to the submarine slope off southwest Taiwan, i.e., not extending to the Taiwan Strait shelf.

Lee et al. (1993) found normal faults cutting the Pliocene–Quaternary strata of the Tainan Basin and considered that they resulted from flexural extension in a modern foreland basin. Angelier et al. (1995) pointed out that downwarping of the lithosphere in the Pliocene–Quaternary produced northeast-trending extension, which reactivated older faults as normal faults. Tandon et al. (1995) and Tandon and Lorenzo (1996) also suggested that the Tainan Basin offshore southwestern Taiwan became a foreland basin in the Pliocene.

STUDY AREA AND DATA

The study area covers the Hsinchu-Miaoli-Taichung region of northwestern Taiwan and adjacent offshore areas (Figs. 1A and 2). The orogenic front in the central parts of the study area is very close to the coastline, so marine seismic data cover the entire foredeep. In general, the quality of seismic data from offshore areas is much better than that of data collected on land. High-quality seismic reflection data in Taiwan Strait provide an excellent opportunity to better understand the structural features of the foredeep.

About 2500 km of industrial, migrated, seismic reflection profiles and four exploratory wells were used for this study (Fig. 2). Most of the seismic profiles were shot and processed by the Chinese Petroleum Corporation between 1976 and 1985. Reprocessing of the data collected from 1976 to 1979 was completed in 1996. The application of shotpoint-gap deconvolution significantly improved the data resolution. FK POWER (frequency-wave number power exponent calculation) and DMO (dip move-out) processes made the details of the sections clearer and improved fault imaging. Nannofossil data and geophysical

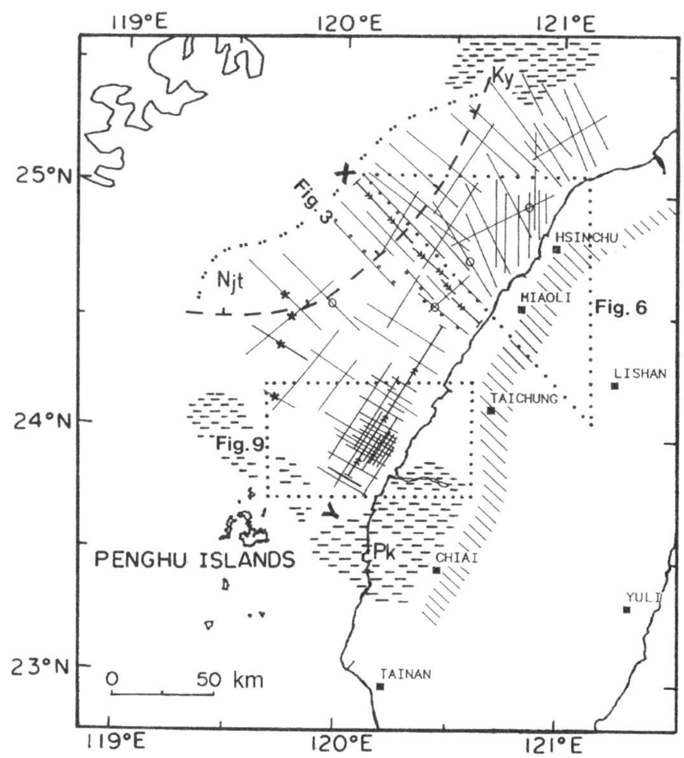

Figure 2. Locations of industrial seismic reflection profiles (thin lines) and four wells (marked by open circles) used in the study area of the western Taiwan foredeep. Long-line hatched area represents approximate western limit of thrust belt of Taiwan. Note that few or no young normal faults can be recognized on the seismic lines marked by asterisks. Ky—Kuanyin basement high, Pk—Peikang basement high, Njt—Nanjihtao Basin.

logging from well boreholes provided approximate age calibration for the seismic profiles.

A set of selected high-quality seismic profiles was used in this study for detailed observations and interpretations of the structures preserved in the thick stratigraphic sections of the foredeep. To ease the recognition of faults, the horizontal-to-vertical proportions were varied, and profiles were displayed as wiggle traces with both peaks and troughs filled in. After selecting the faults on the seismic sections, we matched adjacent seismic lines in map view to determine which fault segments belonged in the same fault plane. Regularly varying dips for fault planes and smoothly changing fault throws and heaves tested our fault correlations.

STRUCTURES OF THE NORTHWESTERN TAIWAN FOREDEEP

Regional unconformities

The Cenozoic strata in the Taiwan Strait comprise three major units—the synrift, postrift passive-margin, and foreland-basin megasequences, designated A, B, and C, respectively—which are bounded by regional unconformities. These three megasequences reflect tectonically distinct stages in the evolution of the western Taiwan foredeep (Figs. 3–5). Three major unconformities can be identified within the seismic profiles and borehole data. The earliest surface separates the Mesozoic basement from the Paleogene sedimentary rocks. The second separates megasequence A from megasequence B and is marked by an angular truncation, which can be correlated to the breakup unconformity associated with the opening of the South China Sea. The youngest unconformity separates megasequence B from megasequence C and is characterized by erosion of the paleoforebulge (Figs. 3–5).

The major unconformity at the base of the Pliocene–Quaternary foreland-basin fill clearly exists in the western parts of the western Taiwan foredeep. This surface is an angular unconformity and extends over a distance of ~100 km. A detailed representative seismic section (Fig. 3) shows that this unconformity changes laterally, toward the orogen, into a conformity or a disconformity with an insignificant stratigraphic gap in the coastal area. The chronostratigraphic gap across the unconformity increases cratonward; the largest gap separates the Holocene sediments from lower Miocene sequences. The characteristics of this major unconformity suggest that it can be regarded as a flexural forebulge unconformity (e.g., Sinclair et al., 1991; Crampton and Allen, 1995) and that it marks the transition from passive-margin to foredeep sedimentation. This unconformity has been cut by a number of young normal faults (see Fig. 3B).

Normal-fault styles and ages

According to structural interpretations in this study, at least two main phases of normal faulting are evident in this region (Table 1; Figs. 3–5). The older phase of faulting cuts Paleogene strata; these faults show considerable throw, averaging 0.5 km but varying up to 2 km, and are interpreted to be related to the rifting of the Chinese continental margin. In the seismic profiles, the ages of the oldest undisplaced horizons above the faults suggest that most of the older normal faults became inactive before the early Miocene. The Paleogene synrift strata usually show thickening across faults, indicating syndepositional motion on these faults. These faults generally strike east and form horsts and grabens in the eastern Taiwan Strait (e.g., Huang et al., 1993). The blocks so formed are oblique to the later contractional structures in Taiwan that generally trend north. Also, these normal faults appear to offset units within the pre-Tertiary basement (Fig. 4).

In contrast, the younger normal faults have generally smaller displacements, ~100 m, and cut the foreland-basin fill (at least through the foreland-basin basal unconformity) and the underlying passive-margin sedimentary sequences. The displacements associated with these faults appear either to be transferred into the basement or to die out at depth (Figs. 4 and 5). The fact that these younger faults are rarely associated with thickening strata or seismically defined facies changes suggests that the initial movement of these faults was mainly after passive-margin deposition. These younger normal faults also predominantly dip east, in contrast to the rift-linked normal faults that dip north or south. Seismic profiles Y and X (Figs. 4 and 5, respectively) are representative of the interpreted seismic lines in the study area (Fig. 2).

The throw of some of the younger normal faults increases downward, and the synrift sequences thicken across the faults. These faults (e.g., the faults marked in profile Y) are considered to have formed during rifting and to have been reactivated during foreland subsidence. Most synorogenic normal faults in this study area, however, are apparently not related to the preexisting structures and seem to have developed independently of the rift-associated fault system.

In the western parts of the dip profiles (e.g., Figs. 3 and 5), Paleocene–Eocene synrift sequences are confined to a half-graben, named the Nanjihtao Basin after Sun (1982, 1985) (Fig. 2). In these dip profiles, only some normal faults reveal obvious linkage to the synrift deposition, probably because of the orientation of the seismic line. The younger normal faults observed in the dip profiles dip mostly toward the southeast and are more deeply buried to the southeast, toward the thrust belt.

Faults were also correlated between interpreted seismic lines in the study area to track their distribution and geometry. From the Coastal Plain to the eastern parts of the Taiwan Strait, over a distance of ~100 km, numerous younger small-displacement normal faults were identified from the seismic data. The majority of these faults dip toward the Taiwan orogen. The normal faults in the eastern parts of the Taiwan Strait generally do not exhibit the features of structural inversion. These normal faults form a set that nearly parallels the most external contractional structures of the orogenic belt (Fig. 6).

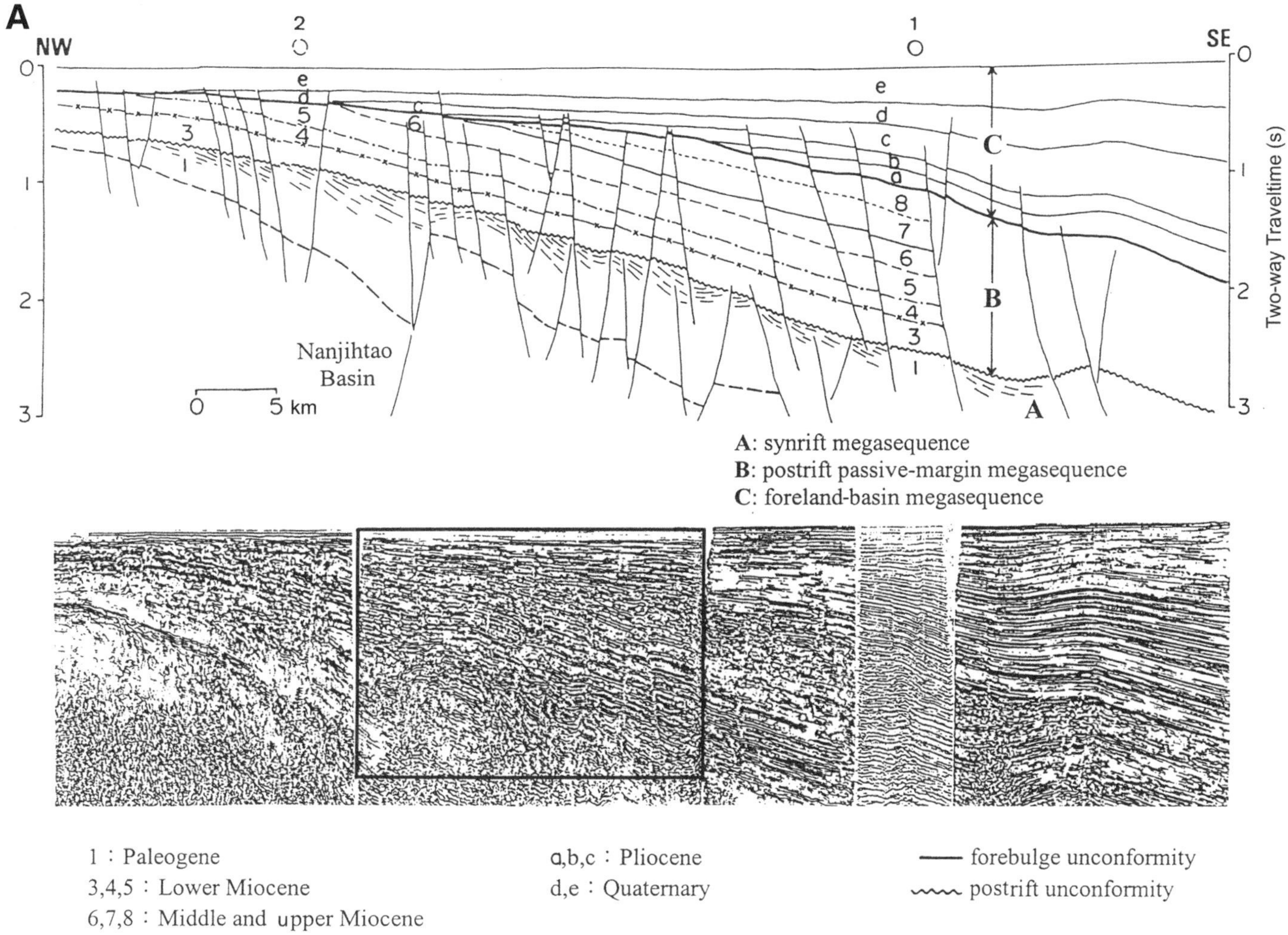

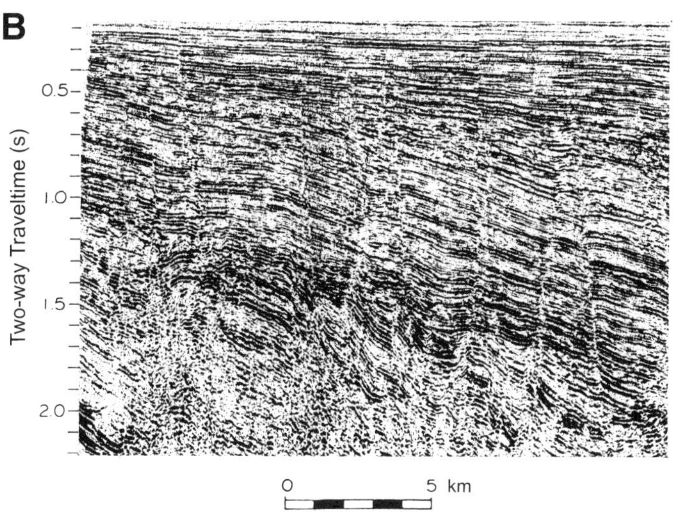

Figure 3. (A) An interpretive cross section (upper panel) and its seismic profile (lower panel), both horizontally compressed (location shown in Fig. 2). The stratigraphic relationships in the middle and distal parts of the western Taiwan foredeep in the Taiwan Strait are illustrated. Cenozoic sedimentary successions consist of three megasequences (synrift, postrift passive margin, and foreland basin). A forebulge unconformity is recognized at the base of the Pliocene–Quaternary sequences that marks the transition from passive-margin to foredeep sedimentation. Borehole 1 (marked by circle), located near the coastal area, indicates an insignificant chronostratigraphic gap across the unconformity. Borehole 2, projected into the profile from the south, shows the largest gap (Quaternary–Holocene strata resting on lower Miocene strata) across the unconformity. This gap progressively decreases toward the thrust belt. Note that this unconformity has been cut by a number of young normal faults. (B) Part of the seismic profile (rectangle in A) is enlarged to show young, well-developed, but relatively small normal faults.

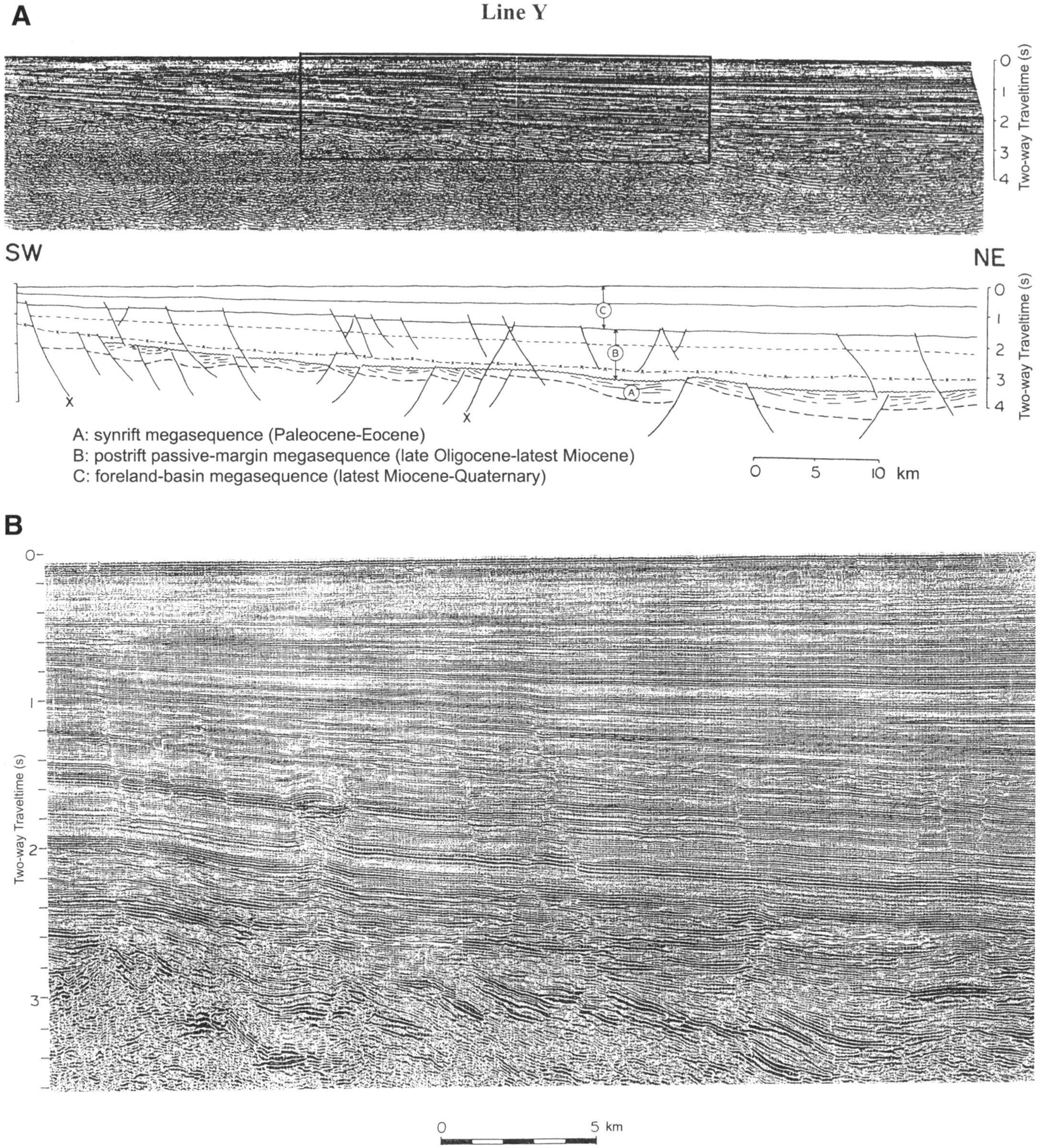

Figure 4. (A) Along-strike seismic profile Y (upper panel) and its borehole-calibrated interpretive section (lower panel) with a different vertical scale, but the same horizontal scale as in Figure 3 (see Fig. 2 for location). Cenozoic sedimentary successions consist of three megasequences (A–C). The breakup unconformity is marked by stratal angularity. Note that synrift normal faults with block-faulting style significantly offset the basement. The normal faults that cut the Miocene are characterized by relatively small displacements and dip mainly toward the Taiwan orogen. Some rift-associated faults were reactivated during the development of the foredeep (e.g., see faults marked by X). (B) Part of seismic profile (rectangle in A) is enlarged to show the young, relatively small normal faults.

Line X

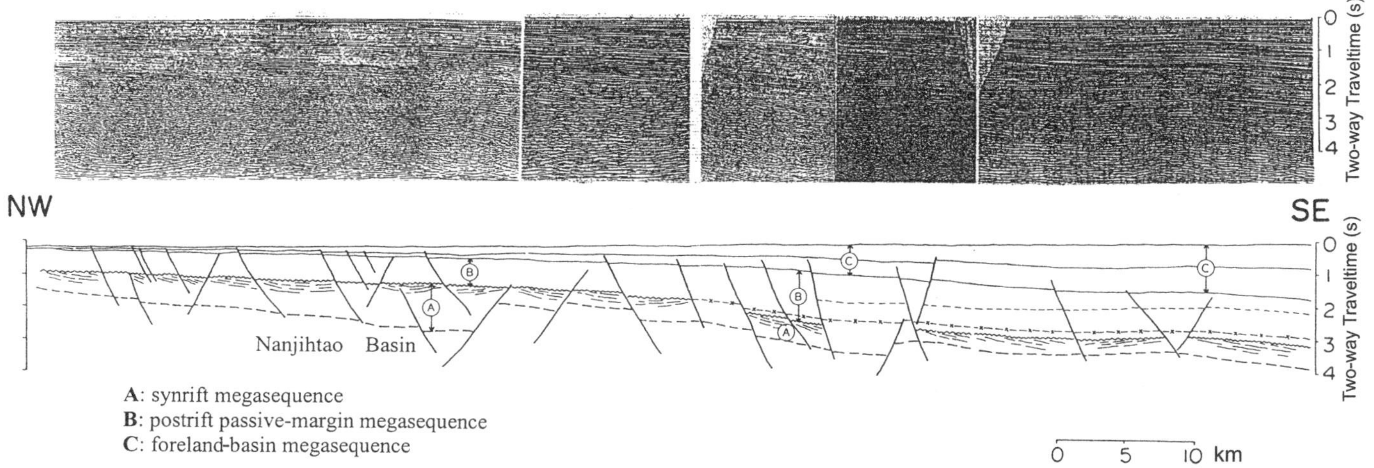

Figure 5. Dip-parallel seismic line X (upper panel) and its interpretive profile (lower panel), consisting of four short lines across the eastern Taiwan Strait (see Fig. 2 for location). Three megasequences (A–C) can be recognized. Note that the displacements along young normal faults are relatively small and that faults predominantly dip toward the thrust belt. The initial movements on these normal faults are mostly dated as Pliocene or Pleistocene.

Yang et al. (1996) also recognized this relationship in the area offshore of Hsinchu.

Few or no younger normal faults can be recognized in the seismic lines located farther than ~80–100 km from the orogen (Fig. 2), which may mean that extensional deformation has not yet occurred in the region beyond the thrust belt.

Figure 6 is a schematic block diagram constructed to show the overall structural features of Cenozoic strata in front of the orogenic belt of Taiwan. Onshore major thrust faults and décollement surfaces are shown in the frontal region of the thrust belt. Offshore, a plan view of the top of the lower Pliocene was adopted to present the patterns of younger faults, which have cut upward into foreland-basin fill.

Some young normal faults are also buried beneath the frontal thrust. For example, a high-angle normal fault in the footwall of the Sani thrust (Fig. 6) has been imaged seismically and penetrated by drilling (Hung and Wiltschko, 1993). In the Miaoli area, strata in the footwall side of the Sani thrust are consistently tilted in a direction opposite to the regional trend (e.g., small structural dips to the west calibrated by borehole dip logs). Previously, these anomalous west dips have been unexplained. They are, however, consistent with the flexural model proposed by Bradley and Kidd (1991).

Observations in the study area lead us to suggest that the younger suite of normal faults was initiated near the thrust front and generally migrated cratonward into the Taiwan Strait as the orogen itself moved westward (e.g., Fig. 7). By using the profile shown in Figure 3 as an example, four seismic horizons were picked within the Pliocene–Quaternary sequences to set limits on the ages of the young normal faults. The horizons were identified by using borehole-controlled paleontologic and/or lithologic data. The horizons are as follows: near the base of the Pliocene (i.e., the top of Kuantaoshan sandstone), the top of lower Pliocene (i.e., the top of Kueichulin Formation), the top of the Pliocene (near the top of the Cholan Formation), and a Pleistocene horizon corresponding to the base of the strata labeled b, c, d, and e in Figure 3. Except for the later inverted zone in the southeast (the Coastal Plain), movement along the young normal faults nearest the thrust belt stopped in the latest Miocene and early Pliocene (Fig. 7). The timing of deactivation of the young faults generally becomes younger toward the Chinese craton, where the latest movements occurred in about the middle Pleistocene.

In summary, extensional faulting, as exhibited in the seismic profiles across the eastern Taiwan Strait, can be grouped into two main stages primarily on the basis of timing of fault deactivation (e.g., Figs. 3–5). The younger extensional deformation in the northwestern Taiwan foredeep was approximately coeval with the collision event to the east on Taiwan Island. This synorogenic extension is interpreted to be related to flexural bending. An older stage of extension appears to be related to the rifting of Chinese continental margin.

Normal-fault reactivation

Compression within the eastern Taiwan Strait and coastal area during the later collisional stage apparently inverted the slip direction along some of the earlier extensional faults into reverse or strike-slip movement. For example, several Holocene high-angle thrust faults near the coastline in the Hsinchu area can be connected to Pliocene normal faults along strike (Figs. 6 and 8) and thus may have originally been normal faults. Some offshore structures immediately adjacent to Tachia may record reactivation of the rift-related grabens (Fig. 6) because the faults

TABLE 1. SUMMARY OF COMPARISONS BETWEEN SYNRIFT AND SYNCOLLISIONAL NORMAL FAULTS IN WESTERN TAIWAN FOREDEEP

	Synrift normal faults	Syncollisional normal faults
Age of initial movement	Eocene to Oligocene	Pliocene to Pleistocene
Age of deactivation	Mostly before early Miocene	Still active
Dip of faults	Horst and graben pattern	Predominantly dip toward orogen
Fault orientation	Nearly east-west (following the rift-basin trend)	Northeast-southwest changing to north-south from north to south (parallel to orogenic front)
Fault displacement	Big (500–2000 m)	Small (average, 100 m)
Offset of basement	Relatively large, evident on seismic profiles	Relatively small, less evident on seismic profiles
Sediments thickening across faults	Evident on seismic profiles	Rarely shown on seismic profiles

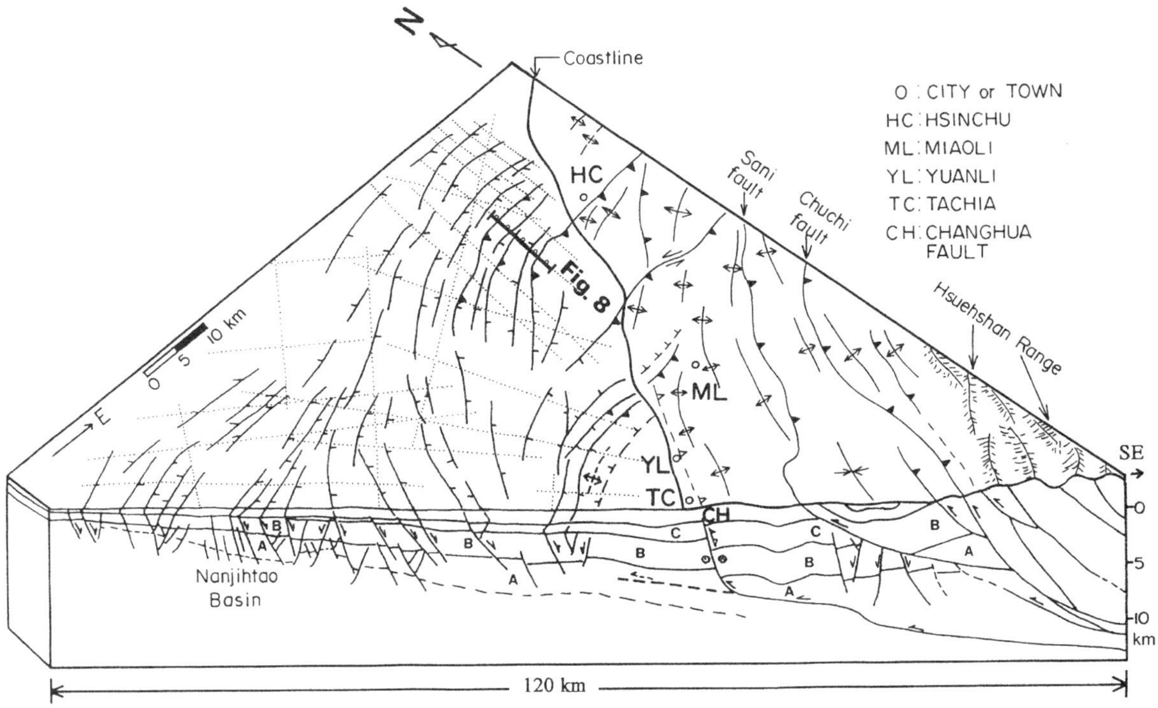

Figure 6. Schematic three-dimensional block diagram showing the deformation of foreland-basin sedimentary units west of the Taiwan fold-and-thrust belts (see Fig. 2 for location). The diagram was constructed by integrating structural interpretations of offshore seismic sections (seismic traces are shown by dotted lines) with published cross sections and maps of surface structures in nearby onshore areas (Lu and Hsu, 1992; Hung and Wiltschko, 1993; Yang et al., 1996). The plan view of the top of the lower Pliocene is presented to show the structural patterns of the younger faults in the offshore area. On a broad scale, the young normal faults in Taiwan Strait strike nearly parallel to the fold axis in the most external zone of the thrust belt. The three Cenozoic megasequences are labeled A, B, and C.

trend obliquely to the orogenic front. Because (1) the Changhua fault in the northern part of the Coastal Plain trends nearly parallel to the orogenic front and (2) the Pliocene strata in the hanging wall thicken, the Holocene thrust movement on the Changhua fault probably represents reactivation of a Pliocene flexural normal fault, but with opposite movement along it. Flexure-induced normal faults that have not been reactivated may be preserved in the stage of tectonic inversion (Fig. 8).

A block diagram (Fig. 6) shows the decreased intensity of inversion both from the external zones of the thrust belt westward to the Taiwan Strait and from the Kuanyin basement high southward to offshore of the area around the city of Tachia; the decreased intensity was also noticed by Huang et al. (1993). This variation of intensity of inversion is probably caused by the southward progression of the collision and by the existence of the Kuanyin basement high that can be considered a relatively rigid indenter.

Flexural normal faults on northern flank of Peikang basement high

In the southern part of the study area, synorogenic normal faults that cut Miocene and younger strata on the northern flank of the Peikang basement high can be recognized in a relatively

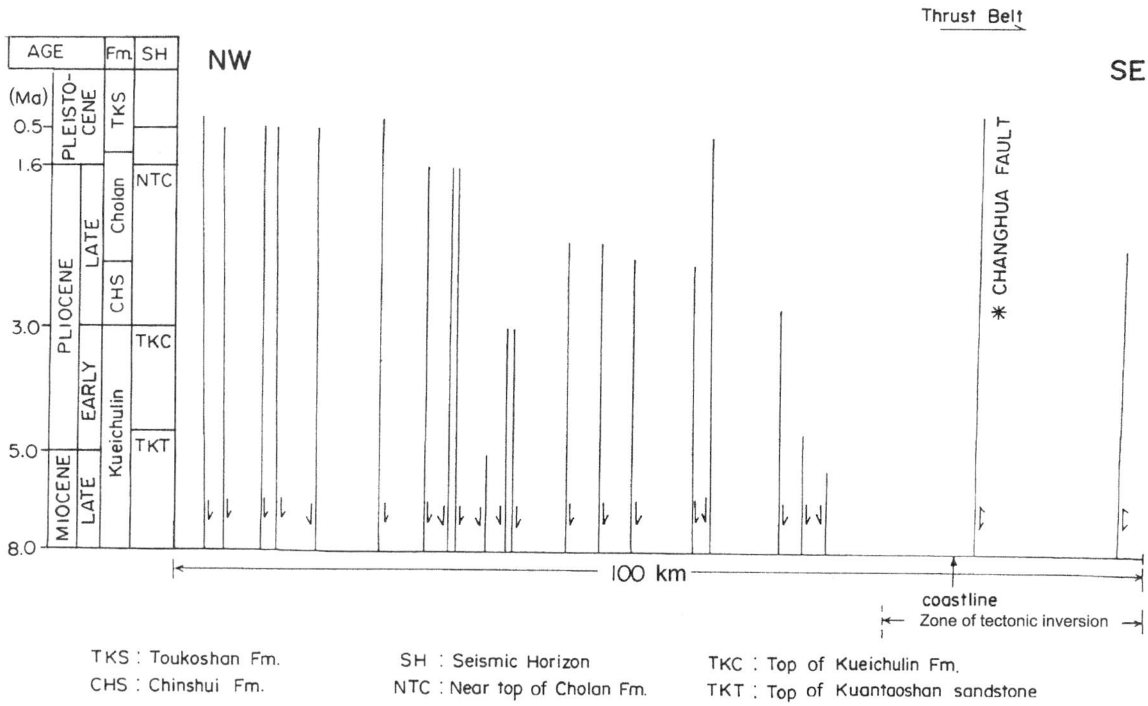

Figure 7. Ages of faults plotted against distance (100 km) from the Coastal Plain to the middle of the Taiwan Strait, by using the fault data in the profile of Figure 3. Note that the normal-fault ages become progressively younger toward the distal part of the western Taiwan foredeep. Movement on the Changhua fault in the Coastal Plain near the thrust belt was inverted from normal slip to thrust slip at around middle Pleistocene time, and the fault became a part of the deformation front. Four reference seismic horizons within the Pliocene–Quaternary sequences correspond to the base of b, c, d, and e in profile of Figure 3.

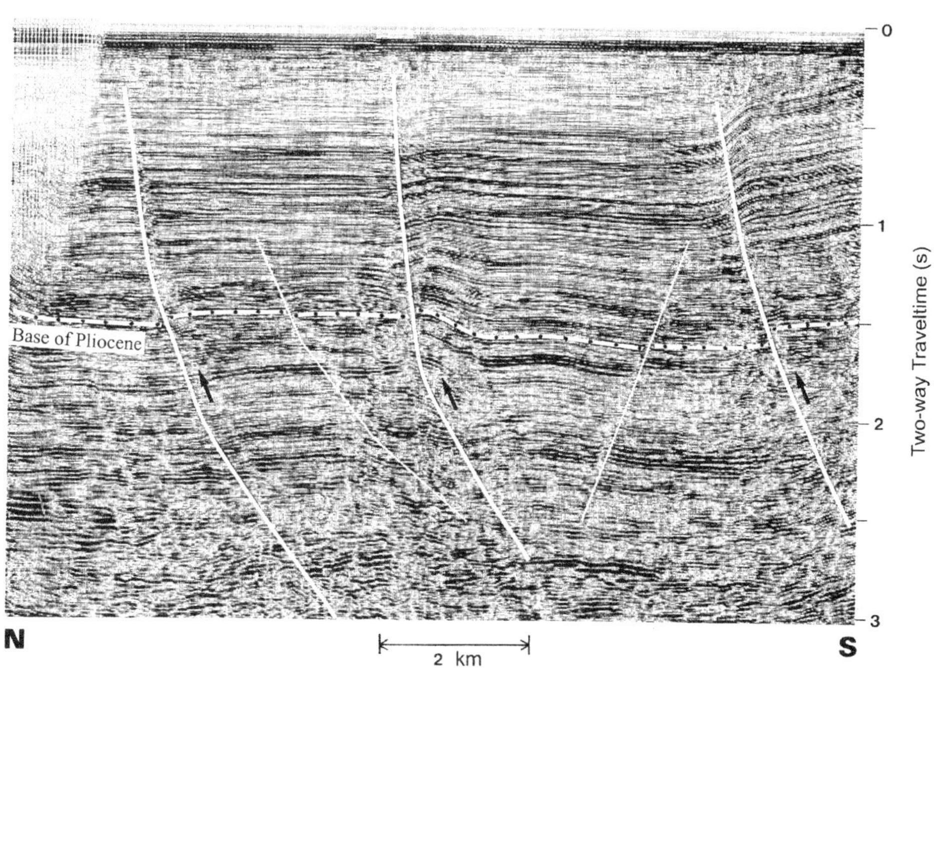

Figure 8. Part of a seismic line near the coastline of the Hsinchu area (see Fig. 6 for location) showing that compression during the later collisional stage has inverted the slip direction along some probably flexure-related, Pliocene normal faults so that they now function as high-angle thrust faults.

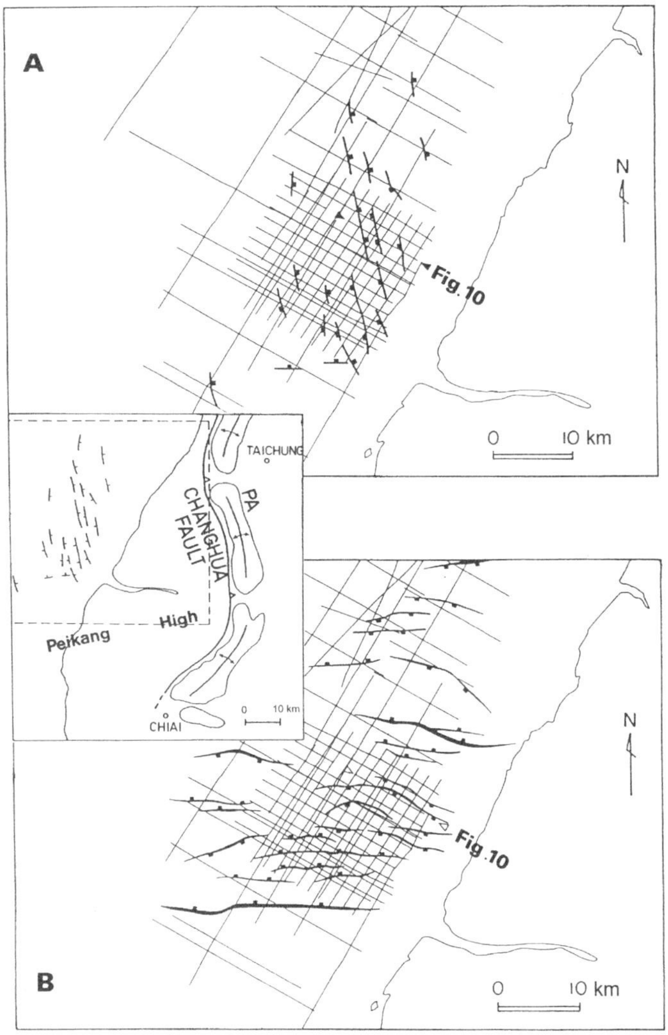

Figure 9. The patterns of fault traces of synorogenic (A) and synrift (B) normal faults on the northern flank of the Peikang basement high (see Fig. 2 for location). The synorogenic normal faults mainly trend north, orthogonal to the synrift faults (east trends) and nearly parallel to the Changhua fault and the associated folding structures in the frontal thrust zone (inset). PA—Pakuashan anticline. Modified after Fuh and Wu (1999).

dense (e.g., 2 km spacing) seismic grid (Fuh and Wu, 1999). Traces of these normal faults trend mainly north, nearly parallel to the strike of Changhua fault and the most external folding structures (e.g., the Pakuashan anticline) of the orogen to the east (Fig. 9A and inset). In contrast, synrift normal faults strike east in this area, consistent with the regional trend of the rift basin (Fig. 9B). In this area, the synorogenic normal faults generally become shorter and rarely cut deeper into the Paleogene synrift strata (Figs. 9A and 10) than similar faults in the Hsinchu and Miaoli areas (e.g., 5 km vs. ~10 km) (Fig. 6). Therefore, we suggest that the magnitude of flexural extension probably decreased in this area on the northern flank of the Peikang basement high.

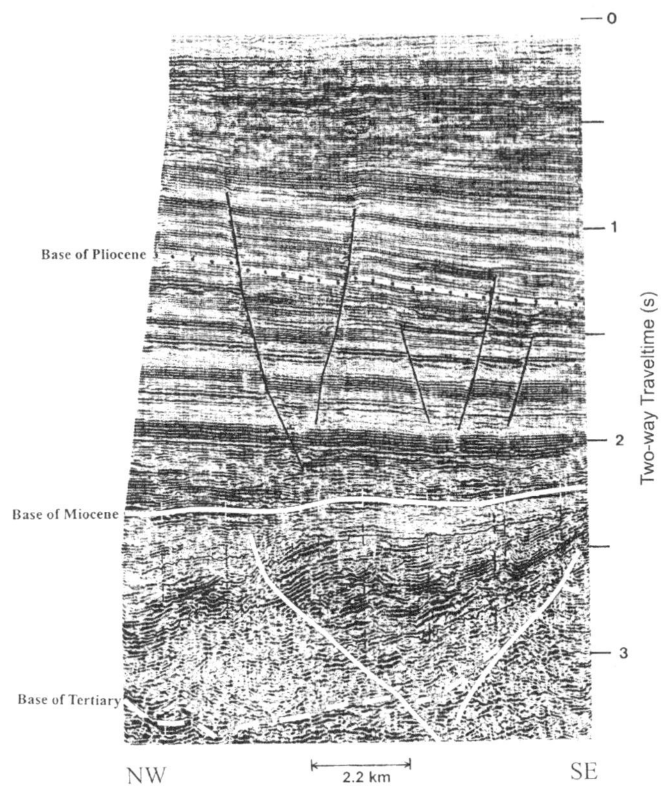

Figure 10. An interpreted seismic section (location shown in Fig. 9) displaying the synorogenic normal faults (black segments) that cut the Miocene and younger strata (but die out at depth) and the synrift normal faults (white segments) that penetrated the Paleogene sedimentary units and pre-Tertiary basement. This section crosses the major synorogenic normal faults recognized in the southern parts of study area.

SUMMARY

The younger normal faults observed in this study are considered to be related to crustal flexure on the basis of the following observations:

1. Displacements are generally dated as Pliocene to Pleistocene. This timing correlates with the major collision event to the east and accelerated subsidence and sedimentation in the foredeep.

2. The faults are largely observed within ~100 km of the thrust front and generally strike parallel to the front. The faults also mainly dip toward the orogen and have relatively small displacements.

3. The magnitude of extension in Taiwan (1.8% to 2.0%) is also comparable with extension magnitudes in the flexed continental margin of Australia in the area of the Timor Trough (~2%) (Jacobson et al., 1979; Karig et al., 1987; Lorenzo et al., 1997, 1998). In contrast, a much greater amount of extension (~20%) was estimated in the rifting regime of the Tainan Basin (Lee et al., 1993).

Finally, igneous intrusions and extrusions are commonly recognized and recorded in the Paleogene–Miocene sequences (e.g., Ting, 1985; Liu and Pan, 1984), but they are conspicuously absent in strata postdating the late Miocene within the basins of western Taiwan. The absence of igneous rocks in the younger strata and their presence in the older strata are facts consistent with our proposal that the younger faults are related to flexure and not continental rifting.

ACKNOWLEDGMENTS

We thank Chinese Petroleum Corporation (CPC, in Taiwan) for permission to publish some selected seismic profiles and borehole information contained in this paper. Tim Byrne critically reviewed the manuscript and provided valuable comments and improvements. Bradley and Kidd's (1991) comment on Taiwan stimulated us to examine and unravel the structures preserved in the thick sedimentary section in the foredeep lying east of the Taiwan Strait shelf. The exploration geophysicists of the Conoco Oil Company engaged in hydrocarbon exploration in the Taiwan Strait provided considerable insights and discussions about basin evolution and structural geology in the region. Further, this paper has benefited in particular from discussions with our colleagues T.W. Shiao, F.F. Huang, and K.M. Yang in CPC and S.K. Hsu and L.Y. Chiao in the Institute of Oceanography, National Taiwan University. An important part of the seismic interpretation was performed on Landmark Graphics workstations at the CPC.

REFERENCES CITED

Angelier, J., Lee, J.C., Chu, H.T., Lu, C.Y., Fournier, M., Hu, J.C., Lin, N.T., Deffentaines, B., Delcaillau, B., Lacombe, O., and Lee, T.Q., 1995, Crustal extension in an active orogen: Taiwan: Active Collision in Taiwan, 3rd Sino-French Symposium, Program and Extended Abstracts, p. 25–32.

Biq, C., 1972, Dual-trench structure in the Taiwan-Luzon region: Proceedings of the Geological Society of China, no. 15, p. 65–75.

Bradley, D.C., and Kidd, W.S.F., 1991, Flexural extension of the upper continental crust in collisional foredeeps: Geological Society of America Bulletin, v. 103, p. 1416–1438.

Brusset, S., Déramond, J., Souquet, P., Mouthereau, F., and Deffontaines, B., 1999, Pro-foreland basin systems linked to Taiwan Mountain building: Active subduction and collision in Southeast Asia: Data and models: International Conference and 4th France-Taiwan Symposium, Program and Extended Abstracts, p. 67.

Chang, S., and Chi, W.R., 1983, Neogene nannoplankton biostratigraphy in Taiwan and the tectonic implications: Petroleum Geology of Taiwan, no. 19, p. 93–147.

Chi, W.R., Namson, J., and Suppe, J., 1981, Stratigraphic record of plate interactions in the Coastal Range of eastern Taiwan: Memoir of the Geological Society of China, v. 4, p. 155–194.

Chou, J.T., 1973, Sedimentology and paleogeography of the upper Cenozoic system of western Taiwan: Petroleum Geology of Taiwan, no. 16, p. 111–143.

Chou, J.T., 1980, Stratigraphy and sedimentology of the Miocene in western Taiwan: Petroleum Geology of Taiwan, no. 17, p. 33–52.

Chou, J.T., and Yang, C.Y., 1986, Characteristics of the sedimentary basins in western Taiwan and their hydrocarbon potential: Petroleum, v. 22, no. 1, p. 2–25 (in Chinese).

Chou, S.C., Teng, L.S., Chung, H.S., and Shiao, T.W., 1994, Geohistory analysis of west Taiwan foreland basin: A preliminary study: Petroleum Geology of Taiwan, no. 29, p. 289–323 (in Chinese).

Chou, Y.-W., 1999, Tectonic framework, flexural uplift history and structural patterns of flexural extension in Western Taiwan Foreland Basin [Ph.D. thesis]: Taipei, National Taiwan University, 125 p. (in Chinese).

Covey, M., 1984, Lithofacies analysis and basin reconstruction, Plio-Pleistocene western Taiwan foredeep: Petroleum Geology of Taiwan, no. 20, p. 53–83.

Covey, M., 1986, The evolution of foreland basins to steady state: Evidence from the western Taiwan foreland basin, *in* Allen, P.A., and Homewood, P., eds., Foreland basins: International Association of Sedimentologists Special Publication 8, p. 77–90.

Crampton, S.L., and Allen, P.A., 1995, Recognition of flexural forebulge unconformities in the geologic record: American Association of Petroleum Geologists Bulletin, v. 79, p. 1495–1514.

Flemings, P.B., and Jordan, T.E., 1989, A synthetic stratigraphic model of foreland basin development: Journal of Geophysical Research, v. 94B, p. 3851–3866.

Fuh, S.C., and Wu, J.C., 1999, Evolution of the faulting in the southern Taihsi basin since early Tertiary and its insight into the conflict of biostratigraphical dating: Geological Society of China, Annual Meeting, Taipei, Abstracts, p. 70–72.

Ho, C.S., 1986, A synthesis of the geologic evolution of Taiwan: Tectonophysics, v. 125, p. 1–16.

Hsu, S.K., Liu, C.S., Shyu, C.T., Liu, S.Y., Sibuet, J.C., Lallemand, S., Wang, C.S., and Reed, D., 1998, New gravity and magnetic anomaly maps in the Taiwan-Luzon region and their preliminary interpretation: TAO, Terrestrial, Atmospheric and Oceanic Sciences, v. 9, no. 3, p. 509–532.

Huang, S.T., Chen, R.C., and Chi, W.R., 1993, Inversion tectonics and evolution of the Northern Taihsi Basin, Taiwan: Petroleum Geology of Taiwan, no. 28, p. 15–46.

Huang, B.S., Chen, K.C., Wang, K.L., and Yen, H.Y., 1998, Velocities of Pn-waves in the Taiwan Strait and its surrounding area from regional earthquakes: TAO, Terrestrial, Atmospheric and Oceanic Sciences, v. 9, no. 3, p. 473–486.

Hung, J.H., and Wiltschko, D.V., 1993, Structure and kinematics of arcuate thrust faults in the Miaoli-Cholan area of western Taiwan: Petroleum Geology of Taiwan, no. 28, p. 59–96.

Jacobson, R.S., Shor, G.G., Kieckhefer, R.M., and Purdy, G.M., 1979, Seismic reflection and refraction studies in the Timor Arc-Trough system and Australian continental shelf: American Association of Petroleum Geologists Memoir 29, p. 209–222.

Karig, D.E., Barber, A.J., Charlton, T.R., Klemperer, S., and Hussong, D.M., 1987, Nature and distribution of deformation across the Banda arc-Australian collision zone at Timor: Geological Society of America Bulletin, v. 98, p. 18–32.

Karner, G.D., and Watts, A.B., 1983, Gravity anomalies and flexure of the lithosphere at mountain ranges: Journal of Geophysical Research, v. 88B, p. 10449–10477.

Lee, P.J., 1963, Lithofacies of the Toukoshan-Cholan Formation of western Taiwan: Proceedings of the Geological Society of China, no. 6, p. 41–50.

Lee, T.Y., Tang, C.H., Ting, J.S., and Hsu, Y.Y., 1993, Sequence stratigraphy of the Tainan Basin, offshore southwestern Taiwan: Petroleum Geology of Taiwan, no. 28, p. 119–158.

Liu, C.H., and Pan, Y.S., 1984, Seismic stratigraphic study on the Tertiary sequences in the Hsinchu Basin, Taiwan: Petroleum Geology of Taiwan, no. 20, p. 97–112.

Lorenzo, J.M., Tandon, K., and O'Brien, G.W., 1997, Neogene flexural reactivation in a modern foreland basin, Timor Sea, N.W. shelf, Australia: American Association of Petroleum Geologists, Annual Convention, Dallas, Texas, Abstracts, p. A72.

Lorenzo, J.M., O'Brien, G.W., Stewart, J., and Tandon, K., 1998, Inelastic yielding and forebulge shape across a modern foreland basin, North West Shelf of Australia, Timor Sea: Geophysical Research Letters, v. 25, p. 1455–1458.

Lu, C.Y., and Hsu, K.J., 1992, Tectonic evolution of the Taiwan mountain belt: Petroleum Geology of Taiwan, no. 27, p. 21–46.

Shiao, T.W., and Teng, L.S., 1991, Flexural tectonics of western Taiwan foreland basin: A preliminary study: Proceedings, 3rd Taiwan Symposium on Geophysics; p. 437–446.

Shih, R.C., Lin, C.H., Lai, H.L., Yeh, Y.H., Huang, B.S., and Yen, H.Y., 1998, Crustal structures across central Taiwan from modeling of the onshore-offshore wide-angle seismic data: TAO, Terrestrial, Atmospheric and Oceanic Sciences, v. 9, p. 317–328.

Sinclair, H.D., Coakley, B.J., Allen, P.A., and Watts, A.B., 1991, Simulation of foreland basin stratigraphy using diffusion model of mountain belt uplift and erosion: An example from the central Alps, Switzerland: Tectonics, v. 10, p. 599–620.

Sun, S.C., 1982, The Tertiary basins of offshore Taiwan: Proceedings, 2nd Asian Conference on Petroleum (ASCOPE), Manila, Philippines, 1981, p. 125–135.

Sun, S.C., 1985, The Cenozoic tectonic evolution of offshore Taiwan: Energy, v. 10, p. 421–432.

Suppe, J., 1981, Mechanics of mountain building and metamorphism in Taiwan: Memoir of the Geological Society of China, v. 4, p. 67–89.

Suppe, J., 1987, The active Taiwan mountain belt, *in* Schaer, J.-P., and Rodgers, J., eds., The anatomy of mountain ranges: Princeton, New Jersey, Princeton University Press, p. 277–293.

Tandon, K., and Lorenzo, J.M., 1996, Comparative study of flexure in incipient continental collision and subducting oceanic lithosphere: Eos (Transactions, American Geophysical Union), v. 77, p. 269.

Tandon, K., Lorenzo, J.M., Boehme, R.S., and Pyne, D.A., 1995, Continental flexural extension in collisional margins: Timor Sea area, North West Shelf of Australia, and offshore southwestern Taiwan: Eos (Transactions, American Geophysical Union), v. 76, p. 613.

Teng, L.S., 1987, Stratigraphic records of the late Cenozoic Penglai orogeny of Taiwan: Acta Geologica Taiwanica, no. 25, p. 205–224.

Ting, J.S., 1985, A study of the Tertiary unconformities, igneous activities and development of sedimentary basins in Taiwan: Chinese Petroleum Corporation, Internal published report, 45 p.

Wang, Y., 1987, Continental margin rifting and Cenozoic tectonics around Taiwan: Memoir of the Geological Society of China, v. 9, p. 227–240.

Yang, K.M., Wu, J.C., Wickham, J.S., Ting, H.H., Wang, J.B., and Chi, W.R., 1996, Transverse structures in Hsinchu and Miaoli areas: Structural mode and evolution in Foothills Belt, northwestern Taiwan: Petroleum Geology of Taiwan, no. 30, p. 111–150.

Yu, H.S., 1993, Contrasting tectonic style of a foredeep with a passive margin: Southwest Taiwan and South China: Petroleum Geology of Taiwan, no. 28, p. 97–118.

Manuscript Accepted by the Society April 12, 2001

Printed in the U.S.A.

Geological Society of America
Special Paper 358
2002

Inversion tectonics of the fold-and-thrust belt, western Taiwan

Chung-I Lee*
Exploration and Development Research Institute, Chinese Petroleum Corporation, Miaoli 36010, Taiwan, Republic of China
Yu-Long Chang[†]
Taiwan Petroleum Exploration Division, Chinese Petroleum Corporation, Miaoli 36010, Taiwan, Republic of China
Mike P. Coward
Geology Department, Imperial College, London SW7 2BP, UK

ABSTRACT

In the Taiwan orogen, Cenozoic passive-margin development on the flank of the Eurasian plate resulted in repeated phases of extension with the formation of faults and associated stratigraphic variations. Extensional faults are found with great lateral extent throughout the fold-and-thrust belt of western Taiwan. Pliocene through present-day shortening in the area, however, has resulted in fold-and-thrust geometries that show a history of interaction between early normal faults and later thrust faults.

We evaluate the structural style and kinematics of contractional deformation in the fold-and-thrust belt of western Taiwan by using an inversion-tectonics model. Four cross sections are presented to illustrate the interaction of normal faults and thrusts in this area. Our study suggests that the contribution of thick-skinned tectonics to the structure of the fold-and-thrust belt of western Taiwan is not less than that of thin-skinned tectonics. The structural style in the fold-and-thrust belt of the Taiwan orogen is related to inversion tectonics.

INTRODUCTION

Research into the foreland fold-and-thrust belt of many contractional mountain systems, including the Canadian Rocky Mountains, has led to the acceptance of thin-skinned tectonics as a mechanism of thrusting (e.g., Rich, 1934; Dahlstrom, 1970; Boyer and Elliott, 1982). Inversion geometries involving reactivation of earlier extensional faults or basement lineaments, however, have also been recognized in many of these foreland fold-and-thrust belts. Thin-skinned thrust tectonics may also often hide inversion geometries at depth (e.g., Gillcrist et al., 1987; Davis and Namson, 1989; de Graciansky et al., 1989; Letouzey, 1990; Welbon and Butler, 1992; Coward, 1994, 1996; Coward et al., 1991). It is important, therefore, to recognize, or test for, basin-inversion and thick-skinned thrust tectonics rather than assume a thin-skinned thrust model.

Since the 1980s, the Chinese Petroleum Corporation has been acquiring new geologic, seismic profiling, and drilling data from the Taiwan foreland basin and fold-and-thrust belt (Fig. 1). In light of these new data, the objective of this study is to reevaluate the structure of the fold-and-thrust belt of western Taiwan not only by using the thin-skinned model but also by incorporating the inversion-tectonics model. This paper illustrates the role of reactivated old normal faults in contractional deformation with four cross sections from the fold-and-thrust belt of the Western Foothills of Taiwan (Fig. 1). In order to compare the difference between the structural interpretation of the thin-skinned model and the results of this study, the cross sections were drawn near, and subparallel to, the sections constructed by Suppe (1980a) and Namson (1981, 1983, 1984).

Our results confirm and build on earlier work that suggested that the Cenozoic basins of western Taiwan have undergone a

*Deceased
[†]Corresponding author. E-mail: 005444@cpc.com.tw.

Lee, C.-I., Chang, Y.-L., and Coward, M.P., 2002, Inversion tectonics of the fold-and-thrust belt, western Taiwan, *in* Byrne, T.B., and Liu, C.-S., eds., Geology and Geophysics of an Arc-Continent Collision, Taiwan: Boulder, Colorado, Geological Society of America Special Paper 358, p. 13–30.

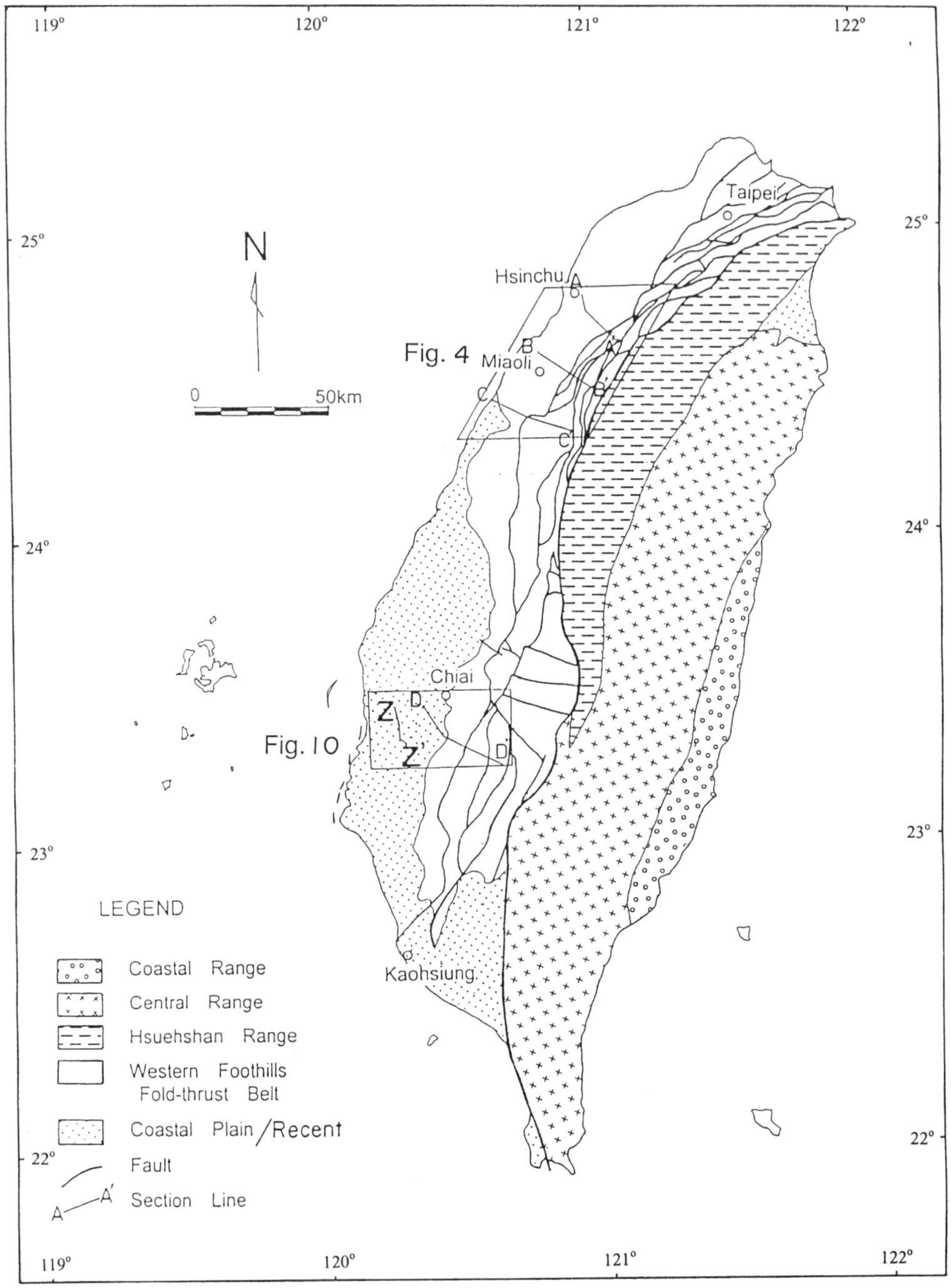

Figure 1. Geographic map of Taiwan showing the Coastal Plain (dotted pattern), fold-and-thrust belt (to the east of the Coastal Plain, unpatterned but faults are shown), cross-section lines, and locations of areas shown in other figures in this study (modified after Ho, 1988).

tectonic inversion with a change from subsidence by extension and normal faulting of Paleogene through Miocene age to uplift by contraction associated with an arc-continent collision since middle Pliocene time (Sun, 1985; Wang, 1994; Chung et al., 1989; Letouzey, 1990; Teng, 1992; Chen and Tang, 1993; Huang et al., 1993; Chi, 1994; Sibuet and Hsu, 1997; Huang, 1998). Analogues to Taiwan basin inversion are present in the frontal regions of many orogenic belts such as the Alps (Gillcrist et al., 1987; Butler, 1989; de Graciansky et al., 1989; Coward et al., 1991; Welbon and Butler, 1992), the Himalayas (Coward, 1994, 1996), and the Moine thrust belt of the Scottish Caledonides (Stewart, 1982; Soper and England, 1995; Butler, 1997).

The stratigraphy of rock units and abbreviations used in cross sections are shown in Figure 2.

Scale Change

my	Age		Hsinchu-Miaoli Area, NW Taiwan		West of Chukou Fault (Chiai-Tainan Area, SW Taiwan)		East of Chukou Fault (Chiai-Tainan Area, SW Taiwan)	
0.2	Pleistocene		Toukoshan Fm	TKS	Liushuang Fm	LS	Liushuang Fm	LS
	Pleistocene		Toukoshan Fm	TKS	Erhchungchi Fm	EC	Erhchungchi Fm	EC
	Pleistocene		Toukoshan Fm	TKS	Kanhsialiao Fm	KH	Yuching Sh	YC
	Pleistocene		Toukoshan Fm	TKS	Kanhsialiao Fm	KH	Chingmien Ss	CM
1.8	Pliocene	Late	Cholan Fm	CL	Liuchungchi Fm	LC	Peiliao Sh	PL
3.0	Pliocene	Late	Cholan Fm	CL	Yunshuichi Fm	YS	Chutouchi Fm	CT
	Pliocene	Early	Chinshui Sh	CS	Niaotsui Fm	NT	Maupu Sh	MP
	Pliocene	Early	Kueichulin Fm (KCL): Yutengping Ss	YTP	Niaotsui Fm	NT	Ailiaochiao Fm	AL
	Pliocene	Early	Kueichulin Fm (KCL): Shihliufen Sh	SLF	Chunlun Fm	CLN	Yenshuikeng Sh	YSK
5.0	Miocene	Late	Kueichulin Fm (KCL): Kuantaoshan Ss	KTS	Chunlun Fm	CLN	Tangenshan Ss	TE
	Miocene	Late	Nanchuang Fm (NC): Shangfuchi Ss	SF	Nanchuang Fm	NC	Nanchuang Fm (NC): Changchihkeng Fm	CC
11.0	Miocene	Middle	Nanchuang Fm (NC): Tungkeng Fm; Hopai Fm	TK; HP	Nanchuang Fm	NC	Nanchuang Fm (NC): Hunghuatzu Fm	HH
	Miocene	Middle					Nanchuang Fm (NC): Sanmin Sh	SM
	Miocene	Middle	Nankang Fm (NK): Kuanyinshan Ss	KY	Nankang Fm (NK): Kuanyinshan Ss	KY	Nankang Fm (NK): *Kuanyinshan Ss*	KY
14.0	Miocene	Early	Nankang Fm (NK): Talu Sh	TL	Nankang Fm (NK): Talu Sh	TL	Nankang Fm (NK): *Talu Sh*	TL
	Miocene	Early	Nankang Fm (NK): Peiliao Fm	PL	Nankang Fm (NK): Peiliao Fm	PL	Nankang Fm (NK): *Peiliao Fm*	PL
	Miocene	Early	Shihti Fm	ST	*Shihti Fm*	ST	*Shihti Fm*	ST
	Miocene	Early	Piling Sh	PLI	*Piling Sh*	PLI	*Piling Sh*	PLI
	Miocene	Early	Mushan Fm	MS	*Mushan Fm*	MS	*Mushan Fm*	MS
23.0	Oligocene	L	*Wuchihshan Fm*	WC	*Wuchihshan Fm*	WC	*Wuchihshan Fm*	WC
	Oligocene	M						
38.0	Oligocene	E						

Figure 2. Stratigraphy of units (and abbreviations) discussed in text or shown in cross sections (modified after Chi, 1994). Italics denote units not exposed.

PROBLEMS IN INTERPRETATION OF THE TAIWAN FOLD-AND-THRUST BELT USING THE THIN-SKINNED DEFORMATION MODEL

Normal faulting of Paleogene through Miocene age is widespread in both the onshore and offshore areas of western Taiwan. Normal-fault structures are best known under the Coastal Plain in the Peikang-Chiai-Tainan area of southwestern Taiwan. Here, a system of east- to east-northeast–striking normal faults is well documented from seismic profiles and well data (Fig. 3). The normal-fault structure extends into the Western Foothills area of Taiwan, but is less obvious because of the sequence of overlying Pliocene–Pleistocene orogenic sedimentary deposits reaching 3 to 4 km in thickness and because of the strong contractional deformation. Wang et al. (1995) described Tertiary normal faults in the East China Sea, to the north of Taiwan, and their slight structural inversion during the Miocene–Pliocene.

Previous studies assumed that thrusting in the fold-and-thrust belt of western Taiwan was relatively shallow, was confined to sedimentary rocks above the basement (which is generally considered to be rocks pre-Miocene in age), and was caused by thin-skinned tectonics (Biq, 1969, 1972; Ho, 1976, 1982). John Suppe and Jay Namson carried out one of the most detailed structural studies of the fold-and-thrust belt in western Taiwan from 1976 to 1986 and used a thin-skinned deformation model to interpret the structures (Namson, 1981, 1982, 1983, 1984; Suppe, 1976, 1980a, 1980b, 1981, 1983; Suppe and Namson, 1979).

During a later stage of their investigations, Suppe (1984, 1986) and Namson (1984) realized the importance of preexisting normal faults in the formation of the fold-and-thrust belt. Suppe's research group also developed a kinematic theory of basement-involved contractional structures and applied it to the structural interpretation of the Rocky Mountains, USA (Narr and Suppe, 1994). Later studies of the gravity modeling (Ellwood et al., 1996) and seismic tomography (Wu et al., 1994, 1997) of Taiwan also suggest basement involvement and reactivation of early normal faults in the fold-and-thrust belt (Suppe, 1986; Chang et al., 1993, 1996; Lee and Chang, 1998;

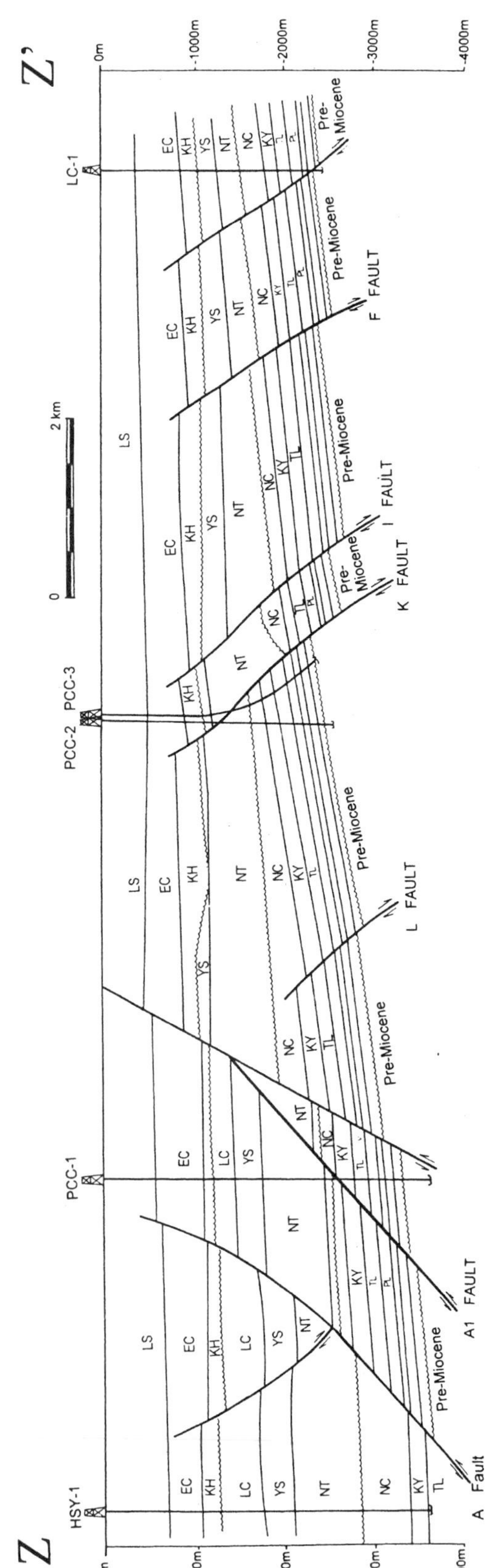

Figure 3. Example of extensional tectonics from the Coastal Plain of the Peikang-Chiai area (modified after Liu and Hsu, 1988; Lin et al., 1997). See Figures 1 and 10 for location of section line. Abbreviations as in Figure 2.

Lee et al., 1993, 1997, 1998; Yang et al., 1996, 1997). Inversion tectonics and fault reactivation may therefore play a significant role in the contractional deformation of Taiwan.

The misinterpretation of folds and thrusts as being related only to thin-skinned shortening without consideration of the possibility of the inversion of a sedimentary basin can have far-reaching effects on the structural and tectonic interpretation of a region, leading to (1) use of the wrong method in section construction, (2) miscalculation of the amount of orogenic shortening, and (3) incorrect assumptions about the nature of structures at depth, both those directly beneath the fold-and-thrust belt and those farther back within the hinterland of the mountain belt. Typical examples include the Sulaiman and Kohat Ranges, Pakistan (Coward, 1994, 1996; Coward and Butler, 1985) and the Moine thrust belt of the Scottish Caledonides (Elliott and Johnson, 1980; Stewart, 1982; Coward, 1988; Soper and England, 1995; Butler, 1984, 1997). It is, therefore, important to consider the concept of inversion tectonics in the interpretation of the geologic evolution of the Taiwan fold-and-thrust belt.

It must be emphasized that almost no balanced section gives a unique solution. It is our experience that oil companies now require several alternative sections and structural models and many require structures to be balanced in three, not just two, dimensions. Many structural sections should contain elements of both thin-skinned and thick-skinned (basement-involved) deformation. Thick-skinned interpretations involving structural inversion and the reactivation of older normal faults may sometimes allow faults to be kinked convex down or convex up, depending on whether the folds are considered to be fault-bend folds or fault-propagation folds, respectively, related to buckling of the hanging-wall strata (Coward, 1996). The trishear model may also be applicable to some fault zones, in which case the fault may not bend at depth. There may be back-rotation of the fault during inversion, and indeed many normal faults rotate their strike as well as their dip during inversion. We must also assume that the footwall may be displaced down the normal fault, i.e., there may be footwall deformation produced by displacement under an irregular fault surface. Finally there may be oblique slip associated with structural inversion. It has been shown that many faults cannot be reactivated unless there is a large component of strike-slip movement (Gillcrist et al., 1987; Letouzey, 1990). Inversion structures should not be considered balanced if examined in only two dimensions. A full three-dimensional structural analysis and restoration program are normally required.

In this paper, the sections presented are not unique solutions; indeed in several areas they will not balance. They aim to show the complexities of the deformation and the various alternative interpretations open to two-dimensional section construction. The sections also incorporate all available data based on surface mapping, seismic profiling, and drilling.

THE HSINCHU-MIAOLI AREA, NORTHWESTERN TAIWAN

The Hsinchu-Miaoli area lies within the northern part of the Western Foothills region of Taiwan (Figs. 1 and 4). The subsurface geology of the area has been studied since 1861 because of the rich oil and gas fields of the region (Chinese Petroleum Corporation, 1972). Most of the surface geology has been mapped at a scale of 1:25 000 and 1:10 000, and abundant subsurface data are available from hundreds of wells and many seismic lines. By using this extensive database and the thin-skinned model, Namson (1981, 1983, 1984) evaluated the structure of the Hsinchu-Miaoli area in a series of detailed cross sections. However, recent drilling and seismic profiling suggest that thin-skinned tectonics cannot fully explain the structure of the area. The structures in this area have been highly influenced by preexisting normal faults that formed from Paleogene to Miocene time (Chang et al., 1993; Lee and Chang, 1998; Lee et al., 1993, 1997, 1998; Yang et al., 1996, 1997).

Cross section A–A′ (Figure 5)

Cross section A–A′ (Fig. 5) is a regional-scale section of the northern Hsinchu-Miaoli area through the Chingtsaohu anticline, Hsincheng fault, Paoshan anticline, Peipu fault, Chutung anticline, Juanchiao fault, Pokungping fault, and Taping anticline. Along the section there exists a series of down-faulted blocks that are bounded by normal faults of pre-Miocene to Miocene age; the basins above these blocks were filled with sediments by early Pliocene time (Lee et al., 1993). The blocks are step-like, and those to the east were offset farther downward along the normal faults before the early Pliocene. This interpretation is subsequently discussed in detail.

In section A–A′, the normal faults are shown accommodating stratigraphic thickness changes. The evidence for these thickness changes is taken from well data shown on the cross section. The normal faults are indicated by abrupt changes in the thickness of the Shihti and Hopai Formations across the Chingtsaohu anticline, Paoshan anticline, and Juanchiao fault.

A normal fault is shown separating the stratigraphic thickness variation across the Chingtsaohu anticline. The thickness of the Shihti Formation changes from 280 m in the CTH-11 and CTH-14 wells to 690 m in the PS-11 well across the Chingtsaohu anticline. A normal fault is also shown separating the stratigraphic thickness variation across the Paoshan anticline. The thickness of the Hopai Formation (equivalent to the Tunkeng Formation and Kuanyinshan Sandstone) changes from 979 m in the PS-4 well to 690 m in the PS-8 well across the Paoshan anticline. A normal fault is indicated by a remarkable change in the thickness of the Hopai Formation from 1030 m in the CTU-25 well to 1730 m in outcrop to the east of the Juanchiao fault.

Chingtsaohu anticline and Hsinchu fault. The Chingtsaohu structure is an asymmetrical anticline with an east-northeast–trending axis. The near-surface shape of the Chingtsaohu anticline is characterized by a relatively steep (30°–60°) north flank, a relatively flat crest, and a more gentle (10°) south flank. The subsurface geometries of the structure and corresponding faults are well defined by data from wells CTH-3, CTH-7, CTH-11, CTH-12, CTH-13, and CTH-14 (Fig. 4). Data from well CTH-14 also show that rocks at least as deep as the Wuchihshan Formation are involved in the folding. There is no evidence of a thin-skinned thrust stack in this area.

The Chingtsaohu anticline is interpreted as involving the reactivation of two normal faults (Fig. 5). The shallow fold is interpreted as being formed by fault-bend folding. A high-angle thrust (Hsinchu fault) cuts the east limb of the deep fold to the middle of the Hopai Formation, where the fault becomes parallel to bedding and causes the sharp kink in the west limb. Limits on the shape of the higher thrust are set by both drilling data and the position of the fold axis at the surface and subsurface. A kink-band fold was produced by shortening of the hanging wall above a high-angle fault within the pre-Wuchihshan rocks, according to a fault-rotation folding model (Coward, 1994, 1996). The high-angle fault is interpreted as a reactivated normal fault to explain the evidence for rifting in this area.

Paoshan anticline and Hsincheng fault. The Paoshan anticline is interpreted as being formed by fault-propagation folding (Fig. 5). The subsurface geometries of the structure and associated faults are well determined by subsurface data from wells PS-3, PS-4, PS-5, PS-6, and PS-8. The entire stratigraphic section in this structure is crosscut by a high-angle thrust fault. A change in the thickness of the Hopai Formation from 979 m in well PS-4 to 1144 m in well PS-8 across the Paoshan anticline suggests that the high-angle thrust fault is a reactivated normal fault. The folding of the hanging wall produced by the displacement along the fault plane is probably related to rotational reactivation and inversion of slip direction along the old normal fault. Cessation of displacement on the fault may have caused the fold to tighten, and a change in inversion kinematics have may occurred, with the growth of two new thrusts. One low-angle thrust was the site of movement within the Hopai Formation. Another, higher-level low-angle thrust, the Hsincheng fault, was the site of movement within the Kueichulin Formation; the fault then stepped up to the lower part of the Cholan Formation and cut to the surface. The old normal fault acted as a buttress (e.g., low-level thrust) or ramp (e.g., high-level thrust) to contractional deformation. The field evidence for the Hsincheng fault is well documented (Huang, 1984; Chang et al., 1998). The fold in the footwall may have been produced by several mechanisms: it may be a fault-bend or fault-propagation fold on a deep-level fault, or it may have been produced as a fault-bend fold by displacement of the footwall down a slightly kinked Hsincheng fault.

Chutung anticline and Peipu fault. The formation of the Chutung structure is related to three faults, a deep, high-angle fault forming the deep fold and two higher-level thrust faults

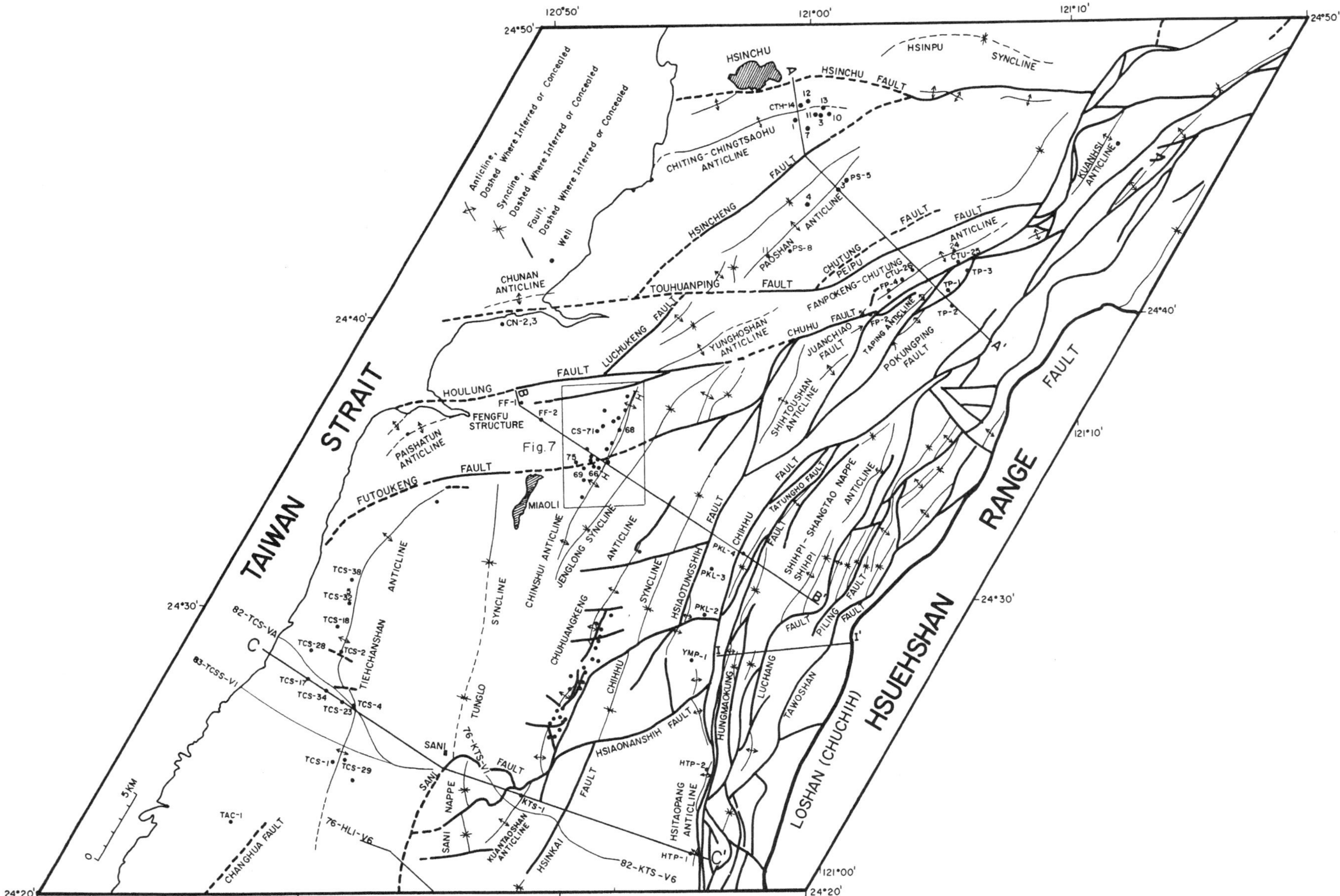

Figure 4. Map of the fold-and-thrust belt of the Hsinchu-Miaoli area, northwestern Taiwan, showing structural elements and the location of wells, seismic lines, and cross sections of this study. See Figure 1 for location of area.

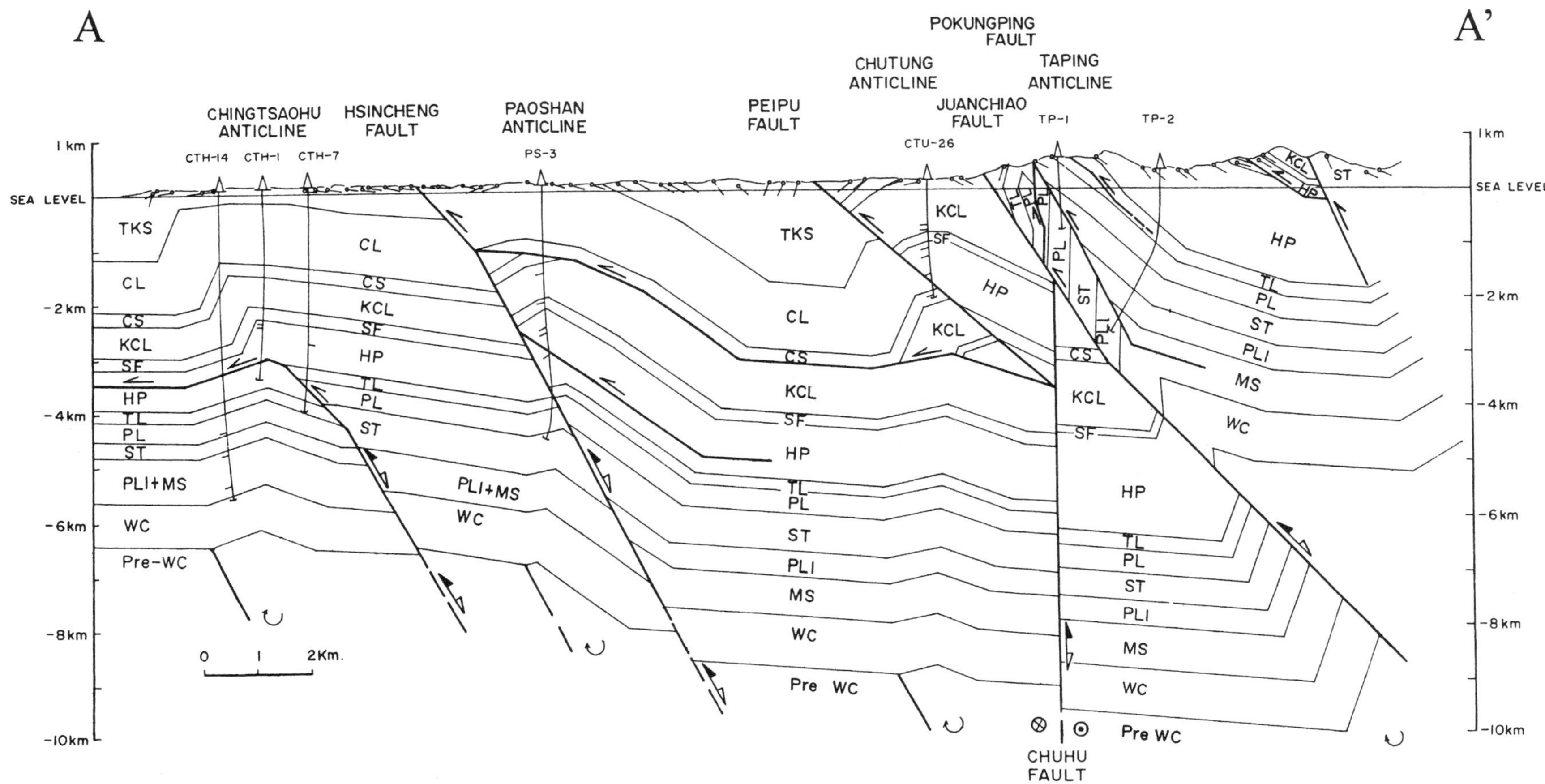

Figure 5. Cross section A–A′ across the northern part of the Hsinchu-Miaoli area, northwestern Taiwan (modified after Lee et al., 1993). See Figures 1 and 4 for location of section line. Abbreviations as in Figure 2.

that complicate the fold near the surface (Fig. 5). The near-surface part of the Chutung anticline is interpreted as having formed by fault-bend folding related to two thrusts. One thrust lies in the Kueichulin Formation and is the eastern part of the Hsincheng thrust. The other thrust, the Peipu fault, cuts to the surface with a dip of about 45°.

Taping anticline and Juanchiao, Chuhu, and Pokungping faults. The Taping structure is interpreted as having formed by fault-propagation folding related to the Juanchiao fault. Three higher-level thrust faults complicate the near-surface part of the Taping anticline (Fig. 5). The Juanchiao fault is interpreted as a reactivated normal fault that overrides the Chuhu fault. Recent drilling of well TP-3 encountered the Juanchiao fault at a depth of 520 m (Cheng and Tseng, 1995). The subsurface geometry of the Pokungping fault is well determined by drilling data from wells TP-1 and TP-2.

There is a significant change in thickness of the Hopai Formation from 1030 m in well CTU-26 to 1730 m in the outcrop across the Juanchiao fault. This fact, combined with seismic profiles and drilling data of well FP-2, indicates that an old, high-angle normal fault with a strike of N75°E exists beneath the Juanchiao fault in this area. This old normal fault crops out in an area to the southwest of the Chutung anticline (with its southern extension as the Fanpokeng anticline) and is known as the Chuhu fault (Lee et al., 1993).

In 1958, well FP-2 was drilled about 100 m east of the Chuhu fault to determine whether the Chuhu fault is a low-angle thrust or a high-angle fault. The well was drilled to a depth of 962.5 m without encountering the fault. This fact, together with surface data and seismic profiles, indicates that the fault is a high-angle fault with a dip of about 85° to the northwest (Huang and Hsu, 1958). The map view of the Chuhu fault (Chinese Petroleum Corporation, 1994) indicates that it has a right-lateral component of separation, whereas the cross section (Fig. 5) indicates normal separation at depth and reverse separations along the fault up the dip of the fault plane. We speculate therefore that the preexisting extensional fault geometry of the Chuhu fault has been inverted by obliquely reversing the sense of motion.

Cross section B–B′ (Figure 6)

Cross section B–B′ (Fig. 6) is a regional-scale section of the central Hsinchu-Miaoli area through the Fengfu anticline, Chinshui anticline, Chuhuangkeng anticline, Hsiaotungshih fault, Pakuali anticline, Chihhu fault, Hungmaokuan fault, and Shihpi anticline.

Fengfu structure and Fengfu fault. The geometries of the Fengfu structure and associated fault, here called the Fengfu fault, are defined by surface mapping, seismic profiles, and wells FF-1 and FF-2. Well FF-1 encounters the Fengfu fault at about 3120 m and records a 350 m repetition of the Hopai Formation. This well is also structurally higher than well FF-2. Separation observed in map view (Chinese Petroleum Corporation, 1994) indicates that the Fengfu fault has a right-lateral component of displacement. We speculate that the Fengfu fault is a reactivated normal fault that has been inverted by obliquely reversing the sense of motion.

Chinshui anticline. The Chinshui anticline is a broad (~8 km × 5 km), slightly asymmetrical structure that trends north-northeast and includes one of the most productive gas fields in Taiwan. The overall structure of the Chinshui anticline is well determined by data from more than 80 wells. The anticline is broken into blocks by two transverse high-angle faults that dip to the north and a nearly strike-parallel thrust fault that is exposed in the northern part of the anticline. This fault dips to the southeast (Figs. 7 and 8) (Chiu and Hsu, 1963; Kuan, 1971; Chinese Petroleum Corporation, 1972).

Surface and subsurface data indicate that the two high-angle faults are reactivated oblique normal faults. First, there is a significant change in the thickness of the Kueichulin Formation across the high-angle faults (Fig. 8). Second, subsurface geometries of the anticline and associated faults are well determined by subsurface data from wells CS-66, CS-69, and CS-75. For example, deep drilling in wells CS-69 and CS-75 reaches the Wuchihshan Formation without intersecting any major thin-skinned thrusts and shows that the Wuchihshan Formation is involved in the folding. Finally, separations in map view (Chinese Petroleum Corporation, 1994) indicate that the faults have a right-lateral component, and separations in cross section (Figs. 6 and 8) show that the fault is a normal fault. We therefore speculate that a preexisting extensional fault has been inverted by oblique reverse slip.

Chuhuangkeng anticline. The Chuhuangkeng anticline trends northeast for ~30 km and has a geometry that is different from most folds in the Western Foothills. It is a nearly symmetric upright chevron fold with limb dips of about 70°–80° at the surface. The anticline is bisected along its axis by steep reverse faults that dip toward the west. The Chuhuangkeng anticline is also divided transversely into separate blocks by normal faults (Fig. 4) (Chinese Petroleum Corporation, 1972, 1994; Wey et al., 1992).

Because the Chuhuangkeng oil and gas field, discovered in 1861 and still in production, is one of the most productive oil and gas fields in Taiwan, more than 140 wells have been drilled in this area (Chinese Petroleum Corporation, 1972, 1994). Results from the deepest wells show that rocks at least as deep as the Wuchihshan Formation are involved in the folding. On the basis of well data near the line of section B–B′, the Chuhuangkeng anticline is interpreted as a kink-band fold formed by shortening of hanging-wall rocks without much displacement on the early normal faults. The formation of the fold is preliminarily inferred to have involved rotational reactivation of a normal fault at depth and corresponding shortening of the hanging wall within the pre-Wuchihshan rocks (Figs. 4 and 6). This deformation is explained by a fault-rotation folding model (Coward, 1994, 1996).

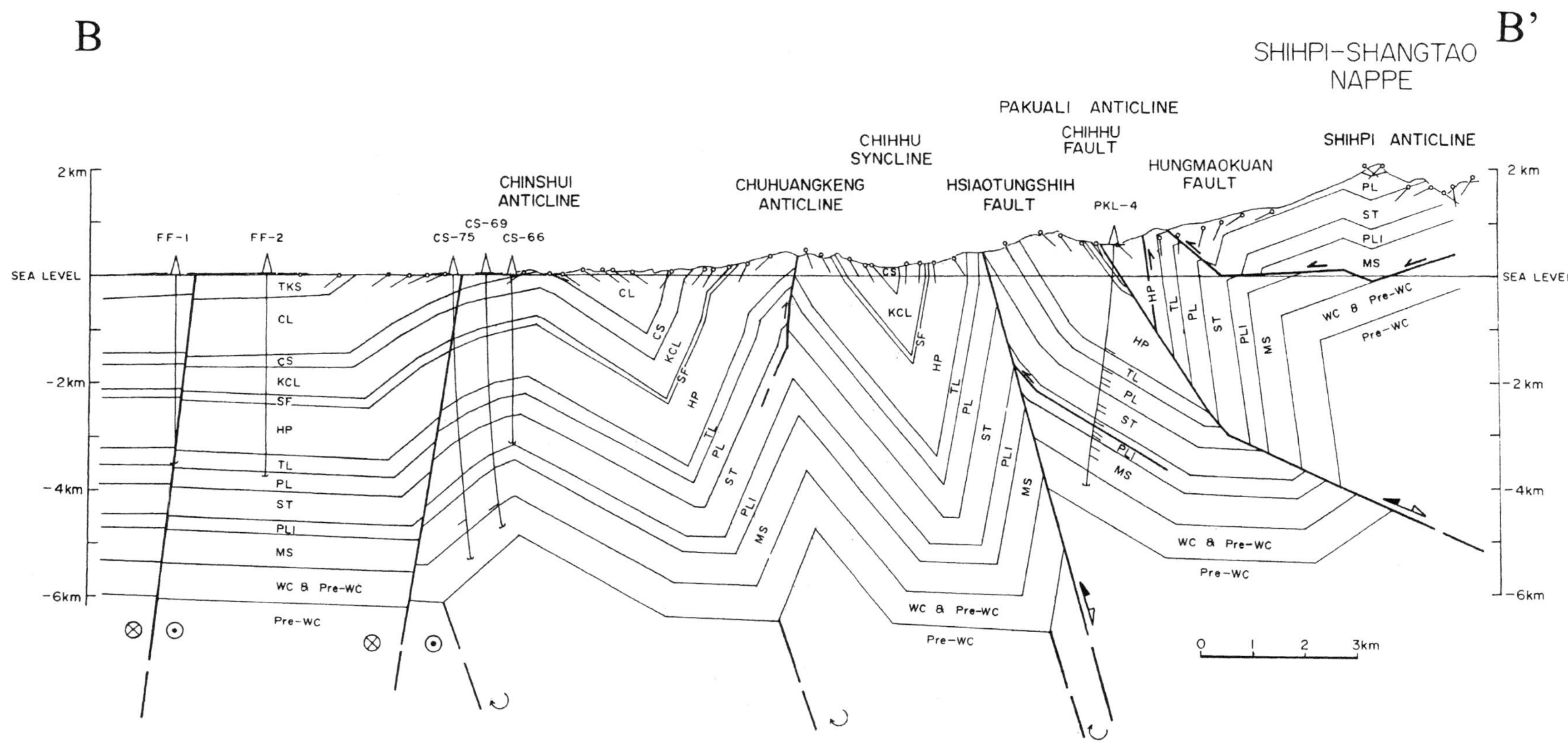

Figure 6. Cross section B–B′ across the central part of the Hsinchu-Miaoli area, northwestern Taiwan. See Figures 1 and 4 for location of section line. Abbreviations as in Figure 2.

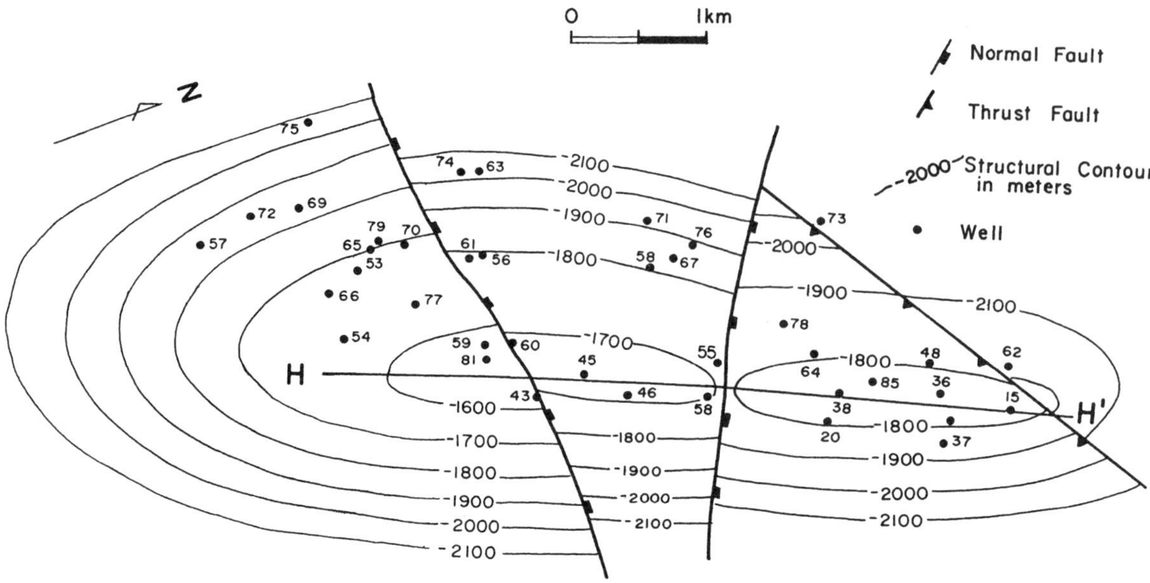

Figure 7. Subsurface structural map of the top of zone 13th within the Talu Shale (see Fig. 8), Chinshui gas field (after Lin et al., 1997; data from Taiwan Petroleum Exploration Division. See Figure 4 for location of area.

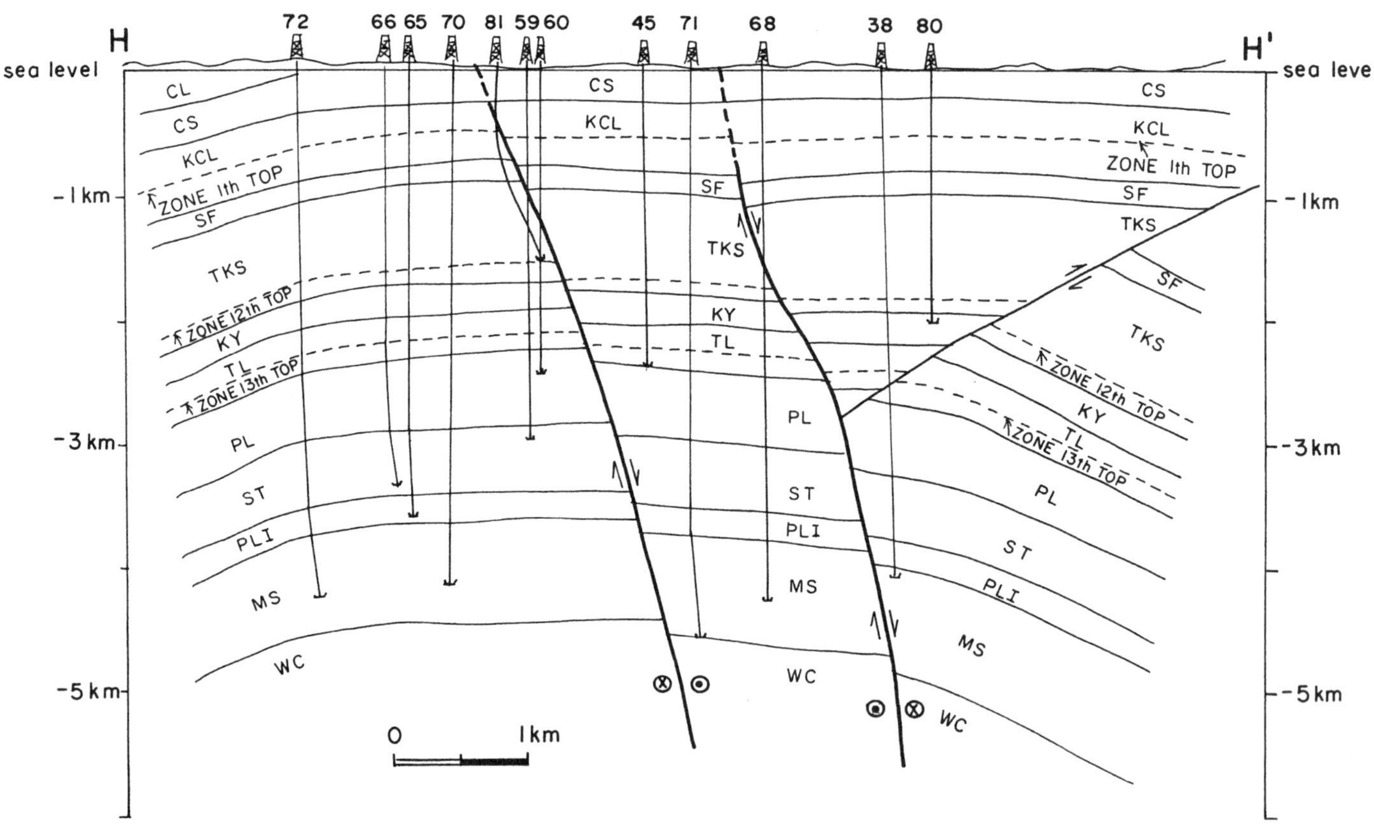

Figure 8. Cross section H–H′ along strike of the Chinshui anticline (modified after Lin et al., 1997; data from Taiwan Petroleum Exploration Division).

Pakuali anticline and Hsiaotungshih fault. On the basis of surface mapping and drilling data from well PKL-4, the Pakuali anticline is interpreted as having formed by fault-propagation folding related to the Hsiaotungshih fault. There is a significant change in thickness of the Hopai Formation from about 780 m to about 1100 m across the Hsiaotungshih fault. Hence, the Hsiaotungshih fault is interpreted as a reactivated normal fault. Its steep dip also suggests an important component of strike-slip deformation. Well PKL-4 encountered the Chihhu fault at 505 m depth.

Shihpi anticline, Shihpi-Shangtao nappe, and Chihhu and Hungmaokuan faults. From surface mapping and drilling data from well PKL-4, the Shihpi anticline is interpreted to be a fault-propagation fold above the Hungmaokuan thrust fault, which roots at the base of the Mushan Formation and cuts to the surface. The hanging wall of the thrust fault is the Shihpi-Shangtao nappe or thrust sheet (Figs. 4 and 6). The Chihhu fault is interpreted as a reactivated normal fault, which cuts through an unnamed syncline between the Pakuali anticline and the Shihpi anticline (Fig. 6).

Cross section C–C′ (Figure 9)

Cross section C–C′ (Figure 9) is a regional-scale section of the southern Hsinchu-Miaoli area through the Tiehchanshan anticline, Tunglo syncline, Sani fault, Kuantaoshan anticline, Hsinkai fault, and Hsitaopang anticline.

Tiehchanshan anticline, Tunglo syncline, and Futoukeng and Hsiaonanshih faults. The Tiehchanshan anticline is a broad gentle fold that is well defined by abundant drilling data, whereas the Tunglo syncline is an asymmetrical fold with a steep-dipping east limb and a more gentle west limb. The Tiehchanshan gas field, which includes these structures, is the most productive gas field in Taiwan. In section C–C′ (Fig. 9), the subsurface geometries of the Tiehchanshan anticline and associated faults are defined by data from wells TCS-4, TCS-17, TCS-23, and TCS-34 and seismic lines 82-TCS-VA, 83-TCSS-V1, 76-KTS-V1, and 82-KTS-V6. For example, well TCS-34 penetrates a normal fault within the Talu Shale at 2585 m and then passes through a normal stratigraphic sequence to a total depth of 5425 m (Hung and Ho, 1979; Chou and Yeh, 1987).

Development of the Tiehchanshan anticline and the Tunglo syncline is interpreted as related to the contractional reactivation of normal faults (Fig. 9). The overall shape of the Tiehchanshan anticline and the Tunglo syncline appears to reflect fault-propagation folding associated with the Futoukeng and Hsiaonanshih faults. Some parts of the Tiehchanshan anticline, however, are complicated by shallow-level faults. Although the existence of the Futoukeng fault is well documented by surface mapping (e.g., Chinese Petroleum Corporation, 1994) and seismic profiling (e.g., Chung, 1996), the specific characteristics of the fault are not completely clear. Chung (1996, Fig. 4.10) interpreted the Tiehchanshan anticline as a fault-bend fold caused by a step in the Futoukeng fault, which was inferred to root at the contact between the basement and overlying sedimentary section. On the basis of an evaluation of both onshore and offshore subsurface data from the Miaoli area, including the drilling data, we prefer to interpret the Futoukeng fault as a reactivated normal fault.

Sani fault and Sani nappe. The Sani fault is a low-angle thrust fault with a geometry that is well defined by surface mapping, seismic profiling, and drilling data (Meng, 1963; Hung and Wiltschko, 1993; Chinese Petroleum Corporation, 1994; Wang et al., 1994). The hanging wall of the Sani thrust—the Sani nappe—is further complicated by shallow-level faults (Fig. 9).

Kuantaoshan anticline and Hsiaonanshih and Hsinkai faults. The subsurface geometries of the Kuantaoshan anticline and associated faults beneath the Sani nappe are well defined by drilling, seismic profiles, and vertical seismic profiling data from well KTS-1. The Kuantaoshan anticline is inferred to be a fault-propagation fold related to displacement on the Hsiaonanshih and Hsinkai faults. The near-surface shape of the Kuantaoshan anticline is complicated by shallow faults, e.g., the Sani thrust.

Outcrop and drilling data (e.g., data from the Tiehchanshan field and wells KTS-1 and HTP-1) also indicate a significant increase in thickness of the Shihti Formation, the Hopai through Kueichulin Formations from the southern part of the Chuhuangkeng anticline to east of the Sani fault (Figs. 4 and 9). This fact, together with surface mapping, drilling data, and seismic profiles (e.g., 76-KTS-V1 and 82-KTS-V6), indicates that a high-angle normal fault with a strike of about N60°E exists beneath the Sani thrust. This early normal fault crops out northeast of the Kuantaoshan anticline and is known as the Hsiaonanshih fault, a right-lateral strike-slip fault separating the Chuhuangkeng anticline from the footwall block of the Sani thrust (Chang et al., 1993; Wang et al., 1994).

Seismic profiling also reveals extensional faults east of the Hsinkai fault and beneath the Sani nappe (e.g., Hung and Wiltschko, 1993; Wang et al., 1994). These faults are especially well illustrated in seismic lines to the south of section C–C′ (e.g., seismic lines 76-HLI-V5, 76-HLI-V6 and 81-TS-V1) (Fig. 4).

Hsitaopang anticline. The Hsitaopang anticline is interpreted as a triangle zone related to fault-bend folding caused by a deep-level thrust that roots in the contact between the Piling Shale and Shihti Formation and steps up to the Sani décollement (Fig. 9). This interpretation is based on surface mapping, drilling and VSP (vertical seismic profiling) data from the KTS-1 well, drilling data from the HTP-1 well, and seismic lines 76-KTS-V1 and 82-KTS-V6.

Summary of observations from northwestern Taiwan

The structural style and kinematics depicted in section C–C′ (Fig. 9) suggest that the contractional deformation possibly involved the following four phases. Phase 1 included formation

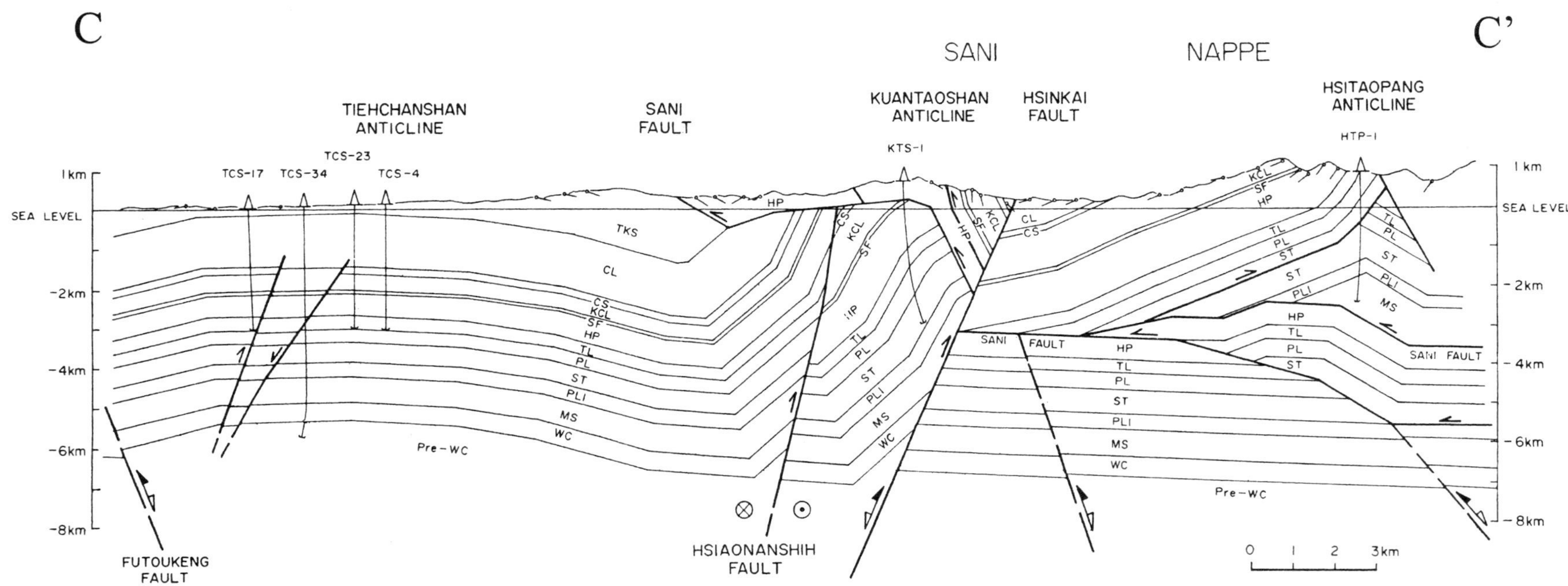

Figure 9. Cross section C–C′ across the southern part of the Hsinchu-Miaoli area, northwestern Taiwan. See Figures 1 and 4 for location of section line. Abbreviations as in Figure 2.

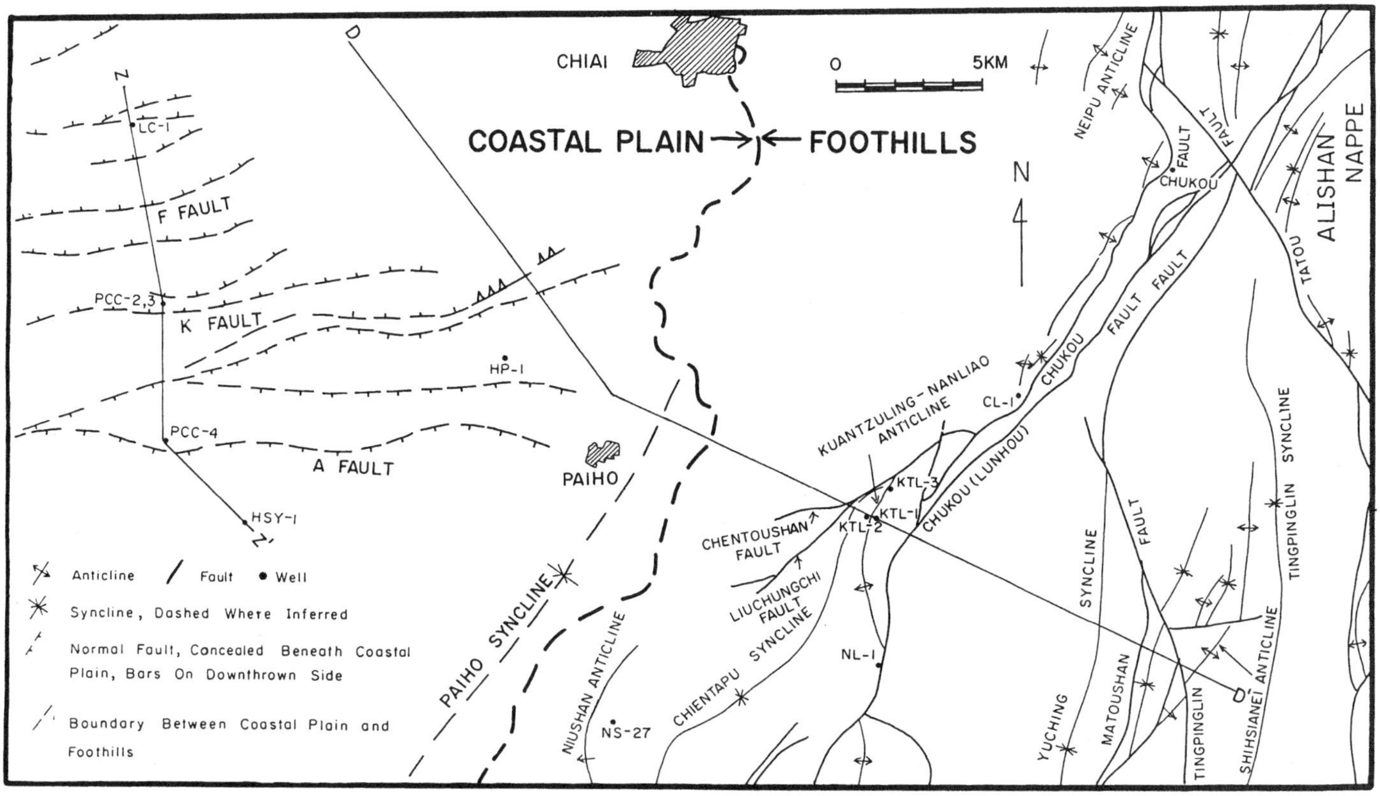

Figure 10. Map of the fold-and-thrust belt of the Chiai-Tainan area, southwestern Taiwan, showing structural elements and the location of wells, seismic lines, and cross sections of this study. See Figure 1 for location of area.

of the Tiehchanshan anticline, Tunglo syncline, and Kuantaoshan anticline by fault-propagation folding associated with the Futoukeng, Hsiaonanshih, and Hsinkai faults. Phase 2 was the emplacement of the Sani nappe. Phase 3 was characterized by deformation of the Sani nappe by uplifting of the Kuantaoshan anticline and propagation of the Hsinkai fault to surface. Phase 4 involved folding of the Hsitaopang structure as a result of the formation of a triangle zone.

The interpretation presented here has some similarities to, but also some distinct differences from, that presented by Chung (1996). The principal differences lie in our emphasis on reactivation and oblique movement along old normal faults and fault-propagation folding and their importance to the observed structural style and kinematics of the contractional structures.

THE CHIAI-TAINAN AREA, SOUTHWESTERN TAIWAN

The foothills of the Chiai-Tainan area (Figs. 1 and 10) form the southern part of the fold-and-thrust belt of western Taiwan. Existing structural interpretations through this area involve thin-skinned tectonics and attribute the structural relief to a series of duplexes and thrust structures in the lower Miocene Mushan Formation to upper Miocene Chunlun Formation (Suppe, 1976, 1980a). However, recent drilling and seismic profiling suggest that the structural relief may not be caused by thin-skinned tectonics but by structural inversion (Wey and Peng, 1982; Hsu and Wey, 1983; Chang et al., 1996).

Cross section D–D′ (Figure 11)

Cross section D–D′ (Fig. 11) is a regional-scale section of the Chiai-Tainan area. The section transects the Coastal Plain and the Western Foothills through the Paiho syncline, Chentoushan fault, Liuchungchih fault, Kuantzuling anticline, Chukou fault, Yuching syncline, Matoushan fault, Tsaoshan syncline, Chutouchi fault, Tingpinglin fault, and Shihsianei anticline. Structural interpretation for the Coastal Plain part of the section is based on drilling, seismic profiling, and gravity data, whereas for the Western Foothills part of the section, the interpretation is based on surface mapping, drilling, and gravity data. Along the section there are a group of subsurface grabens and horsts that are bounded by normal faults of pre-Miocene to Miocene age and the basins thus were formed filled by early Pliocene time (Chang et al., 1996). This interpretation is subsequently discussed in detail.

Paiho syncline. The Paiho syncline is located on the eastern limit of the Coastal Plain, just east of the Kuantzuling anticline, which is a major contractional structure. These two structures correlate with regional-scale gravity lows and highs, respectively (Chang et al., 1996) and represent the deformation front of the Taiwan orogen. The gravity data and additional

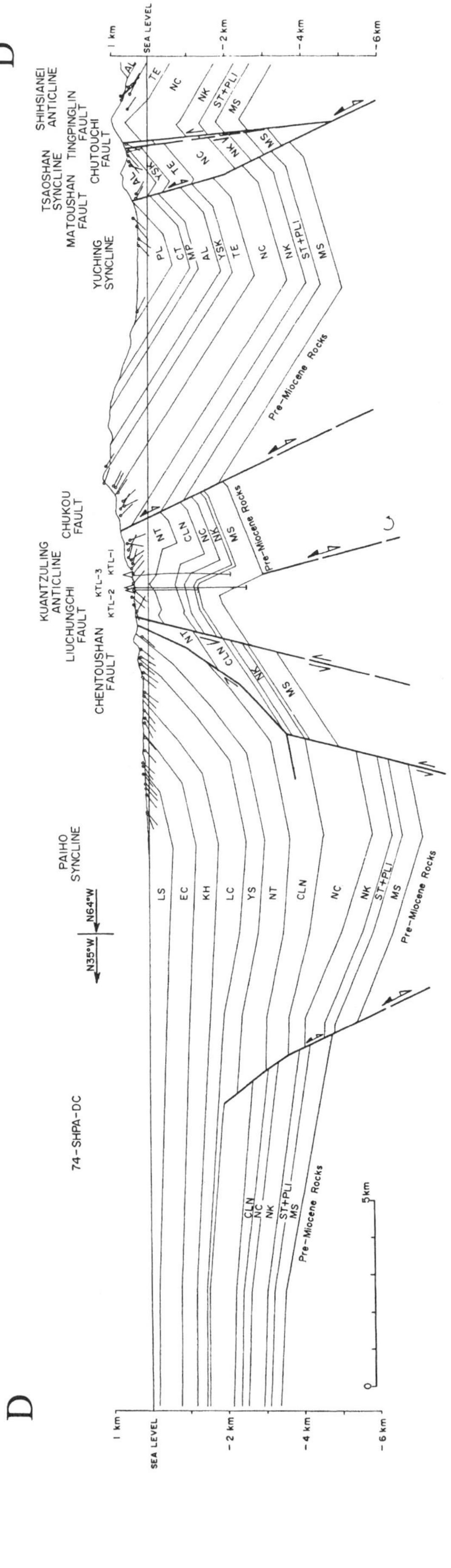

Figure 11. Cross section D–D′ across the Chiai-Tainan area, southwestern Taiwan (modified after Chang et al., 1996). See Figures 1 and 10 for location of section line. Abbreviations as in Figure 2.

seismic reflection data suggest that the Paiho syncline is, at least in part, bounded by high-angle normal faults.

Kuantzuling anticline and Chentoushan and Liuchungchih faults. The Kuantzuling anticline appears to have originated as a basement high or horst. The apparent basement high is defined on its east by a subsurface normal fault, which separates it from the Paiho graben, and on its east side by the Chukou fault. In addition, core drilled into the high is cut by the Liuchungchi fault, which is interpreted as an early normal fault. The Chentoushan fault, which crops out east of the Liuchungchih fault, is interpreted as a high-angle, Pliocene–Pleistocene backthrust that originates from a basal detachment within the Chunglun Formation. An alternative and possibly more radical interpretation is that the Kuantzuling anticline may be a fault-propagation fold above a footwall shortcut to the deeper part of the Chentoushan fault.

Chukou fault and Matoushan fault. The Chukou and Matoushan faults are interpreted to be reactivated normal faults on the basis of an abrupt change in stratigraphic thickness and lithology across them. For example, the thickness of the Miocene Changchikeng Formation changes from 250 m to 750 m across the Chukou fault. In addition, two systems of lithostratigraphic units have been established by geologists of the Chinese Petroleum Corporation to account for sediments of same age but different lithology on the east and west sides of the Chukou fault (Figs. 2, 11, and 12). The Nanchuang Formation also changes thickness from 800 m to 1000 m across the Matoushan fault, suggesting normal displacement.

Restored section D–D′ (Figure 12)

The Mushan through Nanchuang units in cross section D–D′ were restored individually and to equal lengths to illustrate the precontractional fault and basin geometries (Fig. 12). In the restored section, abrupt changes in the thickness of the Nanchuang Formation across faults indicate the existence of extensional fault systems that divide the section into a group of grabens and horsts. This example shows that the Kuantzuling anticline is a basement high bounded by normal faults and the Matoushan fault is a reactivated normal fault. The restored section suggests a total shortening of 4.1 km.

During the Pliocene–Pleistocene orogeny, the Chentoushan thrust formed the important tectonic boundary between intensely deformed rocks of the fold-and-thrust belt and those of the essentially undeformed coastal plain. The intensity of the deformation decreases from east to west along the section. East of the Chentoushan thrust, the old normal faults—including the Chukou and Matoushan faults—were reactivated, and the slip direction along them was inverted. West of the Chentoushan thrust, the graben remains essentially undeformed.

The restored section (Fig. 12) suggests that synsedimentary faulting controlled sedimentation of the Nanchuang Formation in both the east and west directions and that the Kuantzuling basement high was an area of little Nanchuang deposition. In addition, there is a significant increase in the thickness of the

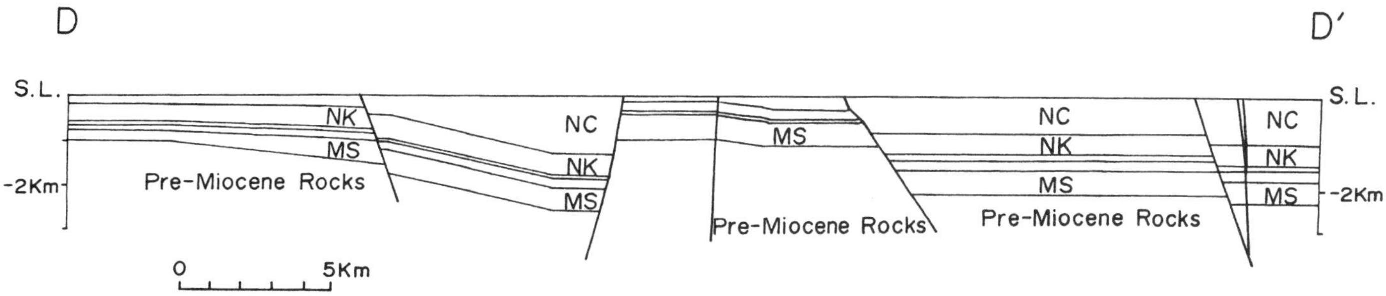

Figure 12. Cross section D–D′ restored by assuming that the top of the upper Miocene Nauchuang Formation was horizontal (after Chang et al., 1996).

Nanchuang Formation from north to south, indicating that the basin deepened to the south (Chang et al., 1996).

DISCUSSION

The Western Foothills of Taiwan provide a good example of a fold-and-thrust belt that previously was interpreted as being thin-skinned but which may warrant reinterpretation as a belt dominated by inversion structures. The problem of identifying inversion tectonics is often aggravated by the development of new, thin-skinned detachments, associated with basin inversion, and by the absence of clear evidence for structural style at depth. Compare, for example, the cross sections shown by Namson (1984) and Suppe (1980a) with the cross sections shown in Figures 5, 6, 9, and 11.

The structural style of contractional deformation in the fold-and-thrust belt of Taiwan appears to involve both thick-skinned and thin-skinned geometries; the major anticlines are interpreted as fault-bend (Suppe, 1983), fault-propagation (Suppe, 1984; Suppe and Medwedeff, 1990; Narr and Suppe, 1994), or fault-rotation (Coward, 1994, 1996) folds. In order to better evaluate the deep crustal structure of the fold-and-thrust belt, however, the structural style and kinematics of the contractional structures in the shallow crust must be understood. Although clearly a difficult task, it is nonetheless highly desirable that every effort be made to establish the depth to regional décollements into which old normal faults merge. This depth may be indicated from the base of seismicity, and the décollement may occur near the boundary between brittle and ductile deformation processes. The décollement is the zone where deformation changes from brittle, fault-related, and block-rotational deformation to more ductile deformation at depth. It is generally not a single surface. On the basis of the evidence for normal-fault reactivation presented here, the décollement is probably deeper than the 8 to 10 km suggested by previous models (Suppe, 1980b, 1981).

In view of the evidence of Paleogene rifting in the footwall rocks of the Taiwan fold-and-thrust belt, a fault-rotation folding process that involves rotation of faulted blocks (Coward, 1994, 1996) is preliminarily used to illustrate reactivation of the old normal faults within pre-Miocene rocks. Our interpretation is by no means definitive. Even though the surface and subsurface data across the study areas provide considerable constraint for structural interpretation, the data are still insufficient to determine a unique solution for either the shallow or deep crustal structure. There might be a number of alternative solutions for the deep crustal structure of the fold-and-thrust belt that are consistent with the observed shallow structural geometry across the area. Additional geophysical data are required to fully evaluate the depth to regional detachments and various models.

In addition, structural cross sections shown in this paper are all two-dimensional and do not take into account any strike-slip component of deformation. Such a component is probably inevitable in many orogenic belts that might appear to be plane strain. In fact, strike-slip faults are common in the Western Foothills of Taiwan (e.g., Tsan and Keng, 1962; Ho, 1967; Chiu, 1971; Huang, 1987; Wey et al., 1992; Hsu et al., 1997; Deffontaines et al., 1997), although most faults appear to have relatively small displacements. The two-dimensional sections presented here should therefore be considered as preliminary. Future, more regional studies of the fold-and-thrust belt should incorporate along-strike motions so that these three-dimensional strains and displacements can then be integrated into the cross sections.

CONCLUSIONS

Extensional faults of Paleogene through Miocene age are found with great lateral extent throughout the fold-and-thrust belt of western Taiwan. Stratigraphic thickening is one of the most useful criteria in identifying the extensional fault geometries. Relatively simple extensional fault geometries commonly have been inverted by an oblique thrust sense of motion resulting from contractional deformation associated with the arc-continent collision. The development of reverse faulting and folding through the inversion of an extensional fault system can result in a thick-skinned fold-and-thrust belt. In addition, low-angle thrust faults associated with regional shortening can also form and result in a thin-skinned fold-and-thrust belt. Our study suggests that in Taiwan, the contribution of the thick-skinned tectonics to the foothills structure is no less than that of the thin-skinned tectonics.

The structural style and kinematics of contractional features of the fold-and-thrust belt of the Taiwan orogen are related to inversion tectonics. The evidence for inversion tectonics includes (1) nonforward propagation of fault displacement, (2) presence of an older basin and the rapid change in thickness or facies of sediments, and (3) an increase in structural relief up the dip of faults.

ACKNOWLEDGMENTS

We thank the Chinese Petroleum Corporation for support to work on this project and for permission to publish this paper. C.-I. Lee expressed appreciation to Robert R. Berg and Biq Chingchang for their inspiration of this study. Figures were prepared by T.Y. Chiu, D.C. Huang, S.C. Lo, C.J. Hsu, Y.L. Wu, Z C. Wu, and S.F. Wu.

REFERENCES CITED

Biq, Chingchang, 1969, Role of gravitational gliding in Taiwan tectogenesis: Bulletin of the Geological Survey of Taiwan, no. 20, p. 1–39.

Biq, Chingchang, 1972, Western Taiwan thrusts, active or inactive?: Acta Geologica Taiwanica, no. 15, p. 69–81.

Boyer, S.E., and Elliott, D., 1982, Thrust systems: American Petroleum Geologists Bulletin, v. 66, p. 1196–1230.

Butler, R.W.H., 1984, Structural evolution of the Moine thrust belt between Loch More and Glendhu, Sutherland: Scottish Journal of Geology, v. 20, p. 161–179.

Butler, R.W.H., 1989, The influence of pre-existing basin structure on thrust system evolution in the Western Alps, *in* Cooper, M.A., and Williams, G.D., eds., Inversion tectonics: Geological Society [London] Special Publication 44, p. 105–122.

Butler, R.W.H., 1997, Late Proterozoic rift faults and basement-cover relationships within the BenMore thrust sheet, NW Scotland: Journal of the Geological Society, London, v. 154, p. 761–764.

Chang, H.C., Lin, C.W., Chen, M.M., and Lu, S.T., 1998, An introduction to the active faults of Taiwan: Central Geological Survey, Ministry of Economic Affairs, Taipei, Taiwan, Republic of China, 103 p.

Chang, Y.L., Mao, E.W., and Tseng, C.S., Lee, C.I., 1993, Application of balanced cross-sections to the foothills area of western Taiwan: Geological Society of China, Annual Meeting, Taipei, Taiwan, Program and Extended Abstracts, p. 15.

Chang, Y.L., Lee, C.I., Lin, C.W., Hsu, C.H., and Mao, E.W., 1996, Inversion tectonics in the fold-thrust belt of the foothills of the Chiai-Tainan area, southwestern Taiwan: Petroleum Geology of Taiwan, no. 30, p. 163–176.

Chen, Y.T., and Tang, S.L., 1993, A review of the petroleum geology of the Neogene basin, offshore Hsinchu-Taichung area, Taiwan: Petroleum Geology of Taiwan, no. 28, p. 289–314 (in Chinese with English abstract).

Cheng, A.B., and Tseng, P.S., 1995, Subsurface geological report of the TP-3 well, the taping structure, Hsinchu: Miaoli, Taiwan, Chinese Petroleum Corporation, Taiwan Petroleum Exploration Division, Unpublished Report, 8 p. (in Chinese).

Chi, W.R., 1994, Basin characterization, tectonic evolution, and hydrocarbon potential of the offshore and onshore Taiwan sedimentary basins: Bulletin of Exploration and Development Research, no. 17, p. 82–114 (in Chinese with English abstract).

Chiu, H.T., 1971, Fold in the northern half of western Taiwan: Petroleum Geology of Taiwan, no. 8, p. 7–19.

Chiu, H.T., and Hsu, C.H., 1963, Subsurface geology of the Chinshui gas field, Miaoli: Petroleum Geology of Taiwan, no. 2, p. 235–269.

Chou, J.T., and Yeh, M.K., 1987, Comparison of the suggested "Szuling Sandstone" in the Tiehchanshan TCS-34 well, Miaoli and the Paileng Formation, Kuoshing, Nantou: Petroleum Geology of Taiwan, no. 23, p. 41–53.

Chow, J., Yuan, J.W., and Yang, K.M., 1986, Geological interpretation of the seismic data in the Houpi area, Tainan: Petroleum Geology of Taiwan, no. 22, p. 27–53.

Chung, H.S., Hu, C.H., and Ting, J.S., 1989, The mode of reverse faults of offshore Hsinchu-Miaoli area: Miaoli, Taiwan, Chinese Petroleum Corporation, Taiwan Petroleum Exploration Division, Unpublished Report, 38 p. (in Chinese).

Chung, K.T., 1996, Foothills structure in the Miaoli area [M.S. thesis]: Chengli, Taiwan, National Central University, 108 p. (in Chinese with English abstract).

Cooper, M.A., and Williams, G.D., 1989, Inversion tectonics: Geological Society [London] Special Publication 44, 375 p.

Coward, M.P., 1988, The Moine thrust and the Scottish Caledonides, *in* Mitra, G., and Wojtal, S., eds., Geometries and mechanisms of thrusting, with special reference to the Appalachians: Geological Society of America Special Paper 222, p. 1–16.

Coward, M.P., 1994, Inversion tectonics, *in* Hancock, P.R., ed., Continental deformation: Oxford, England, Pergamon Press, p. 289–304.

Coward, M.P., 1996, Balancing sections through inverted basins: Geological Society [London] Special Publication 99, p. 51–77.

Coward, M.P., and Butler, R.W.H., 1985, Thrust tectonics and the deep structure of the Pakistan Himalaya: Geology, v. 13, p. 417–420.

Coward, M.P., Gillcrist, R., and Trudgill, B., 1991, Extensional structures and their tectonic inversion in the Western Alps: Geological Society [London] Special Publication 56, p. 93–112.

CPC (Chinese Petroleum Corporation), 1972, Oil and gas exploration and development in Taiwan: Miaoli, Taiwan, Taiwan Petroleum Exploration Division (TPED), 131 p.

CPC (Chinese Petroleum Corporation), 1994, Geologic map of western Taiwan, Miaoli sheet: Miaoli, Taiwan, Taiwan Petroleum Exploration Division (TPED), scale 1:100 000, 1 sheet.

Dahlstom, C.D.A., 1970, Structural geology in the eastern margin of the Canadian Rocky Mountains: Bulletin of Canadian Petroleum Geologists, v. 18, p. 332–406.

Davis, T.L., and Namson, J., 1989, A cross section of the Los Angeles area: Seismically active fold-and-thrust belt, the 1987 Whittier Narrows earthquake, and earthquake hazard: Journal of Geophysical Research, v. 94, p. 9644–9664.

Deffontaines, B., Lacombe, O., Angelier, J., Chu, H.T., Mouthereau, F., Lee, C.T., Deramond, J., Lee, J.F., Yu, M.S., and Liew, P.M., 1997, Quaternary transfer faulting in the Taiwan foothills: Evidence from a multisource approach: Tectonophysics, v. 274, p. 61–82.

de Graciansky, P.C., Dardeau, G., Lemoine, M., and Tricart, P., 1989, The inverted margin of the French Alps and foreland basin inversion, *in* Cooper, M.A., and Williams, G.D., eds., Inversion tectonics: Geological Society [London] Special Publication 44, p. 87–104.

Elliott, D., and Johnson, M.R.W., 1980, Structural evolution in the northern part of the Moine thrust belt, NW Scotland: Transactions of the Royal Society of Edinburgh: Earth Sciences, v. 71, p. 69–96.

Ellwood, A., Wang, C.Y., Teng, L.C., and Yen, H.Y., 1996, Gravimetric examination of thin-skinned detachment vs. basement-involved models for the Taiwan orogen: Journal of the Geological Society of China, v. 39, p. 209–221.

Gillcrist, R., Coward, M.P., and Mugnier, J.L., 1987, Structural inversion, examples from the Alpine foreland and the French Alps: Geodinimica Acta, v. 1, p. 5–34.

Ho, C.S., 1967, Strike-slip faulting in Shantzechiao and Chingshuikeng, northern Taiwan: Proceedings of the Geological Society of China, no.10, p. 83–98.

Ho, C.S., 1976, Foothill tectonics of Taiwan: Bulletin of the Geological Survey of Taiwan, no. 25, p. 9–28.

Ho, C.S., 1982, Tectonic evolution of Taiwan and explanatory text of the tectonic map of Taiwan: Taipei, Taiwan, Republic of China, Central Geological Survey, Ministry of Economic Affairs, 126 p.

Ho, C.S., 1988, An introduction to the geology of Taiwan and explanatory text of the geologic map of Taiwan (second edition): Taipei, Taiwan, Republic of China, Central Geological Survey, Ministry of Economic Affairs, 192 p.

Hsu, C.H., Lee, C.I., Lee, J.F., Chang, Y.L., and Mao, E.W., 1997, NW-SE trending strike-slip faults in Taiwan, and its implications in tectonics: Geological Society of China, Annual Meeting, Tainan, Taiwan, Program and Extended Abstracts, p. 309–312.

Hsu, C.Y., and Wey, S.K., 1983, Structural geology in the Chiai foothills, Taiwan: Petroleum Geology of Taiwan, no. 19, p. 17–28.

Huang, C.S., 1984, Quaternary faults in the coastal area, beneath Hsinchu and Chunan, northwestern Taiwan: Special Publication of the Central Geological Survey, no. 3, p. 103–126.

Huang, C.Y., 1998, Geology and tectonic evolution of Taiwan (in Chinese): Symposium of Geologic Environment and Natural Hazard, Taipei, Taiwan, National Taiwan University, p. 1–16.

Huang, C.T., and Hsu, T.H., 1958, Subsurface geological report of the FP-2 well, the Fangpokeng structure, Hsinchu: Chinese Petroleum Corporation, Taiwan Petroleum Exploration Division, Unpublished Report, 4 p.

Huang, F.W., 1987, Wrench faults in western Taiwan and oil-gas accumulations: Petroleum Geology of Taiwan, no. 23, p. 1–12.

Huang, S.T., Chen, R.C., and Chi, W.R., 1993, Inversion tectonics and evolution of the northern Taihsi basin, Taiwan: Petroleum Geology of Taiwan, no. 28, p. 15–46.

Hung, B.J., and Ho, C.K., 1979, Subsurface geological report of the TCS-34 well, the Tiehchenshan gas field, Miaoli: Chinese Petroleum Corporation, Taiwan Petroleum Exploration Division, Unpublished Report, 15 p. (in Chinese).

Hung, J.H., and Wiltschko, D.V., 1993, Structure and kinematics of arcuate thrust faults in the Miaoli-Cholan area of western Taiwan: Petroleum Geology of Taiwan, no. 28, p. 59–96.

Kuan, M.Y., 1971, Advanced study on the subsurface geology of the Chinshui gas field, Miaoli, Taiwan: Petroleum Geology of Taiwan, no. 8, p. 47–64.

Lee, C.I., Chang, Y.L., Mao, E.W., and Tseng, C.S., 1993, Fault reactivation and structural inversion in the Hsinchu-Miaoli area of northern Taiwan: Petroleum Geology of Taiwan, no. 28, p. 47–58.

Lee, C.I., Hsu, C.H., Coward, M.P., and Chang, Y.L., 1997, Tectonics of the fold-thrust belt of the Taiwan orogen: Tectonics Studies Group (TSG) Annual Meeting, Durham, UK, University of Durham.

Lee, C.I., and Chang, Y.L., 1998, Tectonics of the fold-thrust belt of western Taiwan: Thin-skinned or thick-skinned tectonics: Geological Society of China, Annual Meeting, Chungli, Taiwan, Program and Extended Abstracts, p. 131–132.

Lee, C.I., Chang, Y.L., Coward, M.P., Hsu, C.H., and Chi, W.R., 1998, Inversion tectonics of the fold-thrust belt of western Taiwan: Eos (Transactions, American Geophysical Union), v. 79, no. 24, supplement, paper T52A-12.

Letouzey, J., 1990, Fault reactivation, inversion and fold-thrust belt, *in* Letouzey, J., ed., Petroleum and tectonics in mobile belts: Paris, Techniq, p. 101–127.

Lin, L.H., Chang, C.T., Kuo, C.L., and Chou, T.S., 1997, Geochemistry of hydrocarbon accumulation in western Taiwan: Petroleum Geology of Taiwan, no. 31, p. 169–202.

Liu, C.H., and Hsu, C.Y., 1988, Petroleum exploration on the southwestern plain of Taiwan: Petroleum Geology of Taiwan, no. 24, p. 18–36.

Meng, C.Y., 1963, The Sani overthrust: Petroleum Geology of Taiwan, no. 2, p. 1–20.

Namson, J., 1981, Structure of the western foothills belt, Miaoli-Hsinchu area, Taiwan. 1. Southern part: Petroleum Geology of Taiwan, no. 18, p. 31–51.

Namson, J., 1982, Studies of the structure, stratigraphic record of plate interaction and role of pore-fluid pressure in the active fold-and-thrust belt of Taiwan and a study of manganese deposits from northern California [Ph.D. thesis]: Princeton, New Jersey, Princeton University, 302 p.

Namson, J., 1983, Structure of the western foothills belt, Miaoli-Hsinchu area, Taiwan. 2. Central part: Petroleum Geology of Taiwan, no. 19, p. 51–76.

Namson, J., 1984, Structure of the western foothills belt, Miaoli-Hsinchu area, Taiwan. 3. Northern part: Petroleum Geology of Taiwan, no. 20, p. 35–52.

Narr, W., and Suppe, J., 1994, Kinematics of basement-involved compressive structures: American Journal of Science, v. 294, p. 802–860.

Rich, J.I., 1934, Mechanics of low-angle overthrust faulting as illustrated by Cumberland thrust block, Virginia, Kentucky, and Tennessee: American Association of Petroleum Geologists Bulletin, v. 18, p. 1584–1596.

Sibuet, J.C., and Hsu, S.K., 1997, Geodynamics of the Taiwan arc-arc collision: Tectonophysics, v. 274, p. 221–251.

Soper, N.J., and England, R.W., 1995, Vendian and Riphean rifting in NW Scotland: Journal of the Geological Society, London, v. 152, p. 11–14.

Stewart, A.D., 1982, Late Proterozoic rifting in NW Scotland: The genesis of the Torridonian: Journal of the Geological Society, London, v. 139, p. 413–420.

Sun, S.C., 1985, The Cenozoic tectonic evolution of offshore Taiwan: Energy, v. 10, no. 3/4, p. 421–432.

Suppe, J., 1976, Decollement folding in southwestern Taiwan: Petroleum Geology of Taiwan, no. 13, p. 25–35.

Suppe, J., 1980a, Imbricated structure of western foothills belt south-central Taiwan: Petroleum Geology of Taiwan, no. 17, p. 1–16.

Suppe, J., 1980b, A retrodeformable cross section of northern Taiwan: Proceedings of the Geological Society of China, no. 23, p. 46–55.

Suppe, J., 1981, Mechanics of mountain building and metamorphism in Taiwan: Memoir of the Geological Society of China, v. 4, p. 67–89.

Suppe, J., 1983, Geometry and kinematics of fault bend folding: American Journal of Science, v. 283, p. 684–721.

Suppe, J., 1984, Seismic interpretation of the compressively reactivated normal fault near Hsinchu, western Taiwan: Petroleum Geology of Taiwan, no. 20, p. 85–96.

Suppe, J., 1986, Reactivated normal faults in the western Taiwan fold-and-thrust belt: Memoir of the Geological Society of China, v. 7, p. 187–220.

Suppe, J., and Medwedeff, D.A., 1990, Geometry and kinematics of fault-propagation folding: Eclogae Geologicae Helvetiae, v. 83, p. 409–454.

Suppe, J., and Namson, J., 1979, Fault-bend origin of frontal folds of the western Taiwan fold-and-thrust belt: Petroleum Geology of Taiwan, no. 16, p. 1–18.

Teng, L.S., 1992, Influence of morphotectonic configuration of China continental margin on collision tectonics of Taiwan: Acta Geologica Taiwanica, no. 30, p. 71–78.

Tsan, S.F., and Keng, W.P., 1962, The strike-slip faulting and concurrent or subsequent folding in the Alishan area, Taiwan: Proceedings of the Geological Society of China, v. 5, p. 119–126.

Wang, W.K., Sun, J.H., Chiu, Z.Y., and Tsai, C.J., 1994, Structure and hydrocarbon potential between northern Taichung basin and Hsinlung structure, Miaoli: Miaoli, Taiwan, Chinese Petroleum Corporation, Taiwan Petroleum Exploration Division (TPED), Unpublished Report, 21 p. (in Chinese).

Welbon, A.I., and Butler, R.W.H., 1992, Structural styles in thrust belts developed through rift basin: A view from the western Alps, *in* Larsen, R.M., et al., eds., Structural and tectonic modelling and its application to petroleum geology: Norwegian Petroleum Society Special Publication 1, p. 469–479.

Wey, S.K., and Peng, D.H., 1982, Subsurface geological report of the KTL-3 well, the Kuantzulin structure, Tainan: Miaoli, Taiwan, Chinese Petroleum Corporation, Taiwan Petroleum Exploration Division, Unpublished Report, 26 p.

Wey, S.K., Wu, K.S., Lai, C.Y., and Hsu, J.S., 1992, New prospects on the northern part of the Chuhuangkeng structure, Miaoli, Taiwan: Petroleum Geology of Taiwan, no. 27, p. 199–216.

Wu, F.T., Yeh, Y.H., Rau, R.J., Salzberg, D., Yu, S.B., Liu, C.C., and Lee, T.Q., 1994, The modern orogeny of Taiwan: A review: Taipei, Taiwan, Republic of China, Jubilee Publication in the Commemoration of Director T.Y. Wu on the Occasion of his Retirement, Academic Sinica, p. 45–77 (in Chinese).

Wu, F.T., Rau, R.J., and Salzberg, D., 1997, Taiwan orogeny: Thin-skinned or lithospheric collision?: Tectonophysics, v. 274, p. 191–200.

Yang, K.M., Wu, J.C., Wickham, J.S., Ting, H.H., Wang, J.B., and Chi, W.R., 1996, Transverse structures in Hsinchu and Miaoli areas: Structural mode and evolution in foothills belt, northwestern Taiwan: Petroleum Geology of Taiwan, no. 30, p. 111–150.

Yang, K.M., Ting, H.H., Wu, J.C., and Chi, W.R., 1997, Geological model for complex structure and its implications for hydrocarbon exploration in northwestern Taiwan: Petroleum Geology of Taiwan, no. 31, p. 1–42.

Manuscript Accepted by the Society April 12, 2001

Chung-I Lee
December 15, 1950 to November 20, 1998

Chung-I Lee succumbed to a heart attack on November 20, 1998, shortly after his presentation on inversion tectonics at the Second Cross-Strait Conference on Resource Geology, held at Central University, Chungli, Taiwan.

Lee was born December 15, 1950, in Taiwan and received his B.S. degree in geology from the Chinese Culture University of Taipei in 1974. After passing the entrance examination for graduate school at Taiwan University, he transferred to the University of Illinois at Urbana-Champaign in the United States, where he was supported by a scholarship from the Taiwanese government and where he received an M.S. degree. After working for the Industrial Technology Research Institute of Taiwan, he applied to Texas A&M University and received his Ph.D. degree from there in 1991.

Lee joined the Exploration & Development Research Institute of the Chinese Petroleum Corporation (CPC) as a senior petroleum geologist in 1991. During his work with the CPC, Lee developed the technologies for fracture reservoir evaluation and he successfully employed the petroleum system concept for exploration. By analyzing large volumes of geological and geophysical data, he proposed using inversion tectonics to model the structural evolution and neotectonics of Taiwan. This was a major breakthrough and resulted in rethinking and reevaluating fold and thrust belts both inside and outside of Taiwan.

Lee was a sincere and hardworking scientist. He published a total of 44 technical papers and 18 research reports, and he frequently collaborated with colleagues in academic institutions. His sudden death was a shocking event and a huge loss to the study of the geology of Taiwan, where he was well recognized and well respected. Lee was a very kind man, and he will be dearly missed by his family and his multitude of friends. Lee is survived by his mother and by his wife, Su-Chen Hsu, of Kaoshiung, Taiwan.

Printed in the U.S.A.

Geological Society of America
Special Paper 358
2002

Variations along the strike of the Taiwan thrust belt: Basement control on structural style, wedge geometry, and kinematics

F. Mouthereau
B. Deffontaines
O. Lacombe
J. Angelier
Laboratoire de Tectonique, UMR 7072, Université Pierre et Marie Curie, T.26-25, E1, Case 129, 4 place Jussieu, 75252 Paris Cedex 05, France

ABSTRACT

A model of imbricate thrust wedges based on the conceptual model of the critically tapering wedge is discussed and applied to the case of the Taiwan thrust belt. This model takes into account (1) the occurrence of compressional features located far from the foreland of the orogen and (2) the occurrence of deep-crustal decoupling that allows both the regional stress field to be transmitted in the foreland and the basement to be involved in the orogenic wedge. Accordingly, three different belt fronts are considered, a mountain front, a reactivation front, and a deformation front, based on topographic, kinematic, and mechanical criteria, respectively. The reactivation front located at the outermost reactivated extensional structure displays large curvatures in areas where structural inversion occurs. The mountain front is usually distinct from the reactivation front and localizes the emergence of a shallow décollement. Serial geologic sections of the thrust belt provide strong arguments in favor of the superimposition of deep- and shallow-décollement tectonics at the thrust-belt front in agreement with the model proposed. The along-strike structural changes are usually accompanied by changes in tectonic regimes due to local effects such as frontal contraction and lateral movement in response to indentation by the basement highs. The record of orogenic stresses in the Taiwan Strait allows us to define and locate a deformation front west of the Penghu Islands. Our results suggest that single-minded models based solely on the principles of either thin-skinned or thick-skinned tectonics may be unrealistic in the case of the Taiwan thrust belt.

INTRODUCTION

The Taiwan thrust belt has been taken as a key example for what is usually called thin-skinned tectonics (Suppe, 1976; Namson, 1981). The structure of the Western Foothills units of the Taiwan thrust belt has been investigated and described in terms of balanced cross sections based on the geometric principles of fault-related folds (Suppe and Namson, 1979). Analyses of recent structural data, however, argued in favor of basement-involved tectonics for the foreland fold-and-thrust belt (Lee et al., 1993), and new geophysical works defended the thick-skinned model for the whole thrust belt (Ellwood et al., 1996; Wu et al., 1997). Understanding the structure of the Taiwan thrust belt and the kinematic processes prevailing at depth is still an objective to be met.

The first goal of this paper is to decipher the possible in-

Mouthereau, F., Deffontaines, B., Lacombe, O., and Angelier, J., 2002, Variations along the strike of the Taiwan thrust belt: Basement control on structural style, wedge geometry, and kinematics, *in* Byrne, T.B., and Liu, C.-S., eds., Geology and Geophysics of an Arc-Continent Collision, Taiwan: Boulder, Colorado, Geological Society of America Special Paper 358, p. 31–54.

volvement of the basement in the tectonics of the western frontal units. Second, we aim at relocating the thrust-belt front and determining the type of related structures. In addition, we demonstrate that along-strike variations occur in the wedge geometry. To this purpose, we have first investigated the overall structural framework of the foreland basement and established a new basement map of the western foreland. The accurate location of the thrust-belt front was first approached by a consideration of the nature and the geometry of the thrust wedge that is based on morphostructural and basement-topography analyses. These investigations provided further information on the geometry of structures and their relationship with basement-involved tectonics. On the basis of these considerations, serial geologic cross sections of the frontal thrust units have been constructed. Finally, we have carried out a kinematic analysis of thrust emplacement by means of synthetic paleostress reconstruction.

FORELAND THRUST-BELT FRONT: NATURE AND SIGNIFICANCE

Previous work on thrust-belt fronts

The study of thrust belts has been greatly spurred by the development of thin-skinned tectonics theory, especially the fundamental work of Bally et al. (1966) and Dahlstrom (1969), conducted in the Rocky Mountains of North America. These pioneering works largely contributed to the understanding of the structural framework of thrust systems as a whole (Boyer and Elliott, 1982). Furthermore, because of the increase in petroleum investigation, many structural geologists are focused on the frontal zones of thrust belts such as the one in Taiwan (Suppe and Namson, 1979).

Consequently, thrust-belt fronts (or "mountain fronts") have been described by various terminology. Basically, the terms refer to the topographic boundary of the regional foreland-dipping monocline, elevated above its initial structural level, i.e., the foreland of the thrust belt (Vann et al., 1986). This obvious morphologic frontier was regarded as the result of the emplacement of the outermost thrust. A first attempt at classifying the fronts of thrust belts, in terms of thin-skinned tectonics and based on the belts' geometries, has led geologists to distinguish two main types of fronts: (1) buried thrust fronts and (2) emergent thrust fronts. The occurrence of each of these types depends on geometric and mechanical factors, such as the lateral termination of strata suitable for hosting a décollement and/or the presence of a broad area of weakly strained rocks (Morley, 1986).

Significance of front in the critically tapering wedge model

Improvements in the understanding of thrust-belt mechanics have enabled the fold-and-thrust belt to be modeled. According to the thin-skinned tectonics theory, Chapple (1978) put emphasis on the following key assumption: most fold-and-thrust belts exhibit a basal décollement that gently dips toward the interior of the thrust belt. This basal décollement is usually sited at a relatively weak level, for example, within salt or shale units in the sedimentary cover. Above the basal décollement, compressional deformation occurs whereas in the rigid basement below, the deformation remains limited. These hypotheses led Davis et al. (1983) to describe the mechanical behavior of fold-and-thrust belts and accretionary wedges in terms of critically tapering wedges of Coulomb material (Fig. 1A). According to this model, as the critical taper is achieved, the wedge deforms internally by thrust-sheet imbrication.

In order to apply this model to mountain building, Davis et al. (1983) implicitly considered that the toe of the thrust wedge corresponds to the thrust-belt front defined by structural geologists, i.e., the topographic front. However, the thrust-belt front thus determined in the critical-taper theory (Fig. 1A) is in fact an assemblage of three distinct and fundamental boundaries. First, there is a *topographic boundary*, because one of the two parameters characterizing the critical-taper angle (γ) is the surface-slope angle (Fig. 1A). Second, a *kinematic boundary* is easily defined because as the critical taper is attained, sliding occurs along the basal décollement, resulting in the development of a new frontal thrust. Beyond this limit, i.e., at the toe, the propagation of the wedge ceases. Consequently, the front in the simple model of Davis et al. (1983) could also be viewed as a pin line (the nail in Fig. 1A) that marks the position of no movement. Third, a *mechanical boundary* can be defined between the fold-and-thrust belt hinterland (which deforms internally by faulting and folding) and the undeformed foreland domains (where low stress magnitudes prevail). Therefore, according to the critical-taper theory, defining a front in a fold-and-thrust belt must refer to these three different considerations, and three types of fronts are thus possible: a *mountain front*, a *reactivation front*, and a *deformation front*, based on topographic, kinematic, and mechanical criteria, respectively.

Hereafter, we propose an alternative model, based on the critically tapering wedge theory, in which the different fronts are distinguished and the involvement of the basement is considered (Fig. 1B). This model also takes into account the presence of inversion tectonics—usually documented in the forelands of orogens—as well as imbricate thrust wedges (Fig. 1B).

The upper thrust wedge is restricted to the cover ("thin-skinned" tectonics) and corresponds to the classical steady-state thrust wedge (Fig. 1A) considered by Davis et al. (1983). Its equilibrium depends on a critical-taper angle (γ_1 in Fig. 1B), which is controlled by the dip of the shallow décollement in the cover and the topographic slope (see Fig. 2). This thrust wedge develops by propagation of newly formed frontal thrust sheets toward the foreland; its front, the thrust-wedge front (1) in Figure 1B, corresponds to the outermost limit of the allochthonous units and is equivalent to the topographic front (= the mountain front). However, because the deformation is limited to the upper levels, the thin-skinned tectonics cannot ac-

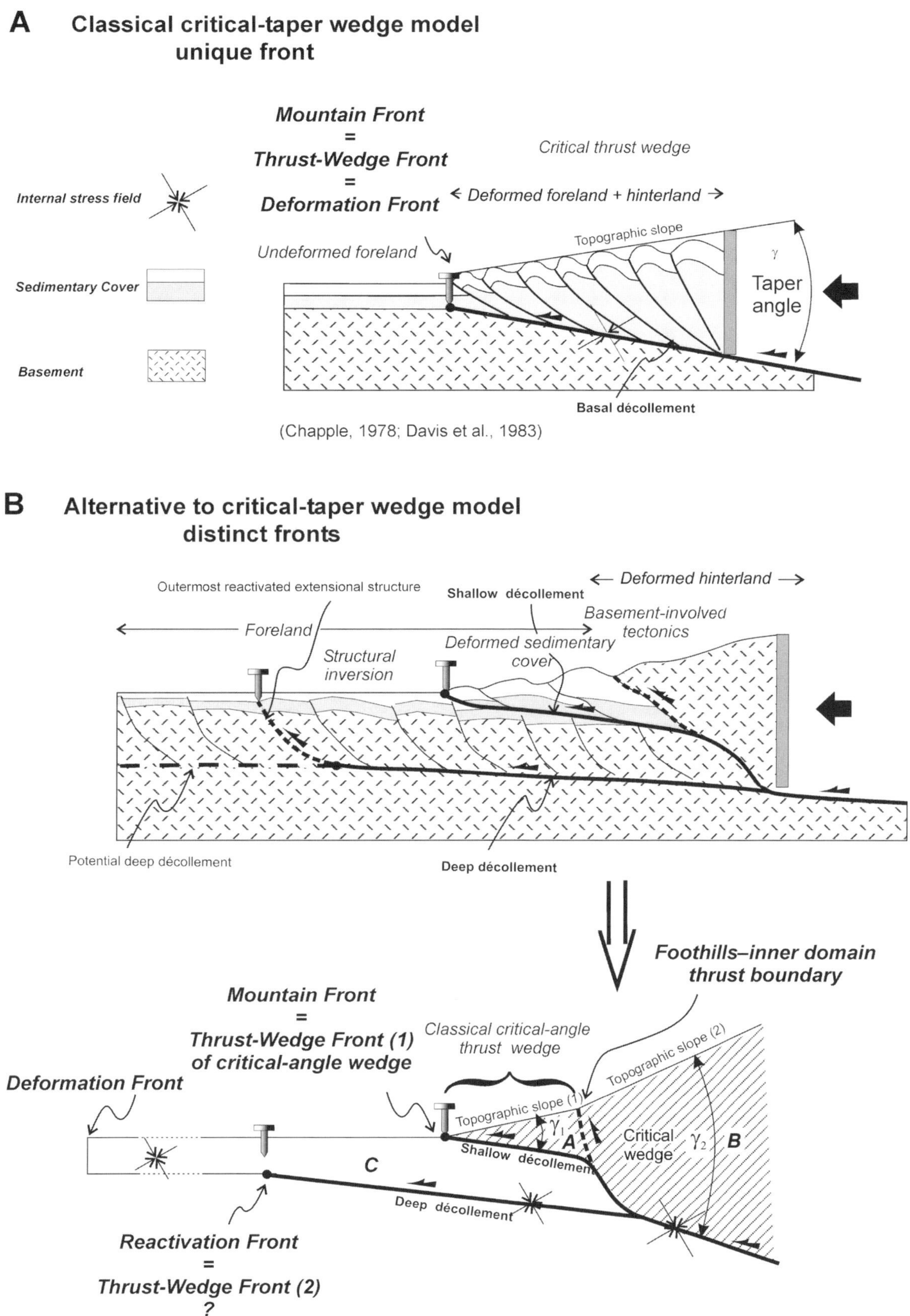

Figure 1. Determination of the different front types of foreland thrust belts on the basis of the décollement tectonics model. (A) Nature of a front in classical critically tapering wedge model after Chapple (1978) and Davis et al. (1983). (B) Alternative to critical-taper model. This alternative considers imbricate thrust wedges bounded by either a shallow décollement (thin-skinned tectonics) or a deep décollement (thick-skinned tectonics) and defines different types of structural fronts: a *mountain front* (the front of the critically tapering wedge), a *reactivation front* (the outermost reactivated preexisting extensional feature), and a *deformation front*. Differentiation of the front types is on the basis of topographic, kinematic, and mechanical criteria. A—shallow thrust wedge, B—inner domain of thrust wedge, C—outer domain of thrust wedge, γ_1—the critical-taper angle controlled by the dip of the shallow décollement, and γ_2—the critical-taper angle controlled by basement structure.

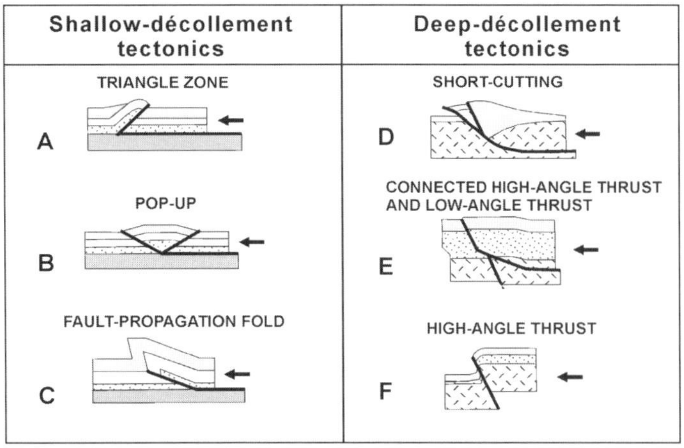

Figure 2. Main fault-related fold structures encountered at the fronts of thrust belts. (Left) Frontal structures associated with shallow-décollement tectonics; newly formed faulting is prominent. (Right) Frontal structures associated with deep-décollement tectonics; reactivation of inherited faults prevails. For details, see text.

count for the propagation of the compressional deformation far in the foreland, even though in particular cases such as the Jura Mountains, the southern Pyrenees foreland, and the Mackenzie Mountains, the occurrence of a widespread and very weak stratigraphic unit that hosts the décollement allows the thrust sheets' propagation to be observed for a significant distance from the mountain front (Vann et al., 1986).

Furthermore, the occurrence of structural inversion over large areas in the forelands of the Pyrenees and the Alps (Roure and Colletta, 1996) as well as the basement-involved tectonics in the inner part of the orogenic wedge in these areas (Roure et al., 1990) led us to consider incorporating a second, deep, basal décollement ("thick-skinned" tectonics) into the alternative to the critically tapering wedge model (Fig. 1B). Above this deep décollement, two domains can be distinguished. The inner domain (B in Fig. 1B) is characterized by a specific critical-taper angle (γ_2 in Fig. 1B) controlled by basement characteristics. As the deep décollement steps up to shallow levels, its leading edge defines the previously mentioned shallow thrust wedge (A in Fig. 1B), which can therefore be regarded as a particular frontal area of a larger critically tapering wedge (hatched area in Fig. 1B) that includes the whole thrust belt (Foothills and inner domain).

In Taiwan, to the west of the mountain front, i.e., to the west of thrust-wedge front 1 in the outer domain (C in Fig. 1B), many preexisting foreland extensional structures (provided that they had appropriate orientation with respect to the transmitted tectonic stresses) have been reactivated or inverted, resulting in the development of inverted basins. A reactivation front can thus be considered for the Taiwan example; this front corresponds to the outermost reactivated extensional structure. It marks the boundary of the activated part of the deep décollement, i.e., the position of no movement (the second outer nail in Fig. 1B). According to this definition, this reactivation front is the external boundary beyond which stresses are not sufficient to create newly formed compressional structures such as folds and faults identifiable at a regional scale. Note that the low-slope and low-strength domain (i.e., the outer domain, C in Fig. 1B) that is subject to orogenic stresses and in which structural inversion occurs above a deep basal décollement also constitutes a thrust wedge limited by the reactivation front (= thrust-wedge front 2 in Fig. 1B). However, as it is not (not yet) at steady state, it therefore cannot be discussed in terms of a critically tapering wedge in the sense of Davis et al. (1983).

Abundant microtectonics studies based on fault-slip data or calcite-twin data (e.g., Letouzey, 1986; Bergerat, 1987; Craddock et al., 1993), as well as additional numerical modeling (Sassi and Faure, 1997), have extensively demonstrated the existence of far-field stresses in the foreland of orogens. In the case of the western foreland of the Taiwan orogen, present-day stress measurements based on borehole breakouts (Suppe et al., 1985) and paleostress reconstructions based on brittle microstructure analysis (Angelier et al., 1990) have revealed that offshore areas far from the mountainous regions have recorded the orogenic stresses responsible for the formation of the orogen. Therefore, a deformation front can be located farther to the west, at least as far west as the vicinity of the Penghu Islands (see Fig. 3). The record of compressional stress regimes in foreland basins does not systematically imply that such areas should be included in the thrust wedge as claimed by some authors (Craddock et al., 1993). In fact, as previously mentioned, the recognition of a thrust wedge is principally related to the occurrence of a basal décollement and not to the sole presence of far-field compression.

Ahead of the thrust-wedge front where the activity of the basal deep décollement has stopped, the part of the foreland subject to an intraplate regional-scale stress field could be viewed as a continually stable province (Fig. 1B). The deformation front could be situated far away from the first orogenic relief: for instance, orogenic stresses have been recorded by microstructures as far as 1700 km away from the Appalachian-Ouachita orogenic belt (Craddock et al., 1993) or more than 700 km away from the Pyrenees (Lacombe et al., 1996). These findings suggest that the "deformation front" can always be defined and that it is obviously distinct from the mountain front and reactivation front (Fig. 1B). The location of the deformation front depends on the spatial attenuation rate of the stress and strain magnitudes. In fact, the maximum horizontal magnitude decreases gradually from the interior of the propagating foreland thrust belt to the stable foreland (Lacombe et al., 1996). Hence, because of the strain release, the stress regimes evolve from compressional to strike-slip, and then to extensional regimes (Letouzey, 1986; Sassi and Faure, 1997). Because this change is in many cases progressive, defining a "deformation front" requires consideration of some arbitrary limit in the amount of deformation.

Summarizing, one is led to define and distinguish three

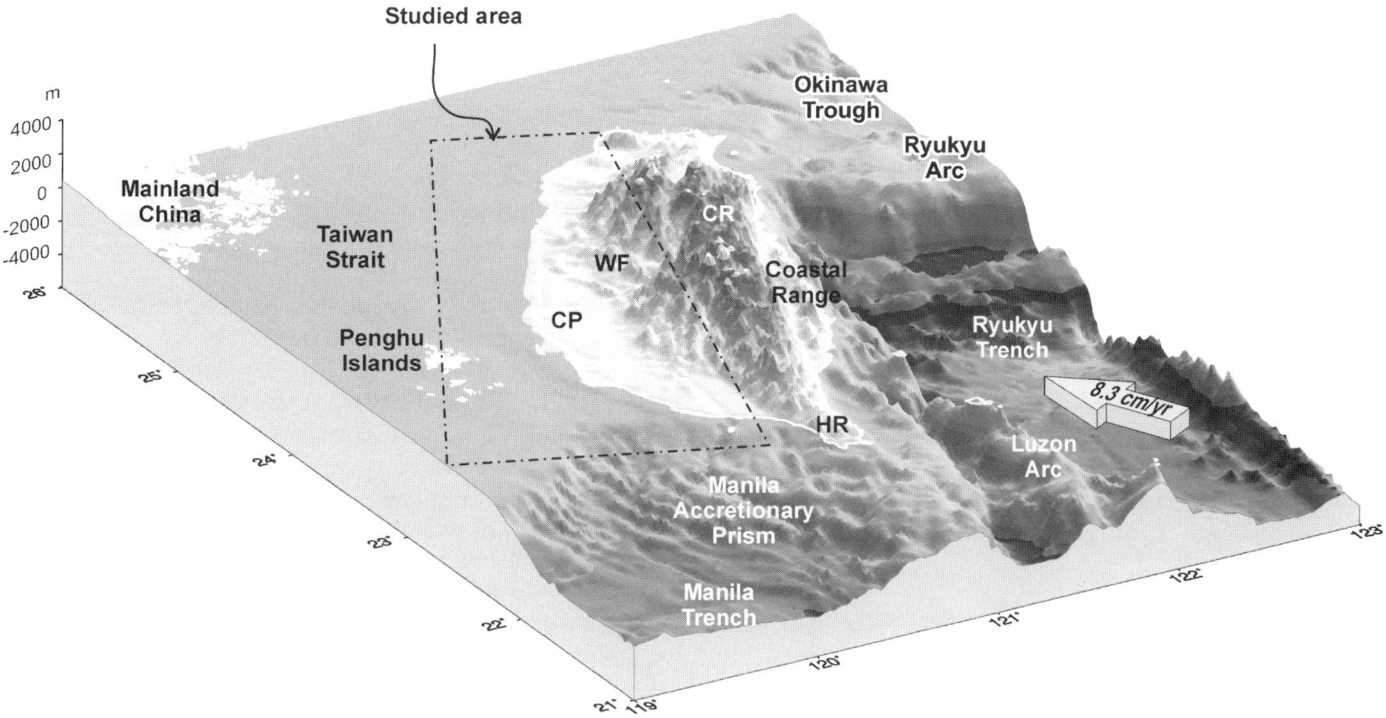

Figure 3. Block diagram of the Taiwan thrust belt in the context of arc-continent collision. Abbreviations: CP—the Coastal Plain, WF—the Western Foothills, CR—the Central Range, HR—the Hengchun Ridge.

fronts (Fig. 1B): the mountain front (the front of the critically tapering wedge), the reactivation front (the outermost reactivated preexisting extensional feature), and the deformation front. These fronts are based on topographic, kinematic, and mechanical criteria, respectively.

TYPES OF FAULT-RELATED FOLDS AT THRUST-BELT FRONTS

Hereafter, we focus on the style of compressional structures encountered at the front of thrust belts with regard to the depth of the basal décollement. Numerous studies in thrust belts have demonstrated that deformation of thrust sheets may reveal major complexities, especially in areas where basement-involved tectonics becomes a significant process in mountain building, such as in the Andes (Winslow, 1981) or in the Rocky Mountains (Dechesne and Mountjoy, 1992). This view has been documented by many authors who have pointed out that preexisting extensional basins were incorporated into thrust belts by structural inversion in which the slip on inherited faults has been reversed from normal to thrust movement, such as in the Canadian Cordillera (McClay et al., 1989) or the Pyrenees (Berastegui et al., 1990). Furthermore, basement-involved tectonics has also been described at the front of a thrust wedge, for instance in the Andes (Zapata and Allmendinger, 1996; Colletta et al., 1997). At the scale of the Taiwan thrust belt, new geophysical evidence, such as gravity anomalies (Ellwood et al., 1996) and seismic tomographic data (Wu et al., 1997), has recently led to discussion of the orogen in terms of basement-involved tectonics. This raised some questions about the mechanics of the orogen. In this paper, we intend to provide confirmation that both thin- and thick-skinned tectonic mechanisms are likely to have contributed to the development of the wedge in both the interior and the frontal zone of the Taiwan thrust belt.

We next focus our structural analyses on a limited area in the vicinity of topographic front. Furthermore, to prevent misunderstanding about the type of tectonic style involved, we chose to refer only to the depth of the basal décollement involved in the deformation under discussion. Consequently, we use the terms "shallow-décollement tectonics" and "deep-décollement tectonics" instead of the terms "thin-skinned tectonics" and "thick-skinned tectonics."

The generic model of a thrust wedge (Fig. 1B) is based on topographic (geometric), kinematic, and mechanical assumptions, but it does not predict the structural styles inside the wedge or at the thrust-belt front. Nevertheless, the type of thrust-related structures in the mountain belt can be divided into two main classes depending on the depth of the décollement (Fig. 2).

Shallow-décollement tectonics

Three main types of frontal structures may usually be identified within the framework of shallow-décollement tectonics.

First, triangle zones (Fig. 2A) are likely to be in a frontal position in most fold-and-thrust belts, mainly because such structures require the presence of mechanically weak units (Cousens and Wiltschko, 1996). Second, pop-up structures that develop above a flat décollement distinguished by low friction (Huiqi et al., 1992) are typically found in accretionary prisms where high pore pressure prevails (Fig. 2B). Note that pop-up structures and triangle zones are frequently associated with buried thrust fronts characterized by a broad zone of low strain (type 1 front of Morley, 1986). Third, fault-propagation folds (Fig. 2C) are common at thrust-belt fronts because in such structures, the foreland-directed displacement is compensated at the tip of the thrust by folding (Suppe and Medwedeff, 1990). These remarkable structures are related to the thrust-wedge front by localizing the basal décollement emergence. Because they are usually subject to significant erosion, they sometimes correspond to a smooth topography in the onshore foreland.

Deep-décollement tectonics

As previously mentioned, reactivation and/or inversion of preexisting extensional structures may be a prominent deformation process in the frontal part of orogen. We distinguish herein three main geometries for frontal thrusts affected by a deep décollement. First, basement shortcutting (Fig. 2D) can generate low-angle thrust faults at the hinge fault of an inverted basin. This geometry has been encountered in nature and reproduced in analogue models (McClay, 1992; Letouzey et al., 1995). A fault-propagation fold and an ancient normal fault could interact, because the décollement zone, necessary for the deformation to propagate, is abruptly terminated (Morley, 1986). Similarly, the nucleation of a high-angle thrust at the upper part of a reactivated normal fault could be expected, as pointed out by Suppe (1986) in northern Taiwan (Fig. 2E). Basement involvement in passive roof thrusting has been also suggested as an alternative model for frontal triangle zones and was described in the Andean front of western Venezuela (Colletta et al., 1997). High-angle thrusting (Fig. 2F), resulting from the simple reactivation of planar normal faults (Cooper and Williams, 1989; Mitra and Islam, 1994) could also be a relevant expression of inversion tectonics in the foreland of thrust belts. Most of these structures have been described and defined in inverted basins (C [the outer domain] in Fig. 1B), where the deformation occurs along a steeply dipping fault characterized by a limited amount of reverse displacement; therefore this geometry does not contribute to significant topographic elevation. However, we can expect to find some of these structures in the vicinity of or associated with the topographic front. Deciphering the particular structural patterns in the front of the Taiwan fold-and-thrust belt will enable us to determine whether shallow-décollement or deep-décollement tectonics is involved at the front of the propagating thrust wedge.

A CASE STUDY: THE WESTERN FOOTHILLS OF THE TAIWAN THRUST BELT

The tectonically active Taiwan mountains are a Neogene thrust belt resulting from the oblique collision between the Philippine Sea plate and the Eurasian plate (Fig. 3). According to the NUVEL-1 model of relative plate motion (De Mets et al., 1990), the present-day convergence of the Philippine Sea plate relative to Eurasia occurs in a N50°W direction at a velocity of ~7 cm/yr (Seno et al., 1993). Results from the Taiwan GPS (Global Positioning System) network surveys, on the other hand, favor a greater velocity, 8.3 cm/yr (Yu et al., 1997). This higher velocity is probably more realistic, because the value of 7 cm/yr neglected the relative motion of the South China block with respect to Eurasia.

In the Taiwan mountains, the initiation of a collision-type tectonic setting has been dated by the beginning of the flexural subsidence in the early Pliocene (Chang and Chi, 1983). The evolution of the collision belt postdated the Oligocene–Miocene continental rifting and spreading associated with the opening of the South China Sea (Ho, 1986). The Taiwan orogen is divided into several tectono-stratigraphic belts, striking N10°–20°E. To the east, the Coastal Range (Fig. 3), mainly composed of Neogene volcanic rocks interbedded with intra-arc Pliocene basin deposits (Huang et al., 1995), is the northern extension of the Luzon arc. It acts as a buttress for the overall thrust belt. Farther west, the rocks exposed in the Central Range belong to either of two metamorphic belts (Tananao Schist, Backbone Range) or a slate belt (Hsüehshan Range). The Central Range results from the exhumation and emplacement of a complex assemblage of ancient sedimentary basins, volcanic deposits, and oceanic materials (Ho, 1986; Teng, 1990). The Western Foothills of Taiwan (Fig. 3) are a typical foreland fold-and-thrust belt, involving east-dipping imbricate sheets being thrust onto the Chinese continental margin.

In detail, the thrust sheets of the Western Foothills are composed of pretectonic Miocene shelf deposits of the Chinese platform secondarily incorporated into the thrust wedge and currently exposed above major thrusts. In addition, Pliocene–Pleistocene synorogenic sediments derived from the erosion of inner domains were deposited and incorporated into the propagating thrust units. To the west, the Coastal Plain consists of alluvial deposits in which compressional structures are nearly insignificant. Offshore southern Taiwan, the Manila accretionary prism is the result of subduction of the South China Sea beneath the Luzon arc.

Since the Pliocene, the emplacement of the thrust belt occurred under an average west-northwest–east-southeast compression deflected to the north in northern areas and to the south in southern ones so that a fan-shaped distribution of trajectories results (Angelier et al., 1986). The recent and present-day stress field interpreted from in situ measurements of borehole breakouts (Suppe et al., 1985), inversion of earthquake focal mechanisms (Yeh et al., 1991), and Holocene faults is also dominated by west-northwest–east-southeast compression, consistent with the present-day direction of convergence.

BASEMENT GEOMETRY IN FRONT OF THE TAIWAN THRUST BELT

Tertiary basins of western Taiwan

The foreland of the Taiwan thrust belt, i.e., the Chinese platform and the Taiwan Coastal Plain, displays numerous Tertiary basins. Two categories of Tertiary basins, mostly Paleogene and Neogene in age, are distinguished on the basis of a widespread early Miocene unconformity (Sun, 1982). First, the lithosphere was stretched in response to the opening of the South China Sea during the Oligocene to the early Miocene. The resulting early Paleogene extensional basins trend northeast and are filled with shelf sediments, tuffs, and lava flows revealing the volcanic activity at this time. Extensional tectonism continued to the early Neogene, but changes of stretching directions occurred with time as revealed by the geodynamic reconstructions and microtectonics data (e.g., Angelier et al., 1990). This period of rifting resulted in the development of a series of northeast-trending horsts and grabens, overprinting the Paleogene basins. The Taiwan orogen developed mainly since 5 Ma, which is the time of the beginning of the flexural subsidence in the foreland (Chang and Chi, 1983). The resulting foredeep basins are superimposed onto the precollisional basins and strike N10°–20°E, nearly perpendicular to the direction of convergence. These basins migrated toward the outer part of the belt and were progressively filled in with synorogenic deposits.

Basement structural framework

Before investigating the subsurface structures of the foreland and discussing the involvement of basement rocks in the compressional deformation, the age, composition, and stratigraphic significance of the so-called "basement" deserve discussion.

First, in the foreland, the major part of the collision-related deposits unconformably overlie the pre-Neogene rocks. This unconformity can be regarded as reflecting a major stratigraphic boundary separating the part of the cover made of synorogenic deposits from the underlying preorogenic rocks. The preorogenic rocks were displaced by normal faulting related to the earlier episodes of extension along the Chinese margin; these rocks behaved in concert with the deeper crust during the subsequent orogenic contraction.

Second, for the Taiwanese petroleum geologists, this unconformity, i.e., the top of the pre-Neogene rocks, is a good seismic reflector that separates the upper sedimentary cover from the acoustic (or seismic) basement (Hsiao, 1974). In the literature, the pre-Neogene rocks have therefore been extensively referred to as the "basement" of the overlying sedimentary cover.

Third, the corresponding pre-Neogene rocks have been drilled in a few places in the vicinity of the Peikang Plain and beneath the Penghu Islands at an average depth of 1000 m (Chiu, 1973). They are Eocene to Cretaceous to Eocene in age and are mainly composed of arkoses and arkosic sandstones; locally some mineral assemblages point to contact metamorphism due to plutonic intrusion (Chiu, 1973). Consequently, even if the pre-Neogene rocks could not strictly be considered as a true crystalline basement, their lithology is nevertheless at least partially crystalline. Therefore, the bottom of the Miocene section can be regarded as a good proxy for the top of the true basement.

In summary, the pre-Neogene rocks might be regarded (1) as an orogenic basement in the sense of tectonic cycles, (2) as an acoustic basement according to the definition of geophysicists, and (3) at least as the lower part of the cover attached to the crystalline basement. Consequently, hereafter, we call "basement" all the pre-Neogene and especially the pre-Miocene undifferentiated rocks.

We have produced a contour map of the pre-Miocene rocks (Fig. 4) based on data collected in the literature for 70 onshore and offshore wells (Chou, 1971; Chou, 1980; Shaw, 1996). This map differs significantly from the previous ones in that it covers the whole frontal part of the foreland thrust belt and the southwestern structural domains. We especially consider the basins and highs, both onshore (Coastal Plain) and offshore (Taiwan Strait). Recognizing basement highs in the frontal Western Foothills will provide indirect evidence for basement-involved tectonics at the belt front.

The basement map clearly highlights the occurrence of two major intrabasinal highs underlying the Taiwan Strait and the Coastal Plain: the offshore Kuanyin high to the north and the partly onshore Peikang high in central-western Taiwan (Fig. 4). In order to characterize the basement topography beneath the foreland and the location of Tertiary basins, we have constructed a longitudinal stratigraphic profile (Fig. 5) based on the available wells drilled in the Coastal Plain and the outer part of the Western Foothills. In this profile, we focus on the stratigraphic characteristics of each province that are relevant in the study of the thrust-belt front. Consequently, we have constructed this profile to fit with the relevant wells rather than considering the orientation of the normal faults that define the basins. In this profile, along-strike variations of basement depth and thicknesses of Neogene deposits are clearly emphasized (Fig. 5). Hence, five different basins could be determined according to their ages and structural frameworks.

The location of the different Neogene basins is likely to be dependent on the occurrence of a basement uplift in the frontal part of the foreland thrust belt (Figs. 4 and 5). The transitions between highs and basins are marked by major boundary faults (labeled in Fig. 5) that are sites of abrupt longitudinal changes in the topography of Neogene and pre-Miocene rocks. These

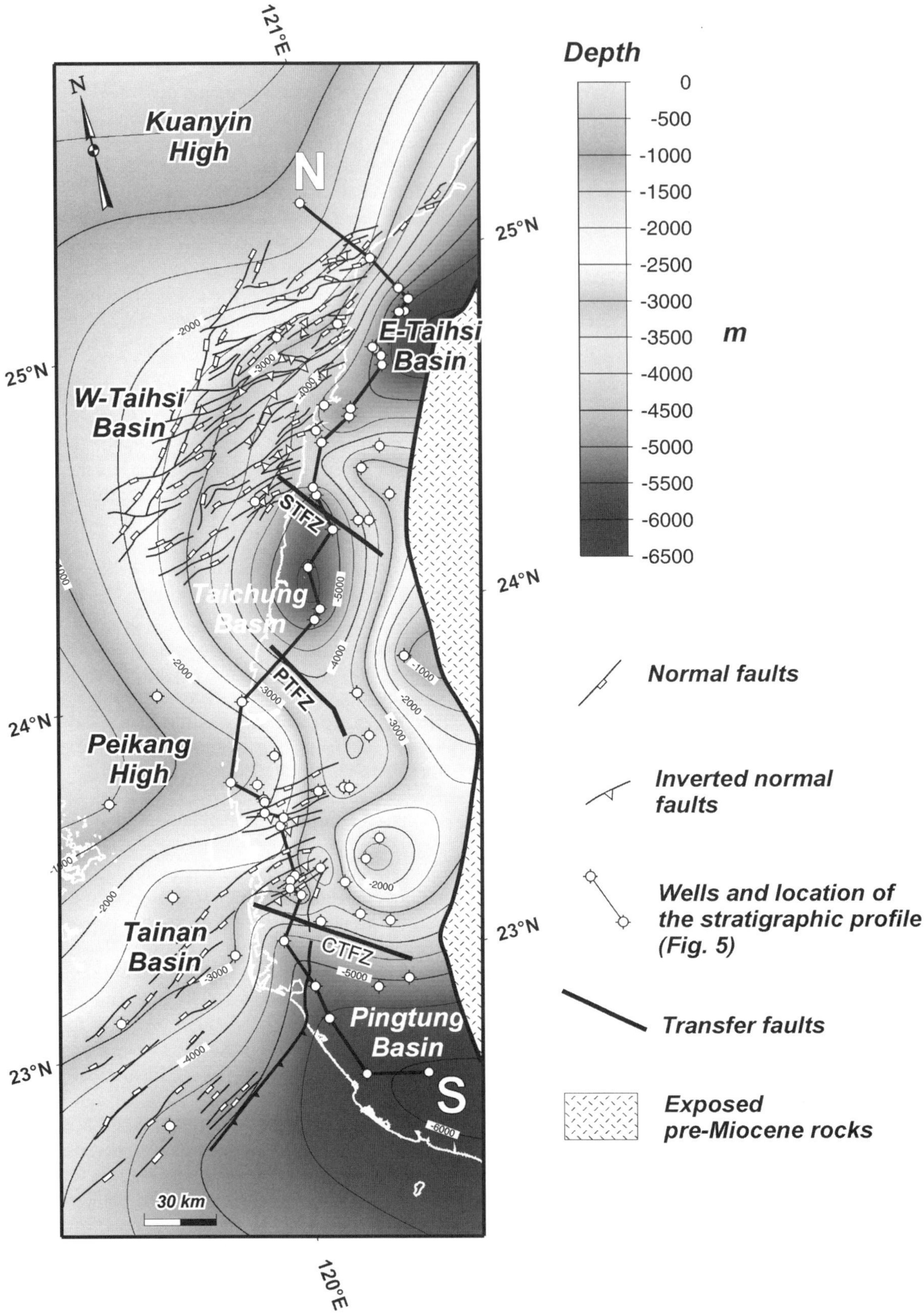

Figure 4. Contoured map of pre-Miocene rocks ("basement") and the structural framework of the Taiwan mountain-belt front. Contour interval is 500 m. CTFZ—Chishan transfer-fault zone, PTFZ—Pakua transfer-fault zone, STFZ—Sanyi transfer-fault zone.

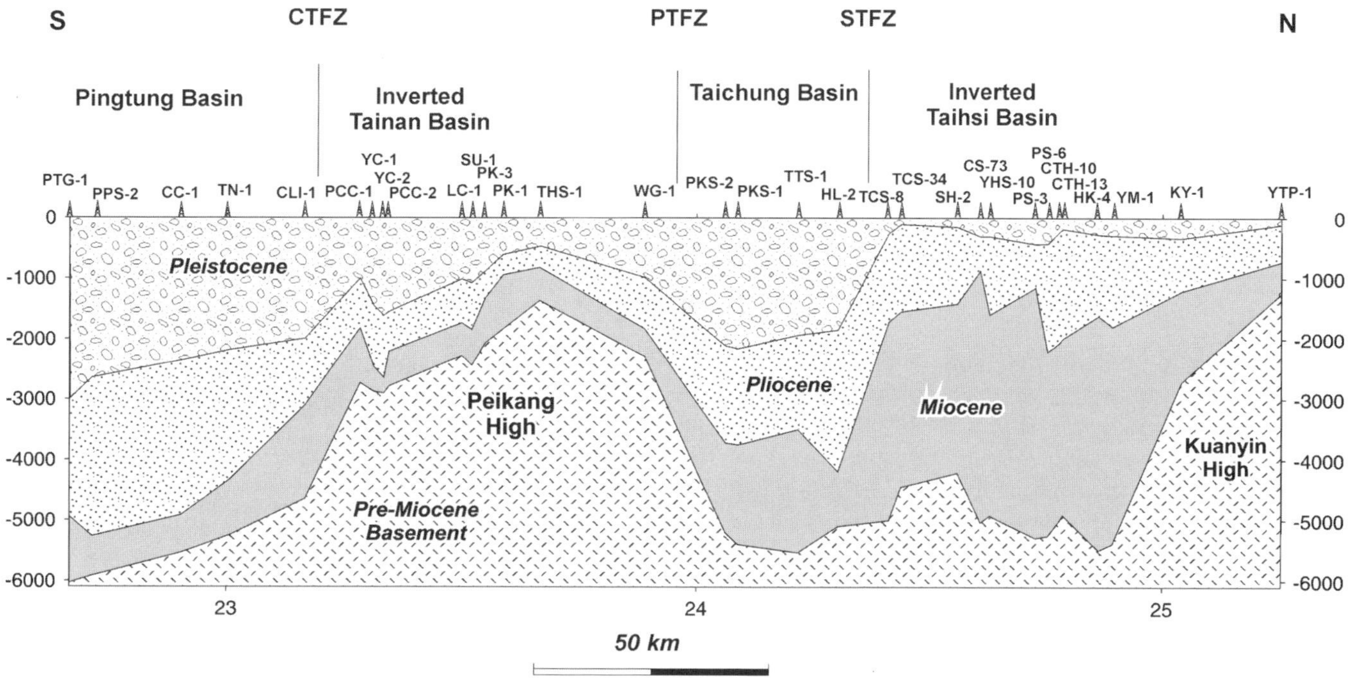

Figure 5. Along-strike stratigraphic profile of the foreland showing the location of basins and highs. Inverted basins are located on the southern edges of the major highs such as (1) the western Taihsi Basin and the Kuanyin high and (2) the Tainan Basin and the Peikang high. The major oblique transfer-fault zones delineate the different parts of the foreland. CTFZ—Chishan transfer-fault zone, PTFZ—Pakua transfer-fault zone, STFZ—Sanyi transfer-fault zone.

major steps in the topography of the pre-Miocene basement are enhanced by the occurrence of major north-northwest–trending transfer faults oblique to the profile (Fig. 5). These faults have been recognized in the cover as major transfer-fault zones on the basis of morphologic analyses and distribution of earthquakes (Deffontaines et al., 1997).

We begin investigating the basement structural framework by distinguishing the precollisional Taihsi and Tainan Basins, which are both located on the southern edge of the main structural highs (Figs. 4 and 5). As revealed by Figure 5, the Taihsi Basin corresponds to the northern Miocene depocenter. Both the pre-Miocene basement of the Taihsi Basin and the entire overlying Neogene sedimentary cover have been uplifted (Fig. 5). The seismic studies of the geologists of the Chinese Petroleum Corporation have extensively demonstrated that the northern Taihsi Basin is an inverted basin (Huang et al., 1993; Shen et al., 1996) in which the pre-Neogene rocks have been offset by numerous high-angle reverse faults with a strike-slip component (Fig. 4). According to the basement map, the Taihsi Basin can be divided into a western inverted basin and an eastern noninverted basin (close to Hsinchu) where the pre-Miocene basement has probably not undergone uplift but rather lateral extrusion accommodated by strike-slip faulting, an important mechanism in this northern region (Lu et al., 1995). Inversion tectonics in northern Taiwan prevails from the southwestern edge of the Kuanyin high area to a basement transfer fault located south of Miaoli (Deffontaines et al., 1997), the Sanyi transfer-fault zone (Fig. 4). Moreover, in the eastern part of the Tainan Basin along the profile shown in Figure 5 (Chang et al., 1983), inversion tectonics have also been demonstrated by the presence of reactivated inherited extensional features and inverted slip direction on originally normal-slip faults that offset both the Neogene sedimentary cover and the pre-Miocene basement in the southern flank of the Peikang High (Fig. 4). Most of the inverted structural features encountered in the foreland of the Taiwan orogen are represented by high-angle reverse faults that offset the pre-Neogene rocks but die out into the basement. The Neogene Taichung and Pingtung Basins show important foredeep subsidence as suggested by the thick series of syntectonic deposits of continental origin (Fig. 5). The development of these basins appears to have been controlled by the two major oblique transfer faults located north and south of the Peikang high, the Pakua transfer-fault zone and the Chishan transfer-fault zone, respectively (Fig. 4 and 5). The northern one is an ancient normal fault that locally controlled the propagation and geometry of frontal thrust sheets (Mouthereau et al., 1999). Evidence of basement uplift east of the Taichung Basin in inner domains of the Western Foothills is suggested in Figure 4.

In summary, major extensional Tertiary basins located beneath the Coastal Plain and offshore have undergone inversion tectonics during the Pliocene–Pleistocene collision. As evidenced by the reactivation of pre-Neogene normal faults and inversion of their sense of motion, the basement is likely to be affected by the inversion. Moreover, the basement-involved tectonics was also controlled by the reactivation of ancient transform faults that have acted as oblique transfer faults during the contractional episode.

LOCATION OF TAIWAN THRUST-WEDGE FRONT, WEDGE GEOMETRY, AND RELATIONSHIP WITH BASEMENT-INVOLVED TECTONICS

Remote-sensing and morphologic analyses have been successfully applied to the Taiwan mountains (Deffontaines et al., 1994; Lee, 1994). In this section, we mainly aim at locating the different fronts by a method based on the previous discussion on the depth of décollement involved and the resulting thrust-wedge geometry. In order to obtain new constraints on the frontal structure and topography, we performed a morphostructural analysis in the outermost part of the Western Foothills and the Coastal Plain. This study includes the use of (Satellite Pour l'Observation de la Terre) SPOT-Panchromatic scenes (10 × 10 m ground resolution) as well as the onshore hill-shading Digital Elevation Model (DEM) of Taiwan (40 × 40 m ground resolution) performed by Taiwanese colleagues. Both these complementary methods have enabled us to recognize small-scale structures and detect small variations in topography. This study was combined with structural information provided by available geologic maps (Chinese Petroleum Corporation, 1974) and complementary field observations and measurements. For offshore investigations, we referred to the published results of the Chinese Petroleum Corporation (Huang et al., 1993).

These combined morphological and structural analyses demonstrate the occurrence of major transfer-fault zones in the Western Foothills (Deffontaines et al., 1997). These transfer faults delineate several structural areas of distinct morphology and thrust-sheet geometry. Therefore, in this study, we distinguish five morphostructurally based regions in the thrust-wedge front from south to north: Tainan (region 1), Chiayi (region 2), Taichung (region 3), Miaoli (region 4), and Hsinchu (region 5). All are named according to the main cities in the region (Fig. 6). From the DEM data, we constructed five topographic sections across the Western Foothills and the Coastal Plain, perpendicular to the direction of the main relief (Fig. 6). Each topographic section is compared with a section of the basement top, which has been calculated from the contoured map (Fig. 4). This indirect method provides constraints and gives a good estimate of the geometry of the thrust wedge, in accordance with the model presented in Figure 1B that allows one to predict whether the basement is involved in the compressional deformation of the considered frontal areas. Our results are summarized in a general structural sketch map of the Western Foothills of Taiwan (Fig. 7).

Tainan area (region 1)

This province is located between the offshore Manila accretionary prism and the Chishan transfer-fault zone (Fig. 6). Preliminary photogeologic studies (Sun, 1964) and analyses of a recent DEM reveal that the frontal fold-and-thrust systems propagate away from the mountainous areas of the Central Range far to the west into the Coastal Plain (Fig. 6). The observation of continuous deformation in the Coastal Plain associated with the low-angle topographic slope and the low-angle, hinterland-dipping, basement-surface slope defines a wedge shape of the sedimentary cover. This result suggests that, instead of deep-décollement tectonics, shallow-décollement tectonics occurs and involves the sedimentary cover (Fig. 6). Locally, in the vicinity of the city of Tainan, the occurrence of a pop-up structure, which is usually found above shallow décollements (Fig. 2), seems to support this conclusion. Consequently, we define a thrust wedge with a low taper angle (Fig. 6), probably consistent with a low-strength area. This interpretation is in agreement with the occurrence of thick series of Pliocene–Pleistocene mudstones in this area, which by definition are characterized by low internal friction. Finally, these results confirm that the basement is not involved in the frontal deformation and that the basal décollement of the thrust wedge is presumably shallow and located above the pre-Miocene basement ("thin-skinned tectonics") in the sedimentary cover.

In summary, the Tainan area exhibits a prominent shallow-décollement tectonics. On the basis of the theoretical model proposed in Figure 1B, the deep basal décollement may be absent or poorly active. In this case, a reactivation front could not be defined whereas a mountain front is observed (Fig. 7).

Chiayi area (region 2)

This region extends from the Chishan transfer-fault zone to the north of the city of Chiayi (Fig. 6). In this region, the frontal thrust sheets exhibit various structural trends ranging from N20°E in the Western Foothills to N60°–70°E in the Coastal Plain (Fig. 7). Furthermore, numerous folds were identified farther west in the Coastal Plain through the use of photogeologic studies (Sun, 1965). Subsurface data (Chang et al., 1983) suggest that the main explanation for the changes in structural trends of frontal units and the occurrence of compressional deformation in the Coastal Plain is the reactivation and/or inversion of N60°–70°E-directed preexisting extensional features belonging to the Tainan Basin. In Figure 6, the topographic section indicates an abrupt increase of the average surface slope with regard to the southern topographic section of the Tainan area. This along-strike topographic change is correlated at depth with the uplift of the basement (Fig. 6). This result suggests the deepening of the basal décollement toward the interior of the belt and thus argues in favor of the internal thickening of the tectonic wedge by basement involvement. Moreover, the presence of inverted structures in the foreland

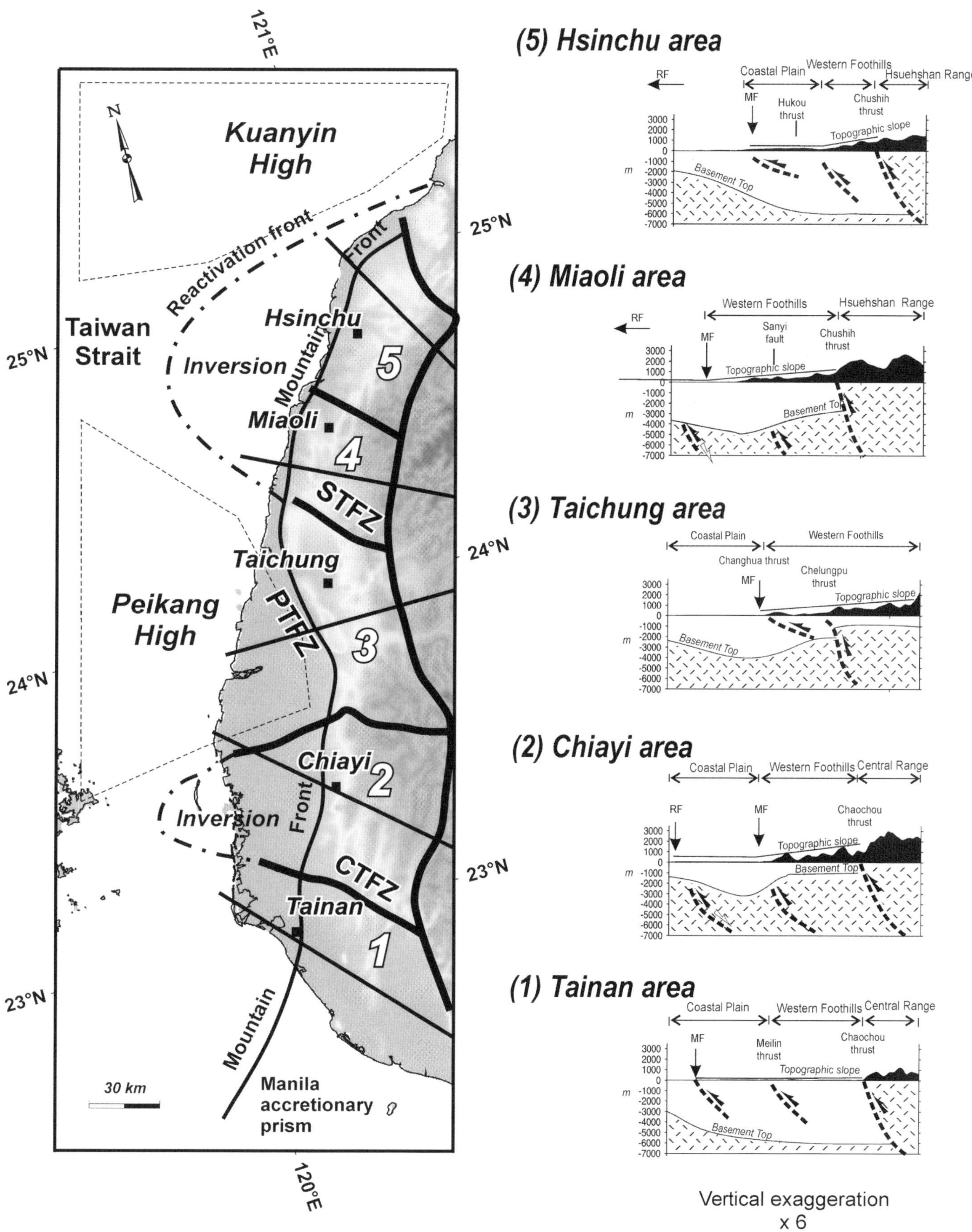

Figure 6. Superimposed profiles of the surface topography and basement top for the five morphostructurally based regions investigated in the Western Foothills. The locations of the mountain front (MF) and reactivation front (RF) are on the basis of considerations developed in the model presented in Figure 1B. Lateral changes in the topographic elevation in the Western Foothills are generally well correlated with basement highs; therefore these sections provide indirect information on the wedge geometry and whether the basement is involved in the frontal tectonics. The thick dashed lines correspond to the major thrusts and inferred frontal thrusts. CTFZ—Chishan transfer-fault zone, PTFZ—Pakua transfer-fault zone, STFZ—Sanyi transfer-fault zone.

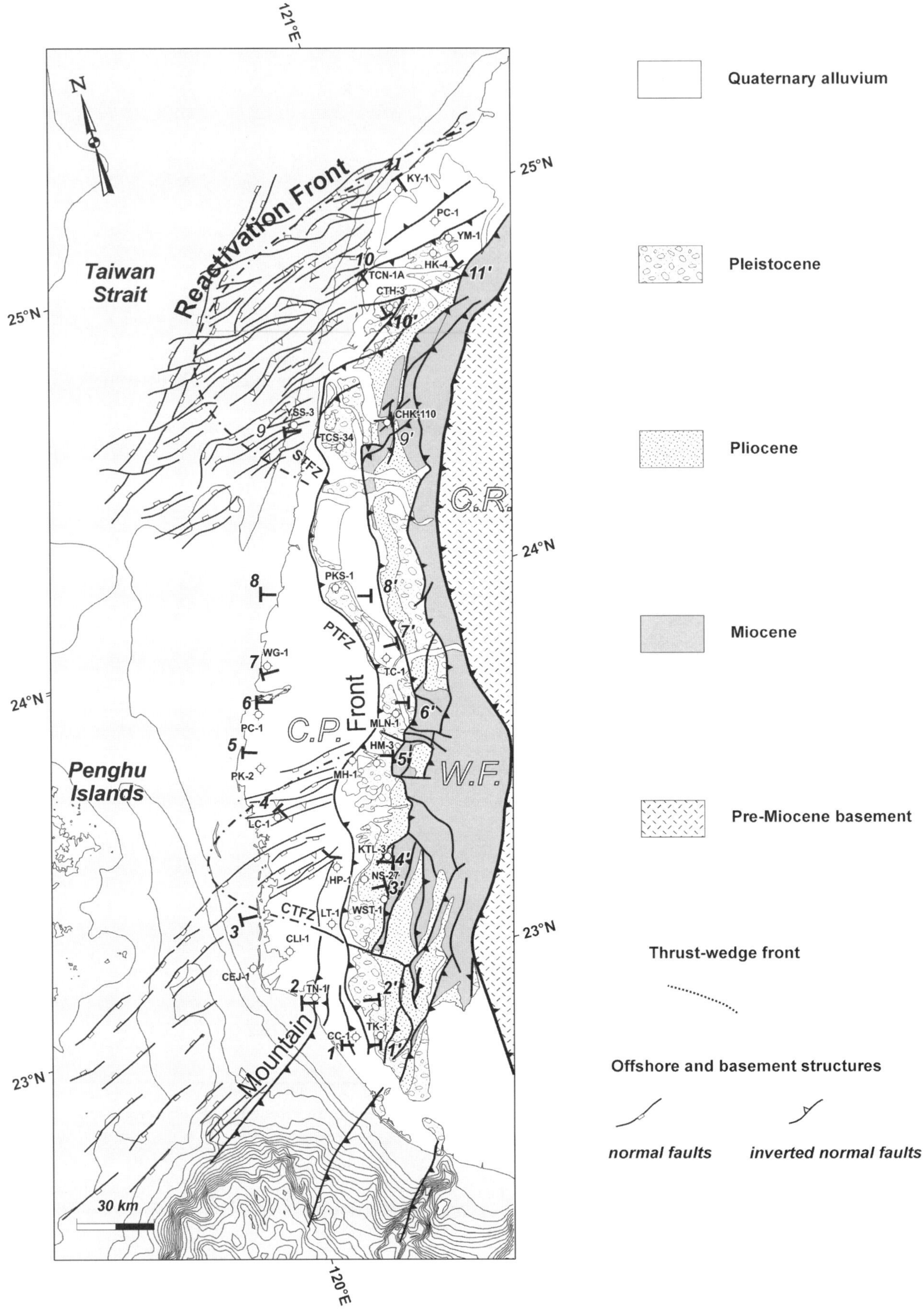

Figure 7. Structural sketch map of the Western Foothills and offshore Taiwan. The Taiwan reactivation front (dash-dot lines) curves between the basement highs and the domains where structural inversion dominates. Numbers give the locations of 11 serial geologic sections shown in Figures 8 to 12. CP—Coastal Plain, CR—Central Range, WF—Western Foothills. CTFZ—Chishan transfer-fault zone, PTFZ—Pakua transfer-fault zone, STFZ—Sanyi transfer-fault zone.

reveals (Fig. 6) that the thrust wedge has propagated westward through the Coastal Plain in favor of the activation of a deep décollement. East of the Chukou-Lunhou thrust, the thrust wedge is characterized by a significant topographic slope, and the major part of the deformation is accommodated by basement involvement that occurs above a deep décollement (Fig. 6). West of this major thrust, the thrust wedge incorporates the inverted Tainan Basin. Consequently, according to the model presented in Figure 1B, imbricate wedges controlled by a deep décollement and limited by a mountain front and a reactivation front must be considered.

In summary, the propagation of the tectonic wedge probably occurs over a deep décollement. We propose to enlarge the thrust wedge through the Coastal Plain and relocate a reactivation front that corresponds to the outermost reactivated extensional structure 20–30 km away from the mountainous areas, which are limited by the Chukou-Lunhou fault (Fig. 7). Furthermore, we notice that, according to these data, a shallow thrust wedge cannot be defined.

Taichung area (region 3)

The central part of the frontal Western Foothills in the vicinity of Taichung is outlined by a remarkable regional-scale curvature (Figs. 6 and 7). In this area, the bending of the thrust-belt front has been interpreted as the result of the activity of a major transverse fold, the Pakua transfer-fault zone (Deffontaines et al., 1997). The analysis of the DEM failed to show any compressional structures west of this transverse fold. This result shows that the thrust-wedge propagation is stopped in the vicinity of the Pakua transfer-fault zone, which thus localizes the mountain front. The occurrence of such oblique structure is due to the basement control, by an oblique inherited normal fault, on the geometry and kinematics of a frontal shallow décollement and associated frontal deformation and thrust emplacement (Mouthereau et al., 1999). The first evidence of basement involvement in the thrust wedge occurs backward in the vicinity of the Chelungpu-Sanyi thrust (Fig. 6). However, in comparison to the previously discussed section in the Chiayi area, the basement uplift in the inner part of the Taichung area is not accompanied by a significant topographic elevation. In contrast, the Western Foothills of the Taichung region display a broad zone having a low-dipping surface slope, which is correlated with an increasing spacing between major thrust sheets (Fig. 6). Such lateral structural changes are usually observed in association with a transfer-fault zone, as has been shown by analogue modeling (Calassou et al., 1993). Moreover, one of the main factors that controls the thrust-wedge propagation is erosion, which is especially important in the frontal part of the Western Foothills, as suggested by the occurrence of thick unconsolidated conglomeratic and sandy syntectonic deposits that fill the Pleistocene depocenter.

In summary, on the basis of the model presented in Figure 1B, we distinguish an inner domain and an outer domain within the critically tapering wedge. One is located to the east of the Chelungpu-Sanyi thrust and is controlled by a deep décollement. As the deep décollement steps up at shallow depth, shallow-décollement tectonics occurs. The mountain front thus defined corresponds to the thrust-wedge front of the critically tapering wedge. It is likely that the deep décollement continues westward, but its activity has probably decreased because of local indentation by the Peikang basement high.

Miaoli area (region 4)

This region is located in the middle of the large curvature of the topographically defined mountain front and corresponds to a transitional structural domain of frontal areas of the Western Foothills where structural trends change northward from predominantly N10°–20°E to N70°E. Both the bending of the front due to the Pakua transfer fault and the transfer faulting along the Sanyi fault transfer motion of the northern thrust sheets with respect to the propagation of the southern thrust sheets blocked onto the Peikang basement high. As a consequence, the related frontal relief delineates the coastline near Miaoli (Figs. 6 and 7).

The structural analysis of the basement (Fig. 4) and further seismic reflection data published by the Chinese Petroleum Corporation (Huang et al., 1993; Shen et al., 1996) have revealed that the offshore western Taihsi Basin was affected by inversion tectonics during the Pliocene–Pleistocene collision. This result implies that the compressional stresses have been transmitted away from the hilly regions in the Chinese platform in favor of the activation of a deep décollement. Consequently, according to the model examined in Figure 1B, we propose to enlarge the interpreted size of the thrust wedge offshore and thus locate the reactivation front about 60 km westward from the mountain front into the Taiwan Strait, in order to include the western edge of the Taihsi Basin in the deformation. To the east of the mountain front, onshore, the profile of the basement top yields no clear evidence of basement involvement at the front (Fig. 6). However, the pre-Miocene rocks exhibit a foreland-dipping slope, which seems to indicate that basement involvement, even though limited, occurred in relationship to the Sanyi thrust activity (Fig. 6).

Moreover, recent reflection seismic profiles (Hung and Wiltschko, 1993) have shown that normal faults have been incorporated into the frontal part of the thrust wedge, implying that basement-involved tectonics is the prominent deformational mechanism in the Miaoli region. This result led us to consider a thrust wedge controlled by a deep décollement in agreement with the model proposed (Fig. 1B). The western area is subject to orogenic stresses producing structural inversion in the foreland. Little reverse displacement occurs, resulting in low topography. East of the Sanyi thrust, the basement is involved in the deformation of the tectonic wedge. On the basis of complete structural study, the reactivation front is extended toward the east and displays a remarkable curvature similar to the shape of the mountain front (Fig. 7).

Hsinchu area (region 5)

The Hsinchu area is the northernmost province investigated in this study. Northward, the structures of the belt front progressively turn and become parallel to the inherited normal faults of the southern edge of the Kuanyin high, trending N60°–70°E (Fig. 4). Accurate topographic data of the DEM highlight the presence of a broad zone of elevated terraces in the Coastal Plain (Figs. 6 and 7). Moreover, these terraces are affected by thrust faulting (Hukou thrust in Fig. 6), demonstrating that compressional deformation has propagated farther in the Coastal Plain from the mountain front in favor of the activity of a shallow décollement. Moreover, the profile of this region reveals the deepening of the basement toward the inner part of the belt. This result shows that basement involvement is absent or limited, which leads us to interpret a typical wedge shape for the sedimentary cover. Such a geometry has been also recognized in the southernmost frontal areas of the belt. As it was noticed earlier, this region is the site of important lateral motion due to escape tectonics (Lu et al., 1995). Consequently, out-of-section transport probably occurs, which makes difficult the observation of reverse motion along faults in the basement.

To conclude, we observe a single wedge of sedimentary cover in this frontal region of northern Taiwan, demonstrating that the deformation is probably controlled by a shallow décollement. Finally, we relocate the thrust-wedge front farther north and close to the shoreline, i.e., distant from the mountain front. The geology of the frontal thrust sheets and the location of the different fronts is summarized in Figure 7. Basically, the Taiwan mountain-belt front (reactivation and mountain fronts) exhibits large curved zones (Fig. 7) that are related to the occurrence of inherited basement reentrants and salients in the foreland and to the location of ancient major oblique-transfer faults reactivated during the Pliocene–Pleistocene compression. The areas of structural inversion located south of the basement highs (Peikang and Kuanyin) are observed in association with basement-involved tectonics near the mountain front.

GEOLOGIC CROSS SECTIONS OF THE BELT FRONT

In order to determine the structure of the frontal units at depth, we have constructed 11 serial geologic sections controlled by 30 Chinese Petroleum Corporation drill holes and field measurements. When available, we have also used several onshore and offshore short seismic profiles, helped by published data (Chen, 1978; Huang et al., 1993; Yang et al., 1996). Each section cuts through the mountain front and is extended in the foreland of the Coastal Plain (see location of sections in Fig. 7). In addition, the geologic sections presented here were constructed by taking into account the conclusions on the thrust-wedge geometry in the preceding section and the location of the different fronts in agreement with the model presented in Figure 1B.

Tainan area (region 1): Sections 1–1′ and 2–2′

On the basis of the previously described results on the thrust-wedge geometry (Fig. 6), we have determined that the deformation in the southern frontal part of the belt is accommodated within a shallow thrust wedge, i.e., in the syntectonic Pliocene–Pleistocene strata deposited in the Pingtung foredeep basin. In this region, the best strata for hosting a décollement is a thick Pliocene–Pleistocene turbiditic formation (2–3 km) composed mainly of mudstones. Moreover, seismic reflection data (Fig. 8) suggest that the development of the frontal Tainan anticline (TN-1 well) is the result of the activity of a shallow thrust that dies out at depth into the Pliocene–Pleistocene mudstones. Both geologic sections constructed in this region (sections 1–1′ and 2–2′) display a series of west-verging imbricate thrust sheets that propagate toward the Coastal Plain (Fig. 8). The deformation occurs above a shallow, gently east-dipping décollement (2–5 km deep) (Fig. 8). These structural patterns match the types of fault-related folds, such as fault-propagation folds (Chungchou anticline, CC-1 well) and pop-up structures (see the line drawing of the Tainan anticline, TN-1 well) (Fig. 8) usually found in shallow-décollement tectonics (Fig. 2). In this region, the front of the shallow thrust wedge (the mountain front) is located at the emergence of the shallow frontal thrusts (Chungchou thrust and Tainan thrust). Different factors could be invoked to account for the thrust wedge to propagate far in the foreland. One is the occurrence of weak internal friction within the thick Pliocene–Pleistocene mudstones, and the other is the presence of overpressured fluids in these deposits, as illustrated by the presence of mud volcanoes in the vicinity of major thrusts. Therefore, the southwestern foreland thrust belt appears to be very similar to structures already described in the Manila accretionary prism (Reed et al., 1992), and the Tainan region may be regarded as the onshore northern extent of the Manila accretionary prism.

Chiayi area (region 2): Sections 3–3′, 4–4′, and 5–5′

The frontal deformation in the Chiayi area is characterized by thrust wedging associated with structural inversion in the outer part and basement-involved tectonics in the inner part. Compared to the previously mentioned southern area (Fig. 8), there are major changes in the thickness of Neogene formations as well as in the age of rock materials involved in the deformation. Moreover, both sedimentary and structural changes are localized across the oblique Chishan transfer-fault zone (Fig. 7). The frontal geologic sections constructed for the Chiayi region (sections 3–3′, 4–4′, and 5–5′) highlight the northward thinning of the Neogene sedimentary deposits onto the southern flank of the Peikang basement high and the increase of the early

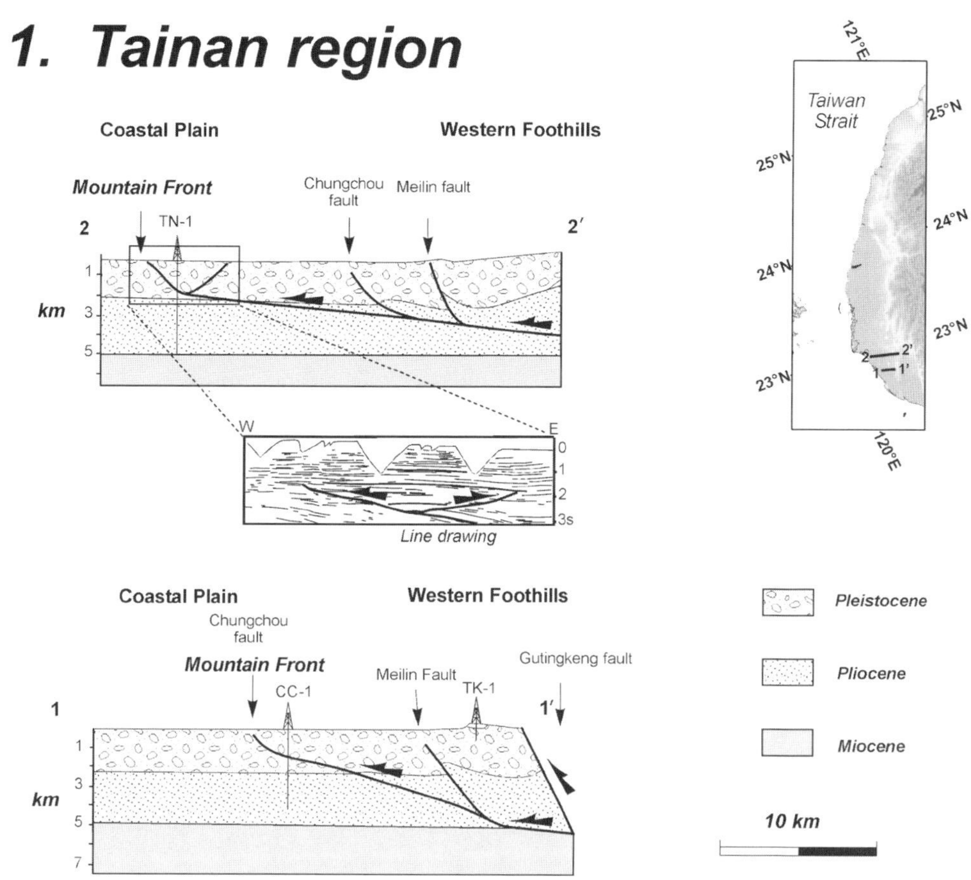

Figure 8. Frontal cross sections in the Tainan area (region 1). These two sections show a dominant shallow-décollement tectonics style affecting a significant thickness of Pliocene–Pleistocene sedimentary deposits. A reactivation front is absent, and the mountain front corresponds to the emergence of the shallow décollement.

Miocene unconformity toward the center of the basement high (Fig. 5). Little structural inversion occurs beneath the Coastal Plain according to section 3–3′ whereas basement involvement becomes much clearer in section 4–4′. With regard to the previous structural interpretations (i.e, based on thin-skinned tectonics; Suppe, 1980; Wiltschko et al., 1997) of the Niushan and Wushantou anticlines (NS-27 and WST-1 wells, section 3–3′) and Kuantaoshan anticline (KTL-3 well, section 4–4′), recent seismic reflection data have led Chang et al. (1996) to reinterpret these frontal structures in terms of simple basement involvement. For instance, seismic reflection profiles across the Niushan anticline (Hsiao, 1974) and Hsiaomei anticline (HM-3 well in section 5–5′) (Wiltschko et al., 1997) have revealed that both the Meilin fault and the Hsiaomei fault are high-angle reverse faults that are probably the result of the reactivation of inherited extensional features. Hereafter, we propose an alternative interpretation for the frontal structures that is based on the superimposition of both a shallow and a deep décollement in the vicinity of the mountain front.

In section 3–3′, the frontal fault-propagation fold geometry of the Niushan anticline (steep forelimb) is interpreted as the result of combined reactivation of an ancient normal fault due to slip along a deep décollement and the abrupt termination of the shallow basal décollement (Fig. 9). This structure has strong similarity with the case described by Suppe (1986) in the northern part of the thrust belt (see also Fig. 2E). The deep décollement extends westward as suggested by the structural inversion in the Coastal Plain. Moreover, it is probably deeper than 9 km. In section 3–3′, the shallow décollement is localized at the base of the Miocene strata (Suppe, 1980) near 5–6 km. The mountain front that corresponds to the emergence of the shallow décollement is distinct from the reactivation front.

To the north, the frontal deformation exhibits structural inversion in the Coastal Plain (Chang et al., 1983), and a broad foreland-dipping syncline (section 4–4′) is observed in the western part of the Western Foothills. Its eastern limb is affected by backthrusting, arguing in favor of the development of a major triangle zone resulting from the imbrication of the pre-Miocene basement beneath the Kuantaoshan anticline (KTL-3 well) (Fig. 9). In the case of superimposed shallow and deep décollement tectonics, and in order to fit with the result of Suppe (1980), a possible alternative to explain the formation of the Kuantaoshan anticline is the result of shortcutting by a reactivated normal fault whose sense of motion has been inverted to thrust slip (Fig. 2D). On the other hand, the deep-décollement surface is located much deeper than commonly assumed at nearly 10–15 km.

Section 5–5′ is located farther north and exhibits the same structural framework as section 4–4′ but structural inversion in the Coastal Plain is absent. The frontal triangle zone evolves

2. Chiayi region

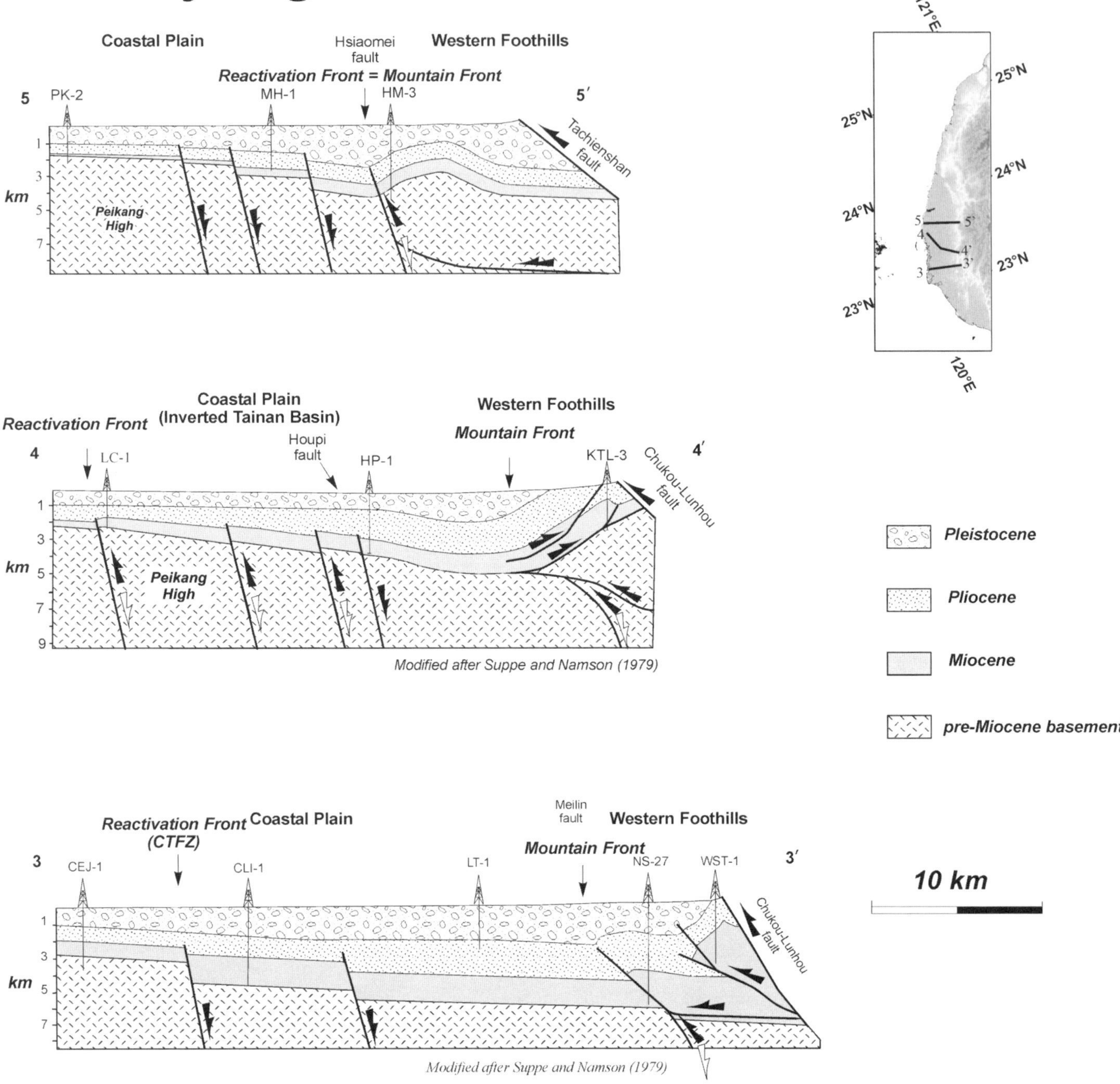

Figure 9. Frontal cross sections in the Chiayi area (region 2). These sections highlight the occurrence of inversion tectonics in the Coastal Plain and basement-involved tectonics in the Western Foothills. Beneath the Chukou-Lunhou fault, the basement involvement occurs by roof thrusting to create a major triangle zone. The reactivation front (the outermost reactivated extensional structure) is distinct from the mountain front. Frontal structures in the vicinity of the mountain front show the superimposition of shallow- and deep-décollement tectonics. CTFZ—Chishan transfer-fault zone.

into a broad anticline in the Hsiaomei anticline (HM-3 well), and the pre-Miocene rocks are also involved in the frontal deformation. Furthermore, the connection between high-angle thrusting (that merges at depth into a deep décollement according to our interpretation) and the shallow décollement (similar to that of section 4–4′) occurs.

Considering results already obtained concerning the geometry of frontal structures in the vicinity of the mountain front (Fig. 6), the structural analysis carried out in the Chiayi area provides support to the interpretation in terms of superimposed shallow-décollement and deep-décollement tectonics.

Taichung area (region 3): Sections 6–6′, 7–7′, and 8–8′

From north to south, the frontal deformation in the Taichung area is controlled by the hinge fault of the Peikang high (Fig. 10). In the cover, this fault has induced the development of a major transfer zone (Mouthereau et al., 1999).

The type of frontal folds in the Taichung area corresponds mainly to fault-propagation folding. In detail, moving southward, the fold limbs flatten, e.g., the Meilin anticline (section 6–6′; MLN-1 well) exhibits a narrow fault-propagation fold with steep limbs (Suppe and Namson, 1979), whereas to the north, the Pakuashan anticline (section 8–8′; PKS-1 well) displays almost flat limbs, as shown in a seismic reflection profile (Chen, 1978). Moreover, the frontal deformation occurs above a shallow décollement (3–4 km deep) that, from north to south, cuts through the Miocene (section 6–6′) and then the Pliocene strata (sections 7–7′ and 8–8′). The atypical, gently dipping limbs shown in section 8–8′ have been probably enhanced by the syndepositional folding when the sedimentation rate became higher than the fold's uplift rate. Effectively, the evolution of fold limbs in this region is characterized by a high deposition rate in the Taichung Basin (Figs. 4 and 5) during the Pleistocene (Mouthereau et al., 1999). In summary, the thrust-wedge front in the Taichung area corresponds to shallow-décollement tectonics, although the thrust location is controlled by a preexisting major normal fault that cuts the basement, as evidenced by a seismic reflection profile (Chen, 1978).

Miaoli area (region 4): Section 9–9′

For many years, the Miaoli and Hsinchu areas of northern Taiwan (Fig. 7) have been investigated by structural geologists in terms of gas and oil exploration potential. The geometry of major anticlines was firstly described as validating the thin-skinned tectonics concept (Suppe and Namson, 1979; Namson, 1981). As a consequence, these previous structural models did not consider the complex array of basement normal faults and the roles of reactivation and structural inversion that have been documented by more recent onshore and offshore seismic reflection profiles (Huang et al., 1993; Shen et al., 1996) (Fig. 7). The geometric study of the thrust wedge reveals that this region exhibits a thick tectonic wedge in which basement is involved and structural inversion occurs (Fig. 6). In the past few years, some authors have entirely reinterpreted this area in terms of thick-skinned tectonics and have omitted any reference to shallow-décollement tectonics (Lee et al., 1993).

The section 9–9′ (Fig. 11) cuts through the frontal broad and symmetrical Tiehchenshan anticline gas field (TCS wells) and, in its eastern part, the tight and symmetrical Chuhuangkeng anticline (CHK-110 well). The latter has been interpreted by Namson (1981) in terms of décollement folding in the shallow-décollement tectonics style. However, evidence of backthrusting in the core of the anticline associated with a foreland-dipping syncline gives constraints on the occurrence of a basement triangle zone. This type of structure has been already encountered in the Chiayi area (section 4–4′; Fig. 9). We propose an alternative hypothesis for the development of the fold that combines shallow and deep décollements. The décollement folding might thus have been generated at the tip of a reactivated normal fault as the result of the activation of a shallow décollement. Some seismic reflection data from the Chinese Petroleum Corporation (personal commun.) reveal that beneath the outermost Tiehchenshan anticline, high-angle thrusts and backthrusts affect the pre-Miocene basement and the sedimentary cover (Fig. 11). Consequently, inversion of deep-seated inherited normal faults could be a mechanism for the development of the outer Tiehchenshan anticline, which localized the mountain front. In summary, even though deep-décollement tectonics is probably the main deformational mechanism in the frontal domain of the Miaoli area, superimposed shallow- and deep-décollement tectonics are likely to have occurred.

Hsinchu area (region 5): Sections 10–10′ and 11–11′

Sections 10–10′ and 11–11′ trend N20°W, perpendicular to the direction of the frontal thrusts. In section 10–10′ (Fig. 12), the relationship between the deep-décollement and the shallow-décollement tectonics is well evidenced. On the basis of seismic sections, Suppe (1986) and Yang et al. (1996) demonstrated that the Chingtsaohu frontal anticline (Fig. 12, CTH-3 well) results from the superimposition of connected low-angle and high-angle reverse faults, in agreement with the model of Figure 2E. Seismic reflection profiles further indicate that the basal décollement lies within the pre-Miocene basement at a depth of 4 km on average. The basal décollement dies out and becomes steeper in the vicinity of the reactivated normal fault. In the same section, the westernmost anticline (TCN-1A well) is interpreted as being the result of the inversion of an inherited extensional feature of the Chinese margin (see Fig. 2F). In section 11–11′, the Yangmei-Hukou anticline (YM-1 and HK-4 wells), which was first interpreted by Suppe and Namson (1979), is the result of fault-bend folding, and the shallow basal décollement lies at 5 km depth. By taking into account these former studies and the occurrence of underlying inherited normal faults bounding the southern part of the Kuanyin basement high, we reinterpret this fold by a similar geometric model that

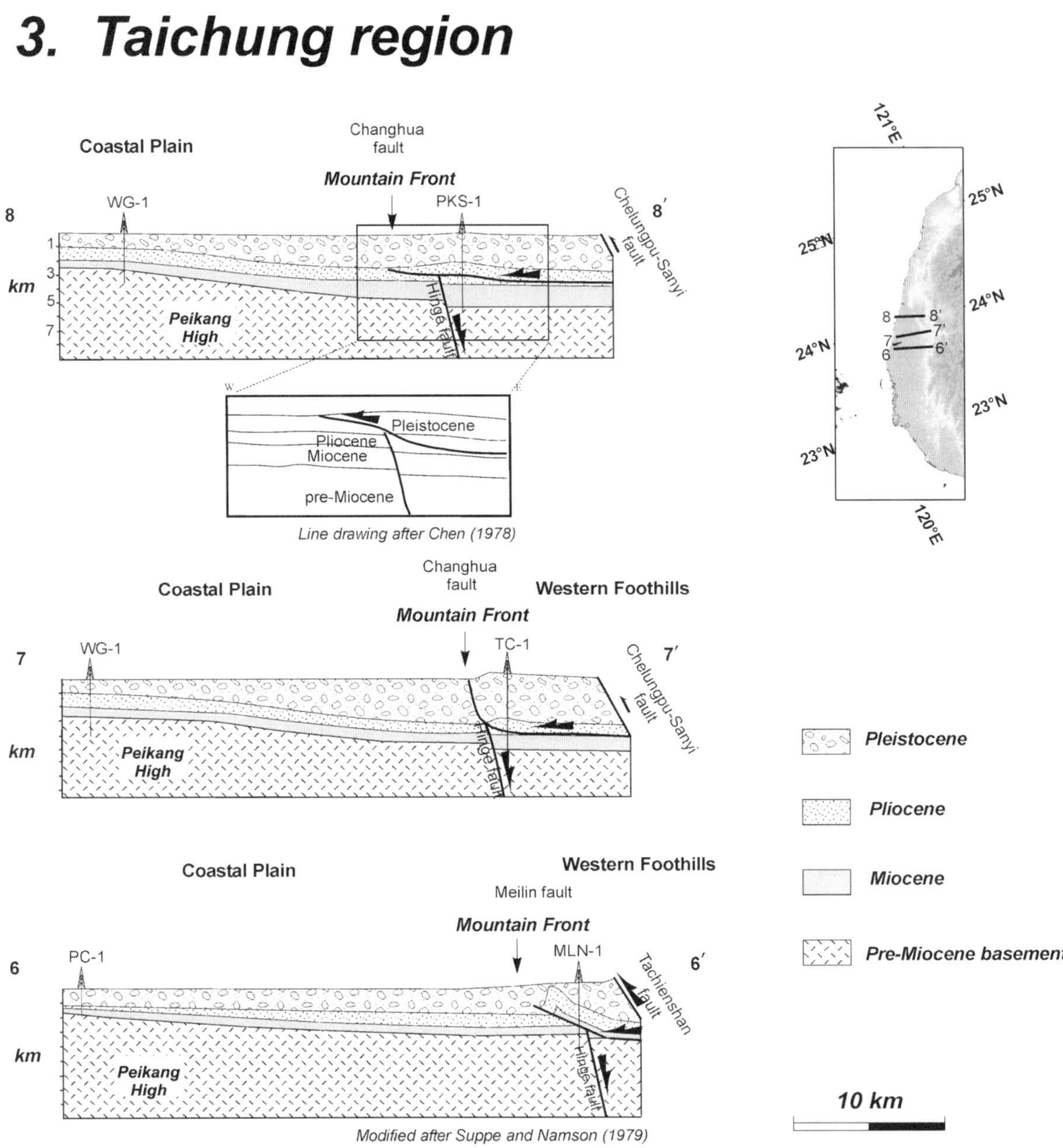

Figure 10. Frontal cross sections in the Taichung area (region 3). The reactivation front is absent. The location of the mountain front (emergence of the shallow décollement) is controlled by a preexisting normal fault that cuts the basement.

was proposed for the Chingtsaohu anticline. Nevertheless, the pre-Miocene basement rocks are probably not involved in the wedge. Therefore, the shallow basal décollement deepens toward the south and displays major steps by localizing ramps above underlying normal faults. Farther west, the thrust wedge propagates in the Coastal Plain above the shallow décollement. A seismic reflection line reveals that the western limb of the Pingchi anticline (PC-1 well) is affected by backthrusting that probably merges at depth with a flat and shallow décollement, thus defining a remarkable frontal triangle zone in the unconsolidated Pleistocene conglomeratic deposits. The western extremity of the triangle zone delineates the mountain front.

To summarize, the compressional structures in this northernmost province exhibit dominant shallow-décollement tectonics. Locally, particular structural styles highlight the superimposition of both shallow- and deep-décollement tectonics, the latter being largely controlled by oblique reactivation of inherited normal faults so that the basement involvement remains limited.

The structural analyses of the serial geologic sections complement the previous investigations of the geometry of the

4. Miaoli area

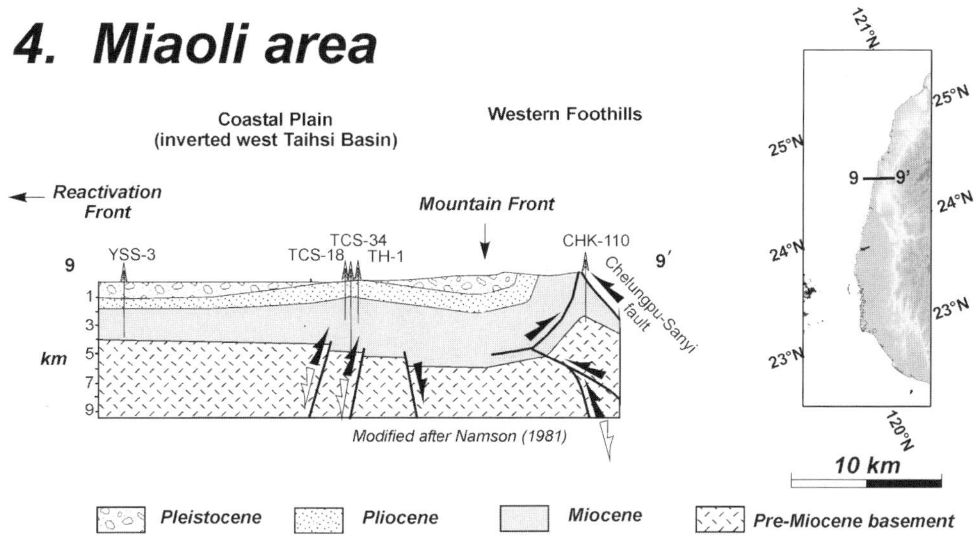

Figure 11. Frontal cross section in the Miaoli area (region 4). This section is marked by structural inversion, i.e., high-angle thrusting along preexisting normal faults. It also shows the basement involvement in the vicinity of the mountain front. In this area, the reactivation front is distinct from the mountain front.

5. Hsinchu area

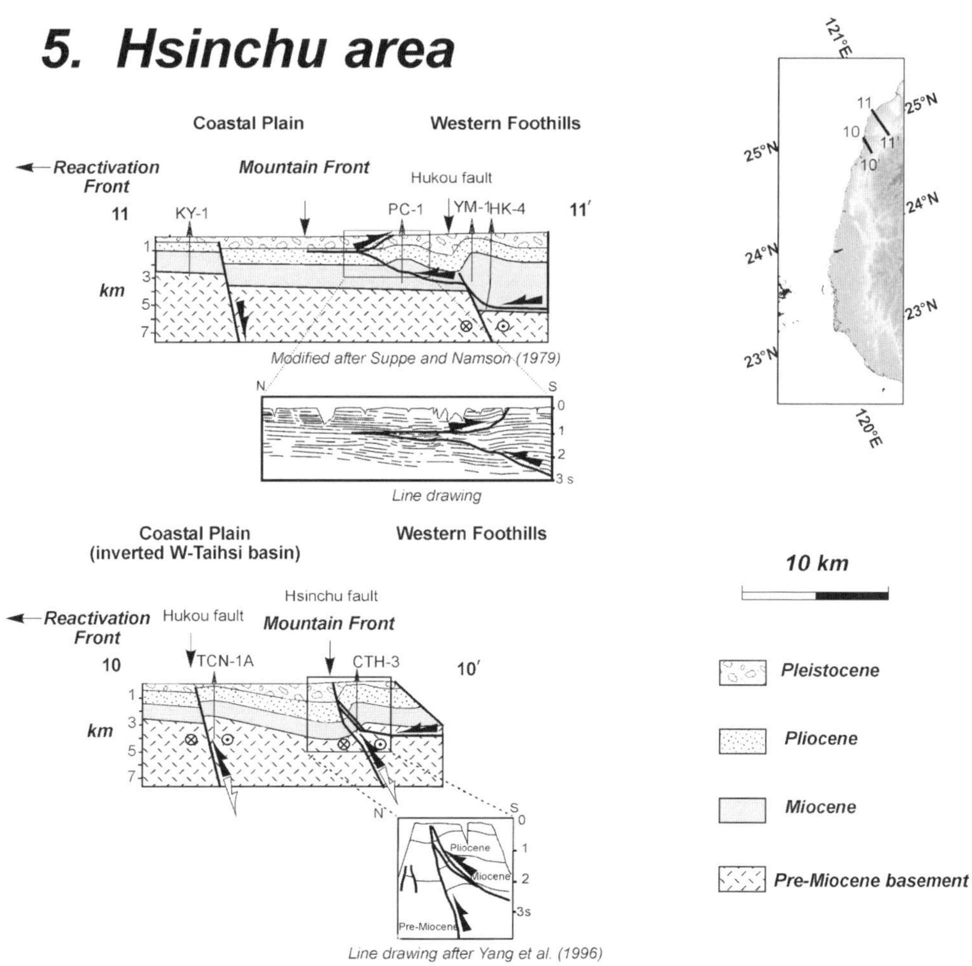

Figure 12. Frontal cross sections in the Hsinchu area (region 5). These sections show the superimposition of shallow- and deep-décollement tectonics. The reactivation front is distinct from the mountain front.

thrust wedge based on study of the surface slope and the topography of the basement. Superimposition of shallow- and deep-décollement tectonics consequently appears as an important tectonic process during the emplacement of frontal thrust sheets.

COMPRESSIONAL TECTONIC REGIMES AT THE TAIWAN THRUST-BELT FRONT

The geometry of the Taiwan thrust-belt frontal zone was highly influenced by the presence of precollisional normal-fault patterns. We consequently attempted to determine the nature and orientation of the tectonic forces prevailing during the Pliocene–Quaternary emplacement of the frontal units. This study is based on kinematic indicators such as striated faults, calcite twins, and impressed pebbles; we collected data as far away as possible from the major faults in the Pliocene–Pleistocene deposits of the outermost thrust units (Angelier et al., 1986; Lacombe et al., 1993, 1997; Hung, 1994; Rocher et al., 1996; Lin and Huang, 1997; this work).

A statistical analysis of the distribution of reconstructed principal paleostress axes (Angelier, 1990) has been performed for each of the regions previously discussed (Fig. 13). A striking result, taking into account the statistical distribution compared with the uncertainties on paleostress reconstructions, is the relative homogeneity of the N120°E orientation of the maximum principal stress axes σ_1. Variations exist, but remain limited considering the along-strike structural complexity (σ_1 distribution diagrams in Fig. 13). Of particular interest is the distribution of tectonic regimes, as a function of the plunges of σ_2 and σ_3 axes (Fig. 13).

The σ_2 axes are predominantly vertical in regions 1 and 3, indicating that strike-slip regimes prevailed in southwestern Taiwan and the Pakuashan area during the thrust emplacement. This geometry may be explained by the obliquity of the trend of the mountain front relative to the regional compression and/or to a decrease in north-south confining stress probably related to incipient lateral extrusion. In addition, the σ_2 and σ_3 axes tend to form a girdle perpendicular to the compression, which argues in favor of stress permutation between pure compressional and strike-slip coeval regimes during the Pleistocene.

In contrast, in regions 2 and 4, steep plunges are predominant for σ_3 axes, suggesting that a purely compressional regime (reverse-faulting type) prevailed during the Pliocene–Pleistocene. This regime is related to the inversion of preexisting features in the Tainan basin at the southern edge of the Peikang and Kuanyin highs, even though some scatter along a girdle in the distribution of σ_2 and σ_3 axes again suggests the possible occurrence of stress permutations.

The data acquired to the north (region 5) are scarce so that no stereo plot is shown. However, they suggest that the dominant regime is of strike-slip type in the northernmost province of the Western Foothills, in relationship to transcurrent faulting along preexisting N60°E faults (Lu et al., 1995).

Thus, the analysis of kinematics markers reveals that (1) regionally σ_1 is approximately parallel to the plate-convergence vector (Fig. 13) and (2) locally strike-slip or reverse-slip compressional regimes dominate, depending on factors such as frontal contraction and lateral movement, in response to indentation by the Peikang and Kuanyin highs.

CONCLUSIONS

The conceptual model of critically tapering wedge (Davis et al., 1983) has been discussed in order (1) to explain the occurrence of inverted extensional features located far into the foreland of an orogen and newly formed microstructures induced by far-field compression related to a deep-crustal décollement, which allows the orogenic stresses to be transmitted within the foreland, and (2) to take into account the basement involvement in the orogenic wedge. On this basis, we propose a model of imbricate thrust wedges.

This model considers three different thrust-belt fronts, a mountain front, a reactivation front, and a deformation front, which are distinguished on the basis of topographic, kinematic, and mechanical criteria (Fig. 1B). This model is applied to the case of western Taiwan (Fig. 14, A and B), and on the basis of these principles, we have redefined and relocated the different fronts.

From south to north, the study of the foreland thrust-belt wedge geometry has been supported by analysis of serial sections in which the basement and the surface topography are compared and discussed (Fig. 6). We have thus distinguished areas where prominent shallow deformation occurs and those where basement involvement dominates. According to this preliminary geometric analysis of western Taiwan, we have shown that a reactivation front exists and corresponds to the termination of the activated part of a deep décollement, i.e., the outermost reactivated extensional structure (Fig. 7). The reactivation front exhibits large curvatures extending offshore in the Taiwan Strait (Fig. 14A). These features are related to the occurrence of inherited basement reentrants and salients in the foreland and to the location of ancient major oblique transfer faults, e.g., the Chishan transfer-fault zone and Sanyi transfer-fault zone, reactivated during the Pliocene–Pleistocene contraction (Fig. 14, A and C). South of basement highs (Peikang and Kuanyin), structural inversion occurs. Such an inversion is commonly found, in the case of the Taiwan mountain-belt front, in association with basement-involved deformation in the inner domains, demonstrating that deep décollement occurs and is continuous beneath the foreland. According to the model shown in Figure 1B, the mountain front delineates the emergence of the shallow décollement and is equivalent not only to the front of a shallow thrust wedge but also to the critical tectonic wedge made of both allochthonous sedimentary and basement units according to the classical critically tapering wedge model of Davis et al. (1983). Because both the present-day stress measurements and the paleostress reconstructions based on brittle microstruc-

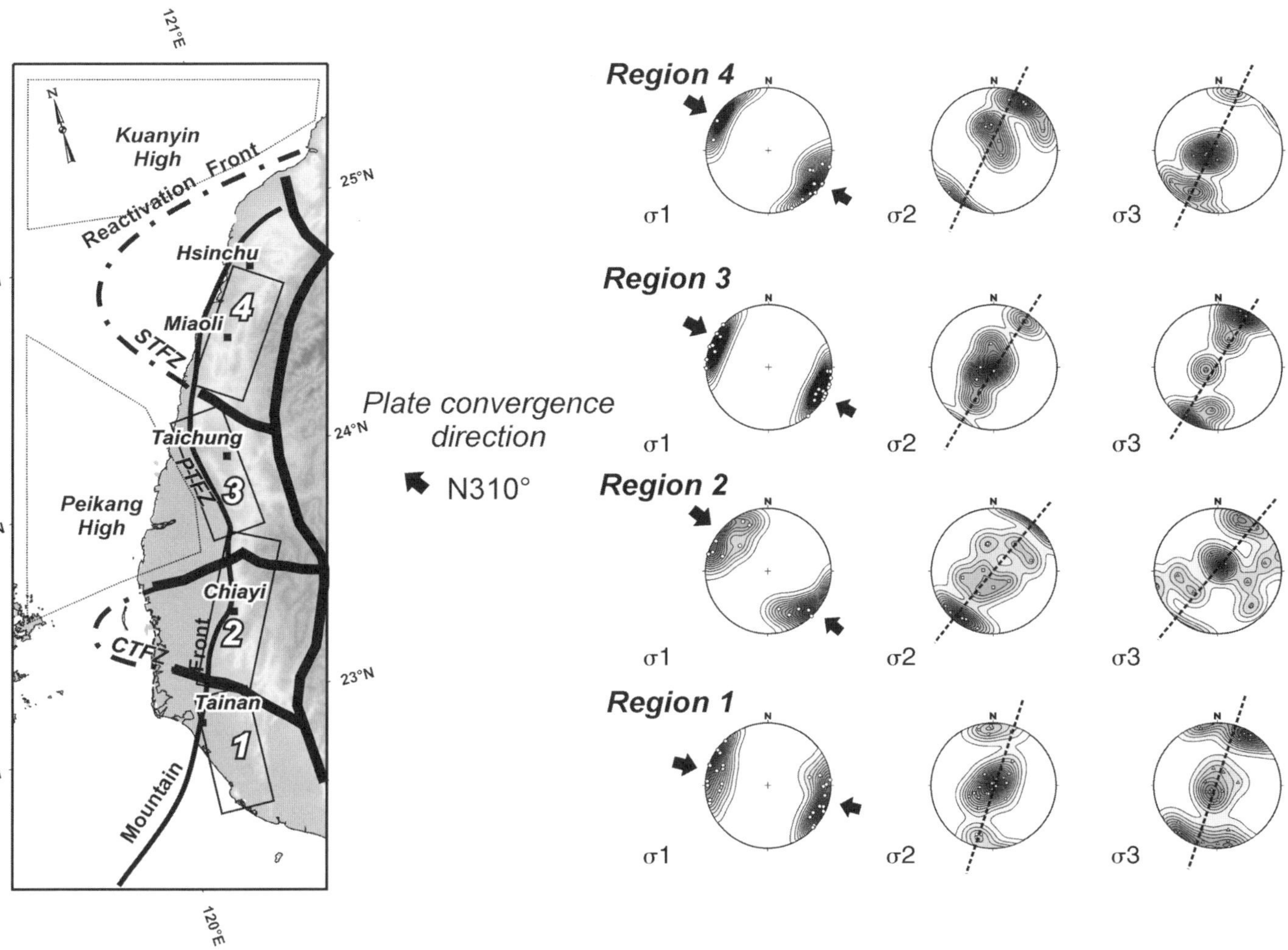

Figure 13. Stereo plots of compiled stress patterns in the studied areas (data are scarce in region 5 so no plot is presented; see text). Each plot represents the statistical distribution of σ_1, σ_2, and σ_3 for a given population. Results are based on original data and others sources (Angelier et al., 1986; Lacombe et al., 1993, 1997; Hung, 1994; Rocher et al., 1996; Lin and Huang, 1997; this work) collected within the frontal part of the thrust belt through the Pliocene–Pleistocene deposits. Dashed lines on σ_2 and σ_3 stereo plots indicate that permutation may have occurred between these two axes. The average compression azimuth is N120°E. CTFZ—Chishan transfer-fault zone, PTFZ—Pakua transfer-fault zone, STFZ—Sanyi transfer-fault zone.

ture analysis have revealed that the orogenic stresses have been recorded in the Taiwan Strait near the Penghu Islands, we define a deformation front located west of the Penghu Islands (Fig. 14A) according to the model proposed in Figure 1B.

At the local scale, the structural styles of the frontal structures have been examined through the construction of serial geologic sections (Figs. 8–12) and provide strong support to the presence of deep- as well as shallow-décollement tectonics. For instance, shallow-décollement structures dominate in southern Taiwan (region 1), as illustrated by pop-up structures and low-angle thrusting. Northward, such shallow-décollement structures are locally superimposed on deeper-décollement structures, which results in the development of basement triangle zones, whereas offshore, simple high-angle thrusting dominates (regions 2 and 4). Preexisting extensional features (regions 2 and 5) might also localize the emergence of shallow décollements.

The along-strike structural changes are followed by variations in the recent kinematics of the deformation in frontal areas, which reflect local effects such as frontal contraction and lateral movement in response to indentation by the Peikang and Kuanyin highs (Fig. 14A).

Investigation of the western Taiwan mountain-belt front has revealed high levels of complexity emphasized by the along-strike variations in the thrust-wedge geometry, structural styles of frontal units, types of décollement-involved tectonics, and faulting mechanisms at the front. This high structural complexity is amplified by the diachronous evolution of tectonic and kinematic domains evolving northward from subduction to collision, in an oblique-convergence setting. Moreover, our re-

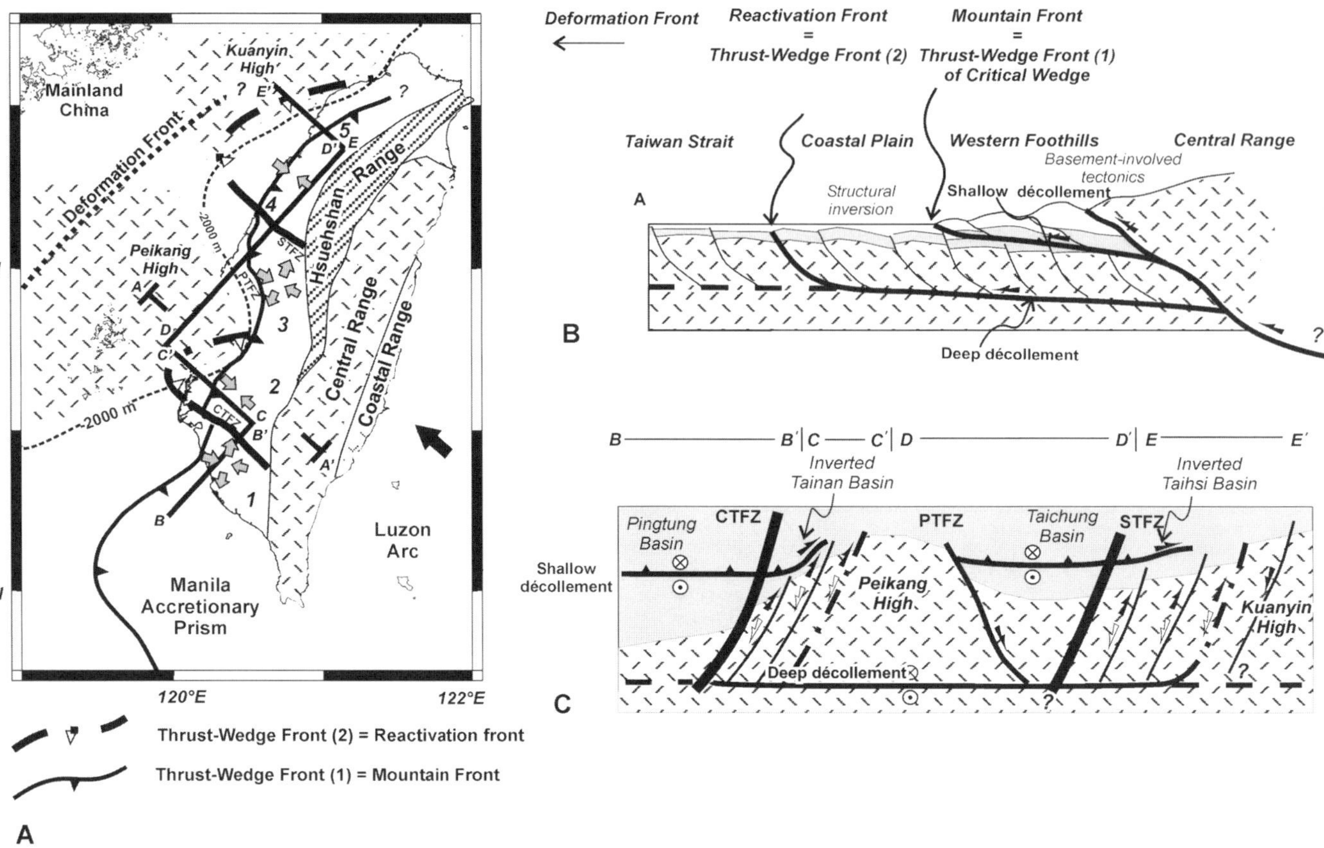

Figure 14. (A) Structural sketch map of the Taiwan thrust belt, showing locations of the different types of thrust-belt fronts in western Taiwan and their relationship with the shape of the pre-Miocene basement (the –2000 m contour). Regions where the reactivation front overlaps the –2000 m contour on the pre-Miocene basement correspond to inverted-basin areas. The along-strike variation of structural style is correlated with along-strike variations in the main stress regimes and demonstrates the evolution from prominent strike-slip regimes, i.e., the Tainan (1) and Taichung (3) areas, to purely compressional regimes especially in areas where basin inversion occurs, i.e., the Chiayi (2) and Miaoli (4) areas. (B) Schematic northwest-southeast cross section of the Western Foothills (example from the Chiayi area) highlighting the superimposition of both shallow- and deep-décollement tectonics according to the model of imbricate thrust wedges presented in Figure 1B. (C) Structural section of the basement showing basins, highs, and transfer faults that compose the basement structural framework. This section outlines the control of the structural inheritance on the deformation style and the kinematics of the thrust sheets. CTFZ—Chishan transfer-fault zone, PTFZ—Pakua transfer-fault zone, STFZ—Sanyi transfer-fault zone.

sults on the Taiwan mountain-belt front suggest that single-minded models such as either thin-skinned or thick-skinned tectonics may be unrealistic in the case of the Taiwan thrust belt and should be considered with care prior to any structural and kinematically consistent reconstruction of reversibly deformable cross sections in the Western Foothills.

ACKNOWLEDGMENTS

This work was supported by the Institut Français à Taipei–National Science Council of Taiwan cooperation framework, the French Ministry of Research and Education, and the Université Pierre et Marie Curie. Help from the Central Geological Survey of Taiwan and the Chinese Petroleum Corporation is gratefully acknowledged. We particularly thank K.M. Yang, S.T. Huang, and J.S. Chen from the Chinese Petroleum Corporation for stimulating discussions. Finally, we are grateful to R. Groshong, D. Cowan, and T. Byrne for their thorough reviews and comments, which have greatly improved the manuscript.

REFERENCES CITED

Angelier, J., Barrier, E., Chu, H.T., 1986, Plate collision and paleostress trajectories in a fold-thrust belt: The Foothills of Taiwan: Tectonophysics, v. 125, p. 161–178.

Angelier, J., 1990, Inversion of field data in fault tectonics to obtain the regional stress. 3. A new rapid direct inversion method by analytical means: Geophysical Journal International, v. 103, p. 363–376.

Angelier, J., Bergerat, F., Chu, H.T., Juang, W.S., and Lu C.Y., 1990, Paleostress analysis as a key to margin extension: The Penghu Islands, South China Sea: Tectonophysics, v. 183, p. 161–176.

Bally, A.W., Gordy, P.L., Stewart, G.A., 1966, Structure, seismic data, and orogenic evolution of southern Canadian Rocky Mountains: Bulletin of Canadian Petroleum Geology, v. 14, p. 337–381.

Berastegui, X., Garcia-Senez, J.M., Losantos, M., 1990, Tectono-sedimentary evolution of the Organya extensional basin (central south Pyrenean units, Spain) during the Lower Cretaceous: Bulletin de la Société Géologique de France, v. 8, p. 251–264.

Bergerat, F., 1987, Stress fields in the European platform at the time of Africa-Eurasia collision: Tectonics, v. 6, p. 99–132.

Boyer, S.E., and Elliott, D., 1982, Thrust systems: American Association of Petroleum Geologists Bulletin, v. 66, no. 9, p. 1196–1230.

Calassou, S., Larroque, C., and Malavieille, J., 1993, Transfer zones of deformation in thrust wedges: An experimental study: Tectonophysics, v. 221, p. 325–344.

Chang, S.L., Hsiao, P.T., Yuan, J., and Chi, W.R., 1983, The Neogene series: Tectonics evolution and petroleum potentialities of southern Taiwan: American Association of Petroleum Geologists Special Bulletin, p. 577–587.

Chang, S.S.L., and Chi, W.R., 1983, Neogene nannoplankton biostratigraphy in Taiwan and the tectonic implications: Petroleum Geology of Taiwan, v. 19, p. 93–147.

Chang, Y.L., Lee, C.I., Lin, C.W., Hsu, C.H., and Mao, E.W., 1996, Inversion tectonics in the Foothills of the Chiayi-Tainan area, southwestern Taiwan: Petroleum Geology of Taiwan, v. 15, p. 199–217.

Chapple, W.M., 1978, Mechanics of thin-skinned fold-and-thrust belts: Geological Society of American Bulletin, v. 89, p. 1189–1198.

Chen, J.S., 1978, A comparative study of the refraction and reflection seismic data obtained on the Changhua Plain to the Peikang Shelf, Taiwan: Petroleum Geology of Taiwan, v. 15, p. 199–217.

Chinese Petroleum Corporation, 1974, Geological maps of Hsinchu, Miaoli, Taichung, Chiayi, Tainan and Kaohsiung areas: Miaoli, Taiwan, Taiwan Petroleum Exploration Division (TPED), scale 1:100 000, 6 sheets.

Chiu, H.T., 1973, Basement rocks under the Neogene formation of west-central Taiwan: Proceedings, Geological Society of China, v. 16, p. 51–58.

Chou, J.T., 1971, A sedimentologic and paleogeographic study of the Neogene formations in the Taichung Region, western Taiwan: Petroleum Geology of Taiwan, v. 9, p. 43–66.

Chou, J.T., 1980, Stratigraphy and sedimentology of the Miocene in western Taiwan: Petroleum Geology of Taiwan, v. 17, p. 33–52.

Cooper, M.A., and Williams, G.D., editors, 1989, Inversion Tectonics: Geological Society [London] Special Publication 44, 376 p.

Cousens, B.A., and Wiltschko, D.V., 1996, The control of mechanical stratigraphy on the formation of triangle zones: Bulletin of Canadian Petroleum Geology, v. 44, no. 2, p. 165–179.

Colletta, B., Roure, F., De Toni, B., Loureiro, D., Passalacqua, H., and Gou, Y., 1997, Tectonic inheritance, crustal architecture and contrasting structural style in the Venezuela Andes: Tectonics, v. 16, no. 5, p. 777–794.

Craddock, J.P., Jackson, M., Van der Pluijm, B.A., Versical, R.T., 1993, Regional shortening fabrics in eastern North America: Far-field stress transmission from the Appalachian-Ouchita orogenic belt: Tectonics, v. 12, no. 1, p. 257–264.

Dahlstrom, C.D.A., 1969, Balanced cross sections: Canadian Journal of Earth Sciences, v. 6, p. 743–757.

Davis, D., Suppe, J., and Dahlen, F.A., 1983. Mechanisms of fold-and-thrust belts and accretionary wedges: Journal of Geophysical Research, v. 88, no. B2, p. 1153–1172.

Dechesne, R.G., and Mountjoy, E.W., 1992, Mutliple decollement at deep levels of the southern Canadian Rocky Mountain Main Ranges, Alberta and British Columbia, *in* Mitra, S., and Fisher, G. W., eds., Structural geology of fold and thrust belts: Baltimore, Maryland, Johns Hopkins University Press, p. 225–238.

Deffontaines, B., Lee, J.C., Angelier, J., Carvalho, J., and Rudant, J.P., 1994, New geomorphic data on the active Taiwan orogen: A multisource approach: Journal of Geophysical Research, v. 99, no. B10, p. 20243–20266.

Deffontaines, B., Lacombe, O., Angelier, J., Chu, H.T., Mouthereau, F., Lee, C.T., Déramond, J., Lee, J.F., Yu, M.S., and Liew, P.M., 1997, Quaternary transfer faulting in the Taiwan foothills: Evidence from a multisource approach: Tectonophysics, v. 274, p. 61–82.

DeMets, C., Gordon, R.G., Argus, D.F., Stein, S., 1990, Current plate motions: Geophysical Journal of International, v. 101, p. 425–478.

Ellwood, A., Wang, C.Y., Teng, L.S., Yen, H.Y., 1996, Gravimetric examination of thin-skinned décollements basement-involved models for the Taiwan orogen: Journal of the Geological Society of China, v. 39, no. 2, p. 209–221.

Ho, C.S., 1986a, A synthesis of the geologic evolution of Taiwan: Tectonophysics, v. 125, p. 1–16.

Hsiao, P.T., 1974, Subsurface geologic study of the Hsinying Coastal Plain area, Taiwan: Petroleum Geology of Taiwan, v. 11, p. 27–39.

Huang, S.T., Chen, R.C., and Chi, W.R., 1993, Inversion tectonics and evolution of the northern Taihsi Basin, Taiwan: Petroleum Geology of Taiwan, no. 28, p. 15–46.

Huang, C.Y., Yuan, P.B., Song, S.R., Lin, C.W., Wang, C., Chen, M.T., Shyu, C.T., Karp, B., 1995, Tectonics of short-lived intra-arc basins in the arc-continent collision terrane of the Coastal Range, eastern Taiwan: Tectonics, v. 14, no. 1, p. 19–38.

Huiqi, L., McClay, K.R., and Powell, D., 1992, Physical models of thrust wedges, *in* McClay, K.R., ed., Thrust tectonics: London, Chapman and Hall, p. 71–81.

Hung, J.H., and Wiltschko, D.V., 1993, Structure and kinematics of arcuate thrust faults in the Miaoli-Cholan area of western Taiwan: Petroleum Geology of Taiwan, no. 28, p. 59–96.

Hung, J.H., 1994, Analysis of deformation fabrics in the Sani thrust sheet and the Chuhuangkeng anticline of western Taiwan: Petroleum Geology of Taiwan, no. 29, p. 105–191.

Lacombe, O., Angelier, J., and Laurent, P., 1993, Les macles de la calcite, marqueurs des compressions récentes dans un orogène actif: L'exemple des calcaires récifaux du Sud de Taiwan: Comptes Rendus de l'Académie des Sciences, Serie 2, Sciences de la Terre et des Planetes: Earth and Planetary Sciences, v. 306, p. 1805–1813.

Lacombe, O., Laurent, P., Rocher, M., 1996, Magnitude de la contrainte déviatorique pyrénéenne dans l'avant-pays nord-pyrénéen: Comptes Rendus de l'Académie des Sciences de Paris, Serie 2a, v. 322, p. 229–235.

Lacombe, O., Angelier, J., Chen, H.W., Deffontaines, B., Chu, H.T., and Rocher, M., 1997, Syndepositional tectonics and extension-compression relationships at the front of the Taiwan collision belt: A case study in the Pleistocene reefal limestones of Kaohsiung (SW Taiwan): Tectonophysics, v. 274, p. 83–96.

Lee, C.I., Chang, Y.L., Mao, E.W., and Tseng, C.S., 1993, Fault reactivation and structural inversion in the Hsinchu-Miaoli area of northern Taiwan: Petroleum Geology of Taiwan, v. 28, p. 47–58.

Lee, J.C., 1994, Structure et deformation active d'un orogene: Taiwan [Ph.D. thesis]: Paris, Université Pierre et Marie Curie, 280 p.

Letouzey, J., 1986, Cenozoic paleo-stress pattern in the Alpine Foreland and structural interpretation in a platform basin: Tectonophysics, v. 132, p. 215–231.

Letouzey, J., Colletta, B., Vially, R., and Chermette, J.C., 1995, Evolution of salt-related structures in compressional settings, *in* Jackson, M.P.A., et al., eds., Salt tectonics: A global perspective: American Association of Petroleum Geologists Memoir 65, p. 41–60.

Lin, C.W., and Huang, M.L., 1997, The fractures and paleostress of deformed pebbles in the conglomerates of the Toukoshan formation, Chiayi-Yuanlin area: Journal of the Geological Society of China, v. 40, no. 1, p. 281–297.

Lu, C.Y., Angelier, J., Chu, H.T., and Lee, J.C., 1995, Contractional, transcurrent, rotational and extensional tectonics: Examples from northern Taiwan: Tectonophysics, v. 246, p. 129–146.

McClay, K.R., Insley, M.W., and Anderton, R., 1989, Inversion of the Kechika Trough, Northeastern British Columbia, Canada, *in* Cooper, M.A., and Williams, G.D., eds., Inversion tectonics: Geological Society [London] Special Publication 44, p. 235–257.

McClay, K.R., 1992, Glossary of thrust tectonics terms, *in* McClay, K.R., ed., Thrust tectonics: London, Chapman and Hall, p. 419–433.

Mitra, S., and Islam, Q.T., 1994, Experimental (clay) models of inversion structures: Tectonophysics, v. 230, p. 211–222.

Morley, C.K., 1986, A classification of thrust fronts: American Association of Petroleum Geologists Bulletin, v. 70, no. 1, p. 12–25.

Mouthereau, F., Lacombe, O., Deffontaines, B., Angelier, J., Chu, H.T., Lee, C.T., 1999, Quaternary transfer faulting and belt front deformation at Pakuashan (western Taiwan): Tectonics, v. 18, no. 2, p. 215–230.

Namson, J., 1981, Structure of the western foothills belt, Miaoli-Hsinchu area, Taiwan. 1. Southern part: Petroleum Geology of Taiwan, v. 18, p. 31–51.

Reed, D.L., Lundberg, N.L., Liu, C.S., and Kuo, B.Y., 1992, Structural relations along the margins of the offshore Taiwan accretionary wedge: Implications for accretion and crustal kinematics: Acta Geologica Taiwanica, no. 30, p. 105–122.

Rocher, M., Lacombe, O., Angelier, J., Chen, H.W., 1996, Mechanical twin sets in calcite as markers of recent collisional events in fold-and-thrust belts: Evidence from the reefal limestones of SW Taiwan: Tectonics, v. 15, no. 5, p. 984–996.

Roure, F., Howell, D.G., Guellec, S., and Casero, P., 1990, Shallow structures induced by deep-seated thrusting, *in* Letouzey, J., ed., Petroleum and tectonics in mobile belts: Paris, Technip, p. 15–30.

Roure, F., and Colletta, B., 1996, Cenozoic inversion structures in the foreland of the Pyrenees and Alps, *in* Ziegler, P.A., and Horvath, F., eds., Peri-Tethys Memoir 2: Structure and prospects of alpine basins and forelands: Memoire Museum National Histoire Naturelle, v. 170, p. 173–209.

Sassi, W., and Faure, J.L., 1997, Role of faults and layer interfaces on the spatial variation of stress regimes in basins: Inferences from numerical modeling: Tectonophysics, v. 266, p. 101–119.

Seno, T., Stein, S., and Gripp, A.E., 1993, A model for the motion of the Philippine Sea Plate consistent with NUVEL-1 and geological data: Journal of Geophysical Research, v. 98, no. B10, p. 17941–17948.

Shaw, C.L., 1996, Stratigraphic correlation and isopachs maps of the Western Taiwan Basin: Terrestrial Atmospheric Ocean, v. 7, no. 3, p. 333–360.

Shen, H.C., Huang, S.T., Tang, C.H., and Hsu, Y.Y., 1996, Geometrical characteristics of structural inversion on the offshore of Miaoli, Taiwan: Petroleum Geology of Taiwan, no. 30, p. 79–110.

Sun, S.C., 1964, Photogeologic study of the Tainan-Kaohsiung coastal plain area, Taiwan: Petroleum Geology of Taiwan, no. 3, p. 39–51.

Sun, S.C., 1965, Geology and petroleum potentialities of the Chingshui-Yuanlin area: Petroleum Geology of Taiwan, no. 4, p. 161–173.

Sun, S.C., 1982, The Tertiary basins of offshore Taiwan: Petroleum geology of Taiwan, *in* Salivar-Sali, A., ed., Petroleum geology of Taiwan, Proceedings, 2nd Asian Conference on Petroleum (ASCOPE): Manila, Philippines, Technical Programme Committee, 1981, p. 125–135.

Suppe, J., 1976, Decollement folding in southwestern Taiwan: Petroleum Geology of Taiwan, no. 13, p. 25–35.

Suppe, J., and Namson, J., 1979, Fault-bend origin of frontal folds of the western Taiwan fold-and-thrust belt: Petroleum Geology of Taiwan, no. 16, p. 1–18.

Suppe, J., 1980, Imbricated structure of western foothills belt, southcentral Taiwan: Petroleum Geology of Taiwan, no. 17, p. 1–16.

Suppe, J., Hu, C.T., Chen, Y.J., 1985, Present-day stress directions in western Taiwan inferred from borehole elongation: Petroleum Geology of Taiwan, no. 21, p. 1–12.

Suppe, J., 1986, Reactivated normal fault in the western Taiwan fold-and-thrust belt: Memoir of the Geological Society of China, v. 7, p. 187–200.

Suppe, J., and Medwedeff, D.A., 1990, Geometry and kinematics of fault-propagation folding: Eclogae Geologicae Helvetica, v. 83, no. 3, p. 409–454.

Teng, L.S., 1990, Geotectonic evolution of late Cenozoic arc-continent collision in Taiwan: Tectonophysics, v. 183, p. 57–76.

Vann, I.R., Graham, R.H., Hayward, A.B., 1986, The structure of mountain fronts: Journal of Structural Geology, v. 8, no. 3/4, p. 215–227.

Wiltschko, D.V., Hung, J.H., Hickman, J.B., Liu, C.C., Yu, S.B., Fang, P., and Bock, Y., 1997, *in* Proceedings, International Conference and Sino-American Symposium on Tectonics of East Asia: Chungli, Taiwan, p. 134.

Winslow, M.A., 1981, Mechanisms for basement shortening in the Andean foreland fold belt of southern South America: Thrust and nappe tectonics: Journal of the Geological Society, London, v. 9, p. 513–528.

Wu, F.T., Rau, R.J., and Salzberg, D., 1997, Taiwan orogeny: Thin-skinned or lithospheric collision: Tectonophysics, v. 274, p. 191–220.

Yang, K.M., Wu, J.C., Wickham, J.S., Ting, H.H., Wang, J.B., Chi, W.R., 1996, Transverse structures in Hsinchu and Miaoli areas: Structural mode and evolution in Foothills Belt, northwestern Taiwan: Petroleum Geology of Taiwan, no. 30, p. 111–150.

Yeh, Y.H., Barrier, E., Lin, C.H., and Angelier, J., 1991, Stress tensors analysis in the Taiwan area from focal mechanisms of earthquakes: Tectonophysics, v. 200, p. 267–280.

Yu, S.B., Chen, H.Y., Kuo, L.C., 1997, Velocity field of GPS stations in the Taiwan area: Tectonophysics, v. 274, p. 41–59.

Zapata, T.R., and Allmendinger, R.W., 1996, Thrust-front zone of the Precordillera, Argentina, a thick-skinned triangle zone: American Association of Petroleum Geologists Bulletin, v. 80, p. 359–381.

Manuscript Accepted by the Society April 12, 2001

Printed in the U.S.A.

Geological Society of America
Special Paper 358
2002

Synchroneity and morphology of Holocene river terraces in the southern Western Foothills, Taiwan: A guide to interpreting and correlating erosional river terraces across growing anticlines

Meng-Long Hsieh*
Peter L.K. Knuepfer[†]
Department of Geological Sciences, Binghamton University, Binghamton, New York 13902-6000, USA

ABSTRACT

We studied river terraces in two large river basins in the southern Western Foothills of Taiwan, and we recognize three major Holocene anticlinal structures that are crossed by the rivers. These structures are interpreted on the basis of converging and diverging relationships of terraces aided by more than 20 radiocarbon dates. Holocene rock-uplift rates at the crests of these anticlines are estimated to be at least 1 cm/yr, as inferred from bedrock-incision rates as great as 20 mm/yr at multiple sites. The valleys where the rivers cross the anticlines tend to develop higher, multilevel, but relatively narrow terraces that converge toward lower, fewer but wider terraces developed away from the anticlines. In addition, just as the terrace surfaces can converge or diverge, so can the "time lines" represented by the terraces. A single terrace thus can develop away from the anticline while multiple terraces are created across the anticline. The converging and diverging relationship of the terraces may not be clearly expressed in the field owing to the low potential for preservation of high narrow terraces across anticlines. However, the contrasting width of terraces or river valleys—wide valleys or terraces in valley reaches that are relatively tectonically stable versus narrow valleys and terraces in valley reaches undergoing rapid uplift—provides an extra criterion in terrace correlation and neotectonic interpretation, providing that bedrock lithology does not change along the river. We used this criterion to inspect river-valley patterns elsewhere in the southern Western Foothills and were able thereby to recognize five additional Holocene folds.

INTRODUCTION

Fluvial terraces are probably the landform most widely used to evaluate tectonic deformation over time scales of hundreds to tens of thousands of years (Keller and Pinter, 1996). Full utilization of terraces in tectonic studies, however, is commonly hindered not only by the uncertain origin of the terraces (Merritts et al., 1994), but also by the subjectivity in spatially correlating terrace-surface remnants to reconstruct fluvial history (e.g., Frye and Leonard, 1954; Ritter et al., 1995). These limitations are especially true where the terraces are laterally discontinuous and their number changes along a valley. Detailed chronologic information certainly helps in correlating such terraces; however, because dates are rarely plentiful, un-

*Present address: Department of Geology, National Taiwan University, Taipei, Taiwan, Republic of China. E-mail: hsieh@gl.ntu.edu.tw.
[†]E-mail: knuepfr@binghamton.edu.

Hsieh, M.-L., and Knuepfer, P.L.K., 2002, Synchroneity and morphology of Holocene river terraces in the southern Western Foothills, Taiwan: A guide to interpreting and correlating erosional river terraces across growing anticlines, *in* Byrne, T.B., and Liu, C.-S., eds., Geology and Geophysics of an Arc-Continent Collision, Taiwan: Boulder, Colorado, Geological Society of America Special Paper 358, p. 55–74.

certainty exists regarding how well the limited chronologic data can resolve the timing of the formation of terraces both at a site and along a river.

One of the reasons why terrace correlation can be difficult is that profiles along the length of a terrace ("long profiles") can converge or diverge with respect to the modern river and with respect to each other. As a consequence, correlations based on the terrace elevation above a certain datum (e.g., modern river bed), as applied in some studies (e.g., Rains and Welch, 1988; Hooke et al., 1990; Amorosi et al., 1996), might be incorrect. Convergence or divergence of terraces in response to tectonic warping or tilting has been well documented in many studies (e.g., Rockwell et al., 1988; Bullard and Lettis, 1993; Willemin and Knuepfer, 1994; Molnar et al., 1994; Merritts et al., 1994; Pazzaglia et al., 1998). Yet very few studies (e.g., Harbor, 1998) have shown explicitly that *two* or *more than two* levels of terraces actually converge with each other to a *single*-level terrace (or a flood plain) or that a *single*-level terrace (or a flood plain) diverges into *more than one* terrace level in response to tectonic processes. In other words, insufficient attention has been paid to the question of how a channel in an actively uplifting river reach can repeatedly incise and laterally erode to form multiple levels of terraces while a channel in an adjacent, less-actively uplifting reach maintains a relatively constant flood-plain level without incision. Such a scenario clearly complicates the possibility of correlating terraces along a river. First, the relative position of a terrace (its number in a local hierarchy) may not be a useful criterion for correlation. Second, the traditional assumption that the correlated terraces constitute a time line (e.g., Amorosi et al., 1996; Pazzaglia et al., 1998) may be arguable, because just as the terrace surfaces can converge or diverge, so can the "time" represented by the terraces, producing diachronous terrace surfaces (e.g., Bull, 1990; Weldon and Stimac, 1994) (Fig. 1). In other words, a terrace surface and its associated deposits along one reach of a river can represent more time of occupation than any one of the terraces that converge to that surface. As a consequence, terraces with different dates may nonetheless have been part of one continuous flood plain during at least some part of their formation (cf. Germanoski and Harvey, 1993) (Fig. 1).

In this paper, we illustrate these concepts of diachronous terrace sequences by using river terraces in the southern Western Foothills of Taiwan (Fig. 2). River terraces, many of them of erosional origin, are ubiquitous in the southern Western Foothills where prominent shortening and uplift (Yu and Chen, 1994; Yu et al., 1997) are producing rapid deformation across faults and folds. These terraces are underlain by weakly resistant mudstone, which minimizes the effect of bedrock on the morphology and sequence of terraces. We focus on three river reaches belonging to two major drainage basins. We interpret that these river reaches cross active anticlinal structures on the basis of the apparent convergence or divergence of terrace sequences and a total of more than 20 radiocarbon dates (mostly Holocene). We find that in association with the converging and

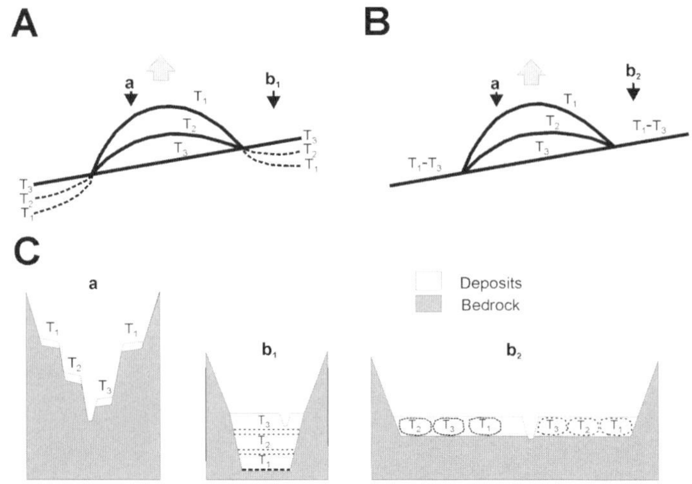

Figure 1. (A and B) Schematic terrace long profiles and (C) cross sections showing hypothetical convergence (terraces at point b) and divergence (terraces at point a) relationships of terraces formed from arbitrary times T_1 to T_3 in response to growth of an anticline. Shaded arrows show axes of uplift; small arrows below a, b_1, and b_2 show locations of diagrammatic sections in C. Bedrock lithology is assumed to be the same along the valley. Two end-member types of terraces may be developed where multiple terrace surfaces diverge across the anticline. In section b_1 (not scaled; the flood plain could be part of a depositional basin), the channel is dominated by aggradation, and younger deposits progressively overlie older ones (cf. Germanoski and Harvey, 1993; Amorosi et al., 1996; Harbor, 1998). This circumstance probably reflects long-term subsidence of the river reach. In section b_2 (wider than section a), the channel is dominated by lateral erosion, with different ages of deposits spreading out on a wide bedrock surface. This circumstance is probably favored in a river reach characterized by long-term tectonic uplift, but with an uplift rate lower than that of the adjacent anticline, and where the uplifting anticline forms a local base level inhibiting upstream incision.

diverging relationships of the terrace surfaces and their formation times, the widths of the terraces also change considerably along the river: narrower terraces and valleys occur in rapidly uplifting reaches, and wider terraces and valleys occur in reaches that are relatively tectonically stable (Fig. 1C). We propose that such a morphologic relationship has genetic implications and should be applicable to other river systems in general. An important implication here is that even where the converging and diverging relationships of terraces are not suitably preserved, the contrast of terrace or valley width may provide a useful criterion in interpreting active structures. We further apply this criterion to search for possible active folds in the rest of the southern Western Foothills.

GEOLOGIC AND GEOMORPHIC BACKGROUND

The island of Taiwan (Fig. 2) is the result of the collision between the Luzon volcanic arc of the Philippine Sea plate and the Chinese continental margin of the Eurasian plate that began at ca. 5–4 Ma (Biq, 1973; Chai, 1972; Bowin et al., 1978; Ho, 1986; Teng, 1990) and continues to the present (Yu and Chen,

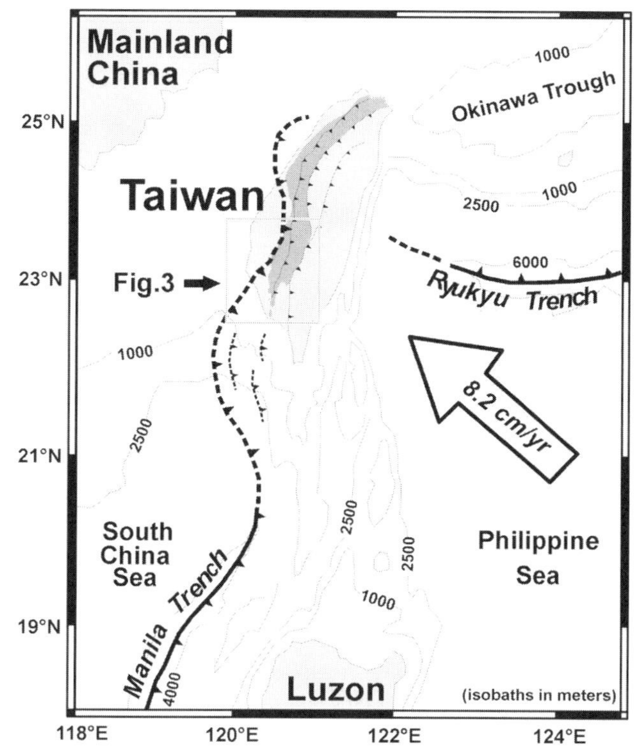

Figure 2. Geodynamic framework of Taiwan and location of the southern Western Foothills (shaded area in box). Open arrow with rate shows the current movement of the Philippine Sea plate relative to the Chinese continental margin (Yu et al., 1997). Thick solid lines with triangle barbs represent subduction zones; barbs are on the hanging wall; the thick dashed line with triangle barbs refers to the deformation front of the collision zone (modified from Ho, 1986).

1994; Yu et al., 1997). The Western Foothills, a hilly and/or mountainous area in western Taiwan, defines the frontal zone of this collisional belt (Fig. 2). It is bounded to the east by the Backbone Range, rising up to 3000 m in elevation, and to the west by the Coastal Plain, which is generally no more than several tens of meters above sea level. The southern part of the Western Foothills is underlain by folded and thrust-faulted Neogene sedimentary rocks (Fig. 3). The sedimentary rocks of the southern Western Foothills consist of preorogenic sandstone and shale of late Miocene age, the sediments of which were deposited on the stable Chinese continental shelf, and synorogenic Pliocene–Pleistocene foreland sediments, dominantly shales and mudstones (Fig. 3). It is these almost homogeneous, fine-grained rocks that underlie the studied river terraces.

The northern part of the southern Western Foothills, underlain by thick Miocene strata, is topographically a massif with elevations up to 2800 m. This northern part is named here the Alishan Highland. The interior of the southern Western Foothills is characterized by a valley-and-ridge topography (Fig. 3) that lowers progressively in elevation from east to west. The ridges in the southern part of the southern Western Foothills are bounded on the west by major reverse faults, and valleys generally follow synclinal structures (Fig. 3). We have concentrated our studies in the basins of two rivers draining this area, the Tsengwen and Erhjen Rivers (Fig. 3). The Tsengwen River originates in the Alishan Highland and flows south mainly between the Chukou and Chutouchi reverse faults (and their associated ranges) until it reaches the Yuching area, where it turns normal to the Chukou fault and crosses it before entering the Coastal Plain (Fig. 3). At the mountain front, the Tsengwen River joins the Tsailiao River, which drains the area between the southern end of the Chukou fault and the Lungchuan fault and is crossed by the northwest-trending Tsochen fault (Fig. 3). The Erhjen River originates in the valley between the Lungchuan and Pingchi faults; it turns across the Lungchuan fault downstream of Chungpu and then generally follows the trend of the Kutingkeng fault before reaching the mountain front (Fig. 3).

Most of the river valleys in the southern Western Foothills that are floored by river terraces and modern flood plains have uniform widths or widen downstream, a feature that is typical of river valleys. There are, however, exceptions. Some of these exceptions (i.e., valley reaches that are significantly wider than those downstream of them) are so distinct that they can be readily recognized from small-scale (e.g., 1:25 000) topographic maps (Fig. 4). For these valleys to be readily recognized at this scale, they must consist of at least several meander bends of the modern channels (or be >1 km in length), and have maximum widths (defined as the combined width of the modern flood plains and major terraces) at least twice as large as the average widths of the valleys downstream from them. Two of these valley reaches are located in the basins of the Tsengwen and Erhjen Rivers and are discussed in detail in this paper. Many of the "abnormally" wide valley reaches in the southern Western Foothills, including those of the Tsengwen and Erhjen Rivers, are either underlain by mapped Pliocene–Pleistocene synclines or are located upstream of mapped anticlines (Fig. 4); their presence thus may have tectonic implications, as are subsequently discussed further.

TERRACE SURVEYING AND CORRELATION

This study employed traditional methods of geomorphic research. River terraces were first mapped from aerial photographs through stereoscopes and plotted onto photograph-based, 1:5000-scale topographic maps issued by the Department of Agriculture and Forestry of Taiwan. The elevation of each terrace surface was directly determined from these maps and referenced to the elevation of the active channel as defined on the maps. Because the maps have a 5 m contour interval, any location that is between two contour lines has an elevation uncertainty of 5 m. The stratigraphy of terrace deposits and the underlying bedrock configuration were then examined in the field. Hand-leveling was used to measure or estimate the thickness of sediments and the height of some specific targets (dated samples, bedrock surfaces) above the local datum (e.g., the

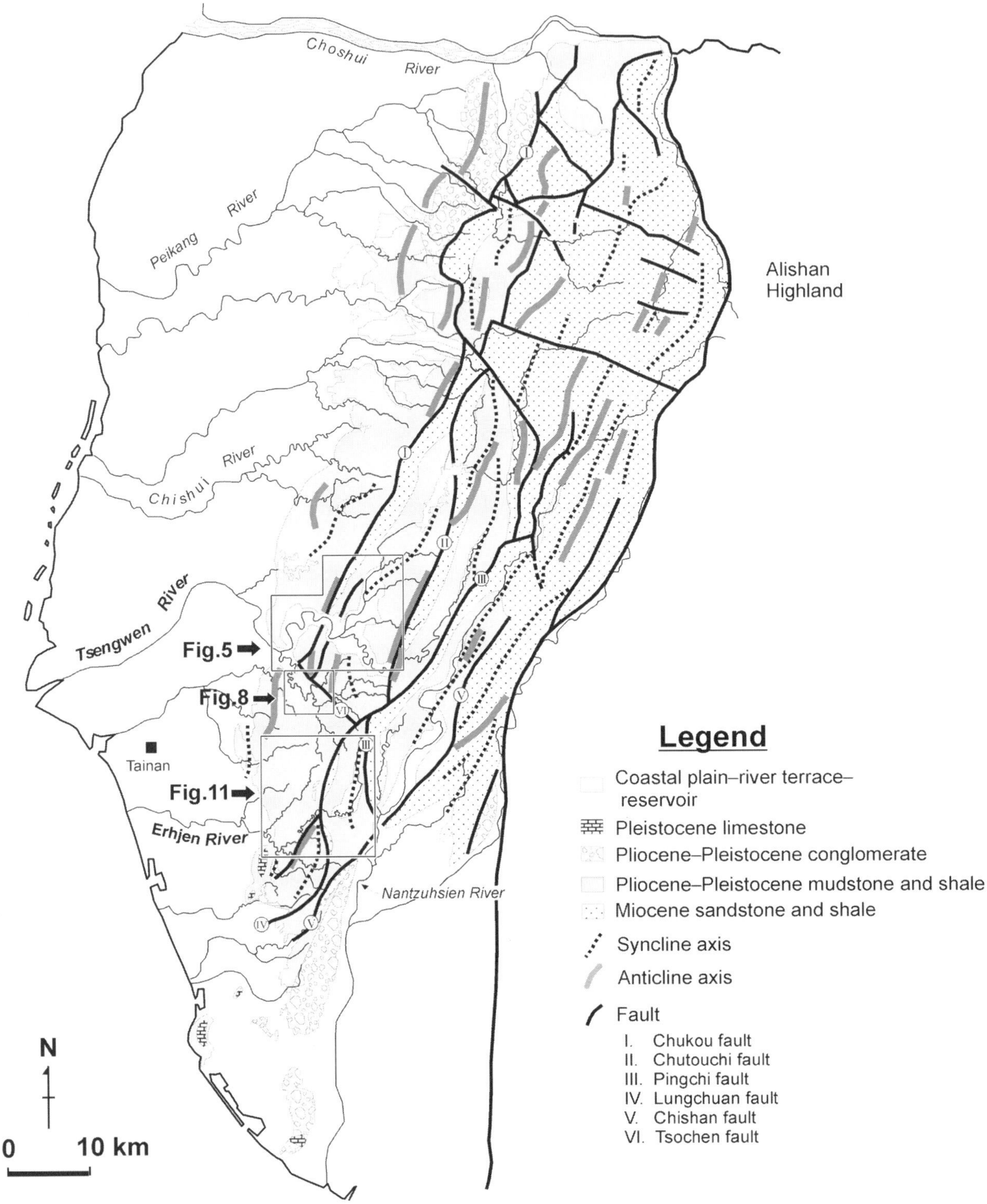

Figure 3. Geology and river systems in the southern Western Foothills (for location, see Fig. 1). Structures and rock units are mainly from Chinese Petroleum Corporation (1971); structures in the downstream Tsengwen River are modified from Chinese Petroleum Corporation (1989). Alishan Highland is northwest of the area illustrated.

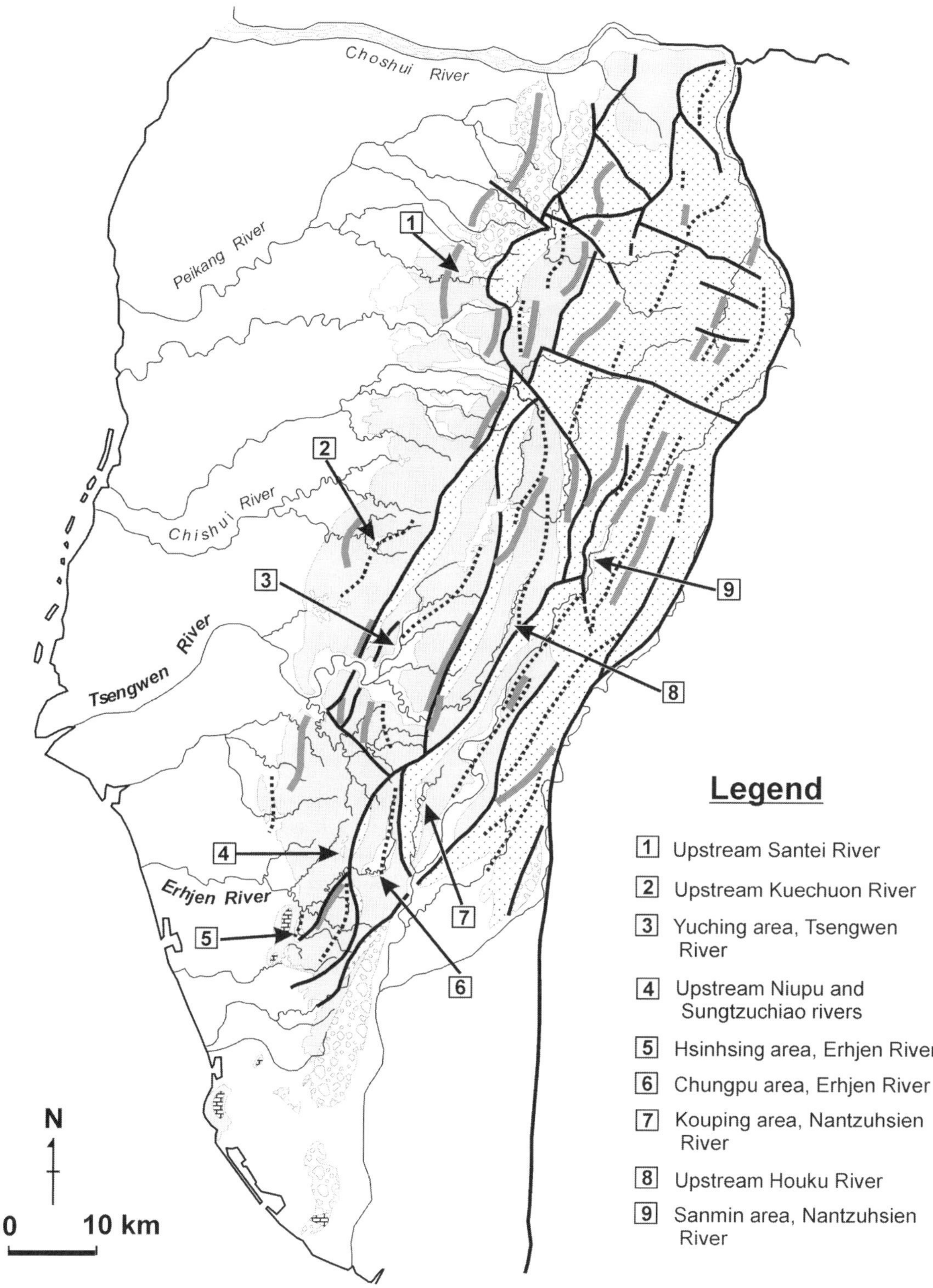

Figure 4. Map of the southern Western Foothills, showing locations of valley reaches (numbered) that are significantly wider than the valleys downstream of them. See text for discussion. Legend as in Figure 3.

modern channel surface). Such measurements are assumed to be more accurate than those determined from the topographic maps and, in most cases, do fall into the uncertainty of the latter. Wood or plant fragments within the terrace deposits were sought for radiocarbon dating. Radiocarbon ages were calibrated to ages in calendar years before present (cal. yr B.P.) by using the program of Stuiver and Reimer (1993).

Our correlation of river terraces involved the following five procedures:

1. We identified nearly continuous terrace surfaces that are interrupted only by tributaries or main-stem channels (terrace YC [for Yuching] along the main-stem Tsengwen River is an example; Fig. 5). It is obvious if a continuous terrace surface thus identified diverges downstream or upstream to more than one terrace surface (e.g., the divergence of YC downstream to YC1, YC2, and YC3; Fig. 5).

2. We initially correlated less continuous terrace remnants. This correlation is based on terrace geometry (pattern and relief) from aerial photographs. For us to have correlated two neighboring terrace remnants, they must have relative relief of less than 5 m and, in most cases, be less than 500 m apart along the valley. Where multiple levels of terraces are close together in both elevation and along-valley position, more than one possible correlation may exist. In this case, we temporarily correlated terrace remnants to result in a smooth long profile, while simultaneously recognizing the meandering character of most of the paleo–flood plains. The correlation ends, however, where a significant position gap (as already defined) occurs between terrace remnants.

3. We used radiocarbon dates to check whether the correlations in step 2 were reasonable. If not, we revised the correlation to seek a consistency between reliable radiocarbon dates and apparent continuity of terrace remnants. However, in interpreting the radiocarbon dates, one must recognize that dates from basal gravel may differ from those obtained from overbank deposits of the same terrace (Merritts et al., 1994).

4. We interpreted the tectonics from the results of steps 1 to 3.

5. We correlated the rest of the terrace remnants according to the inferred tectonic styles.

TERRACES IN THE TSENGWEN, TSAILIAO, AND ERHJEN RIVER BASINS

Terraces of the Tsengwen River in the Yuching area

Yuching Valley. The valley around Yuching is predominantly floored by terrace YC, which is 1.5 km wide and 5–10 m higher than other terraces that are a few meters above the modern channel (Fig. 5). Although flat, this terrace surface is inclined toward the west and is asymmetric in cross section with respect to the valley axis (Figs. 5 and 6, profile B–B′). The thickness of deposits and the underlying bedrock configuration of terrace YC is not clear because of a scarcity of outcrops. The exposed terrace deposits are mainly composed of cobble-sized gravel locally interlayered with cross-stratified sandy beds; limestone boulders as large as 2–3 m in diameter are also observed at the base of some exposures. These coarse sediments are interpreted as channel deposits. Sandy to muddy beds, a few meters in thickness and locally with planar stratification, also commonly overlie these channel deposits; they are probably overbank deposits.

Excavation for new bridge piers west of Yuching exposed terrace YC and its adjacent lower terrace (sample 1 site; Fig. 5). The bedrock strath underlying the lower terrace was about 2 m below the water surface in the active channel. The exposures reveal that this strath level is continuous with the bedrock surface at the base of the sediments on terrace YC, which have a maximum thickness of ~15 m (10 m of channel deposits). We obtained a radiocarbon date, 13 740–14 014 cal. yr B.P. (±1 standard deviation) (sample 1; Fig. 5 and Table 1), from an aggregate of wood fragments within a silty lens in gravel beds within the lower terrace deposits about 4 m above bedrock. We also obtained a date, 3071–3324 cal. yr B.P. (sample 2; Fig. 5 and Table 1), from an aggregate of plant fragments within an overbank mud layer 3 m below the surface of terrace YC nearby.

These two dates appear to be inconsistent. In fact, the other dates we have obtained in the southern Western Foothills from terraces <5 m above the modern channel are all <500 yr B.P. (e.g., samples 7 and 8 in Fig. 5), so sample 1 appears to be too old. One possibility is that it has been reworked from an older deposit, likely the adjacent terrace YC. But the continuity of the strath level with that of terrace YC suggests that the lower terrace may be cut into terrace YC deposits and the dated sample may be related directly to terrace YC. In any event, this date strongly suggests that in this river valley, no significant bedrock incision has occurred since the latest Pleistocene. The 3071–3324 cal. yr B.P. date, on the other hand, indicates that until at least as late as 3 ka, part of terrace YC was still inundated by overbank floods.

Tinghu area. The wide Yuching Valley ends upstream in a gorge. Just downstream from the gorge, the valley around Tinghu is characterized by a series of abandoned meander bends and terraces; the highest one (labeled TH_1 [for Tinghu]) is about 100 m above the modern channel (Figs. 5 and 6, profile A–A′–A″). The river history here thus includes repeated migration and incision of meandering channels as well as channel-bank cutoffs, which is apparently different from the history of the river downstream where the broad terrace YC was developed.

The outcrop in the Tinghu area is very poor. A temple construction exposes coarse gravel beds on the slope of the hill that the major meander bends surround (Fig. 5, loc. A′). These beds, striking east and tilted 10°–20° south, are highly weathered—a distinct contrast to the Holocene terrace deposits exposed elsewhere around the region. We interpret that these gravel beds are fluvial sediments of late Pleistocene age originally deposited in the valley; they are now deformed and serve as "bedrock" for the younger terraces in this river valley.

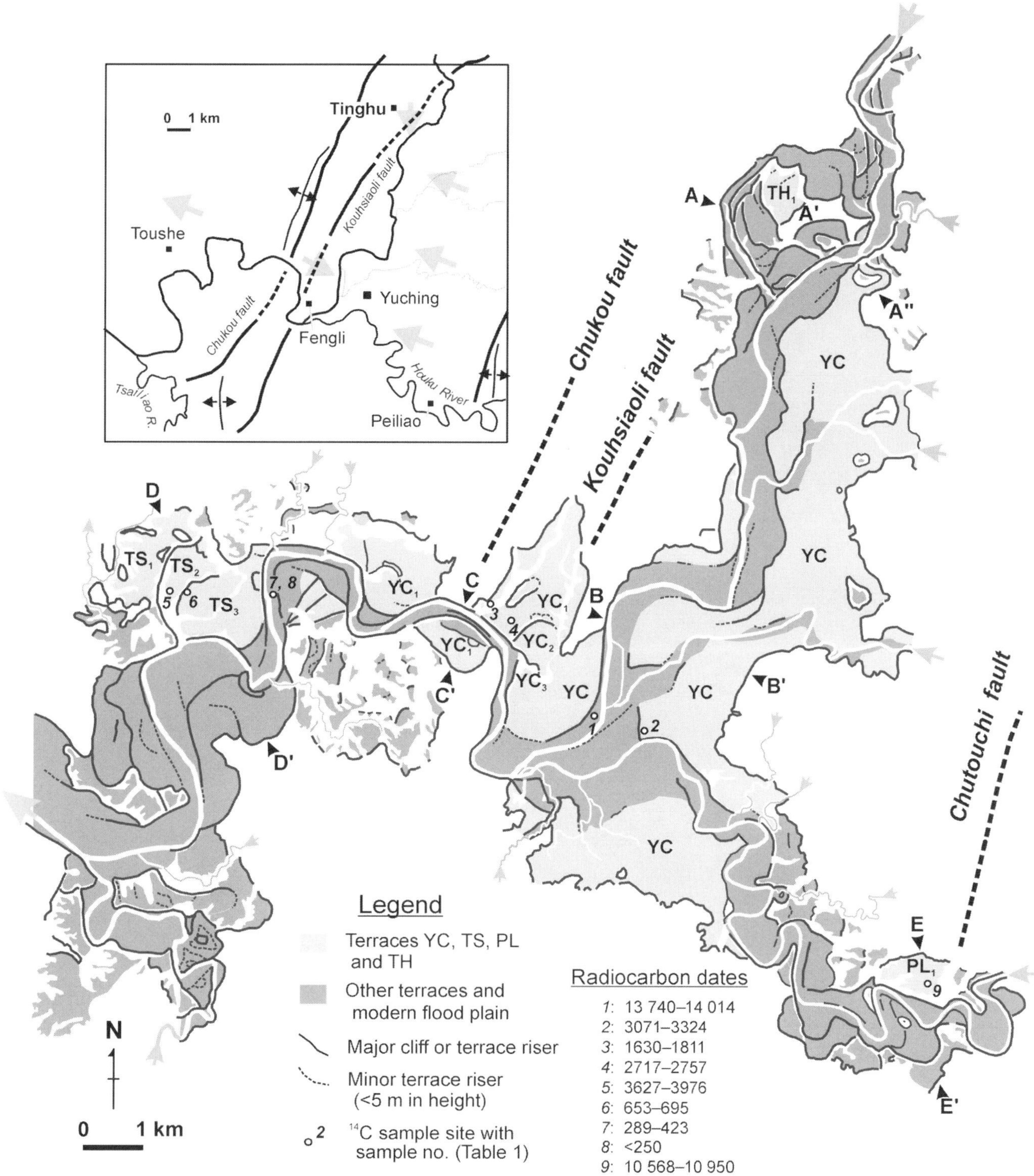

Figure 5. Distribution of river terraces of the Tsengwen River in the Yuching area (for location, see Fig. 3) and radiocarbon dates (ranges shown indicate the $\pm 1\sigma$ span in cal. yr B.P. in this and subsequent figures) and sites. Inset locates Pleistocene structures, major towns, and inferred Holocene structures (light-gray symbols; arrows indicate downdip direction).

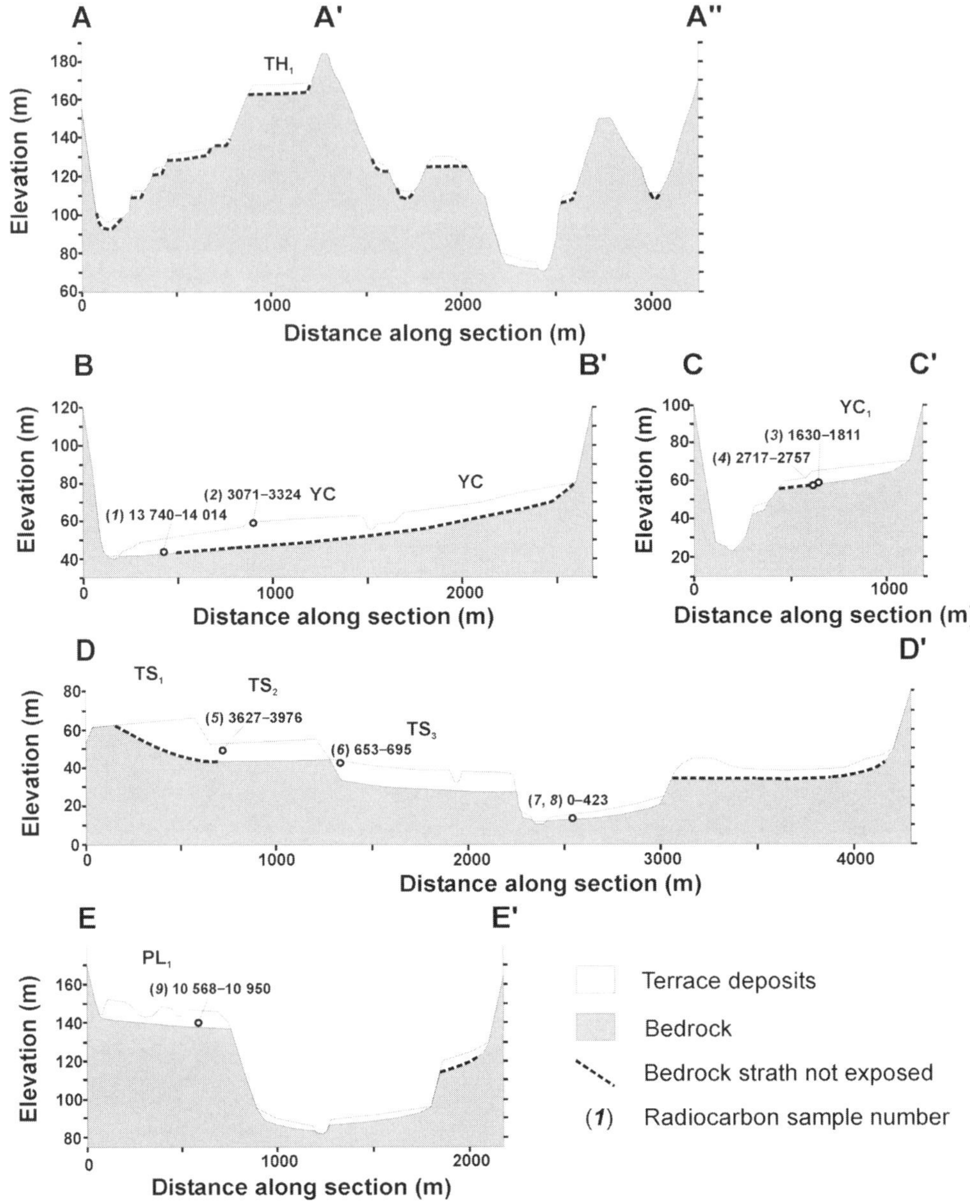

Figure 6. Cross sections of river terraces along the Tsengwen and Houku Rivers (for locations, see Fig. 5). Radiocarbon dates (ranges shown indicate the $\pm 1\sigma$ span in calendar years before present) shown above the sections are projected to their stratigraphic positions within the terrace deposits.

Fengli-Toushe area. The wide Yuching Valley ends downstream where the southward-flowing Tsengwen River turns at a right angle to the west (Fig. 5). Between the villages of Fengli and Toushe, the valley is characterized by terraces that are significantly higher but narrower than terrace YC (Figs. 5 and 6, profile C–C′). The transition from the lower terrace YC upstream to the higher terraces in this area is quite impressive; within less than 1 km, terrace YC rises and diverges downstream into three levels of abandoned meanders (Fig. 5). The highest one, labeled YC_1, can be correlated through much of the valley (Fig. 5). A number of abandoned meanders are preserved on the surface of terrace YC_1. The paleochannel that created this flood plain had a meander pattern out of phase with the modern channel (Fig. 5). All the terraces in this area have straths overlain by fluvial sediments a few meters to more than 10 m thick. These terrace deposits are similar to those of terrace YC; they have coarse gravel beds at the base capped by a few meters of finer-grained sediments (commonly sands and/or silts). The maximum thickness of the gravel beds observed in terrace YC_1 is about 9 m. Two dates are available for this terrace, both derived from combinations of wood fragments within muddy matrix at the base of the gravel beds. One, sampled about 45 m above the modern channel, is 1630–1811 cal. yr B.P. (sample 3; Fig. 5 and Table 1); the other, sampled about 42 m above the modern channel, is 2717–2757 cal. yr B.P. (sample 4; Fig. 5 and Table 1).

Around Toushe where the west-trending valley starts to turn south and widen again, three major terraces are preserved

TABLE 1. RADIOCARBON AGES AND SITES

Sample no.	Lab. no.*	Altitude (m)		^{14}C age (yr B.P.)	Calendar age(s) with 1σ range† (cal. yr B.P.)	Occurrence§	Bedrock incision rate#(mm/yr)	Ref.**
		Sample/ Terrace surf.	Bedrock surf./ modern channel					
Main-stem Tsengwen River								
1	NTU2323	41/50	37/39	11900 ± 70	13740 (13870) 14014	A1	~0	A
2	WK6130	57/60	<55/40	3001 ± 67††	3071 (3170, 3190, 3210) 3324	B2	—	A
3	NTU2322	70/73	70/25	1810 ± 50	1630 (1715) 1811	A1	24.8–27.6	A
4	WK6135	67/75	67/25	2590 ± 60	2717 (2740) 2757	A1	15.2–15.5	A
5	WK5136	47/51	39/13	3520 ± 140††	3627 (3740, 3750, 3770, 3790, 3830) 3976	B2	—	A
6	NTU2324	34/36	24/13	740 ± 55	653 (670) 695	peat	—	A
7	NTU2231	20/23	19/14	280 ± 45	289 (303) 423	A1	11.8–17.3	A
8	NTU2232	20/24	19/14	<250	—	A1	>20	A
Houku River								
9	WK6129	140/147	137/84	9630 ± 60††	10568 (10730, 10890) 10950	B2	4.8–5.0	A
Tsailiao River								
10	NTU2334	41/49	40/25	5490 ± 50	6215 (6291) 6306	B2	2.3–2.4	A
11	NTU2430	44/52	44/26	2320 ± 40	2325 (2340) 2348	B2	7.7–7.8	A
12	NTU2325	54/58	52/24	3150 ± 50	3277 (3363) 3392	A1	7.9–8.3	A
13	NTU2591	51/58	50/24	3210 ± 40	3372 (3399, 3435, 3443) 3465	A1	7.5–7.7	A
14	WK6131	33/37	32/25	587 ± 58††	534 (550) 645	B2	10.8–13.1	A
15	NTU2599	83/89	82/24	7560 ± 50	8315 (8337) 8371	B1	6.9–7.0	A
Erhjen River								
16	NTU2376	72/77	<70/60	870 ± 40	724 (755) 789	B2	—	B
17	WK6132	88/95	<87/60	8580 ± 63††	9482 (9500) 9535	B2	—	B
18	WK4111	76/81	75/15	5060 ± 140	5648 (5760, 5830, 5880) 5935	B1	10.1–10.6	B
19	NTU1337	51/57	50/15	1710 ± 40	1545 (1576, 1590, 1603) 1691	A2	20.7–22.7	B
20	NTU2427	34/38	28/11	1680 ± 40	1530 (1553) 1611	A3	—	B
21	NTU2375	52/59	51/17	2020 ± 40	1897 (1951) 1994	A1	17.1–17.9	B
22	NTU1968	28/38	28/11	2160 ± 50	2062 (2140) 2297	A1	7.4–8.2	C
23	NTU1320	40/47	39/13	2240 ± 40	2150 (2310) 2326	A1	11.2–12.1	D
24	NTU2341	25/37	25/10	2270 ± 50	2156 (2321) 2340	A1	6.4–7.0	B
25	NTU1945	40/47	39/13	2410 ± 40	2350 (2360) 2468	A1	10.5–11.1	C
26	NTU2361	20/35	18/8	2220 ± 50	2141 (2157, 2253, 2320) 2321	A2	4.3–4.7	B
27	NZA6116	41/47	39/13	5155 ± 83††	5763 (5920) 5984	B2	—	B
28	NTU2335	35/41	35/16	1340 ± 50	1194 (1279) 1294	A1	14.7–15.9	B
29	NTU2391	26/32	26/14	1000 ± 45	910 (928) 943	A1	12.7–13.2	B

*Sample number assigned at dating laboratory: NTU = National Taiwan University; WK = the University of Waikato; NZ = Institute of Geological and Nuclear Science, New Zealand.

†Calibrated by using program and calibration curve(s) of Stuiver and Reimer (1993). Value(s) in parentheses is (are) median values.

§A: Wood in mud bed (or matrix) associated with gravelly channel deposits (A1), in sandy or silty channel deposits (A2), or in silty-muddy overbank deposits (A3). B: Plant-fragments, stems, and leaves in silty or muddy channel deposits (B1) or in overbank deposits (B2).

#Calculated by dividing height of bedrock above the modern channel by the calibrated age with 1σ range (only the ages derived from the samples located no higher than 3 m above the bedrock surface are calculated).

**A = this paper; B = Hsieh and Knuepfer (2001); C = Lee et al. (1994); D = Chen (1993).

††Dated by accelerator mass spectrometry.

(Figs. 5 and 6, profile D–D′). The highest one, labeled TS_1 (for Toushe), is fragmented and tilted to the west-northwest by 0.7°–0.9°, as shown by measurements from the topographic map. The middle terrace, labeled TS_2, is also tilted to the northwest but by a smaller angle. It consists of 11–15 m of fluvial sediments above bedrock, including 8 m of muddy sediments in the inner part and 10 m of sandy sediments in the outer part of the terrace overlying 3–5 m of basal gravel. About 8 m above bedrock (~34 m above the modern channel), a combination of plant fragments within a mud bed (probably an overbank deposit) was dated at 3627–3976 cal. yr B.P. (sample 5; Fig. 5 and Table 1) in the inner part of this terrace. The lowest terrace, TS_3, consists of as much as 8 m of coarse gravel and has 3–5 m of sandy-muddy beds on top. About 2 m beneath the surface (10 m above bedrock or ~21 m above the modern channel), multiple layers of peat (with a total thickness of 40 cm) were dated at 653–695 cal. yr B.P. (sample 6; Fig. 5 and Table 1) in the inner edge of this terrace. Either terrace TS_2 or terrace TS_3

could be the downstream continuation of terrace YC_1 (Fig. 7). However, considering that the ages of the basal gravel of terraces TS_2 and TS_3 must be older than the dates (samples 5 and 6, respectively) obtained from their overlying overbank sediments, TS_2 is most likely older than YC_1. Furthermore, the gravel beds in terraces YC_1 and TS_3 are similar in thickness (7–9 m). Thus, terrace YC_1 is most likely correlated with terrace TS_3 (Fig. 7).

Below terrace TS_3, two ages were determined, 289–423 cal. yr B.P. and younger than 250 yr B.P. (samples 7 and 8, respectively; Fig. 5 and Table 1); the samples were obtained from a terrace less than 10 m above the modern channel. These ages are derived from a log (sample 7) and a twig (sample 8), collected within <50 m of each other and both within basal gravel above a bedrock strath. These two dates indicate that these low terraces were formed during the last few hundred years, consistent with the other dates derived from the higher terraces in this river reach (Fig. 6).

Terrace remnants higher than YC_1 and TS_1 are also preserved along the valley between Fengli and Toushe (Fig. 5). They are all severely fragmented or isolated. Still, those south of the modern channel delineate a flight of abandoned meanders, implying progressive incision and lateral migration of the paleo-meander channels. The highest terrace here, now isolated by lower meander remnants, is about 145 m above the modern channel (Fig. 7).

Houku Tributary. The Houku River, a major tributary of the Tsengwen River, flows through the southern edge of the Chutouchi Range (bounded by the Chutouchi fault) and joins the main Tsengwen River near Yuching (Fig. 5). Sequences of complicated abandoned meanders and the irregular long profile of the modern channel hinder detailed correlation of the terraces along this river (Figs. 5 and 7). Nonetheless, the height of these terraces above the modern channel and their number increase upstream (Fig. 7). Near where the Chutouchi fault crosses the valley, an aggregate of plant fragments is dated at 10 568–10 950 cal. yr B.P. from a terrace (labeled PL_1 [for Peiliao]) as high as 65 m above the modern channel (sample 9; Fig. 5 and Table 1). The terrace is capped by about 10 m of sediments with 3 m of gravel at the base; the dated samples were collected from a mud bed (likely deposited in an ephemeral pond on the paleo-flood plain) on top of the gravel beds.

Interpretation of Holocene tectonics. The most prominent tectonic feature that we interpret in this area is an anticline

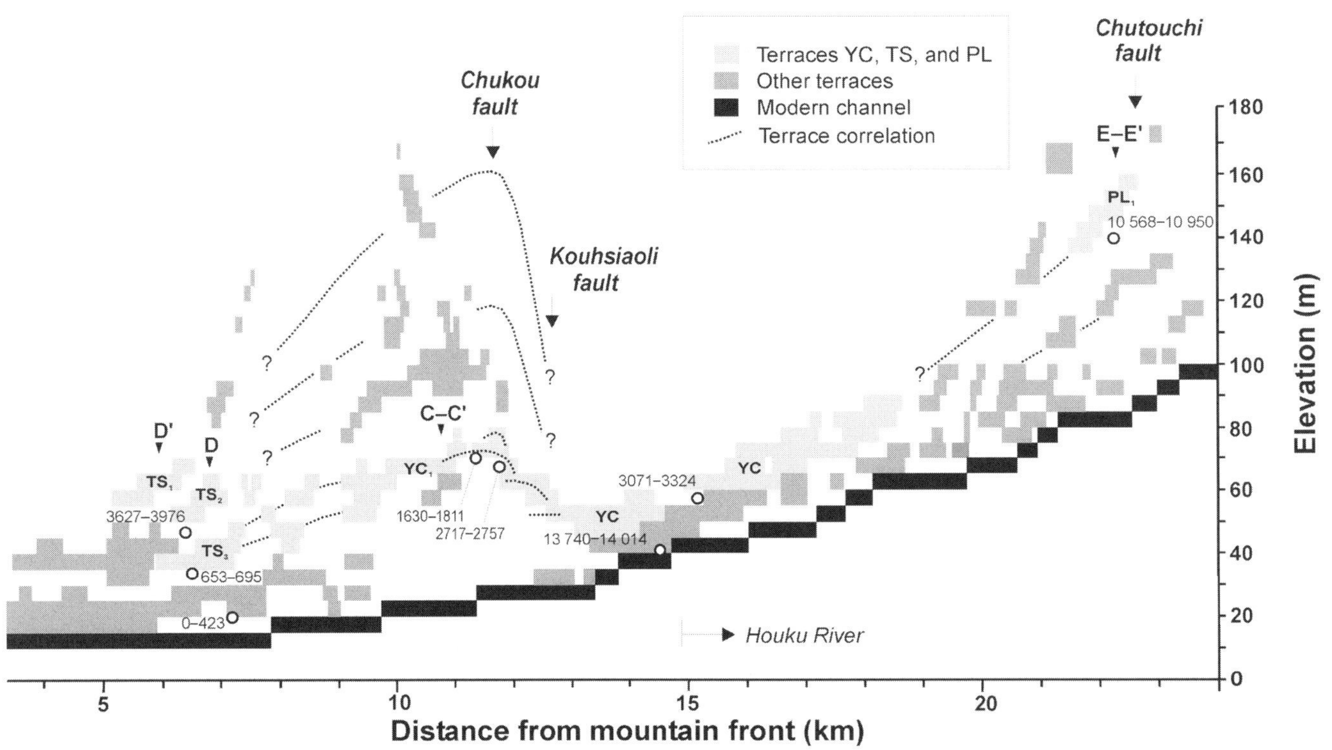

Figure 7. River-terrace long profiles along the main-stem Tsengwen River (downstream of position 14.9 km) and the Houku River (upstream of its junction with the Tsengwen River). Terraces no higher than TS_2, YC_1, YC, and PL_1 are used to define the main valley configuration (Fig. 5) and its midvalley line, to which the modern channel and the terraces are projected. The terraces higher than TS_1 and YC_1 between positions 6.5 and 11.8 km are projected to a straight line parallel to the trend of the valley (N65°W). The elevations of each datum are from 1:5000 topographic maps (with 5 m contour interval). An uncertainty of 5 m is assigned for each elevation reading between two contour lines. Therefore, each projected terrace surface is a rectangle rather than a line. In cases where the projected terraces on both sides of the river partly overlap, only one rectangle covering the projected terraces is shown. Locations of radiocarbon dates shown in proper stratigraphic position.

across the Fengli-Toushe Valley (Fig. 5). The eastern limb of this anticline is evidenced by the downstream rise and divergence of terrace YC into terraces YC_1, YC_2, and YC_3—a feature consistent with the radiocarbon dates (Fig. 5 and 7). The western limb of this anticline is shown by the west-northwestward tilting of terraces TS_1 and TS_2, which is opposed to the paleochannel directions. Terrace TS_1 is also steeper than TS_2, showing the accumulated tilting of the terrace TS_1 surface during formation of TS_2. The long profile for correlated terraces TS_3-YC_1-YC indicates that this anticline is asymmetric, with a sharper eastern limb and a gentler western limb (Fig. 7). The eastern limb is so steep that it apparently does not affect the main-stem Tsengwen River flowing through the Yuching Valley parallel to this anticlinal axis. The crest of this anticline is located about 1 km west of Fengli; the steeper eastern limb is coincident with the mapped trace of the west-dipping Kouhsiaoli fault (Fig. 5). This anticline thus is likely related to active movement of the Kouhsiaoli fault in the shallow subsurface.

We estimated the bedrock-incision rate after the abandonment of the dated terrace YC_1 at the crest of the inferred anticline. We divided the height of the bedrock strath above the modern channel by the dates (of samples 3 and 4) obtained just above the bedrock; this approach yields a rate of 15–28 mm/yr (Table 1). We recognize that the initiation of bedrock incision could have been significantly later than the dates and that the amount of bedrock incision should be greater than the height of bedrock above the modern channel. The rate thus calculated, therefore, is a minimum. Even so, this rate suggests that the rock-uplift rate at the crest of the inferred anticline is at least 1 cm/yr.

We also estimate that the bedrock-incision rate in the Houku River valley downstream of the Chutouchi fault is 4.8–5.0 mm/yr, according to the date (from sample 9) obtained right above the basal gravel (~3 m above bedrock) (Table 1). In contrast, the bedrock-incision rate since the latest Pleistocene at the western side of the Yuching Valley is almost zero (sample 1 and Table 1). We interpret that this contrast in incision rates is due primarily to tectonic tilting. This tilting appears to be widely distributed along the western slope of the Chutouchi Range, as suggested by the asymmetry of the Yuching Valley (and terrace YC) (Fig. 5). There are no large alluvial fans on the eastern side of this valley that could have forced the main channel to its present position. The asymmetry, therefore, is likely the result of tilting (e.g., Leeder and Alexander, 1987).

We interpret that the presence of the higher and multiple-level terraces at Tinghu reflects more active uplift there compared to the situation in the Yuching Valley downstream of Tinghu. We assume that the valley near Tinghu is tilted in the same direction as the south-dipping, highly weathered, pre-Holocene gravel underlying the valley (Fig. 5).

Terraces of the Tsailiao River in the Tsochen area

The river terraces of the Tsailiao River are formed by a complicated sequence of incised meander bends. An abandoned meander valley 2 km in length and up to 1 km in width is preserved around Tsochen (Figs. 3 and 8). It is floored by a terrace (labeled TC [for Tsochen]) with secondary risers (less than 5 m in height) delineating a series of abandoned meander loops (Figs. 8 and 9 profile F–F′). This abandoned valley is distinct; it is relatively intact and much wider than the surrounding multiple levels of terraces (Fig. 8). Given this size, the valley containing terrace TC must have been created by the main Tsailiao River, although the main river now flows south and west of the abandoned valley, separated from it by a local hill (Figs. 8 and 9, profile F–F′). This shift in the channel path probably was due to the river's capture by the north-flowing tributary 4 whose paleovalley has been partly reshaped by the main Tsailiao River since the capture (Fig. 8).

The valley downstream of terrace TC widens and narrows again, exhibiting a downstream-narrowing wedge shape in map view; the modern main channel is generally sited along the valley's southern two sides (Fig. 8). This valley is mainly floored by three terraces labeled TL_1, TL_2, and TL_3 (for Tsailiao) from high to low and from north to south (Fig. 8). Terraces TL_1 and TL_2 are 40–50 m and 30–40 m, respectively, above the modern main-stem channel and are 10–20 m higher than terrace TC upstream from them (Fig. 10). They also form the divides of three closely spaced south-flowing tributaries (tributaries 1–3). One of the remnants of terrace TL_1 also appears to be tilted to the east (shown by an open arrow between tributary 1 and tributary 2 in Fig. 8). The terrace TL_3, an inclining surface 10–25 m above the modern channel, is composed of abandoned meander loops probably from paleo-tributary 1 (based on the meander orientation and size) (Fig. 8). Terraces higher than TL_1 are also preserved north of this wedge-shaped valley reach (Figs. 8 and 9 profile G–G′). They are, however, all severely fragmented and isolated.

The terraces in this area are poorly exposed. Information on sediment characteristics and bedrock configuration relies mainly on highway construction that fortuitously crosses terraces TC and TL_2. Both terraces are underlain by straths. These bedrock surfaces preserve steps and channels with relief as large as 3–4 m. The covering sediments are dominated by planar-stratified silt and mud with a maximum thickness of 8–9 m observed in terrace TC. Basal gravel, mostly pebble- to cobble-sized particles, is mostly less than 0.5 m thick, although channel fills as thick as 2 m are also observed.

Two dates are available for terrace TC, 6215–6306 and 2325–2348 cal. yr B.P. (samples 10 and 11; Fig. 8 and Table 1); both are derived from assemblages of herbaceous leaves and stems in mud beds just above the basal gravel. These dated plant fossils, with delicate fabric, are well preserved in the strata; some of them even stand vertically, preserved in growth position. Thus we interpret these plant remains as deposited in situ, and both ages as representative of the time of deposition. These dates indicate that the formation of the valley containing terrace TC and its associated flood-plain aggradation started no later than 6.3 ka and ended no earlier than 2.3 ka. On the basis of

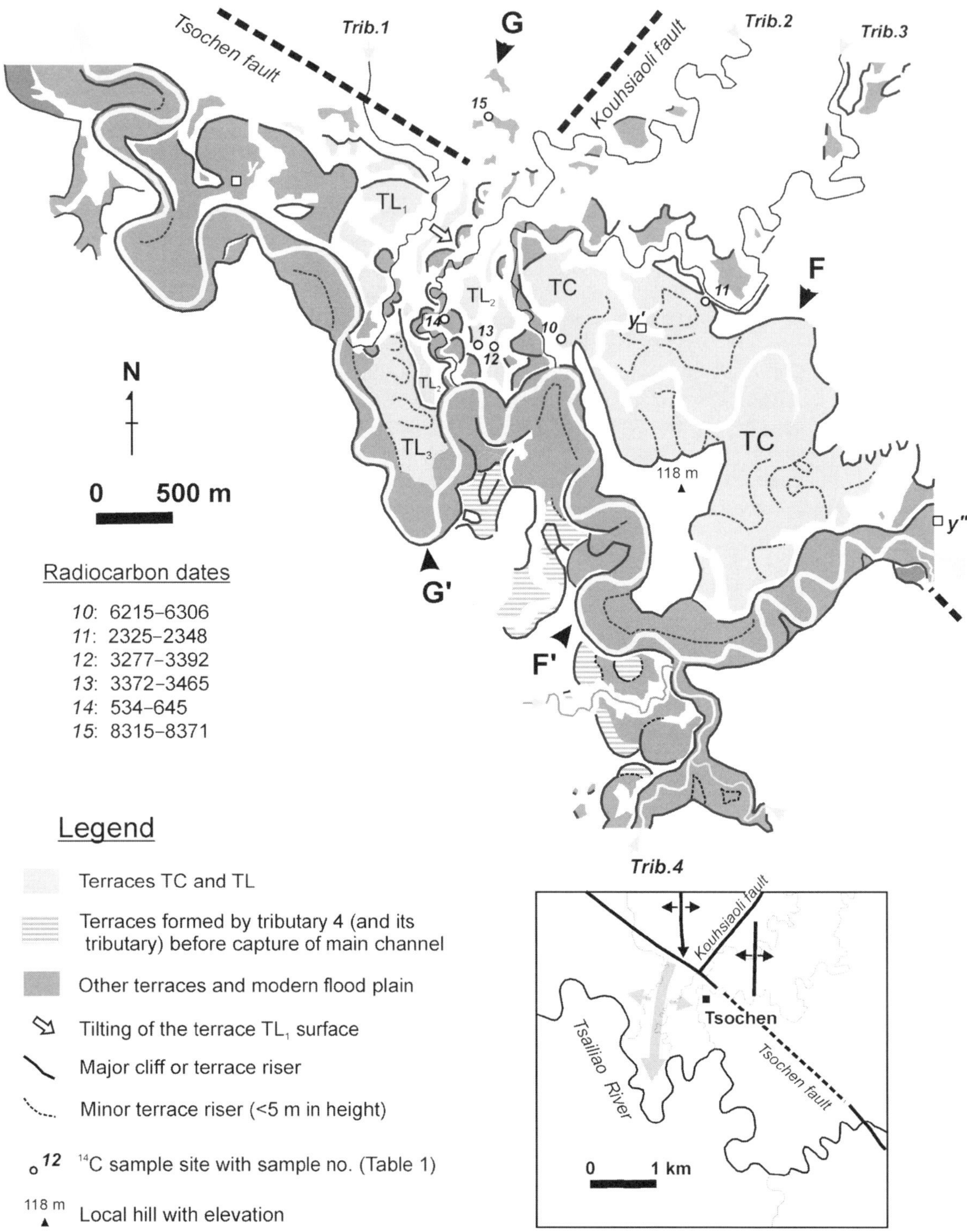

Figure 8. Distribution of river terraces in the Tsochen area of the Tsailiao River (for location, see Fig. 3) and radiocarbon dates and sites. Inset locates Pleistocene structures, the town of Tsochen, and an inferred Holocene anticline (light-gray symbols; arrows indicate downdip direction). Note that no meander loops on the terrace TC surface are offset by the Tsochen fault, and the inferred Holocene anticline coincides with the southward extension of an anticline north of the Tsochen fault.

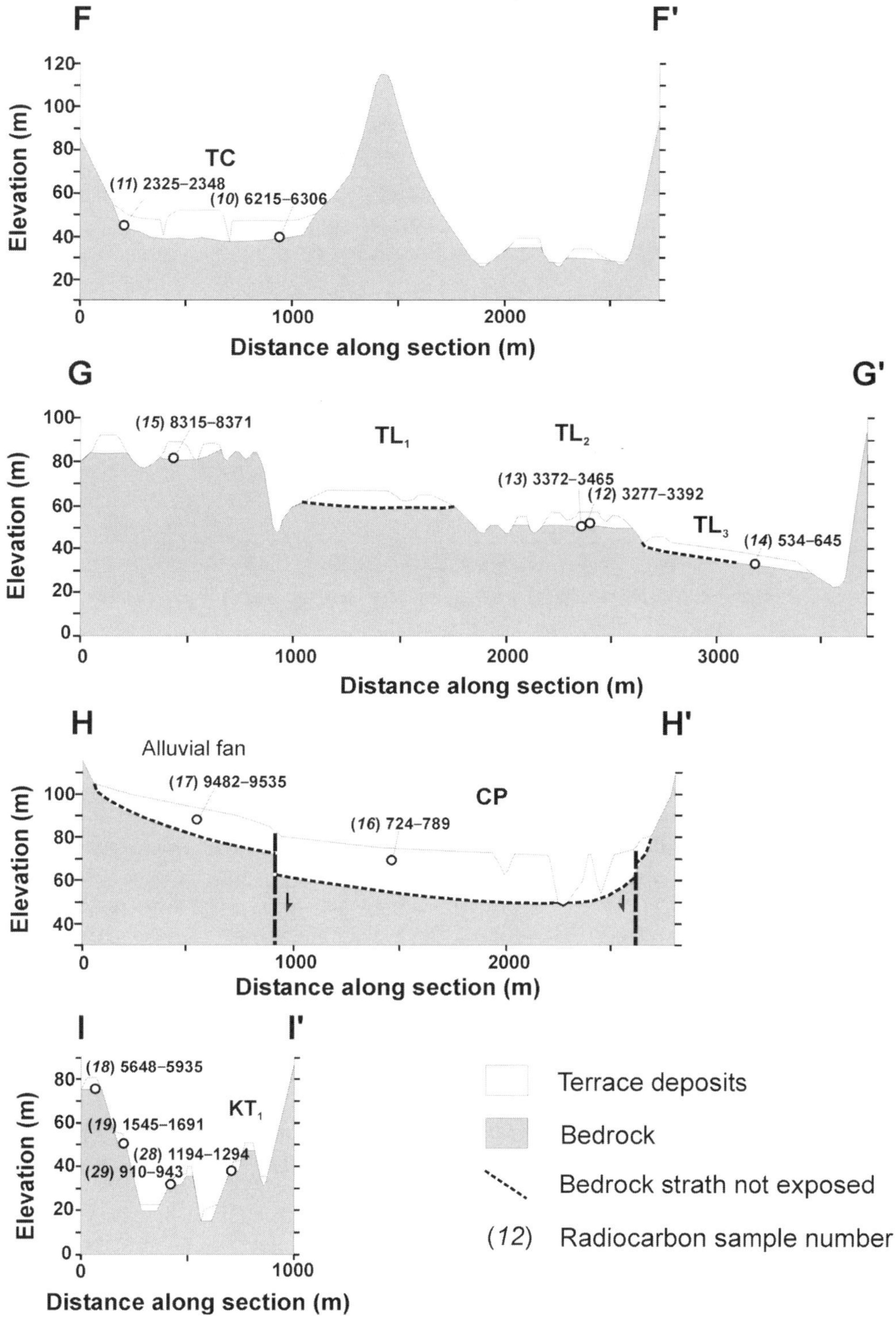

Figure 9. Cross sections of river terraces along the Tsailiao and Erhjen Rivers (for locations, see Figs. 8 and 11, respectively). All the sections are along straight lines except for G–G′, which is along the divide of tributaries 1 and 2 (Fig. 8). Radiocarbon dates shown above the sections are projected to their stratigraphic positions within the terrace deposits.

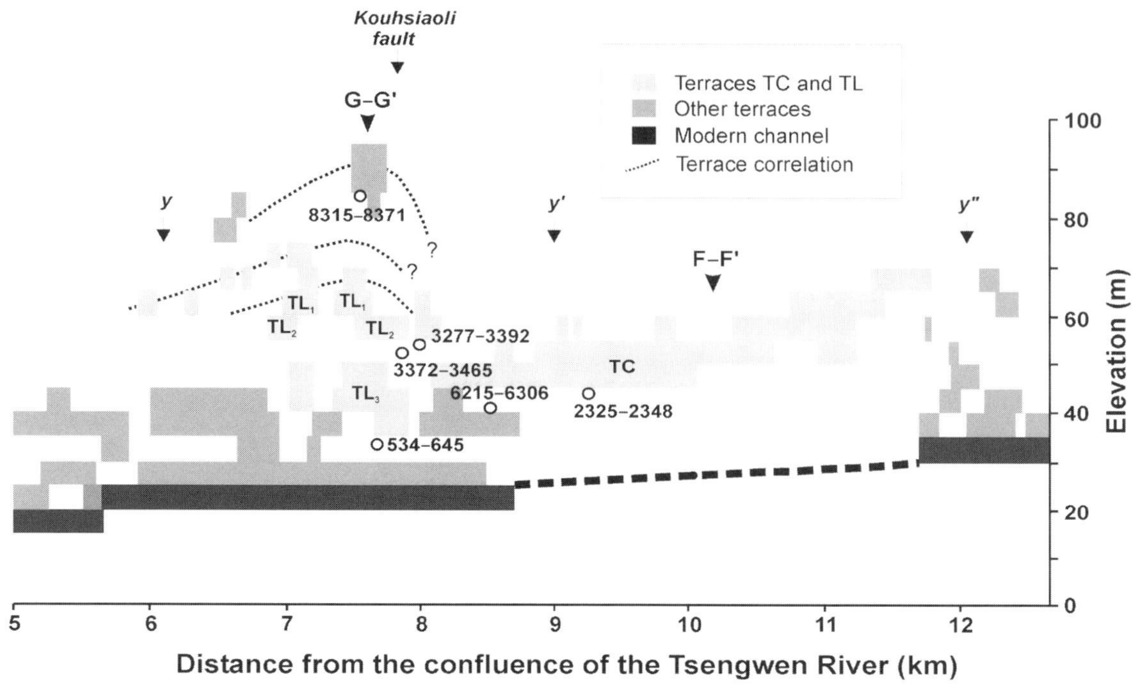

Figure 10. River-terrace long profiles in the Tsochen area. Terraces formed by tributary 4 are not shown. Terraces are projected to a straight line y–y′ (Fig. 8) between positions 6.1 and 9 km and to the midvalley line of the valley defined by terrace TC from points y′ to y″. The modern channel and other terraces that deviate from the valley containing terrace TC are not shown. Other notes as in Figure 7.

the numerous abandoned meanders on this surface, we interpret the channel to have probably meandered across this terrace surface throughout this interval. Two dates also are available for terrace TL_2: 3277–3392 and 3372–3465 cal. yr B.P. (samples 12 and 13; Fig. 8 and Table 1). The younger date was obtained from a log in a mud bed about 2 m above bedrock, and the older date was obtained from an assemblage of twigs in mud matrix associated with channel-fill gravel. We obtained an age of 534–645 cal. yr B.P. (sample 14; Fig. 8 and Table 1) from an aggregate of herbaceous leaves in a terrace of tributary 2, which can be correlated to the lower part of nearby terrace TL_3 because of the similar elevation of the two terraces above the main channel. We also obtained a date, 8315–8371 cal. yr B.P. (sample 15; Fig. 8 and Table 1), from a higher terrace north of terrace TL_1. The dated sample is an aggregate of plant fragments distributed in silty channel deposits a few centimeters to 1 m above bedrock.

Interpretation of Holocene tectonics. The "disappearance" of the wide TC surface downstream, together with the occurrence of the wedge-shaped valley reach, provides information on the active tectonics of the area. The position of the higher terraces (TL_1 and TL_2) that block the downstream continuation of terrace TC may imply that they were initially created by the main-stem channel, but subsequently raised relative to the terrace TC by faulting or warping. This scenario is further supported by the radiocarbon dates, which show that terrace TL_2 (downstream) is synchronous with part of the deposits of terrace TC despite the fact that terrace TL_2 is higher above the channel (Fig. 10). We therefore interpret an anticlinal structure across the wedge-shaped valley reach downstream of terrace TC (Fig. 10). This anticline appears to have a gentler western limb than eastern limb, as TL_1 continues downstream with a lower gradient than the upstream gradient from TL_2 to TC (Fig. 10). In addition, this anticline appears to plunge to the south, as suggested by the southward drop of the terrace sequence from TL_1 to TL_3 and then to the modern flood plain in the wedge-shaped valley reach (Fig. 9, profile G–G′). This inferred south-striking and south-plunging anticline coincides with the southward extension of a mapped anticline north of the Tsochen fault (Figs. 3 and 8). In addition, the steeper eastern limb of this inferred active fold coincides with the southward extension of the mapped Kouhsiaoli fault and thus may also be related to active movement of that fault (Figs. 3 and 8). Given these coincidences, and the fact that no meander loops on the wide terrace TC are offset, we interpret that the Tsochen fault has been inactive during the Holocene (Fig. 8).

We estimated the Holocene bedrock-incision rate from the date obtained from the terrace remnant north of TL_1 to be ~7 mm/yr (sample 15; Table 1). The bedrock-incision rates during the late Holocene calculated from terrace TL_2 (samples 12 and 13) and the lower terrace in tributary 2 (sample 14) range from 7.5 to 13 mm/yr (Table 1). These rates appear to be lower than the rate in the Fengli-Toushe valley of the main Tsengwen River to the north. We interpret that the lower rates may reflect less active tectonic movement of this area compared with that to the north, which is consistent with the southward plunge of the bedrock fold mapped by the Chinese Petroleum Corporation (1989) (Figs. 3 and 8).

Terraces in the Erhjen River basin

The main-stem Erhjen River consists of an abnormally broad upstream valley as wide as 2.5 km, double the widest part of the downstream valley (Figs. 9 and 11, profile H–H′). This upstream valley is also characterized by linear valley walls and terrace risers, also in contrast to the downstream valley that is dominated by incised meanders. This upstream valley is floored mainly by a continuous terrace, labeled CP (for Chungpu), which includes an incised alluvial fan on its western side (Fig. 11). Two other abandoned alluvial fans also are preserved above terrace CP on the western side of the valley; we interpret that they are bounded on the terrace CP surface by a Holocene fault that also offsets parts of the CP surface (Fig. 11) (Hsieh and Knuepfer, 2001). Terrace CP narrows downstream and diverges into two terraces (labeled CP_1 and CP_2) (Fig. 11). The downstream extension of these two terraces, however, is unclear, as the terraces become discontinuous owing to migration and incision of meandering channels (Fig. 11).

The sediment characteristics and bedrock configuration in this wide upstream valley are unclear owing to a scarcity of outcrops. Nonetheless, bedrock is locally exposed along the modern channel; the maximum thickness of the terrace deposits observed in this study is about 20 m. They are dominated by planar-stratified overbank silt and mud. We obtained two dates in this valley. One, 724–789 cal. yr B.P. (sample 16; Fig. 11 and Table 1), is from the upstream part of terrace CP; the sample was collected from an aggregate of herbaceous debris within silty overbank deposits ranging from 5 to 6 m below the terrace surface. The other date, 9482–9535 cal. yr B.P. (sample 17; Fig. 11 and Table 1), is from an aggregate of herbaceous debris in mud (probably overbank deposits) 7 m below an abandoned fan surface bounding terrace CP. These two dates imply that terrace CP may have started forming (and aggrading) by early Holocene time and continued to be occupied by overbank floods until recently.

In contrast to the upstream valley, the downstream part of the Erhjen River consists of a complicated sequence of as many as five levels of terrace remnants (Figs. 9 and 11, profile I–I′). Bedrock is widely exposed in these downstream terraces, with fluvial sediments similar in nature to those exposed in the Tsochen area. We obtained nine dates from terrace remnants (labeled KT_1 [for Kuting]) (Fig. 11) that appear to be correlative on the basis of their geometry from aerial photographs. Eight of these dates, ranging from ca. 1.5 to 2.4 ka (samples 19–26; Fig. 11 and Table 1), were all obtained from wood fragments (trunks or twigs), five of them within basal gravel, two within sandy channel beds, and one within overbank silt. The other date, 5763–5984 cal. yr B.P. (sample 27 and Table 1), was obtained from an aggregate of plant debris within overbank mud. We interpret this as a reworked sample, not only because the date is inconsistent with the other eight dates, but also because the date is similar to another date, 5648–5935 cal. yr B.P. (sample 18), obtained from a nearby terrace remnant higher than correlated terrace KT_1 (Figs. 9 and 11, profile I–I′). The latter date was derived from an aggregate of plant fragments within sandy channel beds on top of basal gravel (or 1 m above bedrock) (Table 1). Given the consistency among the eight dates, we are confident in our initial correlation of terrace KT_1, which led us to correlate another set of discontinuous terrace remnants as the lower terrace KT_2 surface (Fig. 12). The two terraces appear to converge downstream to the Coastal Plain (Fig. 12). We obtained a date, 1194–1294 cal. yr B.P. (sample 28; Fig. 11 and Table 1), from a log in one of the correlated terrace KT_2 remnants, and a date, 910–943 cal. yr B.P. (sample 29; Fig. 11 and Table 1), from an aggregate of twigs in a terrace remnant below terrace KT_2; both of the dated materials are within basal gravel. These dates support the topographic correlation of the terrace KT_2 surface.

Interpretation of Holocene tectonics. The available radiocarbon dates indicate that since the middle Holocene, while the channel in the downstream Erhjen River underwent repeated lateral erosion and downcutting that created multiple levels of terraces, the channel in the upstream valley still slowly aggraded and deposited flood-plain sediments (terrace CP). The terrace remnant that is dated around middle Holocene in age in the downstream valley (Fig. 12) is, however, higher than much of terrace CP upstream. The downstream valley thus must have been raised relative to the upstream valley since the middle Holocene. Further considering the upstream convergence of terraces CP_1 and CP_2 and the apparent downstream convergence of the correlated terraces KT_1 and KT_2, we propose that the downstream valley is crossed by an active anticlinal structure (Figs. 11 and 12). We assume that this anticline has the same trend as a mapped Pleistocene anticline in the same valley and plunges south (Fig. 11).

We estimated bedrock-incision rates of 4–23 mm/yr from the multiple dates (samples 18, 21–26, 28, and 29) obtained from basal gravel of the terraces in the downstream valley (Table 1). The average bedrock-incision rate since the middle Holocene is 10 mm/yr (sample 18) about 2–3 km downstream of the inferred anticline crest. Given these rates, we interpret that the rock-uplift rate at the crest of this anticline is at least 1 cm/yr.

SYNTHESIS AND DISCUSSION

Divergence and convergence of terraces and chronological significance

Downstream divergence and convergence of terrace surfaces, associated with changes in the numbers of local terraces, occur in the river valleys in the southern Western Foothills. The downstream divergence of the single terrace YC on the

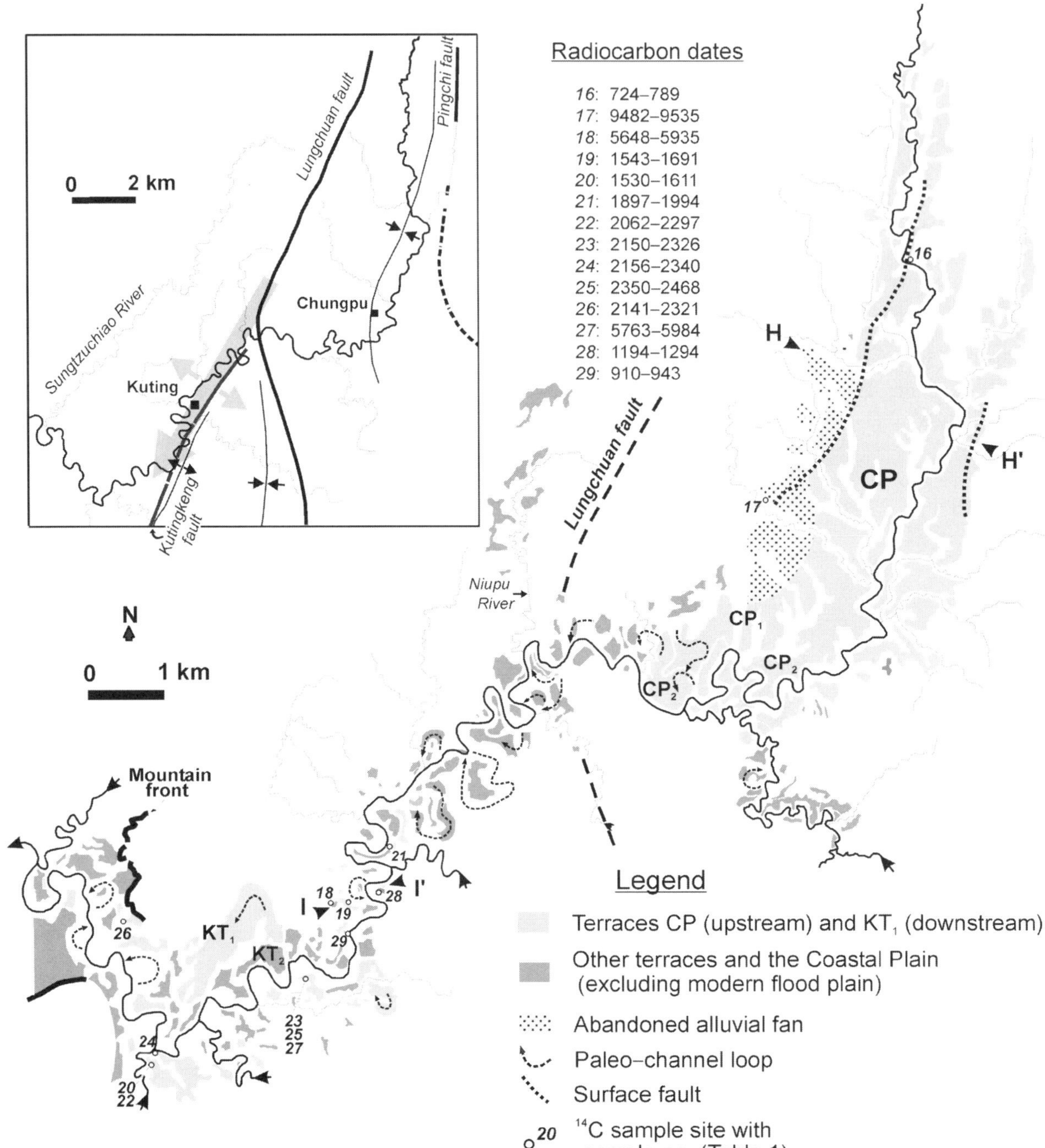

Figure 11. Distribution of river terraces in the main-stem Erhjen River (for location, see Fig. 3), radiocarbon dates and sites. Terraces in the upstream part of the Niupu River are also shown (Fig. 4). Inset locates Pleistocene structures, major towns, and the inferred Holocene anticline (light-gray symbols; arrows indicate downdip direction).

Tsengwen River into multiple terraces is continuous and obvious on aerial photographs; the same is true for the downstream divergence of the single terrace CP on the Erhjen River (Figs. 7 and 12). The downstream convergence of correlated terraces KT_1 and KT_2 on the Erhjen River to form a single terrace on the Coastal Plain is suggested by the apparent geometric continuity of the terrace remnants and by the systematic change in terrace heights and number along the valley; this pattern also applies to the downstream convergence of multiple terraces including PL_1 to form terrace YC on the Houku River (Figs. 7 and 12). These terrace configurations initially led us to interpret active anticlinal structures (because we found no evidence for

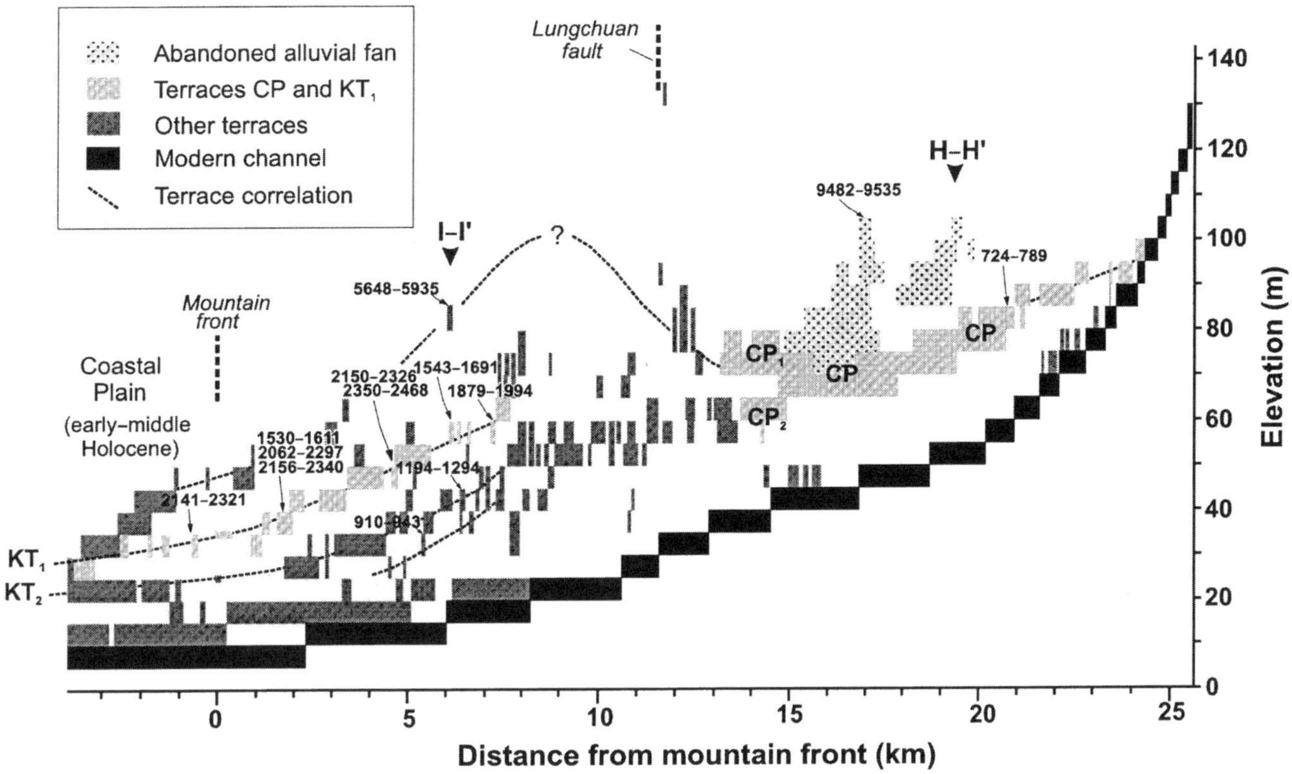

Figure 12. River-terrace long profiles of the main-stem Erhjen River. Terraces are projected to the midvalley line of the valley defined by terraces no higher than terraces CP and KT_1. Only terraces on the right bank of the main channel upstream of position 13 km are shown. The sites of radiocarbon dates are referred to the top of the dated terraces. Other notes as in Figure 7.

surface faulting) across the Fengli-Toushe area (Tsengwen River) and the downstream Erhjen River valley, as well as tilting of the Houku River. These interpretations are further supported by available radiocarbon dates (Figs. 7 and 12). The divergence and convergence of the terraces in the Tsochen area of the Tsailiao River cannot be readily recognized in the topography; however, differential tectonic uplift is suggested by the multiple, higher terraces (TL_1 and TL_2) that "block" the single terrace TC upstream (Fig. 10). This interpretation is further confirmed by radiocarbon dates, which suggest that terraces TL_1 and TL_2 likely converge upstream to terrace TC (Fig. 10).

The available radiocarbon dates conform with the tectonic interpretations based on topographic criteria. They also show two key features. First, the dated terraces in parts of the valleys undergoing what we interpret as more active uplift (YC_1, TS_2, TS_3, PL_1, TL_2, KT_1, KT_2) have developed over shorter time spans than the terraces in parts of the valleys that we interpret as relatively tectonically stable (YC, TC, and CP). Among the former group of terraces, the age of terrace KT_1 in the downstream Erhjen River is best determined. In this ~8-km-long valley reach located in the downstream limb of the inferred anticline, eight dates (1.5–2.4 ka) sampled from either the basal gravel or overbank deposits indicate that the terrace sediments were deposited over a millennium (Fig. 12). The deposition of terrace YC_1 sediments along the downstream Tsengwen River involved a comparable time span, according to the two dates (1630–1811 and 2717–2757 cal. yr B.P.) obtained 0.5 km from each other and limited by the age (3627–3976 cal. yr B.P.) of the higher terrace TS_2 (Fig. 7). Second, except for terrace YC_1, the ages from the terraces in the rapidly uplifting valley reaches (PL_1, TL_2, KT_1, KT_2) all are within the time spans bracketed by the ages of the terraces from which they diverge (YC, TC, and CP).

The combination of both features supports our hypothesis shown in Figure 1 that the formation time of terraces, like the terrace surfaces themselves, can converge or diverge along a valley as a result of active anticlinal structures. This hypothesis requires that while a channel in a more actively uplifting valley underwent repeated incision and lateral erosion to form multiple terraces, the channel in the relatively tectonically stable valley either aggraded or remained at approximately the same level with respect to its flood plains, resulting in a single terrace surface (Fig. 1). In other words, the rivers we have studied must have sufficient stream power *over the long term* (i.e., several thousand years) to keep up with the anticlinal uplift by downcutting through the weakly resistant bedrock, so that the valley reaches upstream of the anticlines can have maintained a stable base level relative to the anticline-affected reaches (cf. Snow and Slingerland, 1990).

Of course, the converging and diverging terrace sequences and their formation times in the studied river reaches are not as ideal as shown in Figure 1. We attribute their nonideality to the low preservation potential of terraces when they become highly elevated across anticlines. Perhaps only very young and wide terraces, such as YC_1, can be preserved as nearly continuous features across much of an anticline after they are uplifted and warped.

We think that the consistency between the geomorphic evidence and available dating is not coincidental, but reflects systematic fluvial behavior accommodating tectonic processes. Given that there commonly is an insufficient number of dates for any terrace sequence as well as the intrinsic uncertainty of these dates, we argue that the convergent or divergent geometry of terrace sequences in a valley can adequately determine the relative times of formation among the terraces. Thus we propose that the timing of complete abandonment of the YC flood plain in the Yuching area must be at least as young as terrace YC_3 (which itself must be later than 1.7 ka—the minimum formation time of the older terrace YC_1 based on available dates), despite the fact that the youngest age available from deposits of terrace YC is ca. 3 ka.

Contrast of terrace widths and controlling factors

A prominent feature associated with converging and diverging relationships of the studied river-terrace surfaces and their formation times is the change of terrace widths along the valley. Surfaces of terraces YC, TC, and CP, with their longer duration of formation, are all wider than any individual terrace surface diverging from them (Figs. 5,8, and 11). Moreover, the widths of the valleys (both up to 2.5 km as measured from the widths of the modern flood plain and major terraces) containing terraces YC and CP are significantly greater than those of the valleys downstream of these terraces where the rivers cross the anticlinal structures (widths of 0.7–2.0 km and 0.3–1.5 km for the valleys of Fengli-Toushe and the downstream part of the Erhjen River, respectively) (Figs. 5 and 11).

This contrast in terrace width, as well as the narrowing of the entire valley across the anticlines in the main-stem Tsengwen and Erhjen Rivers, cannot be explained by contrast in bedrock lithology, which is generally uniform and easily eroded along the rivers as recognized in the field. An increase of channel discharge can be ruled out as the cause of this contrast, because the wider terraces and valleys are all located upstream of their narrower counterparts. A change in sediment load might control changes in terrace and valley width along a river, as it does for alluvial channel width (Schumm, 1968). This possibility seems unlikely for these studied rivers, however, because there are no tributaries appropriately situated to produce a change in sediment-load type or volume, and the deposits are similar on the adjacent wide and narrow terraces.

We therefore propose that the duration of channel occupation of the terrace surfaces is an important factor in governing the contrast in terrace (and valley) width. The positive relationship between the duration of occupation and width of these terraces implies that, other things being equal, the width of a terrace (and valley) depends on how often a channel can contact and erode the valley walls (or bounding terrace risers). As long as a channel remains at a relatively constant level, it can widen its flood plain. On the other hand, once significant channel incision has started, the channel has lost the chance to further widen the surface it previously created. As a result, although the prevailing channel hydraulics in the southern Western Foothills should be capable of widening a flood plain carved into actively growing anticlines, the resulting terraces did not have enough time to obtain widths that are as great as those of the terraces away from the anticlines.

We further consider whether the wider terraces (and valleys) upstream of the inferred active anticlines may be in part a result of flood-plain aggradation on preexisting wide valleys. This concept is possible especially for terraces YC and CP whose sediment thickness and underlying bedrock configuration are in fact not clear (note, at least terrace CP deposits appear to be thicker than those downstream). One possible scenario is that these two valleys were initially intermontane basins, long before the aggradation of Holocene sediments of terraces YC and CP. The presence of the tilted pre-Holocene gravel deposits in the Tinghu area at the upstream end of the Yuching Valley supports this interpretation.

We realize that tectonic tilting normal to the valley trend could be another factor affecting terrace (and valley) width, because it can unidirectionally shift the channel to erode valley walls, which facilitates terrace (and valley) widening as long as no significant channel incision occurs (cf. Leeder and Alexander, 1987). The width of terrace YC thus may be in part attributed to this process, which also may explain why the wedge-shaped valley floored by the combined terraces TL_1-TL_2-TL_3 in the Tsochen area is wider than terrace TC upstream (Fig. 8).

Neotectonic implications in the southern Western Foothills from contrasts in valley widths

Multiple, high but narrow terraces developed in actively uplifting valleys tend to converge toward few, low but wide terraces in relatively tectonically stable valleys—a hypothesis well documented in this study by the three river valleys discussed in detail. The difficulty in using the converging or diverging relationship of terraces in neotectonic studies is that such a relationship may not be preserved, especially owing to the low preservation potential of the high and narrow terraces in actively uplifting valleys. Nonetheless, the contrast of terrace width may still be a useful criterion for interpreting neotectonic style and correlating river terraces. The main-stem Tsengwen and Erhjen Rivers (Figs. 5 and 11) show that not only are individual terraces narrower than the one from which they diverge, but the entire valley across the active anticline can be narrower than that upstream of the structure.

This finding could be particularly useful in recognizing active folding structures from rivers in mountainous areas, as the reversal in downstream widening that is typical of rivers should be obvious. As noted previously, in an attempt to assess whether other rivers in the southern Western Foothills also are tectonically controlled, we examined reaches of river valleys that—like the Tsengwen River and Erhjen River valleys containing terraces YC and CP, respectively—are significantly wider than the valley reaches downstream from them (Fig. 4). These "abnormally" wide valleys are floored by either relatively wide modern flood plains or wide terraces, or both, as are the valleys containing terraces YC and CP. Many terraces in these valleys were probably abandoned during the Holocene, judging from the weakly developed soil profiles in these terrace deposits.

The abnormally wide valley reaches in the Kouping and Sanmin areas (sites 7 and 9, respectively, in Fig. 4) are underlain by less resistant bedrock than that farther downstream. Contrasts in valley width thus should, at least in part, reflect bedrock lithology. The other anomalous valley reaches have bedrock lithology generally the same as that downstream from them. In addition, the valley at Hsinhsing and the upstream valleys of the Kuechuon and Houku Rivers (sites 5, 2, and 8 respectively) are underlain by mapped synclines, and the upstream Santei River valley (site 1) is located upstream of anticlines mapped by the Chinese Petroleum Corporation (1971, 1989) (Fig. 3). The upstream Niupu River valley (site 4), on the western slope of the Lungchuan Ridge, is located upstream of the Holocene anticline previously interpreted in the downstream part of the Erhjen River (Fig. 11).

These "abnormally" wide valleys that are either underlain by synclines or located upstream of anticlines occur in a variety of geomorphic settings. Although most of them are in the upstream part of the river basins, the valley at Hsinhsing (site 5) is not. Although most of the valleys are in the lower hill areas in relatively small drainage basins, the upstream Houku River valley (site 8) drains high mountains. The upstream valleys of the Santei, Kuechuon, and Houku Rivers (sites 1, 2, and 8) are located where the main channels reduce gradient and start accumulating sediments after draining from the steep headwaters, but others are not. The upstream valleys of the Santei and Houku Rivers (sites 1 and 8) also are associated with alluvial fans on the valley sides. In fact, side-valley alluvial fans and channels with gradients that decrease abruptly downstream, as in the upstream Santei River, are common elsewhere in the southern Western Foothills, but they are not associated with valleys as wide and continuous as those shown in Figure 4. The lack of consistent geomorphic factors that may be attributed to the "abnormal" width of these valley reaches increases our confidence that the width of these valleys, like that of the valleys containing terraces YC and CP, is controlled mainly by tectonics.

Most of the terraces in these "abnormally" wide valleys expose bedrock, as is true in the terrace YC reach of the Tsengwen River, the terrace TC reach of the Tsailiao River, and the terrace CP reach of the Erhjen River. This situation implies that even these wide valley reaches have undergone long-term rock uplift, which ensures long-term channel incision into bedrock (Hsieh and Knuepfer, 2001). The tectonic style of the entire southern Western Foothills, therefore, is likely characterized by regional tilting (or flexure) as well as the folding interpreted in this study. The "abnormally" wide valley reaches we observed in the southern Western Foothills are thus probably not inherited, because any preexisting contrast in valley width would be preserved only if the contrast in the tendency for channel incision due to a contrast in tectonic activity along the river can continue.

CONCLUSIONS

This study provides field evidence in the southern Western Foothills to illustrate the development of river terraces across growing anticlines. We recognize three major Holocene anticlinal structures located in the Yuching and Tsochen areas of the Tsengwen River as well as in the downstream Erhjen River. All of them are closely related to nearby Pleistocene structures depicted on the published geologic map. These structures are recognized by converging and diverging river-terrace sequences, with chronologic control provided by more than 20 radiocarbon dates. The Holocene rock-uplift rates at the crests of these anticlines, inferred from channel bedrock-incision rates, are at least 1 cm/yr for the anticline crossing the Tsengwen River downstream of Yuching and for that crossing the downstream Erhjen River, and less for the anticline in the Tsochen area. We highlight the following three findings:

1. Higher, multiple terraces that document the anticlines can converge to a single terrace away from the anticlines. In other words, a channel in a valley reach that is actively uplifting can undergo repeated incision and lateral erosion while a channel in a relatively tectonically stable reach of the river maintains a fairly constant flood-plain level without incision.

2. Just as the terraces can converge or diverge, so can the formation time (time lines represented by the terrace surface[s]) converge or diverge. A single terrace in a relatively tectonically stable valley reach thus can be formed over the time during which multiple terraces in an actively uplifting (and incising) valley reach were created.

3. Along a given river, the terrace surface in a relatively tectonically stable reach is wider than any terrace surface formed in an actively uplifting reach, even if the former is upstream. This finding suggests that, other things being equal, the width of a flood plain is positively related to its duration of formation.

All of these conclusions, in fact, reflect the high tendency of a channel in a narrow, actively uplifting valley to incise, in contrast to the tendency of a channel in a wide, tectonically stable valley. We propose that such conditions should hold in general for other river reaches crossing active anticlines and for time scales longer than the Holocene.

The converging and diverging relationship of terraces may not be suitably preserved owing to the low preservation potential of high and narrow terraces across the actively uplifting valleys. However, the contrasting width of the terraces, or river valleys, may provide a useful clue in terrace correlation and neotectonic interpretation, providing that bedrock lithology does not change along the river. Such an implication is important, as it extends the way terraces can be correlated from two-dimensional long profiles to three dimensions. We further use this criterion to inspect the neotectonic patterns elsewhere in the southern Western Foothills. Five other sites display this character of wide valley reaches upstream of narrow reaches, with no contrast in bedrock lithology. We therefore interpret these narrow-wide valley contrasts as resulting from active folding. The close association between these interpreted Holocene structures and previously mapped Pliocene–Pleistocene ones gives us confidence that this criterion is applicable at least in a tectonically active area like the southern Western Foothills of Taiwan.

ACKNOWLEDGMENTS

We thank Yao-Wen Lee, who first guided Hsieh to the Erhjen River, and Ping-Mei Liew of National Taiwan University for her logistical support and helpful discussions. We greatly appreciate the assistance of Yu-Kao Chen of National Taiwan University, and the Central Geological Survey, Republic of China, who sponsored much of the radiocarbon dating. Kai-Shyuan Shea of the Central Geological Survey, Republic of China, provided one unpublished radiocarbon date. Tung-Sheng Shih of the Central Geological Survey, Republic of China, ably assisted Hsieh in the field. This research was funded largely by National Science Foundation grant EAR93-15041.

REFERENCES CITED

Amorosi, A., Farina, M., Severi, P., Preti, D., Caporale, L., and Di Dio, G., 1996, Genetically related alluvial deposits across active fault zones: An example of alluvial fan-terrace correlation from the upper Quaternary of the southern Po Basin, Italy: Sedimentary Geology, v. 102, p. 275–295.

Biq, C.C., 1973, Kinematic pattern of Taiwan as an example of actual continent-arc collision: United States-Republic of China Cooperative Science Program, Report of the Seminar on Seismology, p. 21–26.

Bowin, C., Lu, R.S., Lee, C.S., and Schouten, H., 1978, Plate convergence and accretion in the Taiwan-Luzon region: American Association of Petroleum Geologists Bulletin, v. 62, p. 1645–1672.

Bull, W.B., 1990, Stream-terrace genesis: Implications for soil development, *in* Knuepfer, P.L.K., and McFadden, L.D., eds., Soils and landscape evolution: Geomorphology, v. 3, p. 351–367.

Bullard, T.F., and Lettis, W.R., 1993, Quaternary fold deformation associated with blind faulting, Los Angeles Basin, California: Journal of Geophysical Research, v. 98, p. 8349–8369.

Chai, B.H.T., 1972, Structure and tectonic evolution of Taiwan: American Journal of Science, v. 272, p. 389–442.

Chen, Y.G., 1993, Sea-level change and neotectonics in the southern part of Taiwan region since late Pleistocene [Ph.D. thesis]: Institute of Geology, National Taiwan University (in Chinese).

Chinese Petroleum Corporation, 1971, Geological map of western Taiwan—the southern half: Chinese Petroleum Corporation, scale 1:200 000, 2 sheets.

Chinese Petroleum Corporation, 1989, Geological map of Tainan: Chinese Petroleum Corporation, scale 1:100 000, 1 sheet.

Frye, J.C., and Leonard, A.R., 1954, Some problems of alluvial terrace mapping: American Journal of Science, v. 252, p. 242–251.

Germanoski, D., and Harvey, M.D., 1993, Asynchronous terrace development in degrading braided channels: Physical Geography, v. 14, p. 16–38.

Harbor, D.J., 1998, Dynamic equilibrium between an active uplift and the Sevier River, Utah: Journal of Geology, v. 106, p. 181–194.

Ho, C.S., 1986, A synthesis of the geologic evolution of Taiwan: Tectonophysics, v. 125, p. 1–16.

Hooke, J.M., Harvey, A.M., Miller, S.Y., and Redmond, C.E., 1990, The chronology and stratigraphy of the alluvial terraces of the River Dane valley, Cheshire, N.W. England: Earth Surface Processes and Landforms, v. 15, p. 717–737.

Hsieh, M.L., and Knuepfer, P.L.K., 2001, Middle-late Holocene river terraces in the Erhjen River basin, southwestern Taiwan: Implications of river response to climate change and active tectonic uplift: Geomorphology, v. 38, p. 337–372.

Keller, E.A., and Pinter, N., 1996, Active tectonics: Earthquakes, uplift, and landscape: Upper Saddle River, New Jersey, Prentice Hall, 337 p.

Lee, Y.W., Chen, Y.G., and Liu, T.K., 1994, Preliminary study of river terraces correlation and their neotectonic significance along the Erhjen-chi, southern Taiwan: Geological Society of China, Annual Meeting, 1994, Program and Abstracts, p. 612–616 (in Chinese).

Leeder, M.R., and Alexander, J., 1987, The origin and tectonic significance of asymmetrical meander-belts: Sedimentology, v. 34, p. 217–226.

Merritts, D.J., Vincent, K.R., and Wohl, E.E., 1994, Long river profiles, tectonism, and eustasy: A guide to interpreting fluvial terraces: Journal of Geophysical Research, v. 99, p. 14031–14050.

Molnar, P. Brown, E.T., Burchfiel, B.C., Deng, Q., Feng, X., Li, J., Raisbeck, G.M., Shi, J., Wu., Zh., Yiou, F., and You, H., 1994, Quaternary climate change and the formation of river terraces across growing anticlines on the north flank of the Tien Shan, China: Journal of Geology, v. 102, p. 583-602.

Pazzaglia, F.J., Gardner, T.W., and Merritts, D.J., 1998, Bedrock fluvial incision and longitudinal profile development over geologic time scales determined by fluvial terraces, *in* Tinkler, K.J., and Wohl, E.E., eds., Rivers over rock: Fluvial processes in bedrock channels: American Geophysical Union Geophysical Monograph 107, p. 207–235.

Rains, B., and Welch, J., 1988, Out-of-phase Holocene terraces in part of the North Saskatchewan River basin, Alberta: Canadian Journal of Earth Sciences, v. 25, p. 454–464.

Ritter, D.F., Kochel, R.C., and Miller, J.R., 1995, Process geomorphology (third edition): Dubuque, Iowa, Wm. C. Brown Publishers, 546 p.

Rockwell, T.K., Keller, E.A., and Dembroff, G.R., 1988, Quaternary rate of folding of the Ventura Avenue anticline, western Transverse Ranges, southern California: Geological Society of America Bulletin, v. 100, p. 850–858.

Schumm, S.A., 1968, River adjustment to altered hydrologic regimen, Murrumbidgee River and paleochannels, Australia: U.S. Geological Survey Professional Paper 598, 65 p.

Snow, R.S., and Slingerland, R.L., 1990, Stream profile adjustment to crustal warping: Nonlinear results from a simple model: Journal of Geology, v. 98, p. 699–708.

Stuiver, M., and Reimer, P.J., 1993, Extended ^{14}C database and revised CALIB 3.0 ^{14}C age calibration program: Radiocarbon, v. 35, p. 215–230.

Teng, L.S., 1990, Geotectonic evolution of the late Cenozoic arc-continent collision in Taiwan: Tectonophysics, v. 183, p. 57–76.

Weldon, R.J., and Stimac, J.P., 1994, The nature of fluvial terraces and their use in climate and tectonic studies: Geological Society of America Abstracts with Programs, v. 26, no. 7, p. A239.

Willemin, J.H., and Knuepfer, P.L.K., 1994, Kinematics of arc-continent collision in the eastern Central Range of Taiwan inferred from geomorphic analysis: Journal of Geophysical Research, v. 99, p. 20267–20280.

Yu, S.B., and Chen, H.Y., 1994, Global positioning system measurements of crustal deformation in the Taiwan arc-continent collision zone: TAO, Terrestrial, Atmospheric and Oceanic Sciences, v. 5, p. 477–498.

Yu, S.B., Chen, H.Y., and Kuo, L.C., 1997, Velocity field of GPS stations in the Taiwan area: Tectonophysics, v. 274, p. 41–59.

Manuscript Accepted by the Society April 12, 2001

Printed in the U.S.A.

Geological Society of America
Special Paper 358
2002

Structure and evolution of the active fold-and-thrust belt of southwestern Taiwan from Global Positioning System analysis

John B. Hickman*
David V. Wiltschko
Department of Geology and Geophysics and Center for Tectonophysics, Texas A&M University, College Station, Texas 77843-3113, USA
Jih-Hao Hung
Institute of Geophysics, National Central University, Chung-Li, 32054, Taiwan, Republic of China
Peng Fang
Yehuda Bock
Institute of Geophysics and Planetary Physics, Scripps Institution of Oceanography, University of California, San Diego, San Diego, California 92037, USA

ABSTRACT

By using GPS (Global Positioning System) and published geologic data, we have investigated the current deformation rates and local kinematic relationships for the southern region of the Taiwan fold-and-thrust belt. The GPS measurements suggest complex motions across the strike of this region. Although there is a general decrease in motion from east to west, significant variations occur across local structures. Deformation across southern Taiwan is neither homogeneous, nor distributed evenly across its surface. The structures and regions that appear to be the most active in this area include the Yuching syncline (limbs converging at 1.8 cm/yr), the vicinity of the Michih transverse fault, and the Tulungwan fault. The Tulungwan fault separates the Western Foothills and Slate Belt and appears to be a nearly vertical fault with some degree of left-lateral motion across it.

INTRODUCTION

Recent mechanical models of fold-and-thrust belts have provided an explanation for many important mechanical features of thrust wedges. For example, these models predict that for a given basal slope and wedge strength, the surface slopes of active fold-and-thrust belts are dependent on the shear strength of the basal glide horizon. A weak basal detachment leads to lower wedge opening angles (Davis et al., 1983; Dahlen et al., 1984; Davis and Engelder, 1985; Willett, 1992). In addition, the opening angle can evolve through time with changes in the strengths of both the basal fault and the rocks making up the wedge (Davis et al., 1983; Dahlen et al., 1984; Willett, 1992; Zhao et al., 1986).

Many of the predictions of these models are in general agreement with measured values from active fold-and-thrust belts (surface slopes, basal fault dip, etc.). Also, results from physical laboratory models (sandbox-type experiments) with materials of known properties (basal shear strength, geometry of the wedge, etc.) approximate those calculated from wedge models. However, the overall composition and structure of the wedge remains generalized. As Woodward (1987) pointed out,

*Present address: Kentucky Geological Survey, University of Kentucky, Lexington, Kentucky 40506-0107, USA.

Hickman, J.B., Wiltschko, D.V., Hung, J.-H., Fang, P., and Bock, Y., 2002, Structure and evolution of the active fold-and-thrust belt of southwestern Taiwan from Global Positioning System analysis, *in* Byrne, T.B., and Liu, C.-S., eds., Geology and Geophysics of an Arc-Continent Collision, Taiwan: Boulder, Colorado, Geological Society of America Special Paper 358, p. 75–92.

a first-order observation of fold-and-thrust belts is that they are composed of discrete thrust sheets, the scale of which is nearly that of the wedge itself. Current wedge models do not describe the behavior of the folds and faults within the wedge. Consequently, data on a much finer scale must be acquired to give further insight on the internal processes of the wedge.

In this paper we present geodetic measurements of surface motions of individual folds and faults within the active thrust wedge of Taiwan. Our purpose, insofar as Taiwan is typical, is to understand both the motions of individual structures, and the sum of their motions within the entire active wedge. We have focused on southwestern Taiwan, the most active part of the island's fold-and-thrust belt, and present three new structural cross sections across this area.

TECTONICS AND GEOLOGY OF SOUTHWESTERN TAIWAN

The island of Taiwan lies ~150 km southeast of mainland China, at the juncture of the Eurasian plate and the Philippine Sea plates (see Fig. 1). To the northeast of the island lies the northwestward-subducting Ryukyu Trench, and to the south lies the eastward-subducting Manila Trench. Relative to the Penghu Islands on the Eurasian plate (Fig. 1), the Philippine Sea plate is moving northwest (306°) at a rate of about 8.2 cm/yr (Yu et al., 1997). The region of contact between the Eurasian and Philippine Sea plates has migrated southward with time at a similar average rate of 84 km/m.y. (Suppe, 1984, although see Byrne and Crespi, 1997). Active deformation within the fold-and-thrust belt has propagated southward with the collision resulting in a relatively inactive northern region and a very active southern region.

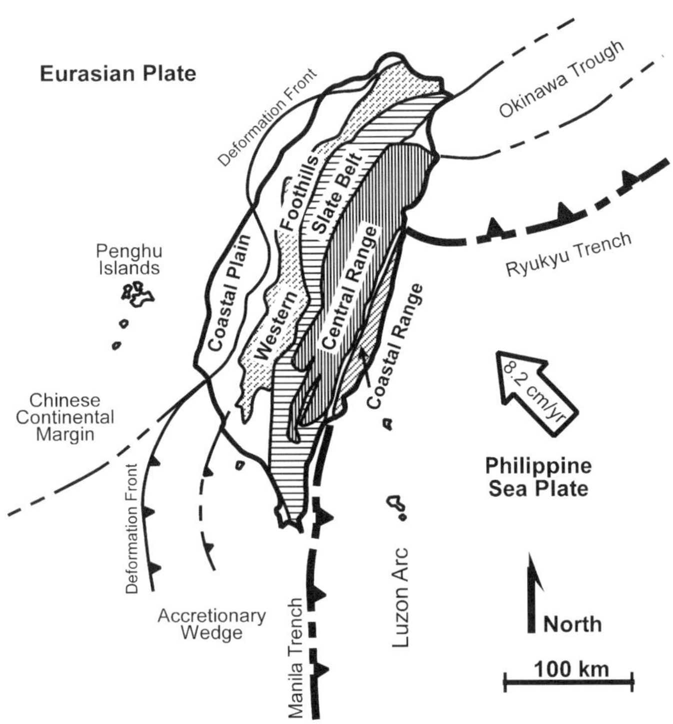

Figure 1. Geodynamic relationships in the Taiwan region. Velocity and direction of Philippine Sea plate motion relative to Penghu Islands shown by the large arrow (after Teng, 1990). Geologic provinces from Ho (1986).

Ho (1986) divided the island into five major geologic provinces (Fig. 1). From west to east these include the Coastal Plain, the Western Foothills, the Slate Belt, the Central Range, and the Coastal Range. The Coastal Plain is a low-relief province composed of Quaternary alluvium and terrace deposits. This province constitutes the foreland and is relatively undeformed. The Western Foothills province extends from a topographic break in the west to the fault-bounded Slate Belt in the east. The westernmost faults and folds of the Taiwan collision and sedimentary rocks as old as Miocene crop out in this belt. The Slate Belt displays high relief and exposes moderately metamorphosed Eocene to Oligocene sedimentary rocks. It is bounded on the west by a series of branching thrust and sinistral oblique-slip faults (Biq, 1989). Owing to the monotony of the lithologies, the regional-scale structure is not well known, but tight folds, reverse faults, and late normal faults have been mapped (Tillman and Byrne, 1995; Crespi, 1996). The Slate Belt can be further divided into the Hsüeshan Range and the Backbone Range. The Central Range is composed mostly of pre-Tertiary highly metamorphosed rocks, including gneisses and schists and some marble. The rocks are also highly folded and foliated. This geologic province is separated from the Slate Belt to the west by a top-to-the-west thrust fault and to the east by the Longitudinal Valley fault.

The Longitudinal Valley is a fault-bounded topographic depression that separates the metamorphic core of the orogen (the Central Range) from the rocks of the colliding island arc (the Coastal Range). The eastern boundary of the valley is marked by a well-mapped, active, east-dipping thrust fault (e.g., Barrier and Angelier, 1986). The western boundary of the valley, however, is not well understood. As a result, the boundary has been variously interpreted to be a strike-slip fault (e.g., Ho, 1986, 1988), a west-dipping thrust fault (Stanley et al., 1981), and a down-to-the-east regional-scale normal fault (Crespi et al., 1996).

The Coastal Range is the remnant of the Neogene Luzon magmatic arc and is considered to be the exposed leading edge of the Philippine Sea plate. The rocks in this region are volcanogenic and flysch-like sedimentary rocks of Miocene through Pleistocene age.

GEODYNAMIC MODELS OF TAIWAN

The tectonic evolution of Taiwan has been a subject of debate for some time. The noncontroversial aspects are that (1) Taiwan is a result of the collision between the north-trending Luzon arc and the northeast-trending Eurasian continental margin, (2) the collision is propagating southward, so that the southern part of the island is now undergoing the majority of the compressional deformation, and (3) the boundary between the Philippine Sea plate and the Eurasian plate lies along the Longitudinal Valley.

More controversial is the deep structure of the collision zone. Suppe (1981, 1984) suggested that Taiwan lies above two opposing crustal-scale detachment faults meeting under the Central Range and forming a double-sided, "thin-skinned" wedge. The rocks beneath these detachments are largely undeformed. Wu et al. (1997), instead, proposed that the entire lithosphere is involved in the collision. In this "thick-skinned" model, both the Eurasian margin and the Philippine Sea plate thicken under the collision zone. This thickening is reflected in the deep-crustal root observed in earthquake focal studies. The buoyancy of the thickened Eurasian margin lithosphere resulted in the uplift of the Central Range.

GEOLOGY OF THE WESTERN FOOTHILLS

Stratigraphy

The rocks exposed in the southern Western Foothills of Taiwan consist of late Oligocene through late Pleistocene sedimentary strata (see Fig. 2). These beds are formed from primarily shelf-margin and continental-slope sediments, covered by relatively thin Quaternary alluvium. Most are shaly sandstones or sandy shales with intermittent sandstone beds (Ho, 1988). Older strata were formed from sediments that were shed off the Eurasian continent. Around the middle Pliocene, the sediment source changed from being west of Taiwan to being the eastern part of Taiwan itself, because of the subaerial erosion of the island and the buildup of the Central Range.

Structures

Map patterns. The structures of the Foothills comprise a series of (1) subparallel folds and east-dipping faults, and (2) oblique structures, mostly faults. The Chukou and Lunhou faults mark the westernmost exposed thrust faults in this area (Fig. 3). The Chukou fault terminates to the north at the Chuchi transverse fault and to the south is inferred to merge or transfer displacement with the Lunhou fault to the south (Lin, 1996; see Fig. 3). The Niushan anticline lies in the footwall of the Chukou-Lunhou thrust system and marks the deformation front of the fold-and-thrust belt. The Yuching syncline occurs in the hanging wall of the Lunhou thrust and merges with the Tingpinglin fault to the north.

The pattern of northeast-trending, subparallel thrust faults continues to the southeast of the Chukou-Lunhou faults as exemplified by, for instance, the Pingchi fault. The Chishan and Tulungwan thrust faults parallel the Pingchi fault; the Tulungwan fault, which both parallels the Pingchi fault and separates the Western Foothills from the Slate Belt to the east, extends southward over 130 km to the southern coast of Taiwan. Biq (1989) suggested that the Slate Belt boundary is a sinistral wrench fault.

Normal faults may be divided into to two categories, (1) those on the Eurasian margin (i.e., the downgoing plate) and those in the fold-and-thrust belt structurally above it. The former are found associated with the Peikang High (Fig. 3), which is considered to be a basement-cored horst-like structure, bounded by normal faults, that is a remnant of the continental rifting that formed the Taiwan Strait. Tertiary strata are thin across the structure and both thicken and become more shale-rich to the south. Numerous small faults within the thrust belt show normal slip (Fig. 3).

The southern Taiwan fold-and-thrust belt is notable for the prevalence of transverse faults (Chinese Petroleum Company, 1986, 1989; Fig. 3). An example is the Michih transverse fault (south of middle of Line A–A′, Fig. 3). This fault strikes northwestward from the Chutouchi fault and crosses the midpoint of the Yuching syncline. The stratigraphic offset is 450 m. Another example is the large transverse fault at the southern termination of the Lunhou and Kouhsiaoli faults (south of the middle of line B–B′, Fig. 3). This left-lateral fault is exposed for 11 km, striking southeastward from the Lunhou fault.

Cross sections. In order to correlate the current motion in the southern part of the Western Foothills west of the Slate Belt with motion on individual structures, a necessary first step is to have a better picture of the structures at depth. We present three cross sections, each about 60 km long, extending to a depth of 15 km below current mean sea level. The locations for these cross sections (see Fig. 3) were chosen on the basis of proximity to both wells and (GPS) monuments. With an approximate spacing of 15 km between lines, they cover the area just off the Peikang High to the north (23°20′N), south to the termination of the Western Foothills at the Pingtung Valley (22°45′N), and from near the coastline in the western Coastal Plain (120°10′E) to the edge of the Slate Belt on the east (120°45′E). The orientations of the lines were chosen so that they lie normal to the strike of the thrust faults, roughly N55°W. They were also designed to be the southern continuation of a companion study (Hung et al., 1999).

Surface dips and stratigraphic contact information were collected primarily from the Chinese Petroleum Company's 1:100 000-scale area map number 5 (Chai-i) and number 6 (Tainan). Apparent dip within the plane of the cross section was calculated from dip data within 1500 m of each line on the map. Subsurface data for each line were collected from Chinese Petroleum Company oil well reports within about 4 km of the lines. These reports provided formation tops, fault depths, bed dips, and bedding thicknesses. Some wells farther than 4 km were used to establish area trends of bedding thicknesses and/or structures.

From surface exposures and well data, it is apparent that there are thickness changes from west to east on each line. The younger (middle Pleistocene) Liushuang Formation thickens to the west and is composed of sediments largely derived from the

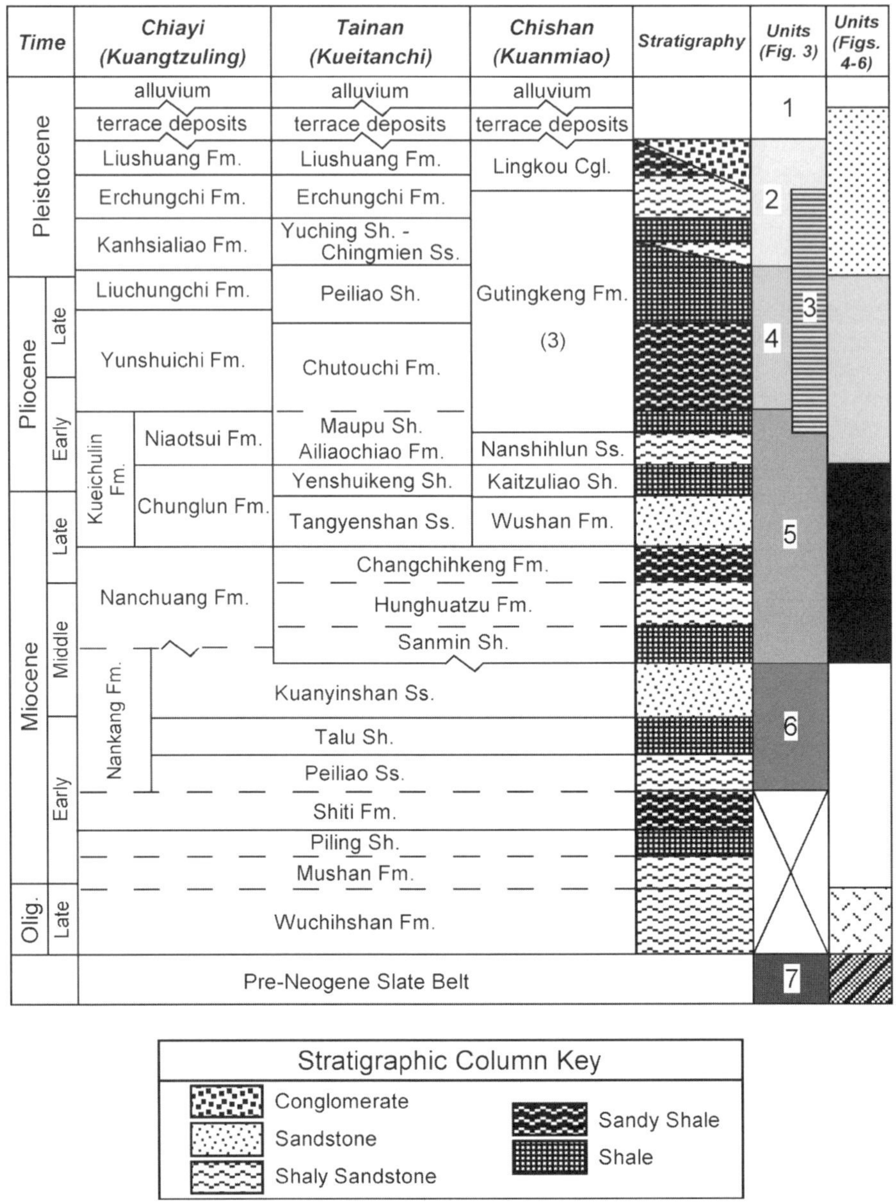

Figure 2. Stratigraphic correlation in southwest Taiwan (after Chi, 1990). The right columns display the patterns for the units used in the geologic map (Fig. 3) and cross sections (Figs. 4–6).

east. The middle to upper Pliocene Yunshuichi Formation thickens dramatically to the east, especially in line B–B′. The older formations (early Miocene) such as the Nankang and Mushan appear to have similar thicknesses throughout the lines. Except for around the Chiali normal faults in Lines B–B′ and C–C′, all the lateral changes in bed thicknesses (1) amount to less than 1° change in dip and (2) take place at more or less a linear rate across the sections.

Thickness variations were also observed between the northern and southern parts of our study area. Miocene beds thicken from line A–A′ to line B–B′ by over 100%. From line B–B′ to line C–C′, the upper Pliocene and lower Pleistocene bed thicknesses increase by almost 100%. This change coincides with an increase in the depth of the basin (Hu et al., 1997) and with moving south off the Peikang High (see Fig. 3).

Construction. We have placed the basal detachment at ~13 000 m (42 620 ft) below sea level to the southeast, rising to ~6000 m below sea level (19 675 ft) to the northwest in all three lines. Although confirmed by neither drilling nor seismic data, this depth trend is the minimum at which a major detachment could exist and provide room for all the necessary stratigraphy exposed at the surface. This same trend of the depth of detachment was used for all three cross sections, regardless of the different stratigraphic intervals in which the detachment occurs. In Line A–A′, the detachment is in middle Oligocene beds, in line B–B′, it is in the upper Oligocene, and in line C–C′, it is in the lower to middle Miocene beds. Placing the detachment at this depth is consistent with the cross sections of an area just to the north where somewhat more subsurface control exists (Hung et al., 1999).

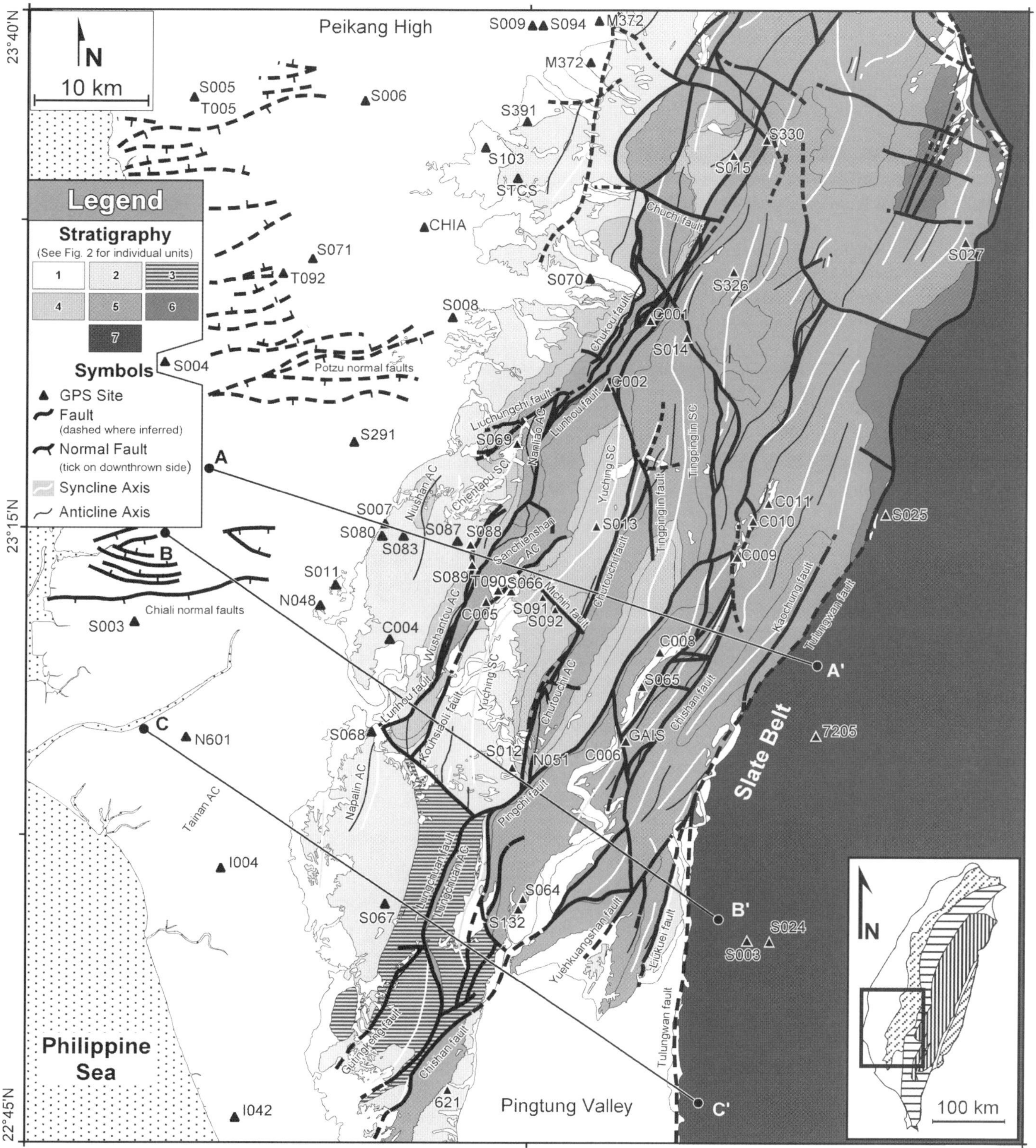

Figure 3. Geologic map of the study area showing site and cross-section locations. Geology from Chinese Petroleum Company Map number 5 (Chia-i). See Figure 2 for stratigraphic names and explanation of patterns.

Line A–A′. The northwesternmost structure crossed by this line (Fig. 4) is the Niushan anticline. This anticline is formed from a duplication of lower Miocene through upper Oligocene strata. This section, consisting of the Talu Shale through the upper Oligocene Wuchihshan Formation, underlies the Kuanyinshan Sandstone below the Niushan anticlinal axis. The elevation and breadth of the Chientapu syncline with respect to the rocks of the Niushan anticline require a bedding-parallel section of the basal fault. At the core of the Niushan anticline is a small fault-propagation fold that effectively thickens the middle Miocene Nanchuang Formation along the Niushan anticlinal axis, as shown by the NS-27 well (located in Fig. 4).

The Sanchienshan anticline is cut by three faults, two of which form a "pop-up" structure. Two of the faults are splays of the Lunhou fault. One splay emerges to the east, placing the upper Miocene Tangyenshan Sandstone against the upper Pliocene Liuchungchi Formation (displacement ≈ 10.4 km). The other splay emerges to the west of the anticlinal axis, but has no appreciable stratigraphic offset. These two splays of the Lunhou fault are interpreted to be southeast-dipping thrust faults that join at ~3200 meters below sea level (mbsl); the eastern fault is the site of the principal displacement. The Kouhsiaoli fault lies to the east of the Sanchienshan anticlinal axis and appears to have about 100 to 200 m of displacement. The Kouhsiaoli fault in outcrop is a backthrust.

The next three fault-related structures have remnants of asymmetrical anticlines present in the hanging walls. The displacements in a thrust sense are 8000 m for the Chutouchi fault, 2900 m for the Pingchi fault, and 6100 m for the Chishan fault. At the point where the Chutouchi fault crops out, the hanging-wall bedding is close to parallel with the fault plane. A 3500-m-long segment of the fault plane within the Sanmin Shale is interpreted as subparallel to bedding to account for this geometry.

The last two faults (Kaochung and Tulungwan faults) have been interpreted as being near-vertical folded thrusts combined with at least some degree of strike-slip motion. The Tulungwan fault delineates the edge of the Slate Belt province.

Line B–B′. The first important structure on the northwest end of this line (Fig. 5) is one of the Chiali normal faults, which lies beneath the Coastal Plain. This fault is apparently a growth structure because the middle to upper Miocene beds have been thickened from about 1300 m northwest of the fault to 3900 m to the southeast. The thicknesses and subhorizontal dips of the upper Pliocene Liuchungchi to the upper Pleistocene Liushuang Formations don't change over this fault (well LT-1).

A westward-tapering section of fault-bounded rock, which at its thickest, includes the Tangyenshan Sandstone to Mushan Formation (see Fig. 2), is interpreted to underlie the central part of this section. Its western end underlies the Wushantou anticline, and its eastern side is beneath the Chutouchi thrust fault. Motion along this imbricated horse has uplifted the region between the Wushantou anticline and the Chutouchi fault. The horse is interpreted as the southern extension of the similar structure described in line A–A′. The Wushantou anticline has also been faulted by the Lunhou fault (3400 m displacement) and Kouhsiaoli fault (1200 m displacement).

The Chutouchi fault and the Pingchi fault (600 m to the east) merge just south of this cross section. The Pingchi fault is the site of 12 km of displacement and has been interpreted as the primary fault, cutting the smaller Chutouchi fault (1700 m of displacement) at 3500 mbsl. Despite the variations of dip in the footwall bedding, we have interpreted a constant dip for the Pingchi fault plane, suggesting out-of-sequence movement along this fault (after the formation of the Chutouchi anticline).

From the crosscutting of the Chishan fault by the Yuehkuangshan fault and from the lateral offsets of the upper Miocene units surrounding the Yuehkuangshan and Liukuei faults,

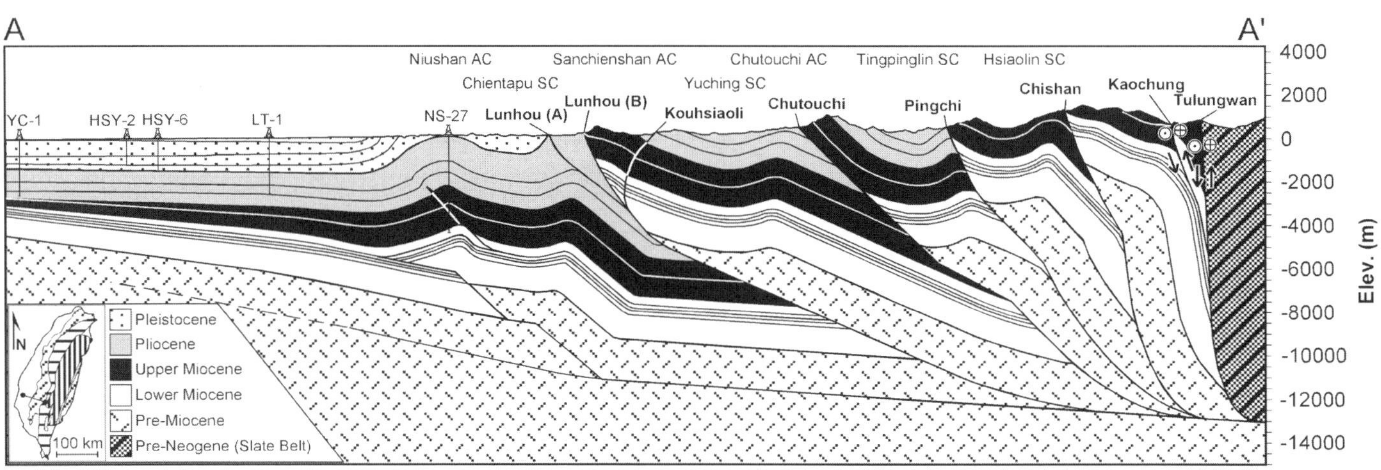

Figure 4. Line A–A′. Cross section through southwestern Taiwan; see inset and Figure 3 for location. Names of major faults are in bold; important wells are shown. AC—anticline, SC—syncline. See Figure 2 for stratigraphy used and pattern key.

the latter two faults have been interpreted as having significant components of strike-slip motion. The Yuehkuangshan fault is interpreted as an oblique left-lateral fault, and the Liukuei is interpreted as an oblique left-lateral reverse fault; little is known about the character of either fault. To the east of the Liukuei fault, the Tulungwan fault separates the Western Foothills from the Slate Belt to the east.

Line C–C′. The cross section C–C′ has the fewest constraints of the three cross sections presented. The first structure apparent on the west at the surface on this line (Fig. 6) is an anticlinal axis (Tainan anticline) that has been interpreted as a detachment fold with accommodation taken up by lower Pliocene units. The ~1 km displacement that caused this folding above the Wushan Formation was a result of movement along the subsurface fault that formed the Lungchuan anticline.

The Lungchuan anticline is interpreted as a fault-propagation fold. The fault has been interpreted to connect the basal Sanmin Shale to the east with the basal Katzuliao Shale to the west, ending in the detachment fold to the northwest. The Lungchuan fault is exposed in the crest of the Lungchuan anticline. This thrust fault has 9.4 km of displacement along the southeast limb of the Lungchuan anticline.

Comparisons to previous cross sections. Suppe (1980) was the first to create detailed, regional-scale geologic cross sections for the Western Foothills of Taiwan. The Niushan line of Suppe (1980) is parallel to, and covers the same structures as, line A–A′ discussed here. The styles of structures within the Niushan line are typical of the other three cross sections of Suppe (1980) through the Western Foothills. With the exception of the dip of the Lunhou fault where it crops out, the near-surface (1 to 2 km depth) structures and bedding are identical between the Niushan line of Suppe (1980) and line A–A′.

In his interpretation, Suppe (1980) used a basal detachment fault at 6000 mbsl below the Chientapu syncline and dipping

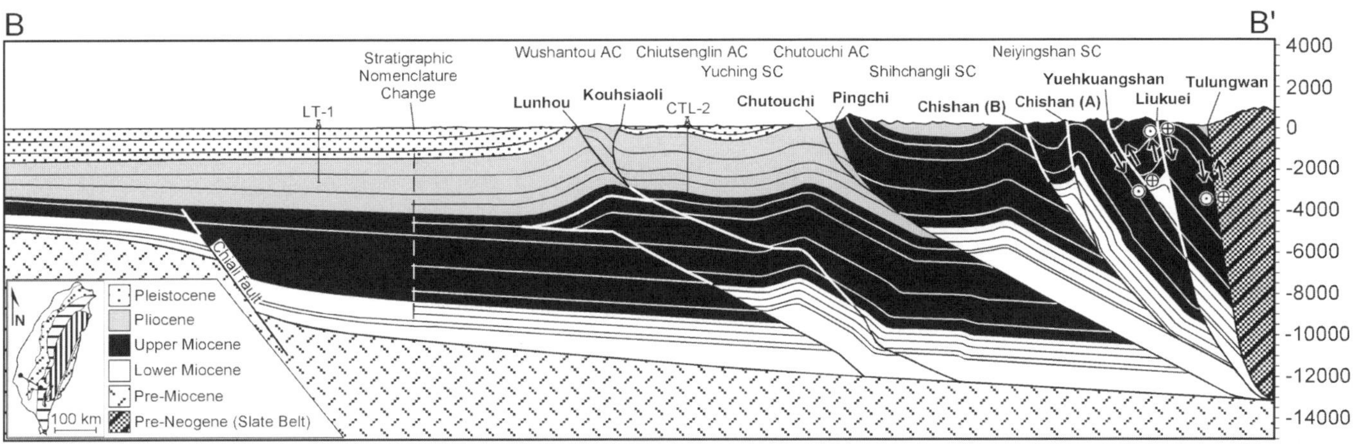

Figure 5. Line B–B′. Cross section through southwestern Taiwan; see inset and Figure 3 for location. Names of major faults are in bold; important wells are shown. AC—anticline, SC—syncline. See Figure 2 for stratigraphy used and pattern key.

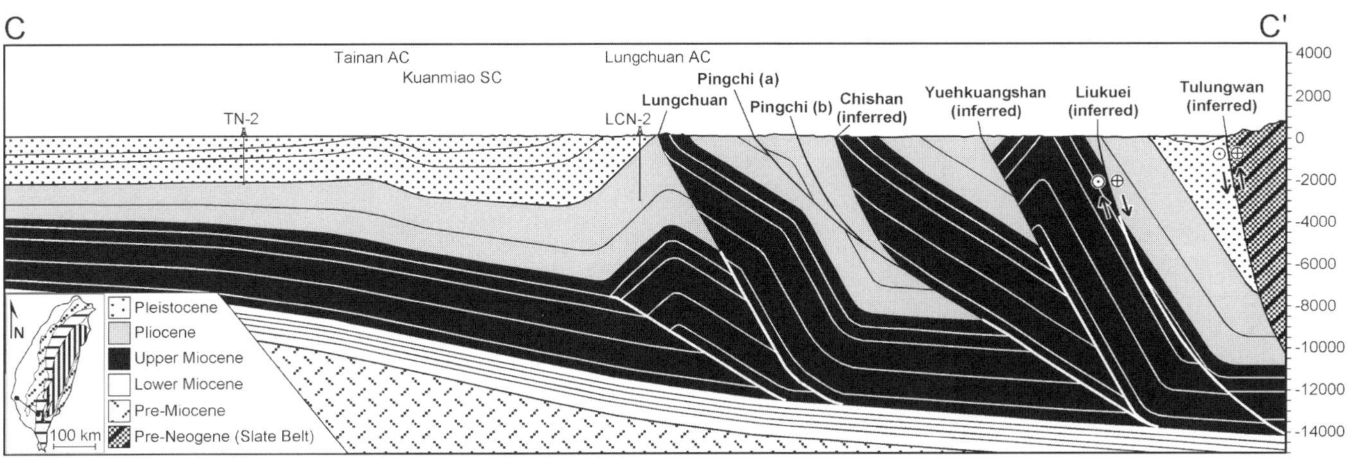

Figure 6. Line C–C′. Cross section through southwestern Taiwan; see inset and Figure 3 for location. Names of major faults are in bold; important wells are shown. AC—anticline, SC—syncline. See Figure 2 for stratigraphy used and pattern key.

6°SE. This fault zone lies along an unconformity between the lower Miocene Mushan Formation and continental Cretaceous rocks. To account for the amount of exhumation needed to expose middle Miocene strata at the surface (about 700 m above sea level), Suppe (1980) interpreted several (10 or more) duplexes of lower to middle Miocene units. Although this geometry does account for the uplift of lower strata, it necessitates a huge amount of displacement (63 km) to form the duplexes.

The same exposed and near-surface structures can be created in a simpler fashion. A deeper basal fault would require fewer, but thicker, ramping units to accomplish the same uplift. In the interpretation presented here, two large ramps are formed above a deeper detachment at 11 kmbsl below the Lunhou fault, as compared to 6.5 kmbsl for the same location in Suppe (1980). These ramps create the same uplift and exhumation, but with only 12 km of displacement. This method has also been used by Hung et al. (1999) to interpret areas to the north within the Western Foothills, with similar results.

SURFACE MOTIONS FROM GPS

The Taiwan GPS Network has been in operation since 1990. The network contains 140 surveying monuments, 9 of which are continuously, or nearly continuously operating, and maintained by the Academia Sinica (Yu and Chen, 1994). Of these 140 sites, 78 are in the southern part of Taiwan where this study is located (see Fig. 7). The velocities are well established (Yu et al., 1997).

For this study, we added 10 monuments across southern Taiwan. The site designation of each of the new sites begins with the letter "C" (Figs. 3, 7). GPS surveys conducted primarily in January and February of 1996 and 1997 used both these new monuments and preexisting ones. Data from several permanent sites maintained by Academia Sinica and the Central Weather Bureau were also incorporated into our solution. Only those permanent-site data that coincided with the times of our field campaigns were used. Our surveys were conducted with survey-grade GPS receivers (Trimble SSI, SSE, and a few SST sites in 1996; Ashtech ZXII) on 4 to 12 sites simultaneously, with an average session occupation time of 8 h (as short as 4 h and as long as 24 h).

Data processing

Data processing and analysis were carried out in two main steps. The first step was to process all regional data session by session by using a weighted least-squares method with the GPS data processing software package GAMIT (e.g., Dong et al., 1998; Feigl et al., 1993) to determine station coordinates. The second step was to combine the sessions by using a Kalman filter approach with the software package GLOBK (e.g., Herring et al., 1990). The session length was defined as 24 h. In the case of more than one observation session within a 24 h period, they were combined as one session.

In the single-session processing, raw observations first underwent an automatic cleaning procedure, which both rejects obvious outliers in the observations and flags data gaps and cycle slips. The cleaned data were then used for forming double differences that became the basic observations for a linear equation system for which the unknown parameters were solved by using a weighted least-squares method. In order to tie the regional network into a global reference frame, a few global sites in the vicinity were also included in the solution. The coordinates of those global sites defined as ITRF94 were tightly defined in the solution. The Scripps Institution of Oceanography final precise orbits and estimate of precision (EOP) consistent with this reference frame were used in the process with tight constraints. In addition to estimating the regional site coordinates, hourly tropospheric delays at each site were also simultaneously estimated. Neill's mapping function (Neill, 1996) was employed in this process by using observations made at elevations as low as 7°. In order to reduce the errors in the vertical component, the standard International GPS Service (IGS) phase-center model was used in the process.

After single-session data reduction was completed, the covariance matrices from all single-session solutions were fed to GLOBK in which the multiday site-coordinate series were treated as continuous variables so that their velocities could be estimated in a consistent manner. In this step, tight constraints were applied to the well-known global sites, as well as orbital and EOP parameters.

Once the velocities of the regional sites were obtained, a strain-rate analysis was performed by carefully grouping together the velocity vectors of a few sites associated with the same geologic feature. By doing so, the true velocity signals of individual structures, or small groups of them, were enhanced while the scatter among individual sites was suppressed. Those data with excessive calculated errors of determinations were omitted from our final interpretations.

Comparison of GPS velocity results and local structures

Horizontal velocity vectors are plotted in Figure 8A, and the preliminary vertical velocities are in Figure 8B. It should be noted that the precision for the vertical velocities is at least three times as large as that for horizontal velocities (e.g., Hofmann-Wellenhof et al., 1994). Figures 3 and 7 show site locations. Site S01R was used as the reference site for velocity calculations in Figure 8. It is situated on the Penghu Liehtao volcanic island group, which is geologically on the Chinese continent.

To study the effects of small-scale structures (faults and folds) on local surface-velocity vectors, five small map areas were created, each with its own reference site. The velocity for this reference site was fixed (set to zero) by subtracting its velocity from all sites on each small map. In this way, specific changes in motion orientation and magnitude could be studied with respect to the individual structures in each area. Changes in the magnitude or direction of velocity vectors over an area

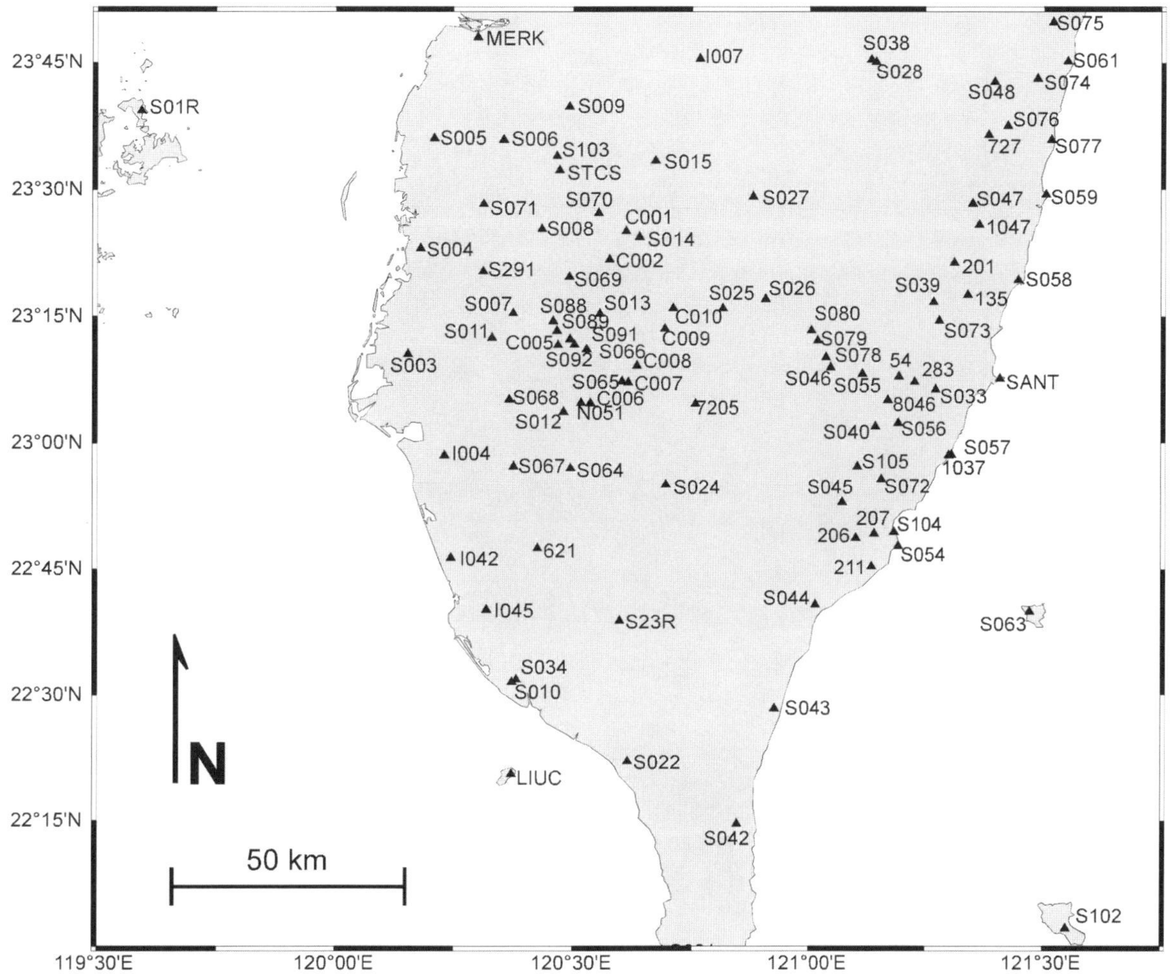

Figure 7. Locations of GPS monuments in southern Taiwan.

are assumed to be from active folding and/or faulting between the stations. To further illustrate changes in the trends of motion, GPS measurements for the entire island have been used to contour the scalar magnitudes of the velocities (Fig. 9).

Map 1. The first map (Fig. 10) contains six sites, whose velocities are displayed with respect to site I004. From this map, it seems clear that the Lungchuan fault and/or Pingchi fault system is active (note change from S064 to S067). The contours closely follow the trend of the Lungchuan-Gutingkeng fault system. However, without more sites, it is difficult to determine what combination of these faults is most active. There also appears to be shortening between S067 and I004. Because there are no mapped faults between these two stations, this contraction is probably being consumed by the growth of the Kuanmiao syncline.

There is a small difference in the horizontal-velocity azimuths of S024 and S064 across the Slate Belt boundary. S064 is moving more rapidly northward than S024, suggesting a small amount of dextral motion across the Tulungwan fault in this area although there are several other faults between S024 and S064.

Map 2. The second map (Fig. 11) contains 14 sites, measured with respect to site S088, which is in the extreme northwest corner. The site velocities suggest that the majority of deformation in this area is accommodated by the growing Yuching syncline. The movements of S066, S091, and S092, all of which are on the same fault block, are toward the axis of the Yuching syncline (between S066 and S091). C005 is moving in a similar fashion with respect to the synclinal axis. Note the possible right-lateral movement along the Chutouchi fault. The anomalous orientations of the movements of sites C007 and C006 are yet to be explained.

Map 3. The third map (Fig. 12) contains 15 sites whose velocities are plotted with respect to site S008. The velocities across the southeastern part of this map display a relatively constant decrease in displacement from south to north. However, in the southwestern part of this area, the gradient in velocity increases around the southern ends of both the Liu-

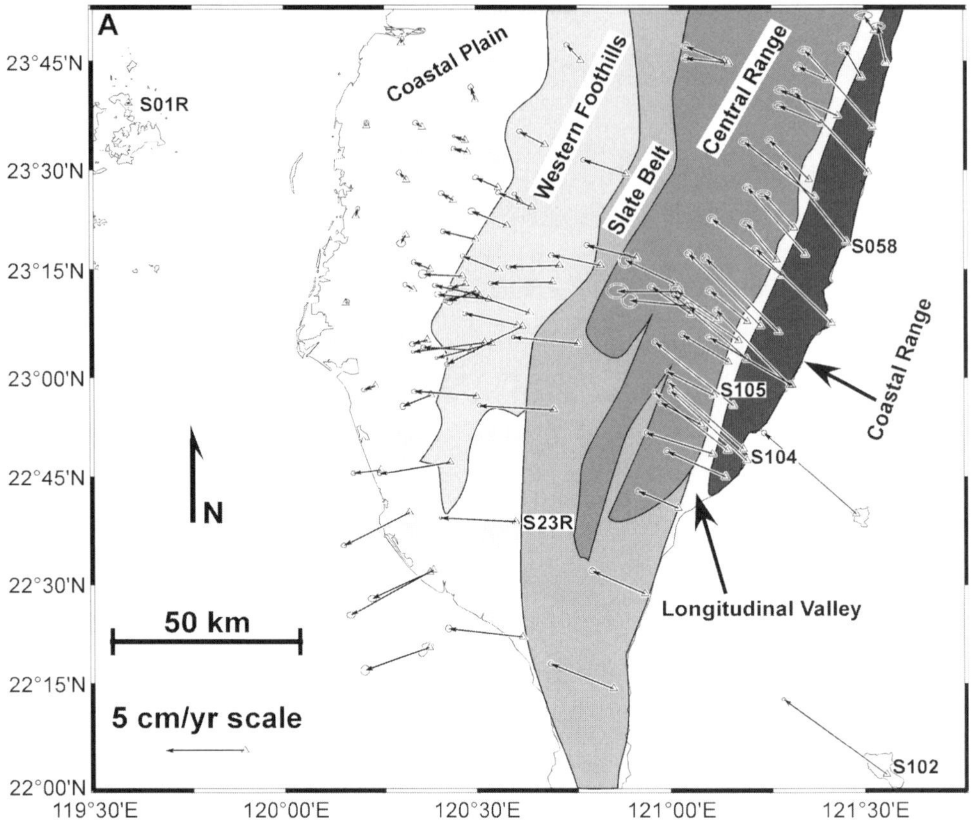

Figure 8. A: Horizontal velocities of GPS stations (open triangles) across southern Taiwan. Measurements relative to Paisha, Penghu (S01R). The 95% confidence ellipse is shown at the tip of each velocity vector. The scale is the same for both the velocity and the error. Only the continuously recording permanent sites are labeled; see Figure 7 for other site labels. GPS data compiled from this work and from Yu et al. (1997). Geologic provinces from Ho (1986). B: Vertical velocities of GPS stations (open triangles) across southern Taiwan. Measurements relative to Paisha, Penghu (S01R). The 95% confidence error bar is shown at the tip of each velocity vector. The scale is the same for both the velocity and the error. All vertical data collected by this study in 1996 and 1997. Geologic provinces from Ho (1986).

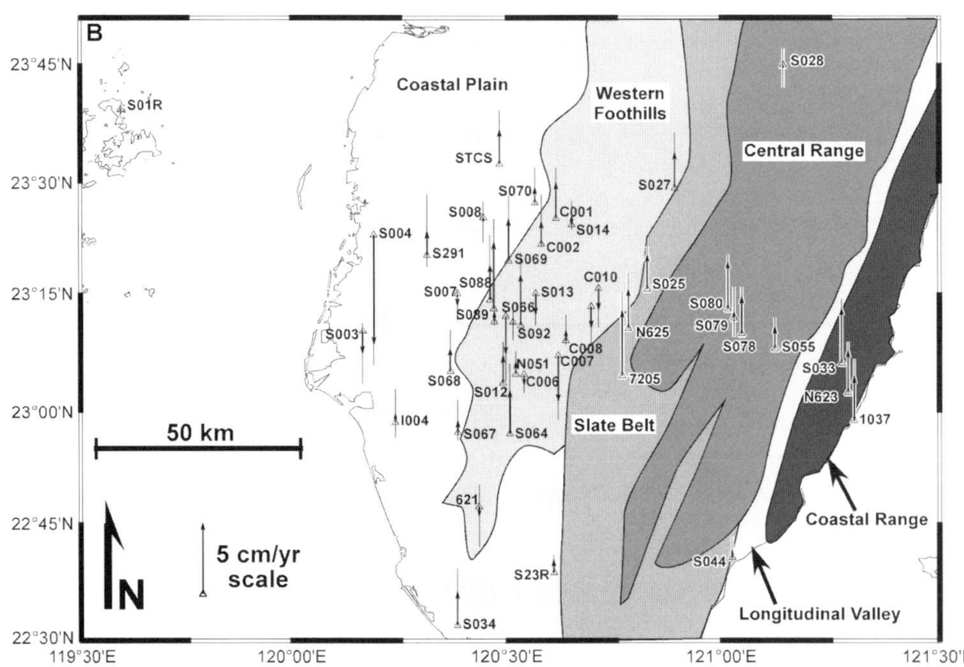

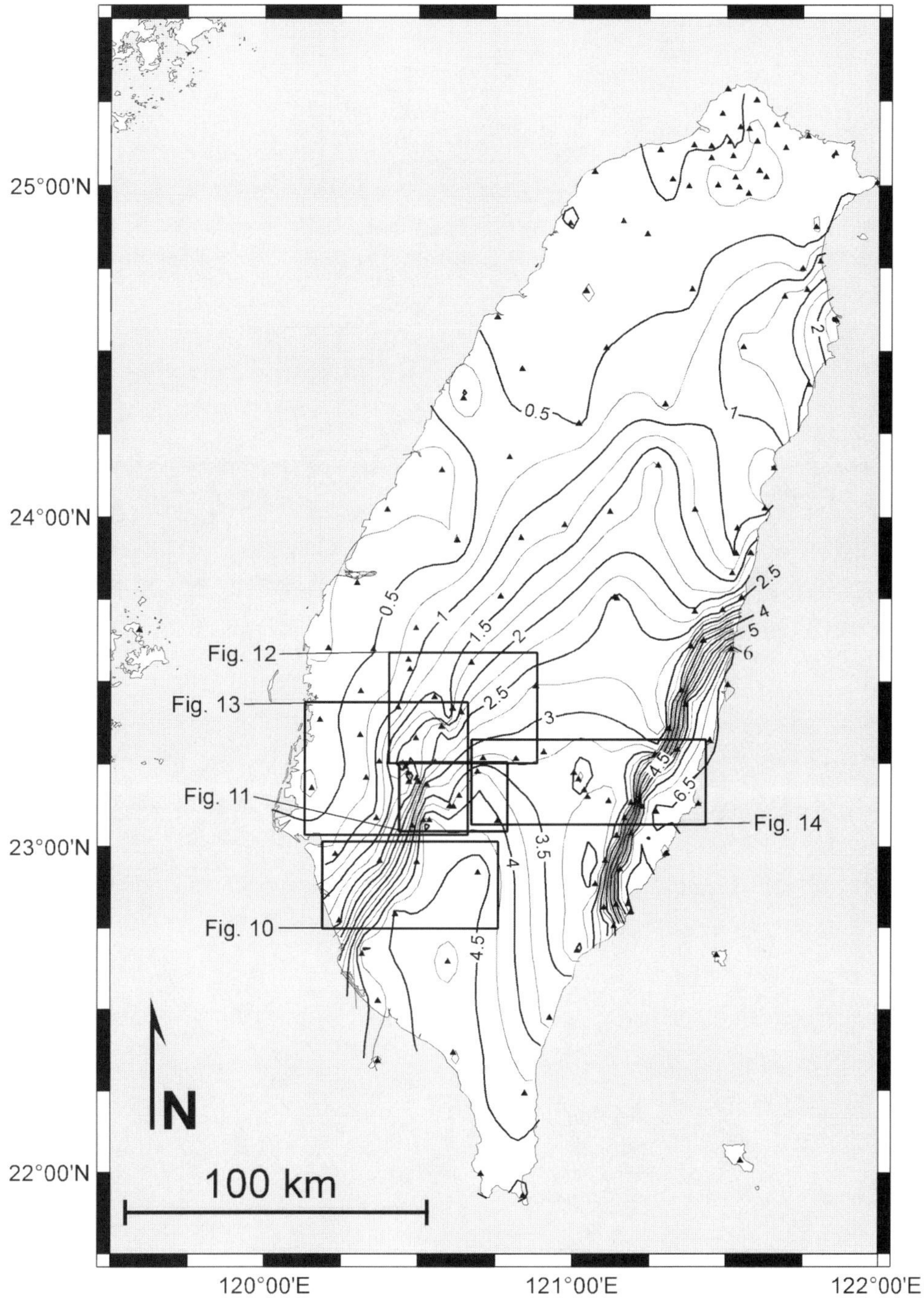

Figure 9. Contoured GPS horizontal velocity magnitudes across Taiwan. A 50% tension factor was applied to the GMT (Generic Mapping Tools; Wessel and Smith, 1998) contouring algorithm within the map area. Contour intervals are 0.25 cm/yr. Data from Yu et al. (1997) as well as our new data. Locations of Figures 10–14 are marked.

chungchi fault and associated folds (Nanliao and Niushan anticlines and the Chientapu syncline). The sites in the vicinity of these folds (S069 and S088) also have high relative uplift rates (Fig. 8B), possibly suggesting movement on a blind part of the Liuchungchi fault.

The difference in direction of motion between S025 and C010 across the Slate Belt boundary suggests sinistral motion, although several structures lie between these two sites (compare S025 and S027). However a large difference in relative uplift rates (about 4 cm/yr, see Fig. 8B) is displayed between sites C008 and 7205 across the Slate Belt boundary in Figure 11. The Slate Belt is being uplifted along a near-vertical fault in this area.

Map 4. The fourth map (Fig. 13) contains 24 sites whose velocities are computed with respect to site S003. The most pronounced structure delineated by the GPS data in this map area is the transfer of displacement from the Chutouchi fault and Yuching syncline to the Liuchungchi and Chukou faults

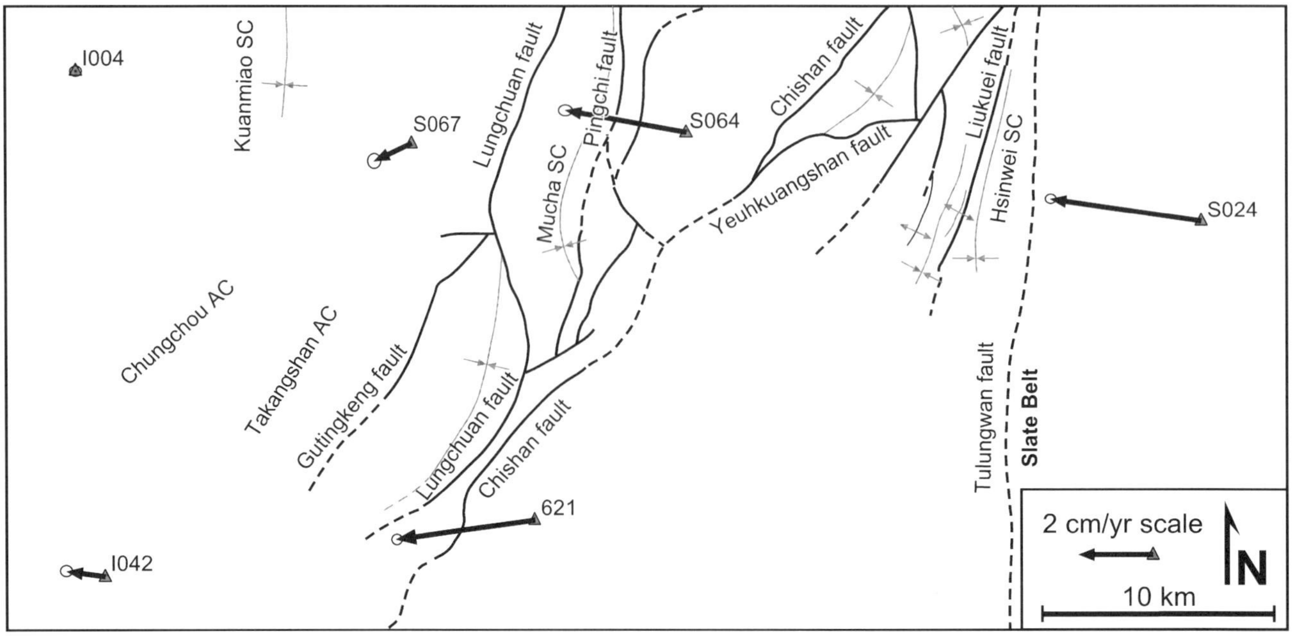

Figure 10. Velocity vectors plotted with respect to site I004 for analysis of local structures (location shown in Fig. 9). Approximate locations of Chungchou and Takangshan anticlines plotted from Chinese Petroleum Company map number 6 (Tainan sheet) (1986). The 95% confidence ellipse is shown at the tip of each velocity vector. The scale is the same for both the velocity and the error.

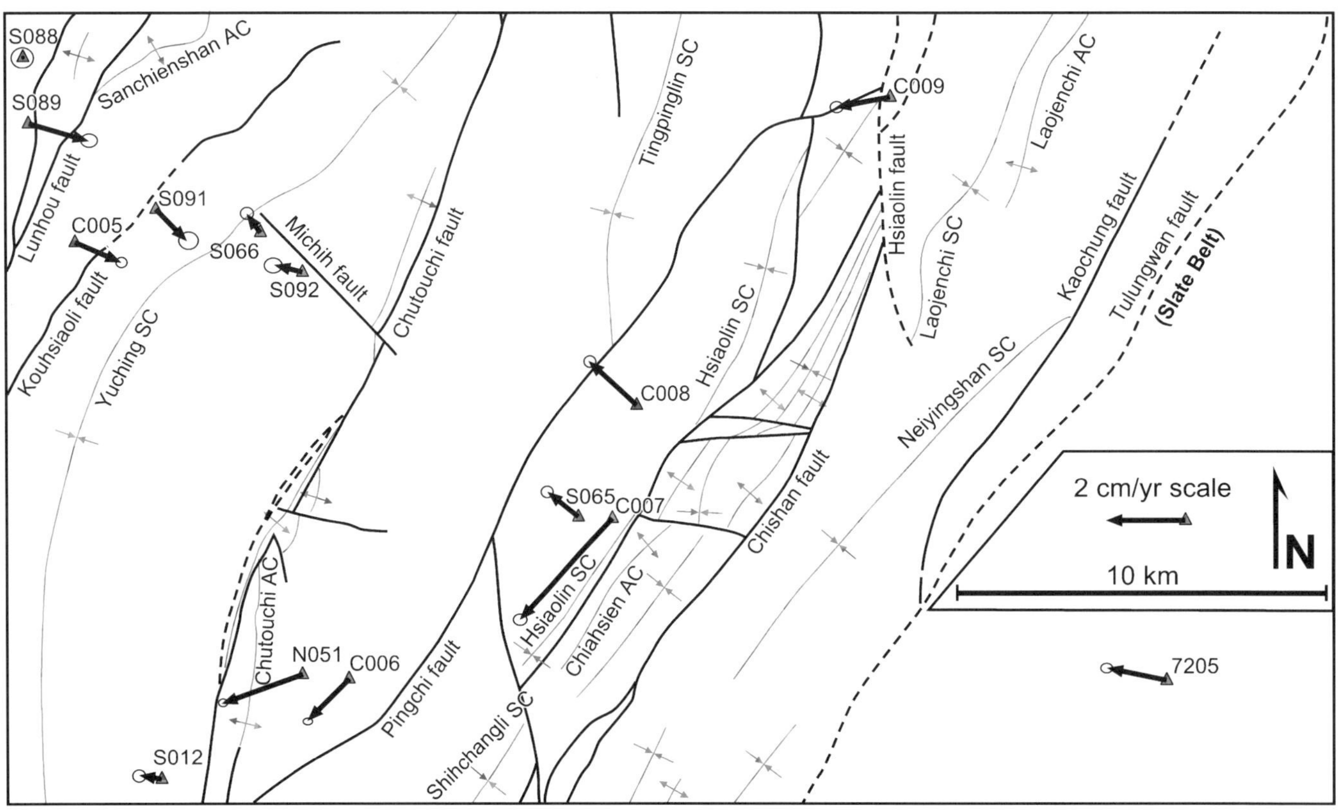

Figure 11. Velocity vectors plotted with respect to site S088 for analysis of local structures (location shown in Fig. 9). The 95% confidence ellipse is shown at the tip of each velocity vector. The scale is the same for both the velocity and the error.

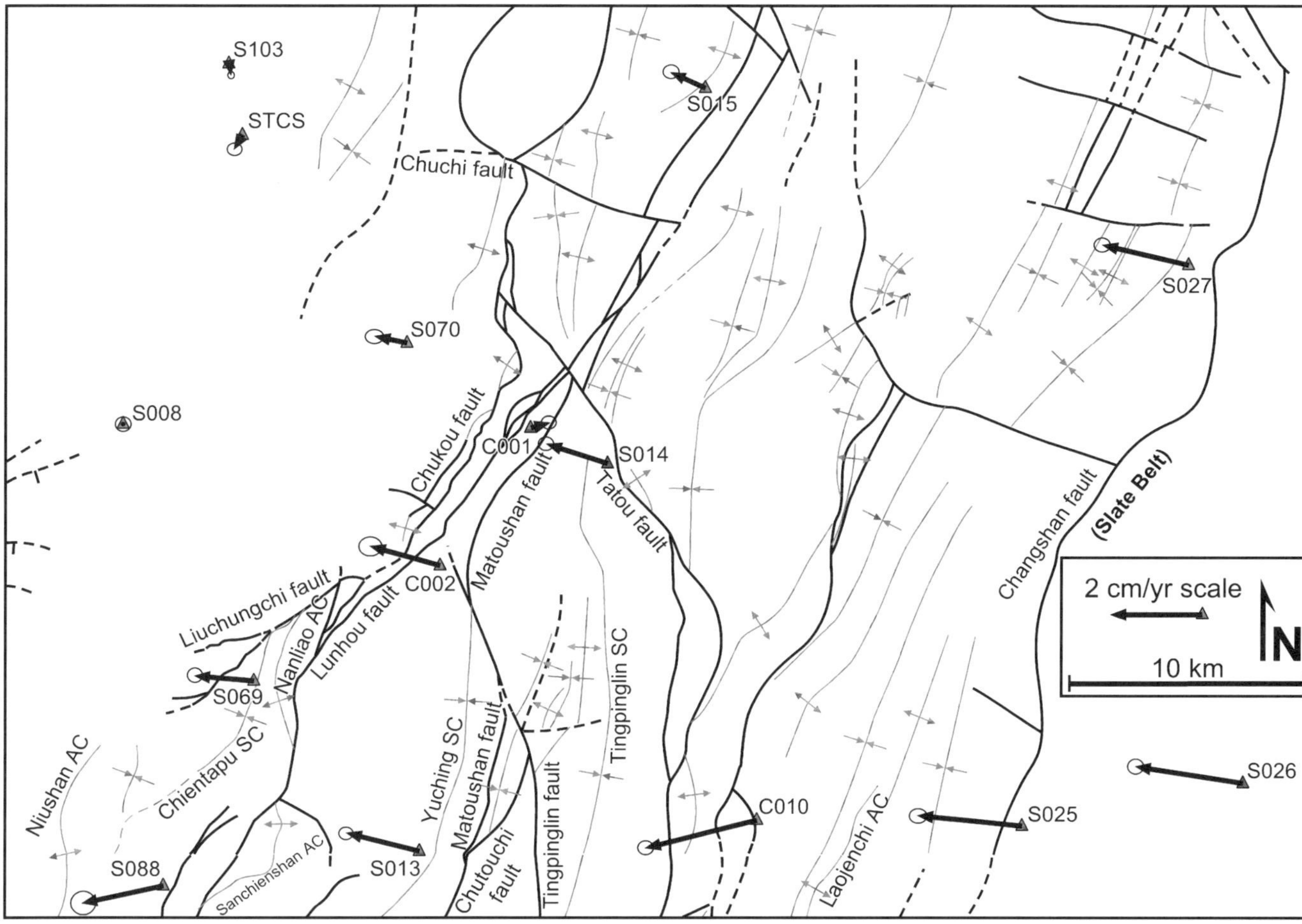

Figure 12. Velocity vectors plotted with respect to site S008 for analysis of local structures (location shown in Fig. 9). The 95% confidence ellipse is shown at the tip of each velocity vector. The scale is the same for both the velocity and the error.

and the Niushan anticline. This transfer of displacement is centered around the Michih sinistral strike-slip fault that originates at the Chutouchi fault and transects the Yuching synclinal axis at 90°, with a mapped left-lateral offset of 350 m. Figure 11 illustrates the motion involved with the growth of the Yuching syncline, and Figure 12 illustrates the motion across the Niushan, Nanliao, and Chientapu folds and the Liuchungchi fault. In Figure 13, it is apparent that these two high-deformation areas are linked through a fault-transfer zone.

There is a high density of sites in this area because of the high level of seismicity that has been associated with the Chukou-Lunhou fault system (Yu et al., 1997). It is possible that this high local seismicity is due, more specifically, to the transfer of displacement and deformation associated with this strike-slip fault system.

The velocities of sites C008 and S065 on the hanging wall of the Pingchi fault are not significantly different in motion perpendicular to strike from those of sites C006 and N051 on the footwall. However, there is a change of about 20° in the orientation of the velocity vector across the fault. Therefore, although the Pingchi fault has had considerable amount of thrust motion along it, the current motion appears to be of a left-lateral nature. However, the Chutouchi fault immediately to the west is still actively thrusting, as shown by the velocity difference between its hanging wall (sites C006 and N051) and its footwall (sites S012 and S092).

The Niushan anticline was interpreted to be actively growing (Fig. 12). Further evidence for this interpretation is displayed in Figure 13, which shows a decrease in velocity between sites S088 and S089 to the east of the Niushan axis and sites S011 and S007 to the west. This westward drop in velocity translates into a convergence of these sites across the axis; as the limbs approach each other, the fold grows.

From the vertical velocity data (Fig. 8B), site S004 appears to be subsiding at a high rate (about 7.7 cm/yr). Although there is a chance that there is reactivated motion along the deeply buried Potzu normal-fault system, it is more probable that this apparent motion is caused by human activity (i.e., over-depletion of shallow aquifers leading to matrix–pore space collapse) in this populous region.

Map 5. This map area extends from the Coastal Range across the Central Range and Slate Belt to the western edge of

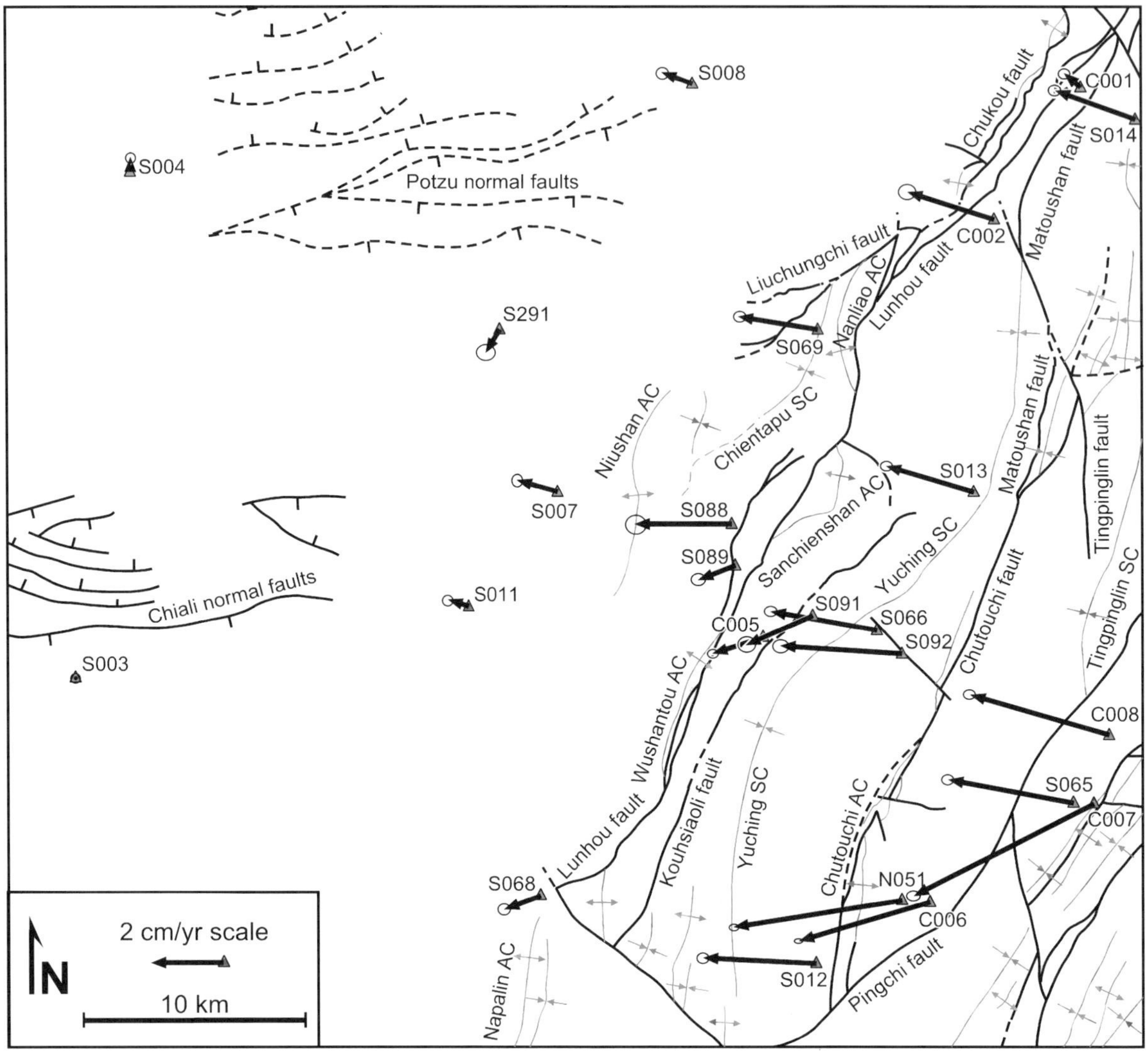

Figure 13. Velocity vectors plotted with respect to site S003 for analysis of local structures (location shown in Fig. 9). The 95% confidence ellipse is shown at the tip of each velocity vector. The scale is the same for both the velocity and the error.

the Western Foothills (Fig. 14). There are 18 sites on this map; motions are plotted with reference to site C010 in the Western Foothills. The plate boundary between the Philippine Sea plate and the Eurasian plate is a left-lateral fault with a large component of west-northwest-verging thrust motion. The sites on the western edge of the Longitudinal Valley have velocity vectors that differ by about 40° and 1.5 to 1.75 cm/yr from those sites on the eastern side.

On the western part of Figure 14, the horizontal velocities of sites S025 and S026 parallel the Slate Belt boundary. Although displaying a large amount of relative uplift (Fig. 8B), these sites are not approaching sites C009 and C010 on the other side of the Tulungwan fault. This result implies that left-lateral oblique motion is occurring along a nearly vertical fault plane. Although site 7205 to the south has a significant westward velocity with respect to site C010, this result may be due to the general increase in westward motion from north to south across the island, not to a lack of left-lateral motion.

DISCUSSION AND PREDICTIONS OF SURFACE MOTIONS

Faults. If no other deformation mechanisms are active within an area, surface velocities will change in magnitude and/or direction abruptly across an active fault. The relative difference in the velocities on both sides of the fault determines the nature of the fault (Fig. 15).

The motion of most faults in southwestern Taiwan is consistent, to the extent that data exist, with the mapped character of the faults. The data density, however, is not sufficient to determine whether most of the large number of individual struc-

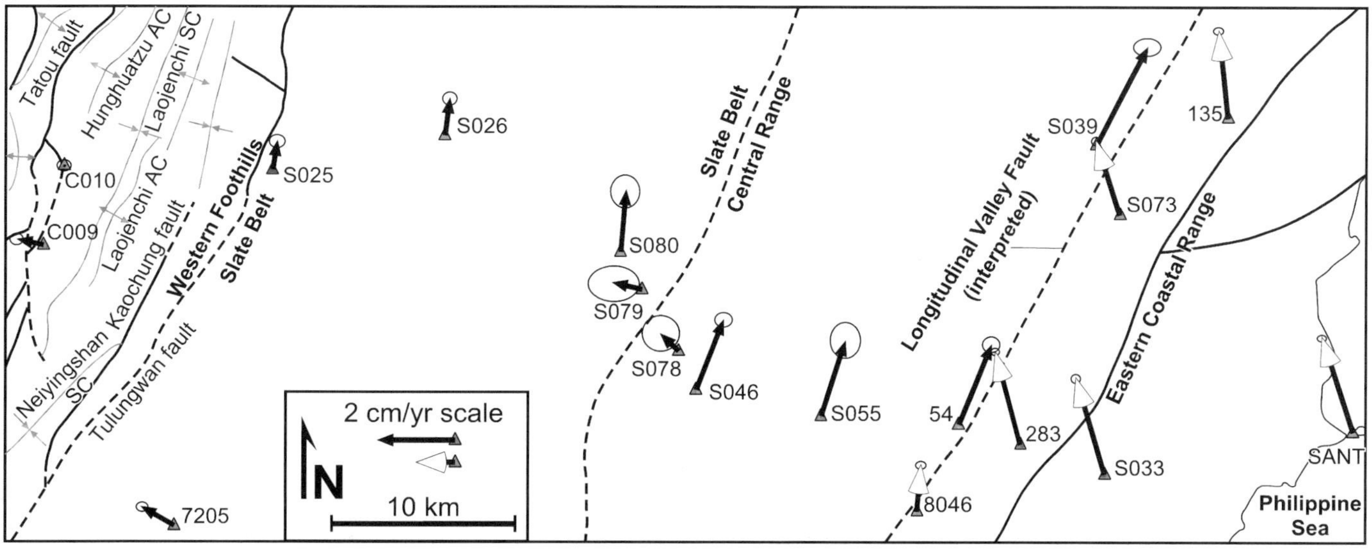

Figure 14. Velocity vectors plotted with respect to site C010 for analysis of local structures (location shown in Fig. 9). Vectors east of the Longitudinal Valley fault (hollow arrows) have been reduced by 50% in order to display their increased magnitudes. The scale is the same for both the velocity and the error.

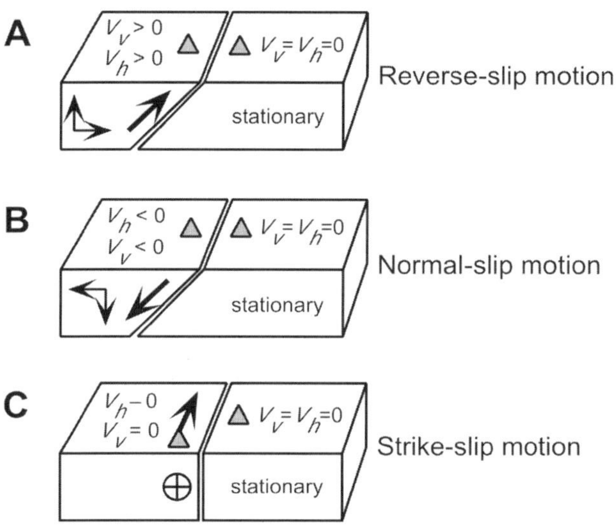

Figure 15. Horizontal (V_h) and vertical (V_v) surface motions of fault-block movements. This analysis assumes that no active internal deformation (folding or shear) is present near the fault plane. Positive vertical and horizontal motions are up and to the right, respectively.

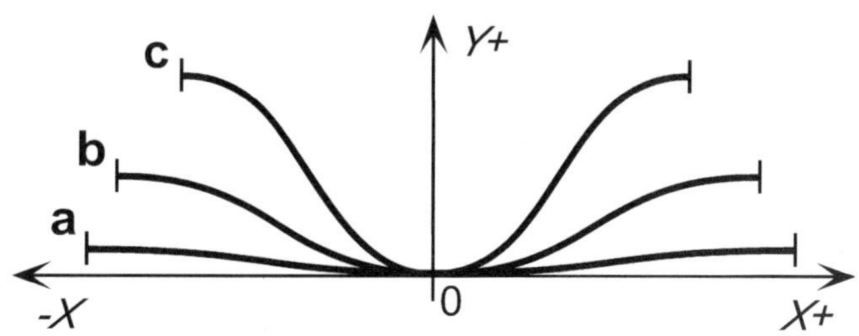

Figure 16. Migration of adjacent anticline axes during synclinal fold growth, under the assumption that no uniform pure or simple shear is applied. Axis of syncline is used as reference. Curves a–c represent progressive development of one folded surface. See text for discussion.

tures are active. The nature of the Western Foothills–Slate Belt boundary is especially problematic in light of data to date. It is a high-angle reverse fault, but the nature of the obliquity is open to interpretation, as was previously mentioned in relation to the individual maps.

Folds. As a simple fold grows by a constant rate of amplitude increase (i.e., constant vertical velocity at the anticlinal axis), the two anticlinal axes approach each other at first rapidly, but then decelerate (Fig. 16). On the other hand, for constant horizontal velocity, the rate of vertical motion at the two endpoints would be at a maximum just after initiation and decrease asymptotically as the fold amplitude reaches a maximum (see Fig. 16, curve c). The limbs of the folds would display similar motions, but of lesser magnitude.

In map view, a growing *detachment fold* in which the forward synclinal axis is held fixed would display maximum horizontal rates in a band in the hinterlandward syncline where horizontal sliding above a basal detachment takes place (see Fig. 17A). Between rear and forward synclinal axes, the horizontal motion decreases across the anticline. Maximum uplift rates would be displayed in a band across the growing anticlinal axis and decrease toward the surrounding synclinal axes. At the base of the foreland limb of the fold, both the horizontal and vertical motion decrease to zero.

In a situation where more than one fold exists in sequence, the horizontal motions become additive across strike. The relative displacement rates would increase with increasing distance from the fixed reference point, across the active folds. Under the assumption that the folds were all active and had the same amplitude, vertical motion across this sequence would oscillate between zero (synclinal axes) and the maximum (anticlinal axes). These patterns are based upon relative movement and would exist regardless of the absolute rate of deformation in the area.

These motions are more complicated in *fault-propagation folds*. These are often asymmetric folds that formed from movement along a propagating blind thrust, causing the overlying strata to deform to accommodate the movement of the hanging wall (see Fig. 17B). The displacement along the fault ramp decreases to zero at the fault tip, beneath the fault-propagation fold. As the hanging wall moves up the growing ramp, the displacement is accommodated above the fault through the shearing of the foreland limb of the fold (Jamison, 1987). Folding of this type causes surface motions that are parallel to the basal detachment-fault surface (horizontal sliding) on the hinterland side of the fold. Material moving up lower parts of the ramp (see Fig. 17B) undergoes uplift accompanying motion toward the foreland. Both vertical and horizontal velocities diminish toward the foreland fold's axial surface. Horizontal and vertical components of motion go to zero at the base of the forelimb, roughly in line with the propagating fault tip (Mitra, 1990). Forelandward of this point, there is no motion associated with this fold-and-fault structure.

The surface motion of the hanging wall of a *fault-bend fold* (Suppe, 1983) as it folds over the ramp will parallel the fault surface (see Fig. 17C). This geometry would give a domain of horizontal motion before the ramp is crossed, a domain of inclined motion above the ramp, and horizontal motion toward the foreland of the ramp (see Fig. 17C). Ramp locations and angles can potentially be determined by the surface vectors.

Given the surface dips and the orientation of the Niushan anticline with respect to the Liuchungchi fault (see Fig. 12), movement in its vicinity may be reflecting the growth of a southwest-plunging fault-propagation fold. The Niushan anticline is the down-plunge extension of the Liuchungchi fault. Other folds are more difficult to interpret. We have interpreted the Yuching syncline as a ramp-related back-limb syncline (Line B–B′, Fig. 5). It is clearly growing. Therefore, the bounding structures must be moving relative to each other. That is, the Chutouchi anticline and footwall of the Chutouchi fault are approaching the Wushantou anticline and immediate hanging wall of the Lunhou fault.

Critical wedge. In critical-wedge models (e.g., Davis et al., 1983), three mechanical states can exist, as shown in Figure 18

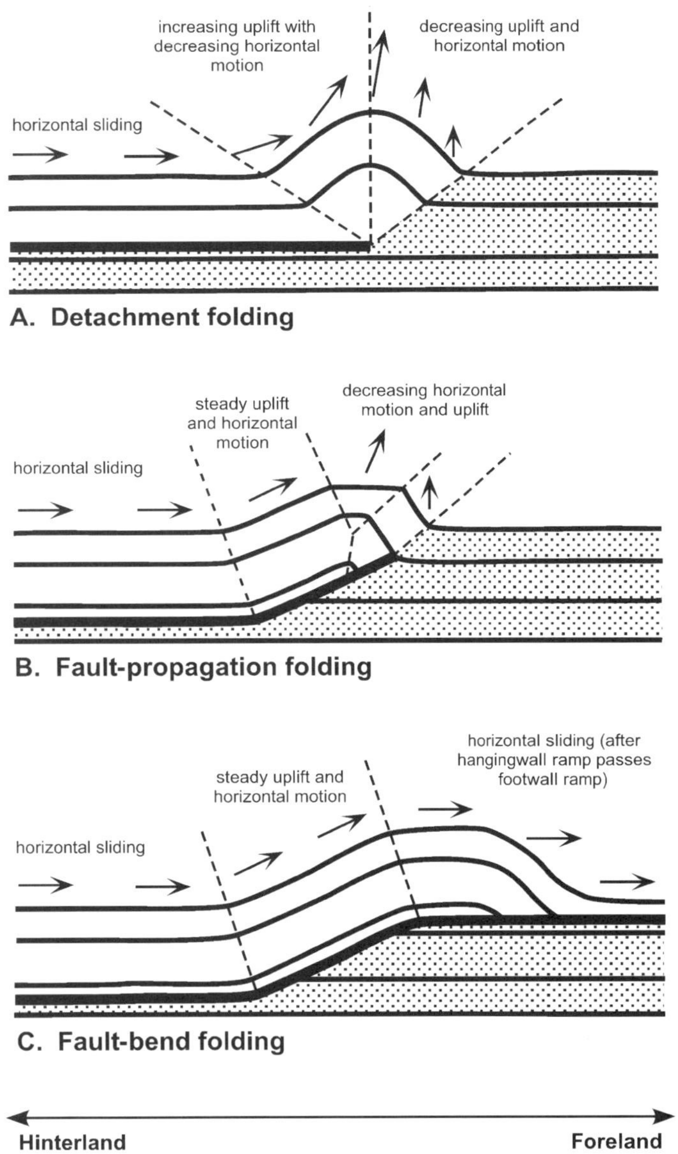

Figure 17. Geometric orientations and surface-motion vectors for the three main types of fold formation mechanisms in fold-and-thrust belts. Several variations of ramp and fold-axis orientations are possible. Shaded regions are undeformed and display no movement.

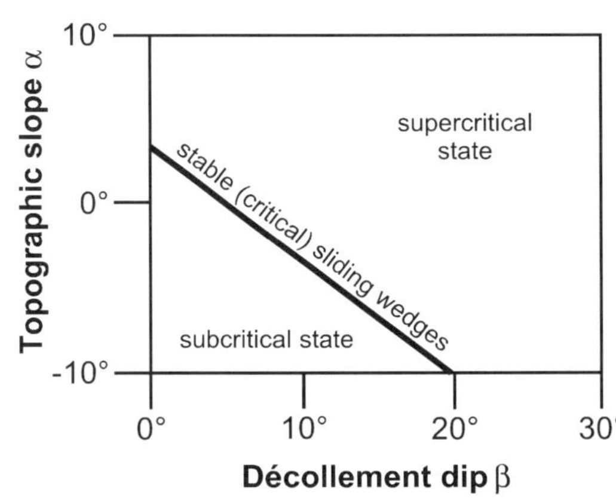

Figure 18. Wedge states as a function of α (topographic slope) and β (basal slope). A plastic wedge in the critical state can slide stably along its entire base. Wedges that are subcritical do not have sufficient topographic slope to overcome basal friction under the entire wedge. Wedges in the supercritical state are gravitationally unstable because their topographic slope is too large. Modified from Willett (1992).

(Willett, 1992). In the critical state, the ratio of the surface slope (α) to the basal slope (β) is such that a plastic wedge can slide stably along the entire length of its basal detachment. This stable state can be maintained in a growing wedge provided that the wedge grows in size without a change in shape. This state would produce surface motions that are both parallel to the basal fault and, in the absence of erosion, of equal magnitude across the entire surface of the wedge (Fig. 19). A wedge in the subcritical state is not at the yield point for sliding motion along its entire length. Consequently, the rear of the wedge moves faster than the toe until shortening has created sufficient thickening to raise the surface slope up to that required for the wedge to enter the critical state. At the toe of a subcritical wedge, no horizontal movement, and little or no uplift would occur at the surface; erosion would enhance this pattern. The rear of the wedge would undergo uplift and horizontal displacement until sufficient thickening had occurred to locally reach a stable, critical state. This zone of movement would then migrate toward the toe until the entire wedge had reached the critical state (Willett, 1992). The surface slope of a wedge that has undergone a weakening of the basal layer, or has accreted material at depth through underplating, can become too large to maintain its topographic slope. As a result, the wedge would enter the supercritical state and become gravitationally unstable. The wedge may undergo either an extensional collapse of the overthickened zone, rapid propagation of the wedge toe (effectively reducing the overall angle by increasing area), or a combination of the two (Platt, 1986; Willett, 1992).

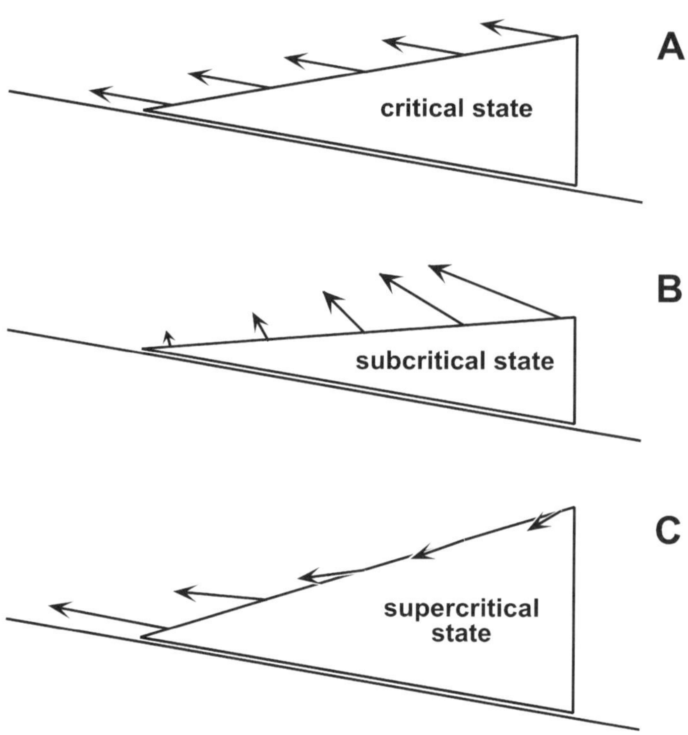

Figure 19. Surface motions of various wedge states relative to the subducting plate. A: A critical wedge has surface motions parallel to the basal fault. B: A subcritical wedge shows reduced motion at the toe and increased surface uplift. C: A wedge has entered a supercritical state in which collapse and forward extension of the wedge occur.

The first-order statement about the current motion of Taiwan is that the general decrease in velocity, relative to Eurasia (site S01R) suggests that the Taiwan wedge is not stably sliding as a block. Instead, the accretionary wedge is subcritical, in the terminology of Willett (1992), with significant internal deformation. Culminations in deformation appear to be occurring in the Longitudinal Valley, i.e., the plate boundary (Yu and Chen, 1994), and in the frontal part of the fold-and-thrust belt associated with the Chukou-Lunhou fault systems.

CONCLUSIONS

The Taiwan wedge is internally complex in both structure and motion. Deformation is neither homogeneous, nor distributed evenly across its surface. Fault movement and/or fold growth is not restricted to a single location. The motion observed resembles that of subcritical wedge growth, i.e., velocities uniformly decreasing with distance toward the foreland. Some of the active structures are the following (from west to east).

1. Liuchungchi fault. This fault is moving on (or near) the southern end. There is an average horizontal velocity increase of 1.25 cm/yr (from west to east) across the fault and its associated folds (Niushan and Nanliao anticlines and the Chientapu syncline). The motions observed along these structures support the active fault-propagation fold interpretation of the cross sections presented here.

2. The Yuching syncline. This fold is actively growing. The limbs of the syncline are both moving toward the axial surface and are being uplifted at a greater rate than the nearby axis.

3. Michih left-lateral fault. There exists a displacement transfer zone between the Chutouchi fault (and Yuching syncline) and the Chukou and Liuchungchi faults (and Niushan, Chientapu, and Nanliao folds), along the Michih left-lateral fault. Displacement south of the strike-slip fault is restricted to a narrow zone around the Yuching syncline and the Chutouchi thrust fault. The southern end of the Chutouchi fault is also active.

4. Lungchuan and Gutingkeng faults. This fault system is the active deformation front to the south. This set of faults continues the narrow zone of deformation south from the Yuching syncline and Chutouchi fault.

5. Slate Belt–Western Foothills boundary. The Slate Belt is rising rapidly along the Western Foothills boundary. The character of the horizontal component of motion appears to change along strike. It is not a simple sinistral strike-slip fault. This low convergence with the Western Foothills suggests that motion across the boundary takes place on a near-vertical fault.

ACKNOWLEDGMENTS

We owe a special debt of gratitude to Yu Shui-Beih of the Institute of Earth Sciences, Academia Sinica, Taiwan, Republic of China, for both providing data from the permanent GPS sites he administers and being of great help throughout our project. Also, thanks to Liu Chi-Ching of the Academia Sinica for loaning us equipment, helping in the field, and sharing data. We also express our thanks to the Central Weather Bureau, Minister of Transportation, Taiwan, Republic of China, for allowing us access to their permanent station data. And we thank Layne Bryant for help with the figures. This work was supported by National Science Foundation grants EAR-9506433 and EAR-9725570 and a grant from the Institute for Pacific Asia, Texas A&M University (to Wiltschko) and National Science Council, Republic of China, NSC86-2116-M-008-008Y (to J.-H. Hung).

REFERENCES CITED

Barrier, E., and Angelier, J., 1986, Active collision in eastern Taiwan: The Coastal Range: Tectonophysics, v. 125, p. 39–72.

Biq, C., 1989, The Yushan-Hsuehshan megashear zone in Taiwan: Proceedings of the Geological Society of China, no. 32, p. 7–20.

Byrne, T. and Crespi, J.M., 1997, Introduction to the geology of Taiwan, *in* Chu, H.-T., and Lee, J.-C., eds., International Conference and Sino-American Symposium on Tectonics of East Asia, Field Guidebook: Taipei, Taiwan, Republic of China, p. 5–10.

Chi, W.R., 1990, Biostratigraphy and correlation of Neogene formations in Taiwan: Miaoli, Taiwan, Republic of China, Chinese Petroleum Company Report.

Chinese Petroleum Company, 1986, Geologic map of western Taiwan, Chiayi Sheet: Miaoli, Taiwan, Taiwan Petroleum Exploration Division (TPED), scale 1:100 000, 1 sheet.

Chinese Petroleum Company, 1989, Geologic map of western Taiwan, Tainan Sheet: Miaoli, Taiwan, Taiwan Petroleum Exploration Division (TPED), scale 1:100 000, 1 sheet.

Crespi, J.M., 1996, Deformation partitioning at shallow crustal levels in the Taiwan arc-continent collision zone: Journal of the Geological Society of China, v. 39, p. 143–150.

Crespi, J.M., Chan, Y.-C., Swaim, M.S., 1996, Synorogenic extension and exhumation of the Taiwan hinterland: Geology, v. 24, p. 247–250.

Davis, D., Suppe, J., and Dahlen F.A., 1983, Mechanics of fold-and-thrust belts and accretionary wedges: Journal of Geophysical Research, v. 88, p. 1153–1172.

Davis, D., and Engelder, T., 1985, The role of salt in fold-and-thrust belts: Tectonophysics, v. 119, p. 67–88.

Dahlen, F.A., Suppe, J., and Davis, D., 1984, Mechanics of fold-and-thrust belts and accretionary wedges: Cohesive coulomb theory: Journal Geophysical Research, v. 89, p. 10087–10101.

Dong, D., Herring, T.A., and King, R.W., 1998, Estimating regional deformation from a combination of space and terrestrial geodetic data: Journal of Geodesy, v. 72, p. 200–214.

Feigl, K.L., Agnew, D.C., Bock, Y., Dong, D., Donnellan, A., Hager, B., Herring, T.A., Jackson, D.D., Jordan, T.H., King, R.W., Larsen, S., Larson, K., Murray, M.H., Shen, Z., and Webb, F.H., 1993, Space geodetic measurement of crustal deformation in central and southern California, 1984–1992: Journal of Geophysical Research, v. 98, p. 21677–21712.

Herring, T.A., Davis, J.L., and Shapiro, I.I., 1990, Geodesy by radio interferometry: The application of Kalman filtering to the analysis of very long baseline interferometry data: Journal of Geophysical Research, v. 95, p. 12561–12581.

Ho, C.S., 1986, A synthesis of the geologic evolution of Taiwan: Tectonophysics, v. 125, p. 1–16.

Ho, C.S., 1988, An introduction to the geology of Taiwan and explanatory text of the geologic map of Taiwan, (second edition): Taipei, Taiwan, Republic of China, Central Geological Survey, Ministry of Economic Affairs, 192 p.

Hofmann-Wellenhof, B., Lichtenegger, H., and Collins, J., 1994, Global Positioning System: Theory and practice: Newark, New Jersey, Springer-Verlag, 355 p.

Hu, J.-C., Angelier, J., and Yu, S.-B., 1997, An interpretation of the active deformation of southern Taiwan based on numerical simulation and GPS studies: Tectonophysics, v. 274, p. 145–169.

Hung, J.-H., Wiltschko, D.V., Huang, H.-C., Hickman, J.H., Fang, P., and Bock, Y., 1999, Structure and motion of the southwestern Taiwan fold and thrust belt: TAO, Terrestrial, Atmospheric and Ocean Sciences, v. 10, p. 543–568.

Jamison, W.R., 1987, Geometric analysis of fold development in overthrust terranes: Journal of Structural Geology, v. 9, p. 207–219.

Lin, H.C., 1996, Structural geology in the Foothills belt of Chianan area, southwestern Taiwan [M.S. thesis]: Chungli, Taiwan, National Central University, 104 p.

Mitra, S., 1990, Fault-propagation folds: Geometry, kinematic evolution, and hydrocarbon traps: American Association of Petroleum Geologists Bulletin, v. 74, p. 921–945.

Neill, A.E., Global mapping functions for the atmospheric delay: Journal of Geophysical Research, v. 101, p. 3227–3246.

Platt, J.P., 1986, Dynamics of orogenic wedges and the uplift of high-pressure metamorphic rocks: Geological Society of America Bulletin, v. 97, p. 1037–1053.

Stanley, R.S., Hill, A.B., Chang, H.C., and Hu, H.N., 1981, A transect through the metamorphic core of the Central Mountains, southern Taiwan: Memoir of the Geological Society of China, v. 4, p. 443–473.

Suppe, J., 1980, Imbricate structure of Western Foothills belt, south-central Taiwan: Petroleum Geology of Taiwan, v. 17, p. 1–16.

Suppe, J., 1981, Mechanics of mountain building and metamorphism in Taiwan: Memoir of the Geological Society of China, v. 4, p. 67–89.

Suppe, J., 1983, Geometry and kinematics of fault-bend folding: American Journal of Science, v. 283, p. 684–721.

Suppe, J., 1984, Kinematics of arc-continent collision, flipping of subduction, and back-arc spreading near Taiwan: Memoir of the Geological Society of China, v. 6, p. 21–34.

Teng, L.S., 1990, Geotectonic evolution of late Cenozoic arc-continent collision in Taiwan: Tectonophysics, v. 183, p. 57–76.

Tillman, K.S., and Byrne, T., 1995, Kinematic analysis of the Taiwan slate belt: Tectonics, v. 14, p. 322–341.

Wessel, P., and Smith, W.H.F., 1998, New, improved version of the generic mapping tools released: EOS (Transactions, American Geophysical Union), v. 79, p. 579.

Willett, S.D., 1992, Dynamic and kinematic growth and change of a Coulomb wedge, *in* McClay, K.R., ed., Thrust tectonics: London, Chapman and Hall, p. 19–31.

Woodward, N.B., 1987, Geological applicability of critical-wedge thrust belt models: Geological Society of America Bulletin, v. 99, p. 827–832.

Wu, F.T., Rau, R.-J., and Salzberg, D., 1997, Taiwan orogeny: Thin-skinned or lithospheric collision?: Tectonophysics, v. 274, p. 191–220.

Yu, S.-B., and Chen, H.-Y., 1994, Global Positioning System measurements of crustal deformation in the Taiwan arc-continent collision zone: TAO, Terrestrial, Atmospheric and Ocean Sciences, v. 5, p. 477–498.

Yu, S.-B., Chen, H.-Y., and Kuo, L.-C., 1997, Velocity field of GPS stations in the Taiwan area: Tectonophysics, v. 274, p. 41–60.

Zhao, W.L., Davis, D.M., Dahlen, F.A., and Suppe, J., 1986, Origin of convex accretionary wedges: Evidence from Barbados: Journal of Geophysical Research, v. 91, p. 10246–10258.

Manuscript Accepted by the Society April 12, 2001

Printed in the U.S.A.

Geological Society of America
Special Paper 358
2002

Taiwan Slate Belt: Insights into the ductile interior of an arc-continent collision

Donald M. Fisher
Department of Geosciences, Pennsylvania State University, University Park, Pennsylvania 16802, USA
Chia-Yu Lu
Institute of Geology, National Taiwan University, Taipei, Taiwan, Republic of China
Hao-Tsu Chu
Central Geological Survey, Taipei, Taiwan, Republic of China

ABSTRACT

The Slate Belt geologic province in Taiwan consists of three structural provinces with markedly different structural styles, histories, and finite strains: the Hsüehshan Range, the western Backbone Range, and the eastern Backbone Range. The Hsüehshan Range is a regional pop-up structure consisting of a biotite-grade core bounded on the west by the east-dipping Chüchih thrust fault and on the east by the west-dipping Lishan fault, a passive-margin growth fault on which the slip direction has been inverted. The western Backbone Range is marked by west-vergent folds, a moderately southeast-dipping cleavage, and a downdip stretching lineation. Pressure shadows and chlorite-mica aggregates record 40%–460% extension during noncoaxial strain histories consistent with top-to-the-west shear where pre-Tertiary basement has been thrust over the Slate Belt. In contrast, the stretching lineation in the eastern Backbone Range plunges gently to the northeast, with a finite elongation of 180%–500%, subparallel to fold axes and the strike of structural fabrics. Thus, from west to east across the Backbone Range, the extension direction changes from downdip to along strike. In the Slate Belt, a comparison of ductile strain indicators with brittle structures, seismicity, and GPS (Global Positioning System) data indicates that there is vertical strain partitioning in the eastern part of Taiwan, whereby the ductile regime has significantly different kinematics than the near-surface brittle regime.

INTRODUCTION

Slate belts record regional patterns of finite strain and strain histories that yield insights into the kinematics associated with mountain building. In this paper, we describe structural histories and regional strain variations within the Slate Belt geologic province of Taiwan and use these data to evaluate the kinematics within the ductilely deformed interior of the collision zone between the Luzon arc and the Chinese passive continental margin. The Taiwan Slate Belt contains many of the incremental and finite-strain markers typical of ancient slate belts, yet the Taiwan slates were deformed and exhumed within an ongoing collision, and samples can be collected along the synorogenic exposure surface. The upper-crustal kinematics of the active collision, based on GPS (Global Positioning System) observations (Yu et al., 1997) and earthquake focal mechanisms (Yeh et al., 1991), can be compared with the ductile kinematics inferred from deformation fabrics and strain indicators. Moreover, the plate velocities are well known; GPS measurements and the NUVEL-1 model of global plate motion predict that

Fisher, D.M., Lu, C.-Y., and Chu, H.-T., 2002, Taiwan Slate Belt: Insights into the ductile interior of an arc-continent collision, *in* Byrne, T.B., and Liu, C.-S., eds., Geology and Geophysics of an Arc-Continent Collision, Taiwan: Boulder, Colorado, Geological Society of America Special Paper 358, p. 93–106.

the Philippine Sea plate moves to the northwest (~305°–310°) at a rate of 70–86 km/m.y. relative to the Eurasian plate (Seno et al., 1993; Yu et al., 1997). This relative plate-motion vector is 15°–20° clockwise with respect to the orthogonal to the mountain belt.

There is also an obliquity between the northeast-trending Chinese margin and the north-trending volcanic arc (Fig. 1). The obliquity has resulted in propagation of the mountain belt to the southwest at a rate of ~55 km/m.y. (Byrne and Crespi, 1997). Because the polarity of subduction reverses between central and northern Taiwan, the collision is no longer active in northernmost Taiwan, and there is a northeast to southwest transition from postcollision to active collision tectonics (Suppe, 1984). Within the mountain belt, stratigraphic thickness variations reflect depositional patterns along the Chinese passive margin, which are largely controlled by normal faults that define a system of basement highs and intervening depocenters filled with growth strata (e.g., Teng et al., 1991) (Fig. 1).

Herein, we evaluate the observed variations in strain within the Taiwan Slate Belt and how these patterns relate to the obliquity of plate motions, the obliquity of the collision, and the passive-margin architecture. We define the Slate Belt as the areas within the Taiwan arc-continent collision that expose penetratively cleaved Tertiary rocks originally deposited along the Chinese passive continental margin (Fig. 2). Given this definition, there are three distinct tectonic and physiographic provinces within the Slate Belt: (1) the Hsüehshan Range in northern and central Taiwan, (2) the western Backbone Range, and (3) the eastern Backbone Range in southern Taiwan (Fig. 2). The structural history recorded by rocks within the Slate Belt reflects only the ongoing arc-continent collision and not Mesozoic deformation and metamorphism that may have occurred in the pre-Tertiary basement (i.e., as recorded in the Tananao Schist).

In the following sections, we describe the stratigraphic framework and structural history for each of the three regions of the Slate Belt. Regional patterns of finite strain and incremental strain are combined to evaluate the kinematics associ-

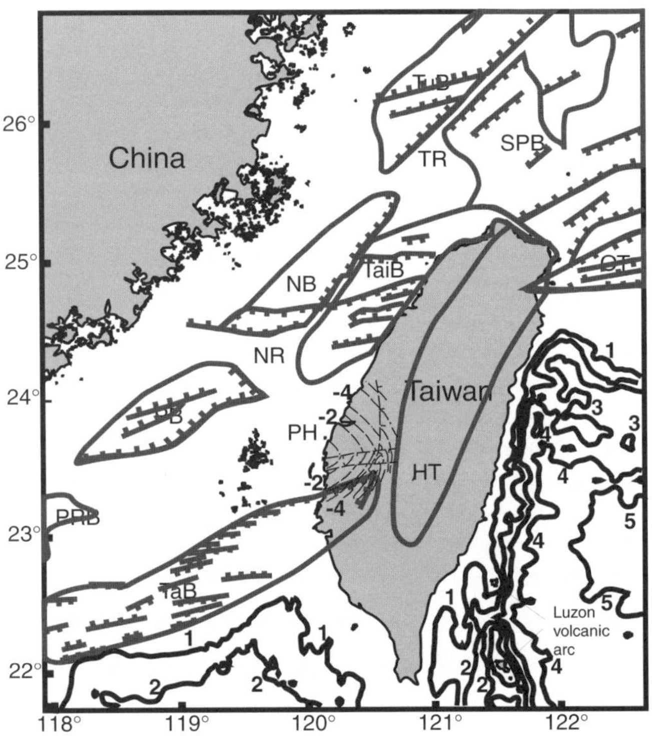

Figure 1. Map showing Taiwan collision between north-trending Luzon volcanic arc and Chinese passive continental margin. Fault-bounded depocenters and basement highs are labeled: HT—Hsüehshan trough; NB—Nanjihto Basin; NR—Nanjihtao Ridge; OT—Okinawa Trough; PB—Peikang Basin; PH—Peikang basement high; PRB—Pearl River Basin; SPB—south Pengchiahsu Basin; TaB—Tainan Basin; TaiB—Taishi Basin; TR—Tungyintao Ridge; TuB—Tungyintao and Tahchentao Basins. Bathymetric contours (in kilometers) are labeled in bold. Contours (in kilometers) of the depth to basement are shown for the Peikang High with light dashed lines. Note the northeast trend of the continental shelf and associated depocenters and the north trend of the Luzon arc south of Taiwan.

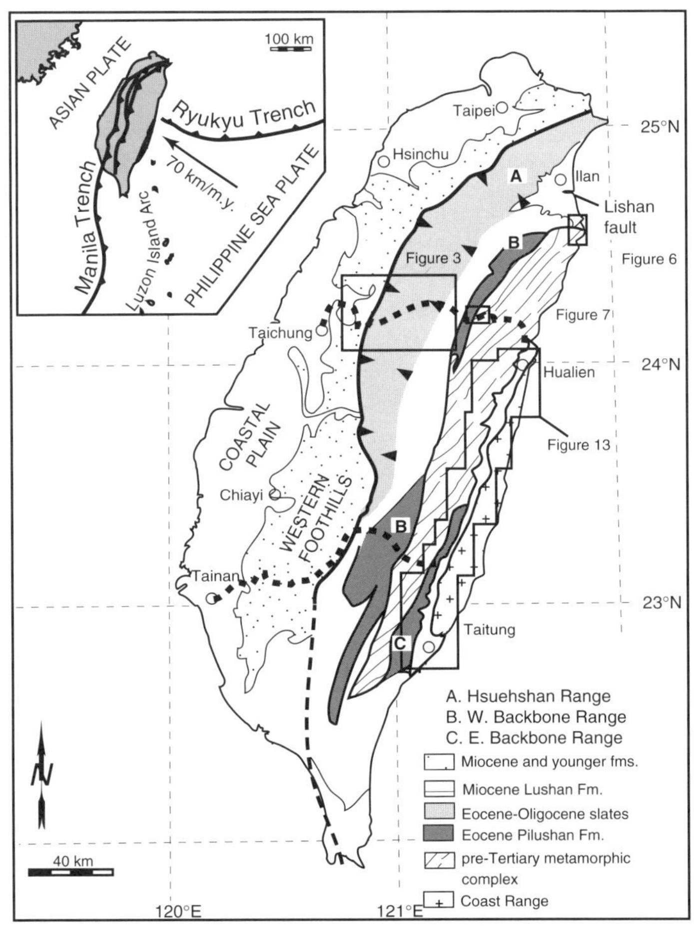

Figure 2. Geologic map of Taiwan showing major tectonostratigraphic units of the Slate Belt. Central and Southern Cross-Island Highways are shown by the dotted lines. Inset shows tectonic framework.

ated with cleavage development. These results are then used to assess the importance of horizontal and vertical strain partitioning across the Taiwan mountain belt in the context of the known plate-convergence vector, the orientations of colliding elements, and the positions of precollision depocenters and intervening basement highs.

HSÜEHSHAN RANGE: A REGIONAL-SCALE POP-UP STRUCTURE

Stratigraphy of the Hsüehshan Range

The Hsüehshan Range is composed of an Eocene to Miocene passive-margin shallow-marine sequence of sandstone, quartzite, and argillite metamorphosed at prehnite-pumpellyite to lower-greenschist facies (Fig. 3) Most of the quartzite layers in this unit retain stratal continuity and can be traced for tens of kilometers. The sequence is characterized by transgressive facies and rift facies (Lu et al., 1991; Sun, 1982; Teng et al., 1991; Wang, 1987) and was deposited as growth strata in a major normal-fault–bounded basin similar to other basins along the Chinese passive margin that are aligned parallel to the trend of the continental shelf (Sun, 1982; Teng et al., 1991) (Fig. 1). Some of these faults have been reactivated during Pleistocene compression in the Western Foothills as reverse faults or tear faults (Bonilla, 1975; Suppe, 1986).

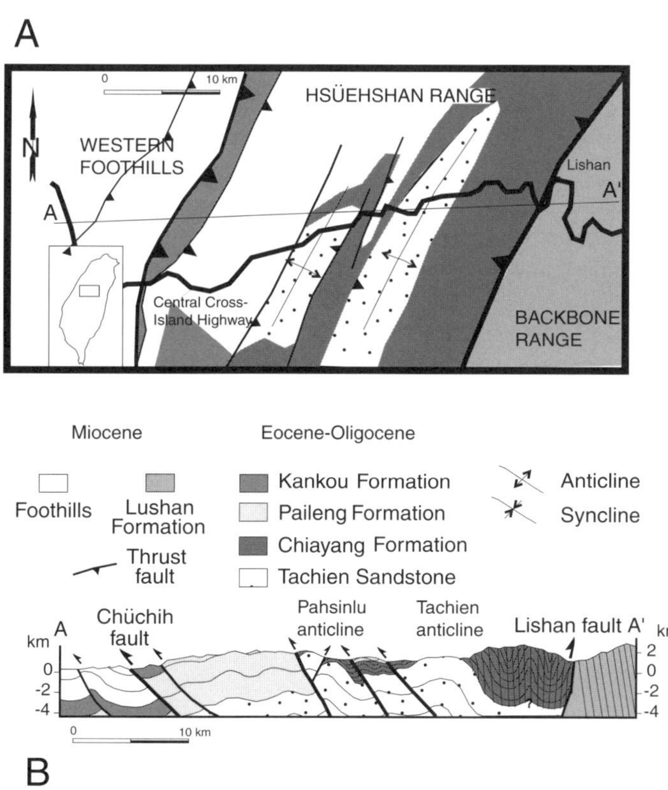

Figure 3. (A) Geologic map of the Lishan fault area in northern Taiwan (after Ho, 1986). (B) Structural cross section across the Lishan fault zone.

The major stratigraphic units in the study area include the Tachien Sandstone and the overlying Chiayang and Paileng Formations (Chen, 1977). The Tachien Sandstone (Chen, 1977) is a thick, coarse-grained, nonfossiliferous sandstone unit (Chen, 1977; Lee, 1987). The Chiayang Formation consists chiefly of black to dark-gray slate intercalated with very fine to fine-grained metasandstone and metasiltstone. Ripple marks are the most abundant sedimentary structures. Although fossils do not allow a definitive age determination, some of the planktonic fossils suggest that deposition of the Chiayang Formation may have extended into the early Aquitanian (or earliest Miocene) (Chang, 1971). The Chiayang Formation thins to the west, where it interfingers with and is completely replaced by the Paileng Formation (Ho, 1988; Lee, 1977). The Paileng Formation is composed of fine- to coarse-grained quartzitic sandstone with minor slaty interbeds. These rocks are nearly void of fossils, but the Paileng Formation is considered to be Eocene–Oligocene or Oligocene in age on the basis of its stratigraphic position (Ho, 1988). In some areas of the Hsüehshan Range, these formations or their stratigraphic equivalents are overlain by Miocene sedimentary deposits whose ages are well determined by fossil data.

Structure of the Hsüehshan Range

The regional structure of the Hsüehshan Range is defined by a culmination that exposes a core of Eocene–Oligocene Tachien Sandstone flanked to the west and east by younger rocks of the Paileng Formation and the Chiayang Formation, respectively (Fig. 3). To the west, the rocks of the Hsüehshan Range are thrust over the Western Foothills along the east-dipping Chüchih fault. To the east, the range is bounded by the steeply dipping Lishan fault. Along the Central Cross-Island Highway, there are two major structural highs that expose Tachien Sandstone—the Tachien anticline and the Pahsinlu anticline. The core of the Tachien anticline exposes rocks with metamorphic biotite porphyroblasts (Clark et al., 1993; Yen, 1973). The isograd associated with the down-section appearance of biotite lies entirely within the massive Tachien Sandstone. This isograd is folded and is observed on both limbs of the fold.

Fold kinematics in the Tachien anticline are characterized by flexural slip; numerous bedding-parallel slickenlined surfaces and slickenlines are oriented downdip and perpendicular to subhorizontal fold axes (Clark et al., 1993). Fibrous steps on bedding-parallel faults indicate a thrust sense of motion on both limbs of anticlines. Thin shale horizons within the Tachien Sandstone are zones of local flexural shear that have resulted in carbonaceous gouge zones (Clark et al., 1993). In the coarse-grained sandstone, the pressure-solution cleavage has a spacing ranging from 0.5 to 5 cm.

Mesoscale folds in the Chiayang and Paileng Formations

are open to tight and upright, with near-vertical, southeast-dipping axial planes and subhorizontal northeast-trending fold axes (Clark et al., 1993; Lu et al., 1997; Tillman and Byrne, 1995; Tsan, 1976). Cleavage in these formations is defined by selvages of carbonaceous material along with parallel preferred orientation of phyllosilicates and elongate quartz grains. The cleavage spacing ranges typically from 0.02 to 0.1 mm and averages 0.05 mm. The cleavage domains are composed of phyllosilicates, whereas the microlithons are composed of quartz, feldspar, muscovite, and minor opaque minerals (Lee, 1987). The cleavage is typically axial planar to mesoscale folds and dips steeply to the southeast. A downdip lineation on cleavage planes is defined by alignment of syntectonic fibrous overgrowths of quartz and secondary phyllosilicates. To the west of the Tachien anticline, faults and shear zones are dominantly east dipping.

Cleavage development occurred late in the structural history of the Hsüehshan Range and resulted in crenulations of bedding-parallel fault surfaces and overgrowths around pyrite framboids, quartz grains, and the biotite porphyroblasts in the core of the Tachien anticline. Thus, the Hsüehshan Range records a history of metamorphism followed by flexural-slip folding, and finally late limb tightening, with flattening and cleavage development. Late brittle structures such as conjugate transcurrent faults separate the mountain range into blocks. These structures indicate a heterogeneous stress distribution, but most localities record a maximum compressive stress direction at high angles to the curvilinear mountain belt and the indenting Peikang basement high (Lu et al., 1997).

The eastern boundary of the Hsüehshan Range is the Lishan fault, a major structure defined by the contact between the Eocene–Oligocene Chiayang Formation of the Hsüehshan Range and the Miocene Lushan Formation of the western Backbone Range (Clark et al., 1993; Lu et al., 1997). Both units are steeply dipping near the contact. The Lishan fault represents a reversal in the regional pattern of eastward-increasing age and metamorphic grade from the weakly metamorphosed rocks of the Coastal Plain to the greenschist-facies slates of the Hsüehshan Range. The fault coincides with a sharp eastward drop in illite crystallinity (Chen, 1981, 1984), consistent with the significantly different thermal histories indicated by fission-track studies (Hsieh, 1990; Liu, 1982, 1988). The fault also marks an abrupt change from the thick Oligocene sequence of the Hsüehshan Range to the absence of Oligocene units in the Backbone Range (Teng et al., 1991). The distribution of Tertiary strata in the Slate Belt is consistent with the interpretation that the Lishan fault acted as a passive-margin growth fault that separated an outer-shelf basement high from a regional middle Tertiary depocenter marked by a Bouguer gravity low (Teng et al., 1991). Thus, the Lishan fault likely originated as a normal fault that juxtaposed the Oligocene strata of the Hsüehshan Range and older rocks of the basement high.

The Hsüehshan Range, as defined by exposure of Eocene–Oligocene rocks west of the Lishan fault, has a constant width from the northern tip of Taiwan southward to the Central Cross-Island Highway. Farther south, the range tapers to a termination near the Southern Cross-Island Highway (Fig. 2). The region over which the width of the Hsüehshan Range decreases is coincident with the southern margin of the Tertiary depocenter and is inboard of the northern flank of the underthrusting Peikang basement high as well as a major embayment in the thrust front of the Western Foothills thrust belt (Figs. 1 and 2). The Tachien and Pahsinlu structural culminations lie in the thickest parts of the depocenter, between the Peikang and Kuanyin basement highs (Lu, 1991).

Incremental and finite strain within the Hsüehshan Range

Pyrite pressure shadows with syntectonic quartz and phyllosilicate fibers are abundant within the Chiayang Formation and shale interbeds within the Tachien Sandstone. These fabrics record information about the shape and orientation of the finite-strain ellipsoid as well as the strain path. In cases where fibers grow continuously without development of an open crack (Durney and Ramsay, 1973; Fisher and Brantley, 1992; Li and Means, 1995; Means and Li, 1995), individual fibers within pyrite pressure shadows track displacement of the matrix with respect to the pyrite and can be used to reconstruct an incremental strain history (Elliott, 1972; Durney and Ramsay, 1973; Beutner and Diegel, 1985; Ellis, 1986; Ramsay and Huber, 1983; Wickham, 1973). In cleavage-perpendicular sections within the Hsüehshan Range, fibers are restricted to the quadrants of incremental extension at the surface of pyrite sphere; peripheral fibers are typically shorter or asymptotic about the central fiber of the pressure shadow (Fig. 4). In cleavage-parallel sections, straight fibers bracket the pyrite; all fibers are of equal length, and the outer margin of the pressure shadow is the same shape as the surface of the pyrite. These observations are consistent with plane strain, with no extension or shortening in the cleavage plane perpendicular to fibers.

Strain histories were evaluated in sections cut perpendicular to cleavage and parallel to the stretching lineation. For most of the Hsüehshan Range, pressure shadows in cleavage-perpendicular sections record coaxial extension parallel to the downdip stretching lineation; fiber bundles are asymptotic about a straight, central fiber (Clark et al., 1993) (Fig. 4). The magnitude of extension (e) varies from $+1.3$ to $+4.5$; some of the highest values are from the western part of the Chiayang Formation where deformation is more concentrated within slate layers interbedded with sandstones. Strain histories in the Hsüehshan Range indicate pure shear, with subhorizontal shortening and near-vertical extension (Clark et al., 1993). Strain magnitudes are not sufficient to explain the near parallelism of bedding and cleavage throughout the Chiayang Formation, an observation that is consistent with the interpretation that the penetrative strain recorded by fibers came in the later stages of flexural-slip folding and limb rotation. Assuming constant-volume, plane-strain deformation, the strain magnitudes re-

Figure 4. Photomicrograph of pressure shadow from the Hsüehshan Range in section cut perpendicular to the steep southeast-dipping cleavage and parallel to the downdip lineation (larger pyrite framboid is ~50 μm in diameter).

Figure 5. (A) Photomicrograph of pressure shadow from near the Lishan fault in section cut perpendicular to cleavage and parallel to the lineation (viewed to the northeast) (pyrite framboid is ~45 μm in diameter). (B) Photomicrograph of pressure shadow in cleavage-parallel section from near the Lishan fault. Note that fibers are straight and bracket the pyrite, indicating no shortening or extension parallel to the finite intermediate stretch direction (i.e., plane strain) (pyrite framboid is ~65 μm in diameter).

corded by fibers indicate >30 km of shortening across the Hsüehshan Range that accumulated by grain-scale deformation during the late flattening of these existing folds.

The Hsüehshan Range also has local regions that display noncoaxial strain histories. For example, a 20-m-wide, east-dipping shear zone between the Tachien and Pahsinlu anticlines shows gradual clockwise rotation of the extension direction in cross sections viewed to the north. On the basis of the presence of other kinematic indicators such as asymmetric folds and shear bands, Tillman and Byrne (1995) interpreted this strain history as a reflection of counterclockwise internal vorticity related to west-directed thrusting.

Near the Lishan fault, pressure shadows record a counterclockwise rotation of the extension direction (Fig. 5A). In cleavage-parallel sections, fiber geometries are indicative of plane strain, with extension parallel to the downdip stretching lineation (Fig. 5B). These results suggest internal vorticity related to northwest-side-up motion on a steeply northwest-dipping reverse fault along which the Oligocene Chiayang Formation is juxtaposed against the Miocene Lushan Formation (Clark et al., 1993). This interpretation is also consistent with the sharp break in thermal histories across the Lishan fault. The Hsüehshan Range is thus best described as a regional pop-up structure—a culmination cored by biotite-grade rocks with west-directed thrusts on the northwest side, an east-directed reverse fault on the southeast side, and a record of penetrative pure shear throughout (Clark et al., 1993). Motion on the Lishan fault—whose hanging wall accumulated a thick sequence of Oligocene sedimentary deposits during normal faulting along the Chinese passive continental margin—was inverted during the arc-continent collision. The pure shear of the Hsüehshan Range is related to shortening of the depocenter between the basement high on the southeast side of the Lishan fault (Tillman and Byrne, 1995) and the Peikang and Kuanyin basement highs to the southwest and northeast.

WESTERN BACKBONE RANGE: A REGION OF TOP-TO-THE-WEST SHEAR

Stratigraphy of the western Backbone Range

The western Backbone Range is generally composed of two rock units: (1) Eocene slates composed mainly of dark gray phyllite with interbeds of whitish to dark-colored metasandstones, limestone lenses, and irregular conglomerate bands and (2) Miocene slates, composed of dark gray argillites or phyllites with occasional interbeds of dark gray sandstone and disseminated marly nodules (Fig. 2). Ho (1986) divided the Backbone Range into two stratigraphic units: the Miocene Lushan Formation and the Eocene Pilushan Formation, but he added that the distinction between these two formations has to be made largely on fossil evidence because there is very little difference in lithology. Both units were metamorphosed under prehnite-pumpellyite- to lower-greenschist-facies conditions (Chen et al., 1983a).

The lateral continuity of sedimentary units in the western Backbone Range is limited, and few beds can be traced for more than 100 m. Fossils in a number of scattered exposures have ages that range widely from the Cretaceous (Yen et al., 1956), Eocene (Chang, 1972), Oligocene (Hashimoto and Matsumaru, 1975), and late Oligocene (Huang, 1980; Lee, 1977) to the Miocene (Chang, 1973, 1974, 1975). Minor occurrences of igneous rocks such as pillow lavas, diorite, and gabbro are observed as well as minor andesitic tuffs indicating synsedimentary arc volcanic activity (Chen, 1991). In many localities along the eastern flank of the Backbone Range, the Lushan Formation, like the Pilushan Formation, directly overlies the Paleozoic or Mesozoic basement, and thus the basement-cover contact must be an unconformity or a fault zone (Chen et al., 1983b; Suppe et al., 1976; Tsan, 1977; Yeh, 1998).

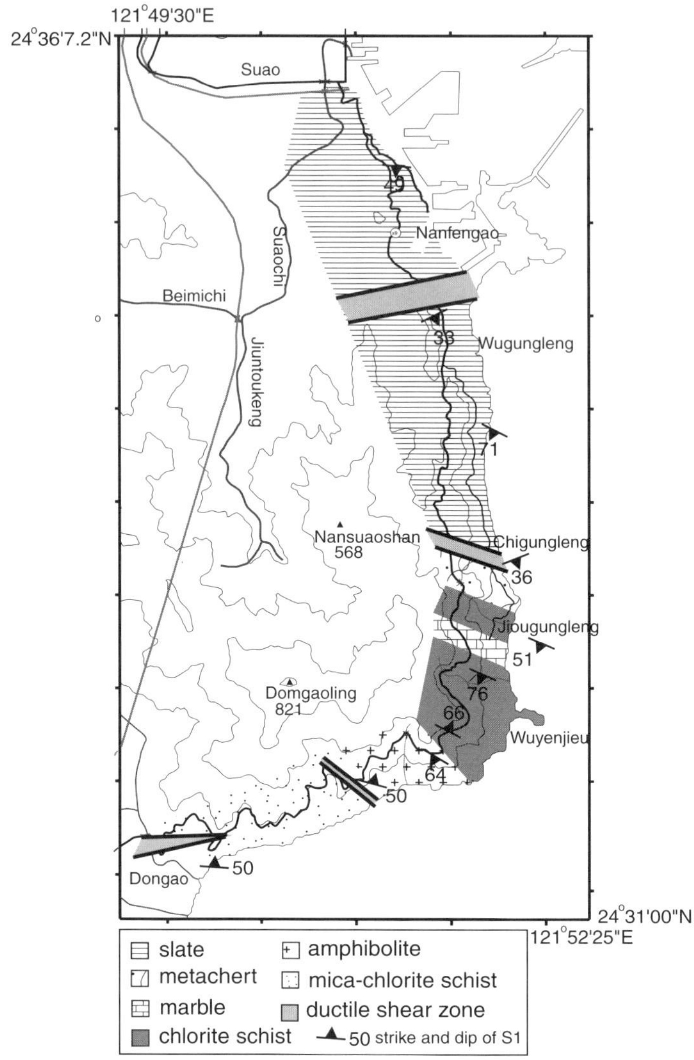

Figure 6. Geologic map of the Suao region, northern Taiwan, showing measurements of orientation of the dominant foliation.

Structure of the western Backbone Range

The western Backbone Range is typified by northwest-vergent folds with moderately southeast-dipping axial planes and northeast-trending fold axes. Folds are tight to isoclinal, and cleavage varies from axial planar to convergent fanning in sandstones and divergent fanning in slates. The cleavage generally dips to the southeast (Figs. 6 and 7). In coarser lithologies, the cleavage is defined by thin layers of insoluble minerals and a planar or anastomosing fabric consisting of phyllosilicate-rich and quartz-rich domains. Finer-grained lithologies have a pervasive, planar slaty cleavage defined by a strong planar preferred orientation of secondary phyllosilicates. The slaty cleavage is associated with a downdip lineation that is parallel to the stretching direction indicated by distributed fibrous overgrowths and elongate chlorite-mica aggregates. Several regional thrust faults have been recognized that postdate peak metamorphism and cleavage development and place the Eocene Pilushan Formation over the Miocene Lushan Formation (Tillman and Byrne, 1996). These out-of-sequence faults have resulted in steepening of the earlier fabrics and have clearly played a significant role in thickening this part of the wedge. From west to east across the western Backbone Range, there is also an appearance of a crenulation cleavage that is steeper than the earlier slaty cleavage.

The eastern boundary of the Slate Belt in the western Backbone Range is the east-dipping contact with the pre-Tertiary metamorphic basement that consists of phyllites, schists, marbles, and quartzites of the Tananao Schist. Suppe et al. (1976) argued that the contact is depositional on basis of the occurrence of overturned Tertiary conglomerate beds along the contact and the observation along the Suao Highway that basement units are truncated at a low angle by the Tertiary strata. There is also very little increase in metamorphic grade across the boundary. More recent structural data from the Suao and Central Cross-Island Highways show that the contact zone has been the site

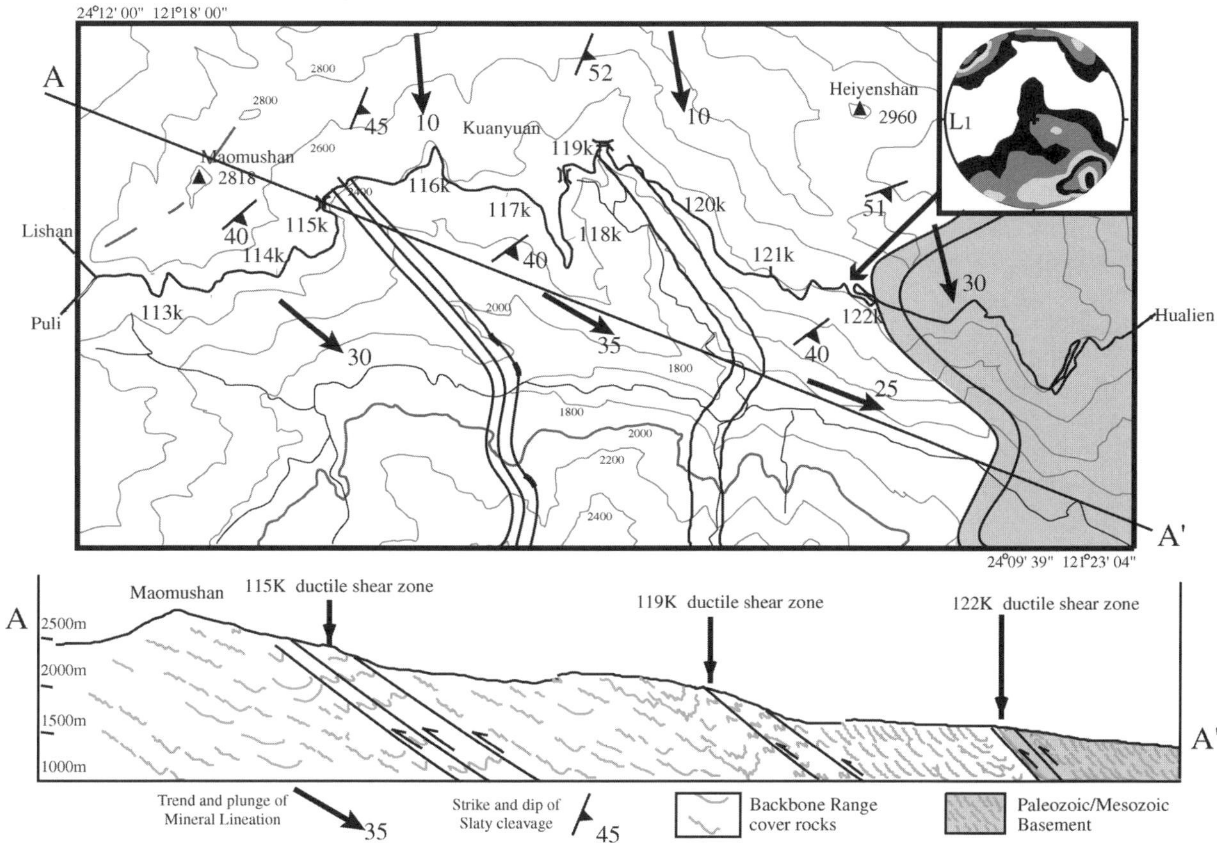

Figure 7. Geologic map of the Central Cross-Island Highway in the vicinity of the basement-cover contact. Contour interval, 200 m. Shear zones are defined by position with respect to the kilometer markers along the highways (labeled as a number plus "k"). C-axis contour plot is shown for an Eocene quartz pebble conglomerate cut perpendicular to cleavage and parallel to the stretching lineation (L1). Contour interval, 2 σ, 300 measurements.

of top-to-the-west overthrusting of pre-Tertiary strata over the Eocene rocks of the Slate Belt. Mylonites within the shear zone display numerous kinematic indicators that show a thrust sense of motion. A quartz *c*-axis fabric from a quartzite along the contact shows a type I cross girdle typical of (001) ⟨100⟩ slip (e.g., Schmid and Casey, 1986), with an asymmetry consistent with top-to-the-northwest thrusting (Fig. 7).

Incremental and finite strain in the western Backbone Range

The western Backbone Range has abundant pyrite pressure shadows in Miocene and Eocene rocks along the Suao, Central, and Southern Cross-Island Highways. There are multiple phases of deformation within the Backbone Range, but fibrous pressure shadows in the slates mainly record the early deformation related to development of the main schistosity. Throughout the western Backbone Range, strain histories associated with this deformation are noncoaxial in sections cut perpendicular to cleavage and parallel to the stretching direction, with a progressive clockwise rotation of the extension direction when viewed to the northeast (Fig. 8A). In cleavage-parallel sections, pressure-shadow geometries are consistent with no strain in the intermediate stretching direction, and extension parallel to the downdip lineation (Fig. 8B). The external rotation recorded by fibers in cleavage-perpendicular sections has been attributed to internal vorticity related to northwest-directed thrusting of pre-Tertiary Tananao Schist over the Tertiary units of the slate (Clark and Fisher, 1995; Clark et al., 1992; Tillman and Byrne, 1995). This interpretation is consistent with other kinematic indicators such as asymmetric boudins, asymmetric folds, and quartz *c*-axis preferred orientations. Along the Southern Cross-Island Highway, Tillman and Byrne (1995) also showed that the magnitude of extension is lower in rocks of the westernmost Backbone Range but that there is an abrupt increase in extension magnitude in the Eocene Pilushan Formation to the east. The low-strain rocks record smaller external rotations and larger angles between bedding and cleavage, a result that also suggests variations in the magnitude of counterclockwise angular shear related to west-directed thrusting.

Chlorite-mica aggregates with large aspect ratios are observed in Eocene–Miocene rocks of the Backbone Range along the Suao and Central and Southern Cross-Island highways (Fig. 9). The basal planes of the phyllosilicates that make up these

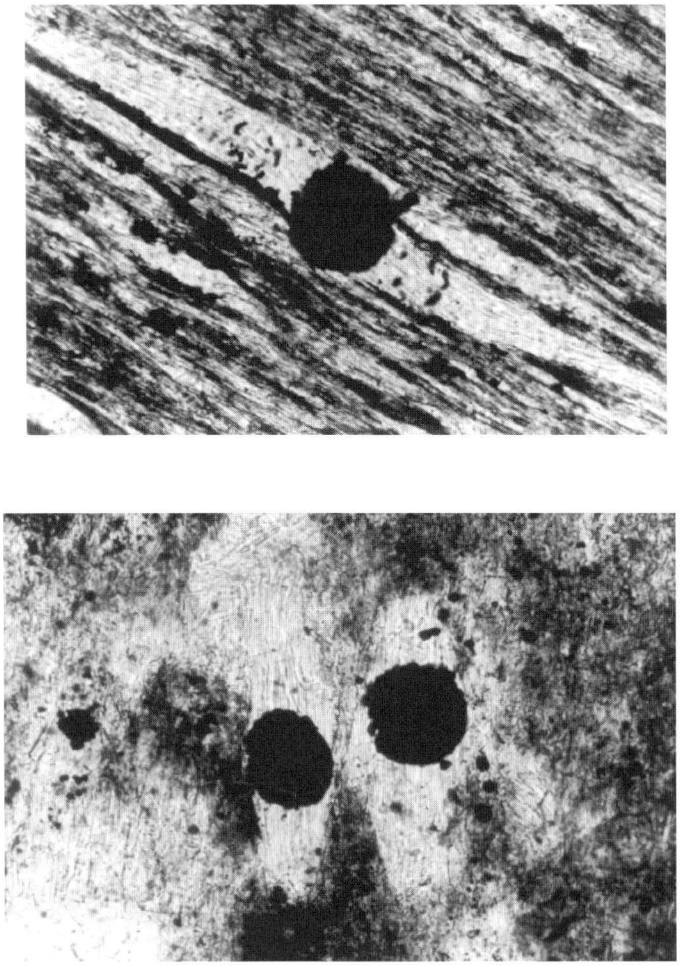

Figure 8. (A) Photomicrograph of a pressure shadow from the Pilushan Formation in the western Backbone Range in a section cut perpendicular to cleavage and parallel to the lineation (viewed to the northeast) (pyrite is ~50 μm in diameter). (B) Photomicrograph of pressure shadows from the western Backbone Range in a section cut parallel to cleavage (pyrite on left is ~150 μm in diameter). Extension is parallel to a downdip mesoscopic stretching lineation. Note that fibers are parallel and fiber bundles show no shortening in the strike direction (horizontal on photomicrograph). Fibers consist of mixtures of quartz and phyllosilicates.

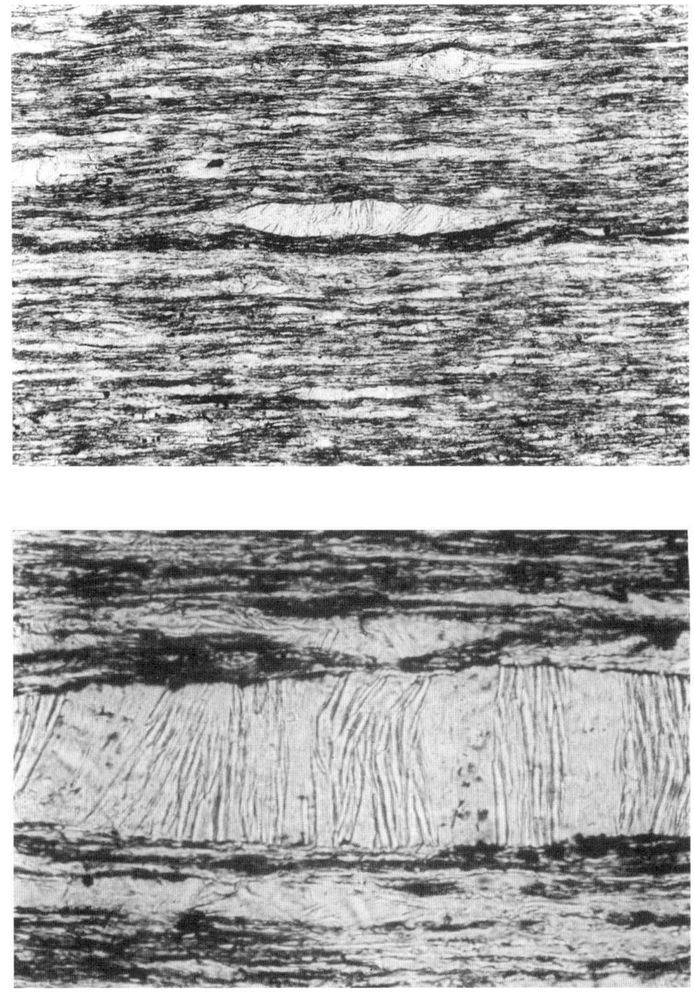

Figure 9. Photomicrographs of chlorite-mica aggregates from the western Backbone Range in sections cut perpendicular to the cleavage and parallel to the lineation. (A) Horizontal field of view is 2.5 mm. (B) Horizontal field of view is 500 μm.

aggregates are typically oriented at a high angle to the cleavage and the long axis of the aggregate. Thus, there is both a shape and a crystallographic preferred orientation for the aggregates. Clark and Fisher (1995) showed that these aggregates are porphyroclasts that grow from detrital seed grains by progressive crack-seal deformation. Chlorite or white mica porphyroclasts have undergone distributed cracking parallel to basal planes, and cracks are filled with a higher-Mg chlorite, mica, and quartz. This crack-seal growth results in changes in the major (*a*), minor (*c*), and intermediate (*b*) axes that define the ellipsoidal shapes of porphyroclasts (Fig. 10). Initially, detrital micas are elongate parallel to the mineral cleavage (001) with $a = b > c$. Growth of aggregates by crack-seal deformation leads to a reversal in the major and minor axes as porphyroclasts become elongate perpendicular to basal planes (i.e., the *z* direction) and parallel to the slaty cleavage (Fig. 10).

There is no growth of chlorite-mica aggregates in the *b* direction, given that pressure shadows indicate plane-strain deformation in the western Backbone Range. Thus, the absolute length of the *b*-axis of porphyroclasts gives an indication of the initial size of porphyroclasts in this direction and a maximum initial dimension for the *a*-axis of porphyroclasts. The observed shape of aggregates thus reflects the strain history in three dimensions and can be used to place constraints on the orientation and magnitude of finite strain. Because these porphyroclasts are stronger than the surrounding fine-grained matrix, the cracks that develop within them are sites of precipitation, and the porphyroclasts undergo volume increase. Because addition of material within cracks produces extension perpendicular to basal planes of phyllosilicates, the porphyroclasts grow initially par-

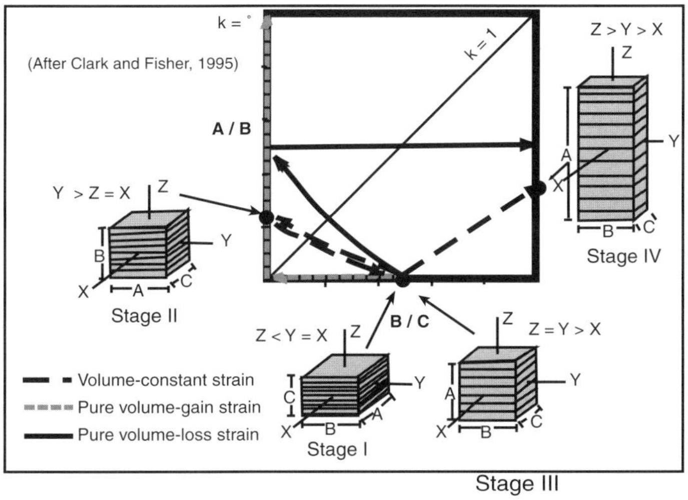

Figure 10. A schematic Flinn diagram showing the paths that a grain with initial dimensions $a = b > c$ would follow for volume-constant strain (dashed line), pure volume loss (solid bold line), and pure volume gain (gray line); x, y, and z refer to the crystallographic axes of chlorite-mica aggregates, with z perpendicular to basal (001) planes. Block diagrams show aggregate dimensions relative to these crystallographic axes at different points in the history.

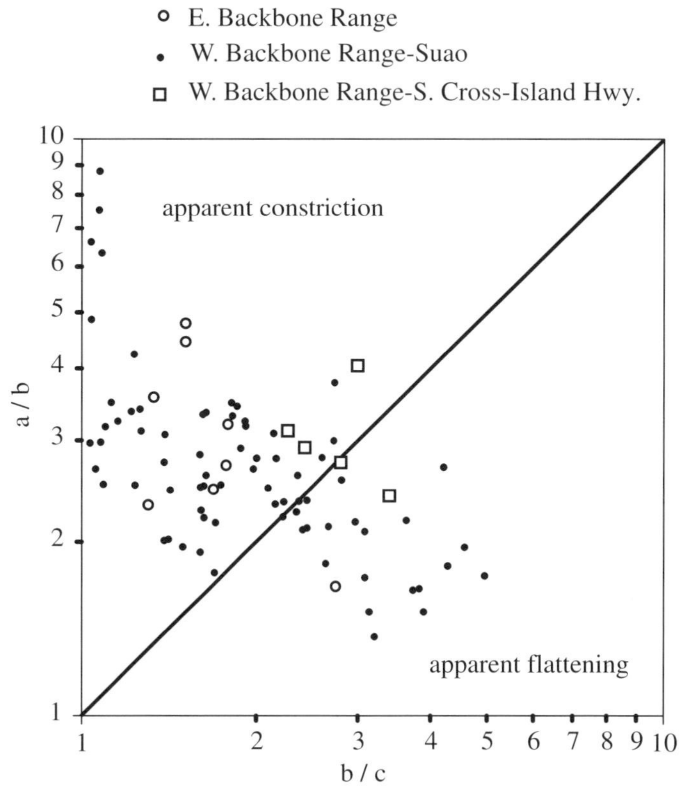

Figure 11. Flinn diagram depicting the three-dimensional shape of aggregates in the western (solid dots) and eastern (open dots) Backbone Range.

allel to the c-axis and later parallel to the a-axis when the dimension perpendicular to basal planes (z) exceeds the dimensions of the other two porphyroclast axes (Fig. 10).

Chlorite-mica aggregates in the western Backbone Range show a variety of shapes from prolate to oblate, but in general the long axis of aggregates is parallel to the stretching lineation as well as the finite elongation direction as determined by pyrite pressure shadows (Fig. 11). Values of a/b for porphyroclasts range from 2.4 to 4.0 along the Southern Cross-Island Highway and from 1.5 to 8.8 along the Central Cross-Island and Suao Highways, with higher values within shear zones (Fig. 12, A and B). The axial ratio b/c varies from 1.0 to 5.0 (Fig. 11).

Clark and Fisher (1995) argued that this range of oblate and prolate shapes reflects a strain history that involves an overall volume increase for porphyroclasts due to growth in the z direction, with some shortening along the c-axis due to dissolution at the clast margins. The range of observed porphyroclast shapes is not consistent with pure volume-increase or constant-volume deformation of porphyroclasts (Fig. 10). The final shape (i.e., prolate or oblate) is a function of the initial shape and the amount of porphyroclast volume increase. Given an initial porphyroclast shape of $a = b > c$ and a deformation history that involves growth parallel to z, a sample with prolate porphyroclasts can be explained relative to samples with oblate porphyroclasts in terms of a greater increase in volume or a lower initial b/c value for detrital seed grains. Most of the observed three-dimensional shapes of porphyroclasts require increases in volume for reasonable estimates of initial shape. Thus, the more competent porphyroclasts underwent crack-seal growth and a spinning coaxial deformation while the surrounding ductile matrix underwent grain-scale diffusive mass transfer and internal vorticity related to top-to-the-northwest thrusting.

The magnitude of finite elongation (e) recorded by fibrous overgrowths in the western Backbone Range is +0.4 to +4.6, with lower values to the west. These values are comparable to the a/b ratios indicated by chlorite-mica aggregates. Throughout the western Backbone Range, the direction of maximum finite elongation as determined by both pressure shadows and chlorite-mica aggregates lies approximately parallel to the downdip stretching lineation with a moderate plunge to the southwest. The incremental extension direction for the last increment of deformation is near vertical or more steeply plunging to the southeast than the finite stretching direction. There is a locally developed crenulation cleavage with a trace on cross sections that is parallel to the final extension increment. The angle between this final incremental extension direction and the slaty cleavage is greater in samples with higher strain (Tillman and Byrne, 1995), an observation that supports a model of top-to-the-west shear. The magnitude of the strain in the Slate Belt is greatest in the eastern part of the western Backbone Range near the shear zone at the thrust contact between the pre-Tertiary basement and the Eocene slates of the Pilushan Formation.

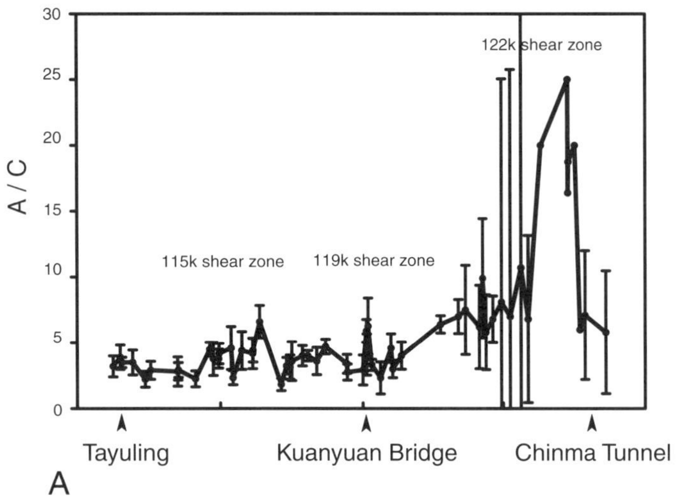

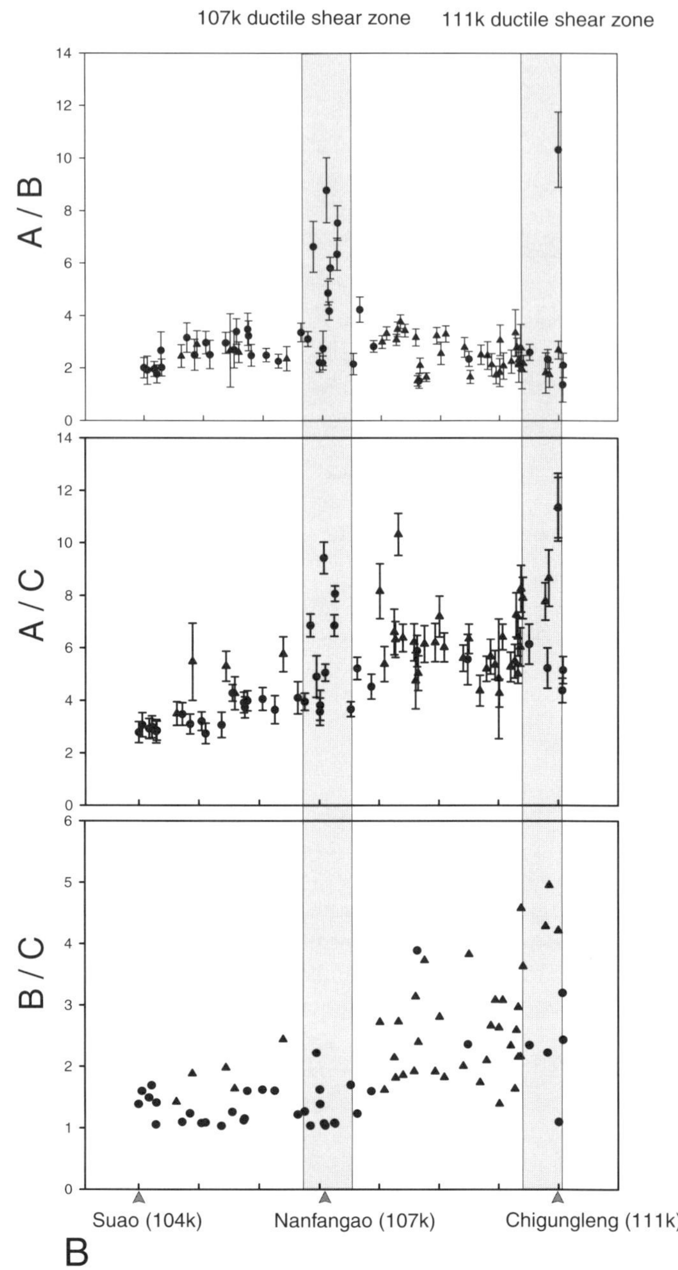

Figure 12. Axial ratios of chlorite-mica aggregates from the western Backbone Range along the (A) Suao and (B) Central Cross-Island Highways. Axial ratios are higher in shear zones. Shear zones are defined by position with respect to the kilometer markers along the highways.

EASTERN BACKBONE RANGE: A REGION OF OROGEN-PARALLEL EXTENSION

Stratigraphy of the eastern Backbone Range

Eocene slates of the Pilushan Formation are exposed in the eastern Backbone Range south of the Southern Cross-Island Highway (Figs. 2 and 13). There is a northeast to southwest change from deeper to shallower structural levels in the eastern Backbone Range with pre-Tertiary basement in the northeast, the Eocene Pilushan Formation in the middle, and finally Miocene rocks in the southwest. The structural observations in this paper refer to the Pilushan Formation.

Structure of the eastern Backbone Range

The eastern Backbone Range is marked by development of west-vergent mesoscale folds, an axial-planar slaty cleavage, and a stretching lineation that plunges gently to the northeast and is defined by distributed fibrous overgrowths and elongate chlorite-mica aggregates (Fisher, 1999). The slaty cleavage is locally folded, with gentle, northeast-plunging fold axes and an associated crenulation cleavage that varies in orientation from steep to moderately west dipping. Mesoscale secondary folds are concentrated in the short limbs of larger folds and are less common within more prevalent southeast- or east-dipping slaty-cleavage panels. Both folding-related structures are crosscut by west-dipping, brittle normal-fault zones that are spaced tens of meters apart. These fault zones are intensely veined and are sites of active hot springs.

To the north along the mountain front, a west-dipping crenulation cleavage, gentle northeast-plunging lineation, and system of late normal faults have been observed in basement rocks. The pervasive early foliation varies from east dipping in the south to west dipping in the north (Fig. 13). This rotation of the early foliation from west to east and from south to north has been attributed to (1) southward propagation of a regional east-vergent overturned backfold within the retrowedge of the collision (Clark and Fisher, 1995; Clark et al., 1992; Stanley et al., 1981) or (2) rotation along west-dipping normal faults (Crespi et al., 1996). Structural lineations such as fold axes, stretching lineations, and later crenulation axes display a gentle northeast plunge along the length of the mountain front in both the Eocene slates in the south and the pre-Tertiary basement rocks in the north. Thus, later folding events are not likely to

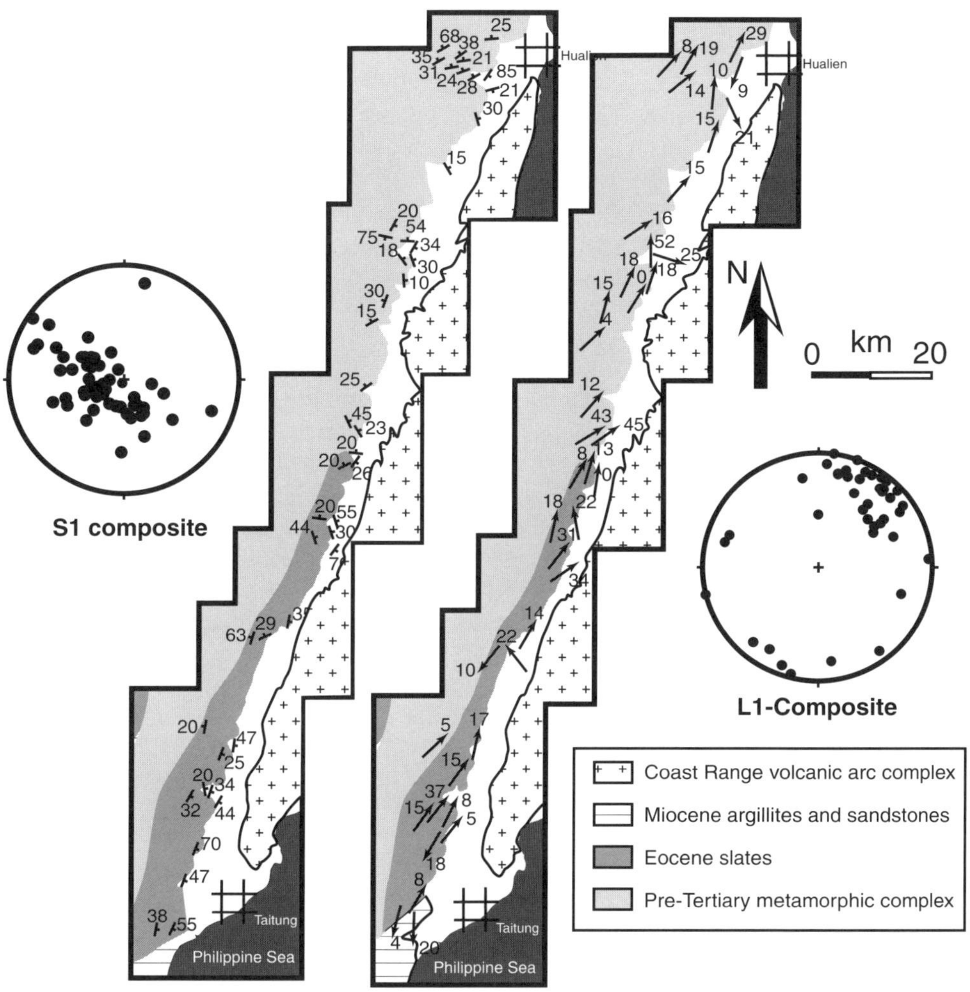

Figure 13. Geologic map of the eastern Backbone Range showing orientations of the cleavage and stretching lineation.

have resulted in significant reorientation of the early stretching lineation as suggested by Faure et al. (1991). The consistency in orientation of early fabrics across the contact between the Eocene slates in the south and the pre-Tertiary basement in the north indicates that the polyphase deformation characteristic of the basement in the eastern Backbone Range is due to the variations in regional kinematics during the arc-continent collision.

Incremental and finite strain in the eastern Backbone Range

Eocene slates of the eastern Backbone Range contain pyrite pressure shadows that record histories of three-dimensional strain with noncoaxial strain in both cleavage-perpendicular and cleavage-parallel sections (Fig. 14, A and B) (Fisher, 1999). These observations indicate that the deformation is not characterized by plane strain, and a single section cannot be used to characterize a complete strain history. Nevertheless, a combination of observations for orthogonal sections depicts many aspects of the strain history associated with the development of the cleavage and lineation. In sections cut perpendicular to cleavage and parallel to the northeast-trending, gently plunging lineation, fibers generally depict a counterclockwise rotation of the extension direction when viewed downward, consistent with highly oblique, southwest-directed thrusting with a component of right-lateral shear (Fisher, 1999). The (minimum) estimates for the extension (e) based on these incomplete histories are $+1.5$ to $+4.5$.

Cumulative-strain histories for cleavage-parallel sections are generally noncoaxial (Fig. 14B); the direction of maximum stretching varies in cleavage-parallel sections. In all cases, the fibers bracket the pyrite spheres, and there is little variation in the width of the pressure shadow from the earliest to latest fiber segments, indicating that there is no *incremental* stretching in the intermediate stretching direction.

The chlorite-mica aggregates generally have an ellipsoidal shape, and in y and z sections, the median values of a/b, a/c, and b/c are 1.8–5.0, 3.0–9.8, and 1.3–2.9, respectively (Fisher, 1999). Measurements of the porphyroclast dimensions display normal or positively skewed distributions with ratios of a/b and b/c that plot in the field of prolate ellipsoids on a Flinn diagram (Fig. 11).

The a/b values from cleavage-parallel sections provide a minimum estimate of the maximum stretch of 1.8–5.0 for these

A

B

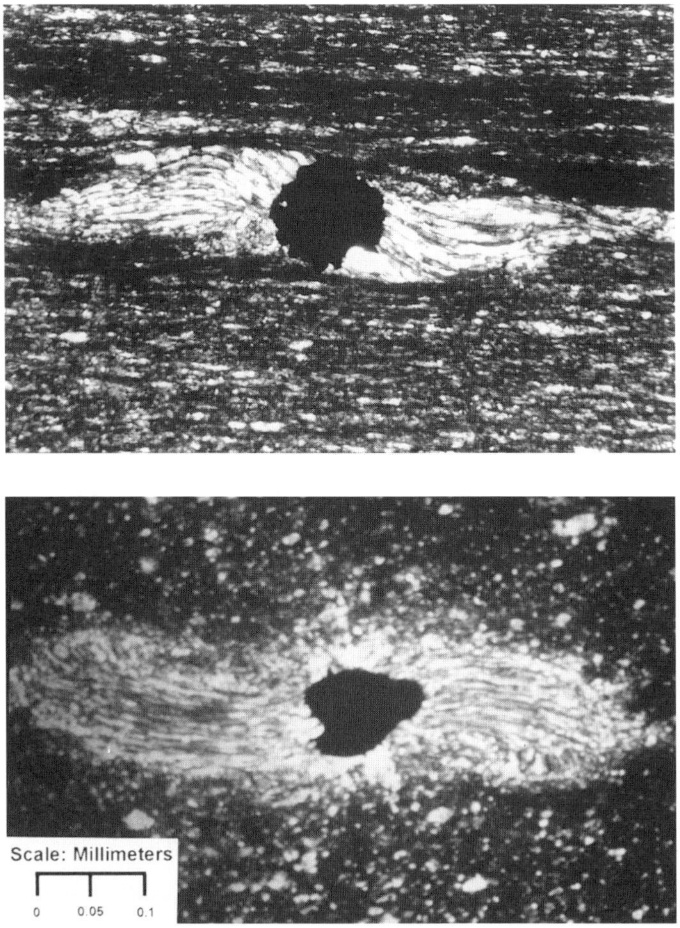

Figure 14. Fibrous pyrite pressure shadows from the slates of the eastern Backbone Range in sections cut (A) perpendicular to the cleavage and parallel to the stretching lineation (viewed toward down on gently dipping sections) and (B) parallel to the slaty cleavage (horizontal on photomicrograph is parallel to strike). Note that the strain history is noncoaxial but adjacent curved fibers within pressure shadows are parallel along their length, suggesting no incremental shortening or extension perpendicular to the fibers in this section. The finite deformation is not plane strain, but there is no intermediate incremental stretch throughout the history. In this case, there is an early event of downdip extension but most of the extension is along strike, nearly parallel to the mesoscopic stretching lineation.

samples. This estimate is a minimum because detrital grains are likely to have initially been elongate parallel to the basal plane (i.e., the *b*-axis). These results indicate that, in the Eocene rocks of the eastern Backbone Range, there is at least 180%–500% extension in a direction that plunges gently parallel to fold axes and the strike of the mountain belt. This noncoaxial deformation is the earliest recognized event in these rocks. The folds related to this deformation are consistently west vergent, and the thrust faults are consistently east dipping; these structures presumably were developed as the older sedimentary rocks of the continental margin were structurally underplated and incorporated into the west-facing collisional prow edge.

CONCLUSIONS

The Slate Belt in Taiwan, as defined by penetratively cleaved Tertiary strata, consists of three structural provinces with distinct structural styles, structural histories, finite strain, and incremental strain histories. The Hsüehshan Range is a regional pop-up structure consisting of a culmination that exposes a biotite-grade core bounded on the west by the east-dipping Chüchih thrust fault and on the east by the west-dipping Lishan fault, an inverted passive-margin growth fault. These faults place Eocene–Oligocene rocks of the Hsüehshan Range over Miocene rocks of the Western Foothills and Backbone Range, respectively. Fibrous pressure shadows record plane strain associated with development of a near-vertical cleavage and downdip stretching lineation, with 130%–450% of vertical coaxial extension consistent with pure shear during the late flattening of upright, flexural slip folds. Noncoaxial strain histories consistent with west-directed thrusting are recorded within east-dipping shear zones of the western Hsüehshan Range, and strain histories consistent with backthrusting are observed near the Lishan fault.

Deformation of the Hsüehshan Range was primarily influenced by the three-dimensional passive-margin architecture and distribution of depocenters. The variations in width, orientation, and exposed structural level of the Hsüehshan Range can be explained in terms of the position of the Peikang and Kuanyin basement highs. Thus, the Hsüehshan Range is a Tertiary depocenter that was squeezed between an outer-shelf basement high and two converging basement highs at the northwest and southeast margins of the range. The outer-shelf high acted as a mechanical backstop at the boundary between pre-Tertiary basement rocks and passive-margin sedimentary rocks (Tillman and Byrne, 1995). The difference in strength across the boundary would decrease as strain accumulated in the Hsüehshan Range because pure shear in conjunction with development of a slaty anisotropy is a strain-hardening path (Fisher, 1990).

The western Backbone Range is marked by west-vergent folds with northeast-trending axes, a moderately southeast-dipping axial-planar or fanning cleavage, and a downdip stretching lineation. Pressure shadows record noncoaxial strain histories consistent with top-to-the-west shear, with an eastward-increasing magnitude of elongation toward a shear zone where pre-Tertiary basement is thrust over the Slate Belt. Pressure shadows and chlorite-mica aggregates record 40%–460% extension, with highest values within discrete shear zones. The lack of any sharp metamorphic break across the contact between pre-Tertiary basement and the Slate Belt reflects the importance of late Tertiary metamorphism and deformation during juxtaposition of the basement rocks and the Tertiary cover.

The eastern Backbone Range exposes Tertiary rocks with slaty cleavage in southern Taiwan. These rocks display southwest-vergent folds with northeast-trending axes and a southeast-dipping axial-planar or fanning cleavage. In contrast to the western Backbone Range, the stretching lineation plunges gently to the northeast, with a finite elongation of 180%–500%

subparallel to fold axes and the structural grain of the mountain belt. From west to east across the Backbone Range, there is a change from downdip extension typical of most of the mountain belt to along-strike extension within the eastern Backbone Range. Variations in orientation of the lineation between the eastern Backbone Range and the western Backbone Range reflect significant spatial variations in the orientation of finite strain associated with slaty-cleavage formation and are not due to post-slaty-cleavage folding.

Indicators of shallow brittle deformation across the Taiwan mountain belt are consistent with strain partitioning, in which the obliquity between the convergence direction and the strike of the mountain belt has been partitioned into components of orthogonal shortening (in the western Backbone Range) and sinistral transpression or strike slip (in the eastern Backbone Range). Sinistral shear is not consistent with the kinematics of Eocene slates within the eastern Backbone Range as recorded by pressure shadows and other kinematic indicators. The recent near-surface velocity distribution in the eastern Backbone Range may not reflect the velocity gradients that produced the ductile fabrics of the eastern Backbone Range, given that these fabrics developed at depth within the west-facing prow edge and may have been advected eastward with respect to the geometry of the double-sided wedge during continued shortening and exhumation.

Lateral extrusion of ductile rocks (e.g., Gomez et al., 1998; Jones et al., 1997) during oblique collision is consistent with the noncoaxial strain histories that show highly oblique southwest-directed thrusting with a component of right-lateral shear. Lateral extrusion is only possible given certain boundary conditions. In the case of Taiwan, lateral or oblique extrusion of ductile continental basement to the southwest may be favored by the steep topographic gradient to the southwest in the transition from the collision zone to the precollision zone.

Brittle structures and low-temperature fabrics that overprint the ductile strain indicators in the eastern Backbone Range depict kinematics that differ from the earlier strain history, with prevalent left-lateral strike-slip faults and normal faults. Similarly, seismicity in the eastern Backbone Range is characterized by focal mechanisms consistent with left-lateral slip along the range (Yeh et al., 1991). The horizontal velocities from GPS studies show a distinctive pattern across southern Taiwan, with motion orthogonal to the deformation front in the Western Foothills and oblique to the front in the eastern Backbone Range (Yu et al., 1997). The corresponding velocity gradients indicate shortening perpendicular to the mountain belt on the northwest side of the Backbone Range, similar to the inferences from ductile-strain indicators in the western Backbone and Hsüehshan Ranges. The southern Backbone Range records extension perpendicular to the mountain belt, and oblique strain axes in the eastern Backbone Range are consistent with left-lateral shear. These results are consistent with observed brittle structures but are not compatible with the earlier strain history. Thus, rocks of the eastern Backbone Range record a history of vertical-strain partitioning whereby the ductile regime had significantly different kinematics than the near-surface brittle regime.

The Slate Belt in the western and eastern Backbone Range shows the same general structural patterns and inferred kinematics as the adjacent pre-Tertiary basement. These similarities include the orientations of the main schistosity and stretching lineation as well as later crenulation cleavages. Thus, the polyphase deformation history characteristic of most of the Backbone Range is predominantly a reflection of the ongoing arc-continent collision and is not restricted to pre-Tertiary rocks. The range of available kinematic indicators and strain markers in the Slate Belt makes it ideal for unraveling the ductile strain associated with mountain-building in Taiwan.

ACKNOWLEDGMENTS

We thank A. Goldstein and D. Cowan for constructive reviews. The research was supported by National Science Foundation grant EAR-9628352.

REFERENCES CITED

Beutner, E.C., and Diegel, F.A., 1985, Determination of fold kinematics from syntectonic fibers in pressure shadows, Martinsburg slate, New Jersey: American Journal of Science, v. 285, p. 16–50.

Bonilla, M.G., 1975, A review of recently active faults in Taiwan: U.S. Geological Survey Open-File Report 75-0041, p. 41–72.

Byrne, T., and Crespi, J., 1997, Kinematics of the Taiwan arc-continent collision and implications for orogenic processes, *in* International Conference and Sino-American Symposium on Tectonics of East Asia: Taipei, Taiwan, p. 38.

Chang, L.S., 1971, A biostratigraphic study of the so-called slate formation in Taiwan based on smaller foraminifera. 1. The E-W Cross-Island Highway: Proceedings of the Geological Society of China, v. 14, p. 45–61.

Chang, L.S., 1972, Eocene/Miocene hiatus and N conglomerate in the Central Range of Taiwan: Proceedings of the Geological Society of China, v. 15, p. 93–98.

Chang, L.S., 1973, A biostratigraphic study of the so-called Slate Formation in Taiwan based on smaller foramanifera. 3. Sankuan Hsuiluan area along the upper courses of the Tanshuiho and the Yulochi: Proceedings of the Geological Society of China, v. 16, p. 69–84.

Chang, L.S., 1974, A biostratigraphic study of the so-called Slate Formation in Taiwan based on smaller foraminifera. 4. Northernmost part of the Central Taiwan: Proceedings of the Geological Society of China, v. 17, p. 85–94.

Chang, L.S., 1975, Biostratigraphy of Taiwan: Geology and Paleontology of Southeast Asia, v. 15, p. 337–361.

Chen, C.-H., 1981, Change of X-ray diffraction pattern of K-micas in the metapelites of Taiwan and its implication for metamorphic grade: Memoir of the Geological Society of China, v. 4, p. 475–490.

Chen, C.-H., 1984, Determination of lower greenschist facies boundary by K-mica—Chlorite crystallinity in the Central Range, Taiwan: Proceedings of the Geological Society of China, v. 27, p. 41–53.

Chen, C.-H., 1977, Some stratigraphic problems of the Hsüehshan Range of Taiwan: Proceedings of the Geological Society of China, v. 20, p. 61–70.

Chen, C.-H., Chu, H.T., Liou, J.G., Ernst, W.G., 1983a, Explanatory notes for the metamorphic facies map of Taiwan: Special Publication of the Central Geological Survey, 32 p.

Chen, C-H., Chu, H.T., and Chuang, T.Y., 1983b, Some structural problems of the Central Range of Taiwan: Bulletin of the Central Geological Survey, v. 2, p. 1–16 (in Chinese).

Chen, P.Y., 1991, Basaltic-andesitic volcanic rocks from the areas of Changshihchiao and Hsiangyang, Southern E-W Cross Island Highway, Taiwan: Special Publication of the Central Geological Survey, v. 5, p. 127–159.

Clark, M.B., and Fisher, D.M., 1995, Strain partitioning and crack-seal growth of chlorite-muscovite aggregates during progressive noncoaxial strain: An example from the Slate Belt of Taiwan: Journal of Structural Geology, v. 17, p. 461–474.

Clark, M.B., Fisher, D.M., and Lu, C.-Y., 1992, Strain variations in the Eocene and older rocks exposed along the central and southern cross-island highways, Taiwan: Acta Geologica Taiwanica, v. 30, p. 1–10.

Clark, M.B., Fisher, D.M., Lu, C.-Y., and Chen, C.-H., 1993, Kinematic analyses of the Hsüehshan Range, Taiwan: A large-scale pop-up structure: Tectonics, v. 12, p. 205–217.

Crespi, J.M., Chan, Y.-C., and Swaim, M.S., 1996, Synorogenic extension and exhumation of the Taiwan hinterland: Geology, v. 24, p. 247–250.

Durney, D.W., and Ramsay, J.G., 1973, Incremental strains measured by syntectonic crystal growths, *in* DeJong, K.A., and Scholten, R., eds., Gravity and tectonics: New York, John Wiley and Sons, p. 67–96.

Elliott, D., 1972, Deformation paths in structural geology: Geological Society of America Bulletin, v. 83, p. 2621–2635.

Ellis, M., 1986, The determination of progressive deformation histories from antitaxial syntectonic fibres: Journal of Structural Geology, v. 8, p. 701–709.

Faure, M., Lu, C.-Y., and Chu, H.-T., 1991, Ductile deformation and Miocene nappe-stacking in Taiwan related to motion of the Philippine Sea Plate: Tectonophysics, v. 198, p. 95–105.

Fisher, D., and Brantley, S., 1992, Models of quartz vein and overgrowth formation: Fluid flow in an ancient accretionary wedge: Journal of Geophysical Research, v. 97, p. 20043–20063.

Fisher, D.M., 1990, Orientation history and rheology of slates: Journal of Structural Geology, v. 12, p. 483–498.

Fisher, D.M., 1999, Orogen-parallel extension in the eastern Central Range of Taiwan: Journal of the Geological Society of China, v. 42, p. 41–58.

Gomez, F., Allmendinger, R., Barazangi, M., Er-Raji, A., and Dahmani, M., 1998, Crustal shortening and vertical strain partitioning in the Middle Atlas Mountains of Morocco: Tectonics, v. 17, p. 520–533.

Hashimoto, W., and Matsumaru, K., 1975, On the Lepidoczclina bearing limestone exposed at the southern cross mountain highway, Taiwan: Contributions to the Geology and Paleontology of Southeast Asia, v. 16, p. 103–116.

Ho, C.S., 1988, An introduction to the geology of Taiwan (second edition) and explanatory text of the geologic map of Taiwan: Taipai, Taiwan, Republic of China, Central Geological Survey, Ministry of Economic Affairs, 192 p.

Hsieh, X.L., 1990, Fission track dating of zircons from several east-west cross sections of Taiwan Island [M.S. thesis]: Taipai, National Taiwan University, 134 p.

Huang, T.C., 1980, Calcareous nannofossils from the slate terrane west of Yakou, Southern Cross-Island Highway, Taiwan: Petroleum Geology of Taiwan, v. 17, p. 59–74.

Jones, R.R., Holdsworth, R.E., and Bailey, W., 1997, Lateral extrusion in transpression zones: The importance of boundary conditions: Journal of Structural Geology, v. 19, p. 1201–1218.

Lee, C.T., 1977, Structural geology of Likuan region, South Cross-Island Highway (in Chinese) [M.S. thesis]: Taipai, National Taiwan University, 71 p.

Lee, J.F., 1987, Preliminary study on Paleo-stress of Tachien sandstone, Chishan to Techi (in Chinese) [M.S. thesis]: Taipai, National Taiwan University, 78 p.

Li, T., and Means, W.D., 1995, Experimental antitaxial growth of fibrous crystals. 1. Technique, fiber character, and implications: Geological Society of America Abstracts with Programs, v. 27, no. 6, p. A70.

Liu, T.K., 1982, Tectonic implications of fission track ages from the Central Range, Taiwan: Proceedings of the Geological Society of China, v. 25, p. 22–37.

Liu, T.K., 1988, Fission track dating of the Hsüehshan Range: Thermal record due to arc-continent collision in Taiwan: Acta Geologica Taiwanica, v. 26, p. 279–290.

Lu, C.-Y., Chu, H.-T., and Lee, J.-C., 1997, Structural evolution in the Hsüehshan Range, Taiwan: Journal of the Geological Society of China, v. 40, p. 261–279.

Lu, C.Y., Lee, J.C., and Lee, J.F., 1991, Extensional and compressional tectonics in central Taiwan, *in* Cosgrove, J., and Jones, M., eds., Neotectonics and resources: London, Belhaven Press, p. 85–92.

Means, W.D., and Li, T., 1995, Experimental antitaxial growth of fibrous crystals. 2. Internal structures: Geological Society of America Abstracts with Programs, v. 27, no. 6, p. A70.

Ramsay, J.G., and Huber, M.I., 1983, The techniques of modern structural geology, Strain analysis, Volume 1: London, Academic Press, 307 p.

Schmid, S.M., and Casey, M., 1986, Complete fabric analysis of some commonly observed quartz c-axis patterns, *in* Hobbs, B.E., and Heard, H.C., eds., Mineral and rock deformation, laboratory studies; the Paterson volume: American Geophysical Union Geophysical Monograph 36, p. 263–286.

Seno, T., Stein, S., and Gripp, A.E., 1993, A model for the motion of the Philippine Sea plate consistent with NUVEL-1 and geological data: Journal of Geophysical Research, v. 98, p. 17941–17948.

Stanley, R.S., Hill, L.B., Chang, H.C., and Hu, H.N., 1981, A transect through the metamorphic core of the Central Mountains, Taiwan: Memoir of the Geological Society of China, v. 4, p. 443–473.

Sun, S.C., 1982, The Tertiary basins of offshore Taiwan, *in* Salivar-Sali, A., ed., Petroleum geology of Taiwan, Proceedings, 2nd Asian Conference on Petroleum (ASCOPE): Manila, Philippines, Technical Programme Committee, 1981, p. 125–135.

Suppe, J., 1984, Kinematics of arc-continent collision, flipping of subduction, and backarc spreading near Taiwan: Memoir of the Geological Society of China, v. 6, p. 21–24.

Suppe, J., 1986, Reactivated normal faults in the western Taiwan fold and thrust belt: Memoir of the Geological Society of China, v. 7, p. 187–200.

Suppe, J., Wang, Y., Liou, J.G., and Ernst, W.G., 1976, Observation of some contacts between basement and Cenozoic cover in the Central Mountains, Taiwan: Proceedings of the Geological Society of China, v. 19, p. 59–70.

Teng, L.S., Wang, Y., Tang, C.-H., Huang, C.-Y., Huang, T.-C., Yü, M.-S., and Ke, A., 1991, Tectonic aspects of the Paleogene depositional basin of northern Taiwan: Proceedings of the Geological Society of China, v. 34, p. 313–336.

Tillman, K., and Byrne, T., 1995, Kinematic analysis of the Taiwan Slate Belt: Tectonics, v. 14, p. 322–341.

Tillman, K., and Byrne, T.B., 1996, Out-of-sequence thrusting in the Taiwan Slate Belt: Journal of the Geological Society of China, v. 39, p. 189–208.

Tsan, S.F., 1976, The folding and block movement of the Hsüehshan Range, Taiwan: Bulletin of the Geological Survey of Taiwan, v. 25, p. 29–34.

Tsan, S.F., 1977, Remarks on the Suao section of the Central Range of Taiwan: Memoir of the Geological Society of China, v. 2, p. 141–145.

Wang, Y., 1987, Continental rifting and Cenozoic tectonics around Taiwan: Memoir of the Geological Society of China, v. 9, p. 227–240.

Wickham, J.S., 1973, An estimate of strain increments in a naturally deformed carbonate rock: American Journal of Science, v. 273, p. 23–47.

Yeh, E.C., 1998, The ductile shear deformation and structural evolution of the Penglai Orogeny at Suao-Donao area in northeastern Taiwan [M.S. thesis]: Taipai, Taiwan, National Taiwan University, 123 p.

Yeh, Y.-H., Barrier, E., Lin, C.H., and Angelier, J., 1991, Stress tensor analysis in the Taiwan area from focal mechanisms of earthquakes: Tectonophysics, v. 200, p. 267–280.

Yen, T.P., 1973, The Eocene sandstones in the Hsüehshan Range terrain, northern Taiwan: Proceedings of the Geological Society of China, v. 16, p. 97–110.

Yen, T.P., Sheng, C.C., Keng, W.P., and Yang, Y.T., 1956, Some problems on the Mesozoic formation of Taiwan: Bulletin of the Geological Survey of Taiwan, no. 8, p. 1–14.

Yu, S.B., Chen, H.Y., Kuo, L.C., 1997, Velocity field GPS stations in the Taiwan area: Tectonophysics, v. 274, p. 41–59.

Manuscript Accepted by the Society April 12, 2001

Printed in the U.S.A.

Geological Society of America
Special Paper 358
2002

Lateral extrusion in a transpressional collision zone: An example from the pre-Tertiary metamorphic basement of Taiwan

Michael H. Pulver
Jean M. Crespi
Timothy B. Byrne
Department of Geology and Geophysics, University of Connecticut, 354 Mansfield Road, Storrs, Connecticut 06269-2045, USA

ABSTRACT

Analysis of the geometry and kinematics of structural fabrics in the pre-Tertiary metamorphic basement of east-central Taiwan provides evidence for transpression and lateral extrusion during oblique collision. Four structural domains are recognized on the basis of systematic variations in the dip direction of S1 and the trend of L1. Domains I, IIa, and IIb contain a northeast-striking S1 and a northeast- or southwest-plunging L1, raking 30°–45° on S1. In contrast, domain III, at the western edge of the pre-Tertiary metamorphic basement, has an approximately downdip L1, similar to the Slate Belt. Microscale kinematic indicators from domains I, IIa, and IIb yield a consistent, left-lateral, strike-slip component of oblique motion. Analyses of microstructures in the Chipan granitic gneiss in domain I and the geometry of mesoscale folds in phyllitic rocks in domain IIa provide qualitative evidence for constrictional strain. Evaluation of potential primary configurations of structural domains, combined with geologic and tectonic information, suggests that the pre-Tertiary metamorphic basement and the Slate Belt formed as a southeast-dipping package; post-S1 block rotation of domain IIa is interpreted to have produced the present arrangement of domains. The pre-Tertiary metamorphic basement was deformed within a zone of left-lateral transpression. The shallow plunge of the L1 lineation and the evidence for constrictional strain in domains I, IIa, and IIb suggest that the orogen-perpendicular component of oblique convergence was accommodated by lateral extrusion. Significant vertical thickening may have been prevented by the presence of a shallowly east-dipping rigid arc lid overlying the pre-Tertiary metamorphic basement.

INTRODUCTION

Oblique convergence results in zones of transpression where plate motion is accommodated by complex, three-dimensional deformation. Transpression is commonly manifested at shallow structural levels by coeval thrust and strike-slip faults, the thrust faults accommodating the component of motion oriented orthogonal to the plate margin and the strike-slip faults accommodating the component of motion oriented parallel to the plate margin (Mount and Suppe, 1987; Oldow et al., 1990; McCaffrey, 1992). The understanding of transpressional deformation at deep structural levels has advanced through two types of studies: (1) forward modeling of finite strain and strain paths using specific boundary conditions (Sanderson and Marchini, 1984; Robin and Cruden, 1994; Tikoff and Teyssier, 1994; Teyssier et al., 1995; Jones et al., 1997; Fossen and Tikoff, 1998; Teys-

Pulver, M.H., Crespi, J.M., and Byrne, T.B., 2002, Lateral extrusion in a transpressional collision zone: An example from the pre-Tertiary metamorphic basement of Taiwan, *in* Byrne, T.B., and Liu, C.-S., eds., Geology and Geophysics of an Arc-Continent Collision, Taiwan: Boulder, Colorado, Geological Society of America Special Paper 358, p. 107–120.

sier and Tikoff, 1999) and (2) structural analysis of deeply eroded, ancient margins for which the boundary conditions are poorly known (Gillotti and Hull, 1993; Dias and Ribeiro, 1994; Wallis, 1995; Northrup and Burchfiel, 1996).

Taiwan is a modern example of an oblique collision. The orogenic belt, which began to form by Pliocene time (Chi et al., 1981), is active, and plate-motion studies (Seno, 1977; Seno and Maruyama, 1984) reveal about 20° of obliquity between the plate-convergence vector and the normal to the plate margin in the vicinity of Taiwan. Moreover, high erosion rates have resulted in the exposure of high-grade metamorphic rocks in the Taiwan hinterland. Taiwan, therefore, presents an opportunity to study ductile deformation in a zone of transpression where the geometry of plate convergence is well known.

In this study, we present field and microstructural evidence for subhorizontal extension oblique to the trend of the orogen and for constrictional strain in the pre-Tertiary metamorphic basement of Taiwan. These findings contrast with the structural geometry of the Slate Belt, which is described by approximately downdip or orogen-perpendicular extension and plane strain. We argue that the distinct structural geometry and kinematics of the metamorphic basement can be explained best by the dominance of lateral extrusion over vertical thickening and, to a lesser degree, by the partitioning of orogen-parallel simple shear during oblique collision rather than by the complex superposition of several distinct tectonic events.

TECTONIC AND GEOLOGIC SETTING OF TAIWAN

The Taiwan orogenic belt is a result of the ongoing collision of the Luzon volcanic arc with the Eurasian passive continental margin (Chai, 1972; Ho, 1986). Analyses of changes in sedimentation rate with time indicate that the collision began by the Pliocene (Chi et al., 1981). The plate margin in the Taiwan region is inferred to lie along the Longitudinal Valley, which trends 020° and separates accreted volcanic-arc rocks from deformed continental basement and overlying cover rocks of the passive margin. The relative plate-convergence vector in the Taiwan region is calculated by using poles of rotation presented in Seno (1977) and trends 310°. The present-day data, therefore, indicate about 20° of obliquity between the relative plate-convergence vector and the normal to the plate margin (Fig. 1A). In addition, the colliding constituents, i.e., the Luzon arc and Eurasian margin, trend obliquely to one another, which results in the southward propagation of the collision (Suppe, 1981) and the exposure of older, more deeply eroded parts of the orogenic belt in the north.

The Taiwan orogenic belt can be divided into five major, north-northeast-trending tectonostratigraphic units: the Coastal Plain, Western Foothills, Slate Belt, pre-Tertiary metamorphic basement, and Coastal Range (Fig. 1B) (Ho, 1986). In general, the units increase in age and metamorphic grade from west to east, with the exception of the Coastal Range. The Coastal Range consists of weakly metamorphosed volcanic rocks and forearc sedimentary deposits and is interpreted to be the accreted segment of the Luzon arc (Ho, 1986). The Western Foothills consists of a fold-and-thrust belt that, together with the Slate Belt and pre-Tertiary metamorphic basement, forms the accretionary complex that developed in response to the collision (Ho, 1986).

GEOLOGY OF THE PRE-TERTIARY METAMORPHIC BASEMENT ALONG THE CENTRAL CROSS-ISLAND HIGHWAY

The pre-Tertiary metamorphic basement exposed along the Central Cross-Island Highway in east-central Taiwan is composed of three main rock types; these are divided into four lithologic units (Fig. 2). The ages of the units are poorly determined, although all are considered Mesozoic or older (Yen, 1960; Chai, 1972). The two oldest lithologic units are composed primarily of marble and are interpreted to be late Paleozoic in age on the basis of a fusulinid identification (Yen, 1953). Pb-Pb isotopic dating yields a 166 Ma age, which provides a minimum age for the marble (Jahn, 1988). Two distinct units are recognized. In the Yentzuko area, the marble is a massive cliff-former (the Chiuchu Formation of Wang Lee et al. [1982]). In other parts of the transect, the marble is typically interlayered with quartz-mica phyllites (the Changchun Formation of Wang Lee et al. [1982]). The marble is completely recrystallized and lacks any relict primary textures.

Phyllites (the Tienhsiang Formation of Wang Lee et al. [1982]) lie structurally above and are, therefore, considered to be younger than the marbles. This interpretation is supported by a report of dinoflagellates consistent with a Jurassic to Cretaceous age for the unit (Chen, 1989). Mineral assemblages are typically quartz + chlorite + carbonate + albite + biotite + epidote + sphene ± actinolite ± tremolite and are interpreted to be the result of greenschist-facies metamorphism of carbonate-rich sedimentary and volcanic protoliths.

The Chipan granitic gneiss is the youngest lithologic unit in the pre-Tertiary metamorphic basement. Evidence for an intrusive relationship with the marbles includes locally preserved, interfingering contacts and calc-silicate mineralization in marble enclaves (Lo and Wang Lee, 1981a). $^{40}Ar/^{39}Ar$ hornblende and U-Pb dating indicates a crystallization age of 85–90 Ma (Jahn et al., 1986; Lo and Onstott, 1995) for the granite.

All of the units show predominantly greenschist-facies mineral assemblages, which are interpreted to be the result of the Pliocene–Holocene orogeny (Lo and Wang Lee, 1981b; Wang Lee et al., 1982; Ernst, 1983; Lo and Onstott, 1995). Metamorphic grade increases gradually across the pre-Tertiary metamorphic basement from lower-greenschist facies in the west to upper-greenschist facies in the east (Ernst, 1983; Liou and Ernst, 1984). Relicts of amphibolite-facies mineral assemblages are interpreted to be related to the Mesozoic Nanao orogeny when the pre-Tertiary basement was the high-temperature, low-pressure member of a paired metamorphic belt on the mar-

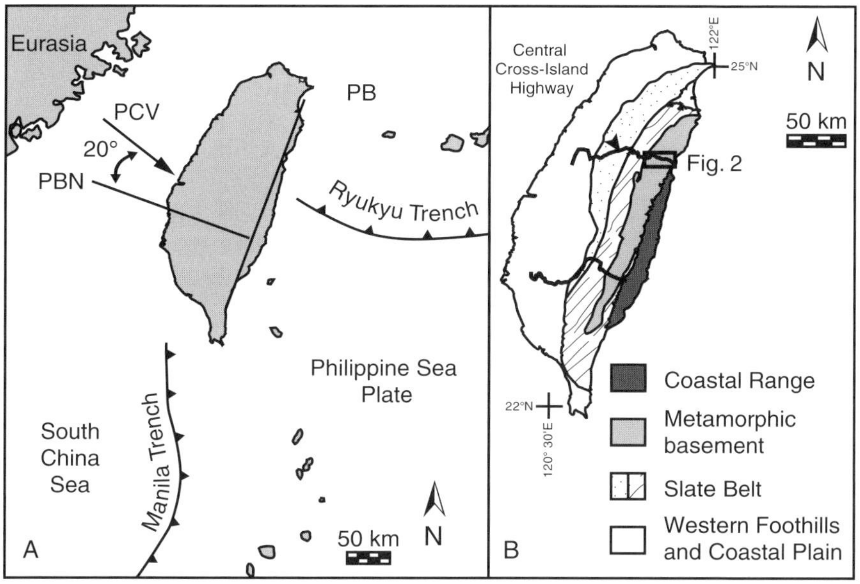

Figure 1. (A) Map of regional tectonic setting of Taiwan. PCV—plate-convergence vector, PB—plate boundary, PBN—normal to plate boundary. (B) Tectonostratigraphic map of Taiwan after Ho (1988). Location of Figure 2 is shown by box.

gin of the Eurasian continental plate (Lo and Wang Lee, 1981a; Wang Lee et al., 1982; Liou and Ernst, 1984; Jahn et al., 1986).

Lo and Onstott (1995) showed that the pre-Tertiary metamorphic basement records only two tectonometamorphic events: (1) amphibolite-facies metamorphism during the Mesozoic Nanao orogeny and (2) greenschist-facies metamorphism during the Pliocene–Holocene orogeny. They further demonstrated that $^{40}Ar/^{39}Ar$ dates that fall between these two events reflect mixing of Ar released from minerals that formed during these two events and not a third orogenic event.

STRUCTURAL ANALYSIS OF THE PRE-TERTIARY METAMORPHIC BASEMENT ALONG THE CENTRAL CROSS-ISLAND HIGHWAY

This research focused on understanding the geometry and kinematics of structural fabrics in the pre-Tertiary metamorphic basement along the Central Cross-Island Highway. Field work consisted of the systematic measurement of S1 and L1 orientations as well as the orientation and interlimb angle of mesoscale folds. In addition, oriented samples were collected for microstructural and kinematic analyses. A qualitative assessment of the state of strain in the metamorphic basement was achieved through (1) microstructural analysis of the Chipan granitic gneiss and (2) examination of the orientation and tightness of folds. The systematic variation in the orientation of S1 and L1 across the transect was analyzed to determine the primary orientation of structures during D1 deformation.

The pre-Tertiary metamorphic basement along the Central Cross-Island Highway is dominated by a single, penetrative fabric. S1, a continuous foliation, strikes northeast and contains a mineral and stretching lineation (L1) that has a rake of 30°–45° on S1 throughout most of the transect. Primary sedimentary and igneous layering has been strongly transposed by S1 and can only be recognized locally as intrafolial hooks and marble lenses in phyllitic rocks. In the western part of the transect near Pilu, however, primary layering (S0) can be recognized as folded, quartz-rich layers that are cut by an axial-planar foliation (S1) defined by the preferred orientation of fine-grained muscovite. In the Tienhsiang area, a second, widely spaced (up to 10 cm) crenulation cleavage (S2) is present locally.

Although S1 and L1 exist throughout the pre-Tertiary metamorphic basement, their relative intensity shows minor variation according to rock type. The phyllites and interlayered marbles are the most strongly foliated. The marbles in the Yentzuko area, in contrast, are relatively massive. Fabric development in the Chipan granitic gneiss is heterogeneous (Clark, 1994); in general, however, the granitic gneiss is an L > S tectonite with a strongly developed L1 and a weakly developed S1.

The minerals that constitute the D1 fabric vary according to rock type. In the Chipan granitic gneiss, L1 and S1 are defined by the preferred orientation of biotite, white mica, and elongate aggregates of coarse-grained (2–5 mm in diameter) quartz. In marbles and carbonate-rich phyllites, the shape preferred orientation of carbonate grains characterizes the D1 fabric. In the metavolcanic phyllites, S1 and L1 are defined by the preferred orientation of chlorite, actinolite, biotite, and rarely white mica. In each rock type, the D1 fabric is defined by greenschist-facies minerals that are interpreted to be related to the Pliocene–Holocene orogeny (Wang Lee et al., 1982; Ernst, 1983; Liou and Ernst, 1984; Lo and Onstott, 1995).

Determination of the deformation mechanisms allows the characterization of the environmental conditions during the formation of the D1 fabric. In the Chipan granitic gneiss, coarse quartz grains exhibit patchy, undulose extinction, subgrains, and irregular grain boundaries indicative of deformation in the dislocation-creep regime. In contrast, bent or offset twin lamellae and strong alteration show that deformation of feldspar

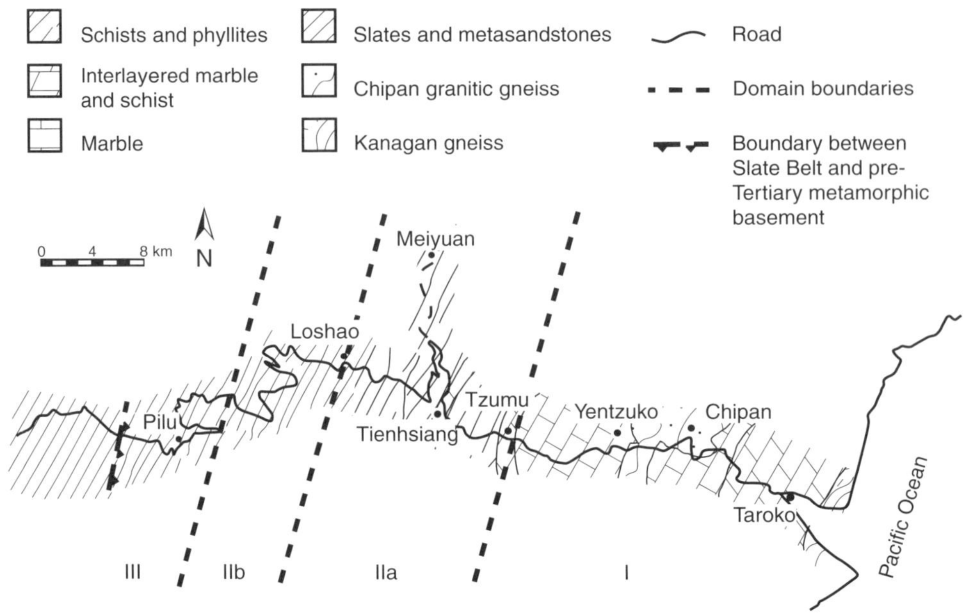

Figure 2. Simplified map of geology in vicinity of Central Cross-Island Highway, east-central Taiwan, after Lu and Wang Lee (1986).

grains was dominated by internal microfracturing and cataclasis. Large, igneous biotite grains contain kink bands that deform the (001) plane. Carbonate grains in marbles are typically large (0.2–1.0 mm) and equidimensional, most likely as a result of grain-boundary-area reduction that minimizes internal free energy by decreasing the total surface area of grain boundaries (Passchier and Trouw, 1996). Carbonate-rich phyllites contain boudinaged dolomite porphyroclasts within a calcite-rich matrix; both carbonates are twinned. Calcite strain shadows on dolomite porphyroclasts do not show evidence of internal deformation, suggesting precipitation of material during deformation. The microstructures displayed by the different minerals in the pre-Tertiary metamorphic basement are consistent with deformation at ~300–400 °C.

Foliation, lineation, and fold geometries

Although the S1 foliation strikes northeast and the L1 lineation has a shallow to moderate plunge throughout most of the study transect, there are systematic variations in the dip direction of S1 and the trend of L1. These differences have allowed the division of the study transect into four domains, each with an internally uniform S1 and L1 orientation (Fig. 3).

The domain boundaries mark relatively gradual transitions in fabric orientation and, except for the boundary between domains I and IIa, are not associated with major structural discontinuities. Near the boundary between domains I and IIa in the Tzumu area, a complex zone of high strain containing sheath folds and superposed folds exists and has been described by Lu and Wang Lee (1986). The general increase in metamorphic grade from lower-greenschist facies in the west to upper-greenschist facies in the east (Ernst, 1983; Liou and Ernst, 1984) also implies that there are no post–peak-metamorphism faults with significant vertical displacement.

Domain I shows the largest internal variation in S1 orientation: the poles to S1 define a π-circle with a subhorizontal π-axis trending 035. Overall, however, the domain is dominated by gentle southeast dips for S1. The mean orientation of S1 is 012, 31 ESE. L1 maintains a consistent shallow plunge to the northeast throughout domain I; the mean orientation of L1 is 045, 19.

To the west in domain IIa, the orientation of structural fabrics changes significantly. S1 dips moderately to the northwest, and L1 plunges gently to the southwest. Mean orientations of S1 and L1 are 219, 45 NW and 243, 21, respectively.

At Loshao, the orientation of S1 changes from the northwest dip characteristic of domain IIa through vertical to a steep southeast dip, which it maintains throughout domain IIb. Concomitant with the change in orientation of S1, L1 shifts to a moderate south-southwest plunge. In domain IIb, the mean orientations of S1 and L1 are 033, 68 SE and 194, 39, respectively.

Near Pilu, L1 changes to approximately downdip. The orientation of S1, however, remains about constant. Domain III is characterized by a southeast-dipping S1 and a nearly downdip L1, the mean orientations of S1 and L1 being 024, 68 SE and 143, 56, respectively. The change from an oblique to downdip lineation between domains IIb and III is gradual in the field and broadly overlaps the mapped location of the contact between the Slate Belt and the pre-Tertiary metamorphic basement (Fig. 2 of Lu et al., 1997).

The hinge lines of mesoscale folds do not show a variation in orientation and are subhorizontal and northeast- or southwest-trending throughout the study transect. The mean trend and plunge of fold hinge lines is 215, 01 in domain I and 224, 04 in domain IIa. The folds generally deform marble lenses, inter-

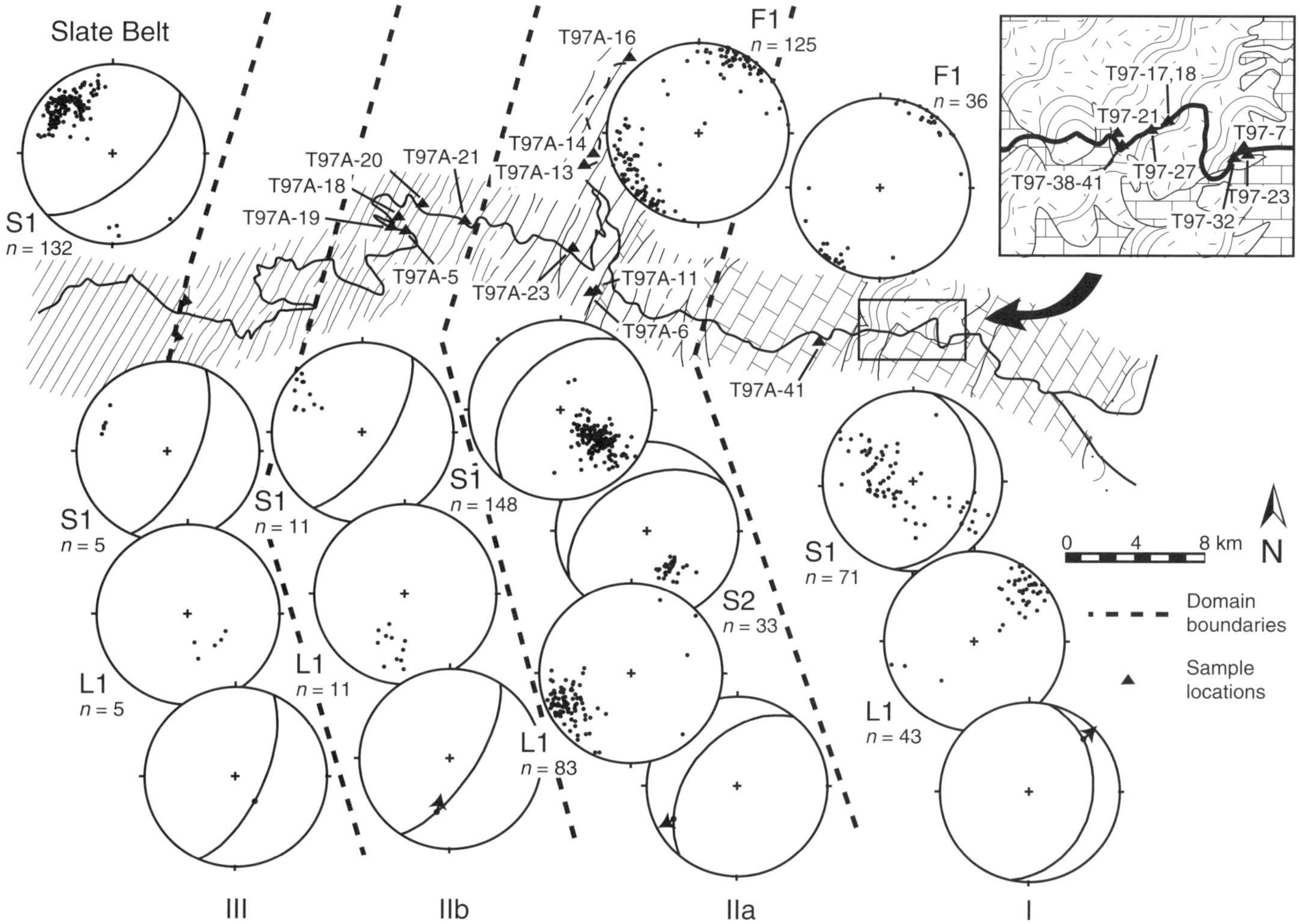

Figure 3. Orientation of structural elements in different domains. Stereoplots of S1 and S2 show poles to foliations; great circle represents mean orientation in each domain. F1 stereoplots show orientation of fold hinge lines. Lowermost stereoplot for each domain shows mean L1, mean S1, and sense of shear, except for domain III, where no shear-sense indicators were observed. Locations of samples that contain kinematic indicators are shown on transect. Inset shows locations of samples used for microstructural analysis of strain in Chipan granitic gneiss. Stereoplots are equal-area, lower-hemisphere projections.

preted as primary lithologic layering (S0), within the phyllites. A wide variety of fold geometries is present, ranging from tight, west-vergent, asymmetric, recumbent folds to open, symmetric folds. The S1 foliation is axial planar to the majority of the folds.

Previous workers identified five generations of folds on the basis of locally preserved overprinting relationships (Lu and Wang Lee, 1986). However, multiple fold generations may form as a result of flow heterogeneities within a single progressive deformation (Skjernaa, 1989). Relating deformation phases to metamorphic events can be useful for determining the significance of superposed folds and associated foliations. The foliations associated with the generations of folds previously recognized in the pre-Tertiary metamorphic basement are all defined by greenschist-facies minerals. Because the different fold generations formed at similar metamorphic conditions, we suggest that they are more likely the result of a single progressive deformation rather than several distinct deformation phases or tectonic events.

Kinematic analysis

Previous work on the kinematics of D1 deformation in the pre-Tertiary metamorphic basement along the Central Cross-Island Highway has been conducted by Chan (1990) and Clark (1994) who focused on the Chipan granitic gneiss in domain I. Kinematic indicators in the granitic gneiss include S-C fabrics, strain shadows around feldspar porphyroclasts, and foliation geometries in centimeter- to decimeter-wide shear zones. These indicators consistently yield left-lateral, top-to-the-northeast sense of shear (Chan, 1990; Clark, 1994).

This study builds on this previous work and describes kinematic indicators in marble of domain I, calc-phyllites and calcareous slates of domain IIa, and mafic phyllonites of domain IIb. No shear-sense indicators were observed in domain III. Kinematic indicators were observed in thin sections cut perpendicular to the S1 foliation and parallel to the L1 lineation (*XZ* sections). All sections are described viewing down, regardless of fabric orientation. In other words, samples with a southeast-dipping S1 and northeast-plunging L1 (domain I) are described viewing down to the southeast; samples with a northwest-dipping S1 and southwest-plunging L1 (domain IIa) are described viewing down to the northwest; and samples with a southeast-dipping S1 and south-southwest-plunging L1 (domain IIb) are described viewing down to the northeast.

A sample of strongly foliated and lineated marble from the Yentzuko area of domain I contains an oblique foliation defined by the shape preferred orientation of carbonate grains (Fig. 4A). The obliquity of the shape preferred orientation of carbonate grains relative to the S1 foliation indicates a left-lateral, top-to-the-northeast sense of shear, consistent with the observations of Chan (1990) and Clark (1994).

Kinematic indicators in calc-phyllites and calcareous slates of domain IIa include asymmetric strain shadows, oblique fo-

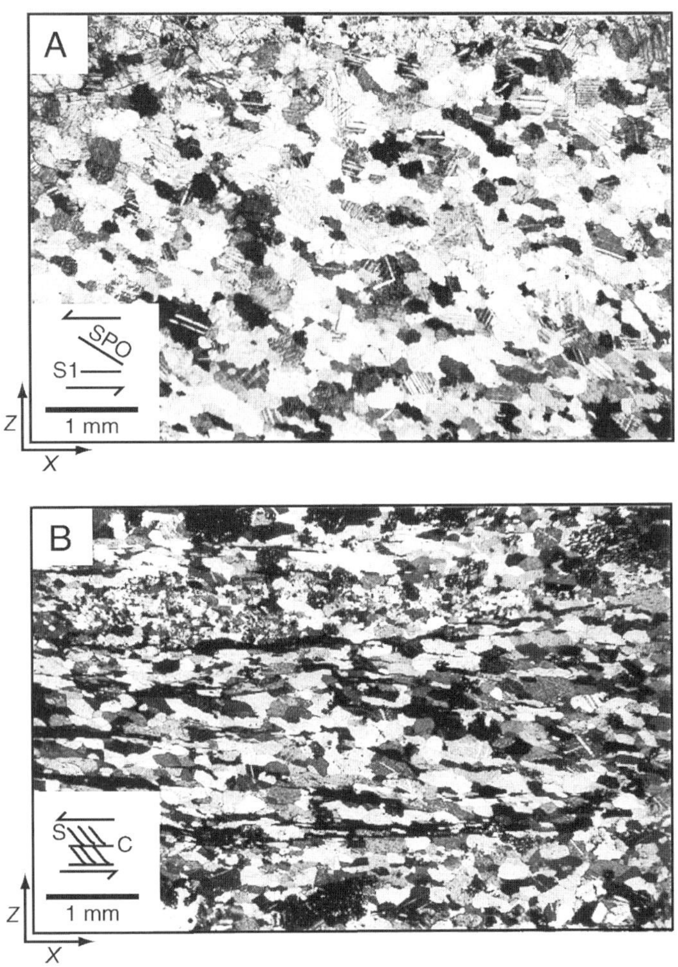

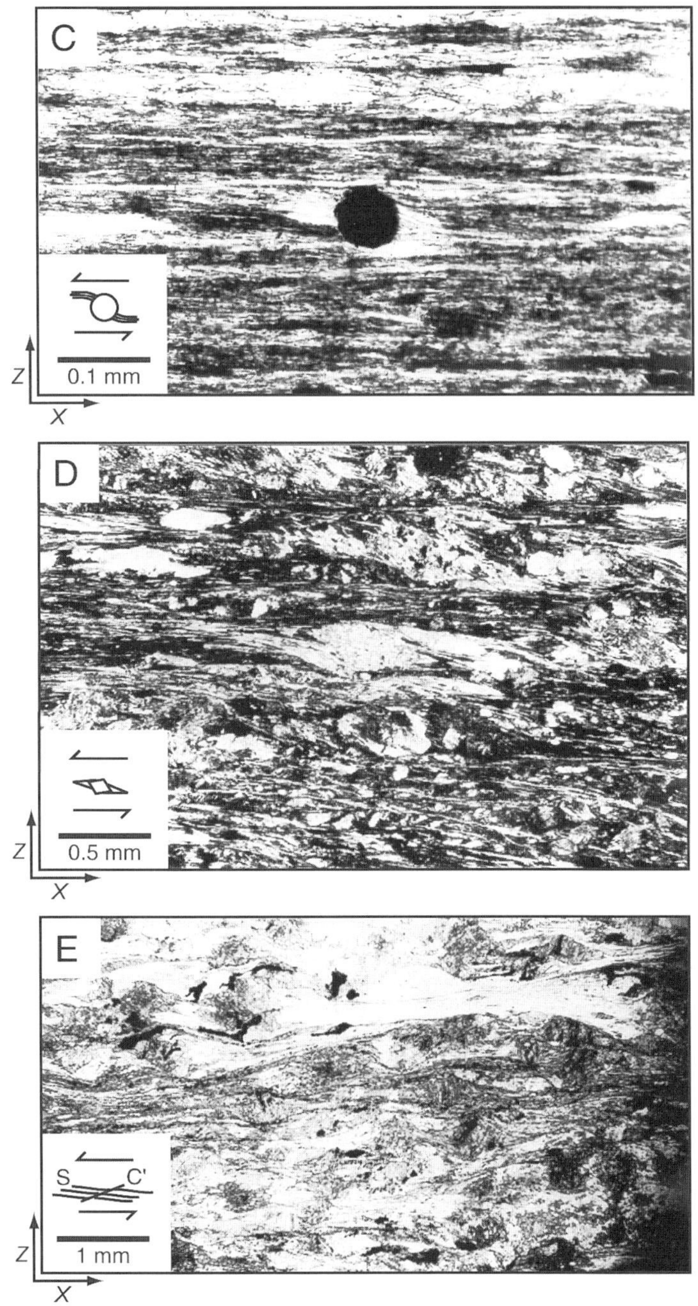

Figure 4. (A) Marble with oblique foliation defined by shape preferred orientation (SPO) of carbonate grains. S1 is horizontal, and oblique foliation trends from upper left to lower right. Sample T97A-41. Cross-polarized light. (B) S-C fabric in calc-phyllite. C planes are horizontal and defined by chlorite layers. S planes are defined by shape preferred orientation of carbonate grains. Sample T97A-14. Cross-polarized light. (C) Curved muscovite and biotite strain fringes around pyrite framboid. Sample T97A-13. Plane-polarized light. (D) Asymmetric actinolite and chlorite strain shadows around calcic clino-amphibole porphyroclast. Sample T97A-18a. Cross-polarized light. (E) Dolomite boudins offset across extensional shear band. Shear band cuts S1 at a low angle and trends from upper right to lower left. Sample T97A-19. Plane-polarized light. All photographs view down.

liations defined by the shape preferred orientation of dynamically recrystallized quartz grains, and S-C fabrics; these indicators consistently yield a left-lateral, top-to-the-southwest sense of shear. An example of a strongly developed S-C mylonite fabric is shown in Figure 4B. Discrete layers of chlorite define C planes that separate domains of carbonate grains with an oblique shape preferred orientation (S planes); the asymmetry of the S and C planes indicates noncoaxial flow and left-lateral, top-to-the-southwest sense of shear. In the same sample, asymmetric chlorite strain shadows around plagioclase porphyroclasts, oblique foliations defined by the shape preferred orientation of dynamically recrystallized, 20–40 µm quartz grains in a quartz-rich layer, and rotated epidote porphyroclasts with asymmetric carbonate strain shadows also yield left-lateral, top-to-the-southwest sense of shear. An example from a slate sample of a pyrite framboid with curved muscovite and biotite strain fringes is shown in Figure 4C. The fibers in the strain fringes are interpreted to be antitaxial because the fibers and the core object consist of different minerals. In antitaxial strain fringes, the youngest fibers, which are inferred to lie parallel to the maximum instantaneous stretching axis, are adjacent to the core object (Ramsay and Huber, 1983). The fibers in the strain fringes display clockwise curvature when traced from their distal ends toward the core object, and the youngest fibers lie at an angle to S1. The curvature and obliquity of the fibers indicate noncoaxial flow and a left-lateral, top-to-the-southwest sense of shear.

Left-lateral, top-to-the-northeast motion is observed throughout domain IIb. An example of a mafic phyllonite with asymmetric chlorite and actinolite strain shadows around a calcic clino-amphibole porphyroclast is shown in Figure 4D. The asymmetry of the strain shadows relative to S1 indicates a left-lateral, top-to-the-northeast sense of shear. Figure 4E shows a strongly deformed carbonate-rich phyllonite with extensional shear bands at a low angle to S1. Two halves of a boudinaged dolomite porphyroclast are offset across a shear band and are connected by dynamically recrystallized quartz grains with a strong lattice preferred orientation. The asymmetry of the shear bands relative to S1 and the offset of the dolomite boudins indicate noncoaxial flow and left-lateral, top-to-the-northeast sense of shear.

In summary, domains I, IIa, and IIb are all characterized by oblique motion, the strike-slip component being left-lateral. The three domains, however, differ in their dip-slip component. The dip-slip component is top-to-the-northeast and normal sense in domain I, top-to-the-southwest and normal sense in domain IIa, and top-to-the-northeast and thrust sense in domain IIb. The change in sense of the dip-slip component is associated with the domain change in the orientations of S1 and L1.

Qualitative strain analysis

A study of microstructures in the Chipan granitic gneiss in domain I and an analysis of fold geometries in domain IIa were conducted to estimate qualitatively the state of finite strain in the pre-Tertiary metamorphic basement. Although the methods do not provide quantitative finite-strain data, they do allow characterization of the shape of the finite strain ellipsoid.

Granitic gneiss microstructure. Fabric development in a rock is directly related to the geometry and magnitude of finite strain (Ramsay and Huber, 1983); detailed study of fabric geometry can, therefore, provide qualitative information on finite strain. The goal of the study of the Chipan granitic gneiss was to determine the state of strain associated with the L > S fabric.

The state of finite strain depends on the value of e_2, the extension (e) along the intermediate axis of the finite strain ellipsoid [$e = (l_f - l_0)/l_0$] (Ramsay and Huber, 1983). A positive value for e_2 indicates oblate or flattening strain, whereas a negative value indicates prolate or constrictional strain. Grain-scale deformation features were analyzed in the Chipan granitic gneiss to assess qualitatively the state of finite strain. This method was chosen to avoid the assumptions inherent in quantitative strain analysis of ductilely deformed, polymineralic rocks.

Microstructures were examined in thin sections cut parallel to each of the three principal planes of the finite strain ellipsoid. The S1 foliation and L1 lineation were used as approximations of the *XY* plane and *X*-axis, respectively. We assume that the rock underwent bulk extension along the *X*-axis and bulk shortening along the *Z*-axis. If grain-scale structures reflect the bulk strain, then it is possible to determine the microstructures that accommodated extension along the *X*-axis and shortening along the *Z*-axis in thin sections cut parallel to the *XY* and *XZ* planes of the finite strain ellipsoid. The presence or absence of similar microstructures in *YZ* sections can be used to determine whether there was a change in length parallel to the *Y*-axis and, if so, the sign of the change in length (e_2). This approach only determines qualitatively whether the finite strain ellipsoid is a flattening, plane strain, or constrictional type; it provides no information on the magnitude of finite strain or volume strain. Fletcher and Bartley (1994) also used microstructural observations to assess qualitatively the strain geometry in granodioritic and granitic gneisses, and they obtained results consistent with those from R_f/ϕ analyses of deformed primary quartz grains in the granodioritic gneiss and the shapes of cataclasized and altered garnet porphyroclasts in the granitic gneiss.

Eleven samples of the Chipan granitic gneiss were used for strain analysis. Weakly deformed rocks or rocks with evidence of post-D1 deformation such as semibrittle shear zones or an east-southeast-trending stretching lineation that overprints L1 were not used. Microstructures that indicate extension along the *X*-axis include microboudinaged feldspar grains, fibrous quartz infills within extensional fractures, and quartz strain shadows around feldspar porphyroclasts (Fig. 5A). Folds, microfaults within feldspar grains, and the preferred orientation of phyllosilicates are interpreted as evidence for shortening along the *Z*-axis (Fig. 5B). In general, a strong preferred orientation of minerals was observed in *XZ* and, to a lesser degree, in *XY* sections.

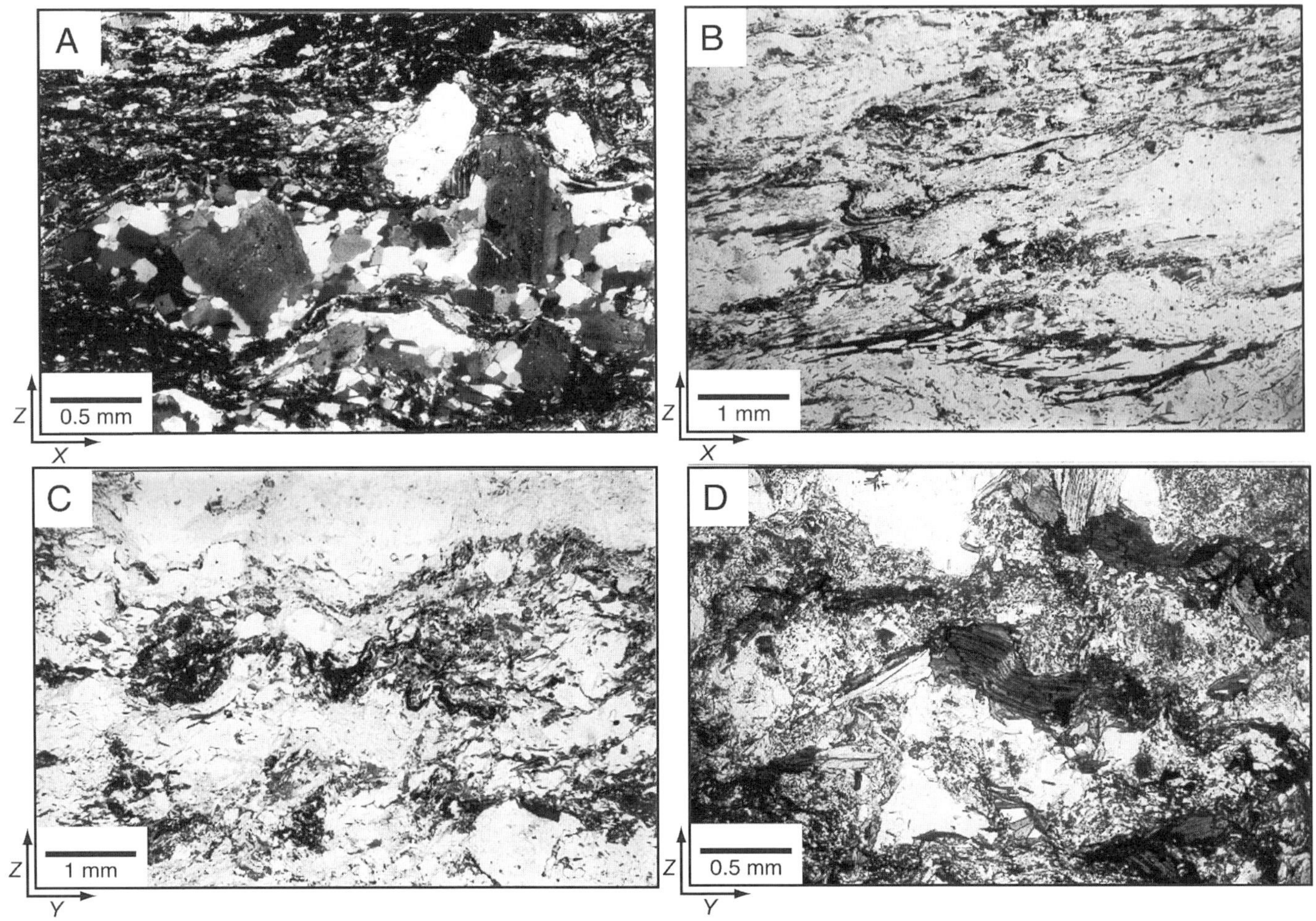

Figure 5. (A) Microboudinage of feldspar accommodating extension. *XZ* section. Sample T97-8. Cross-polarized light. (B) Microfolds accommodating shortening. *XZ* section. Sample T97-41. Plane-polarized light. (C) Microfolds accommodating shortening parallel to *Y*-axis. *YZ* section. Sample T97-39. Plane-polarized light. (D) Kinked porphyroblastic biotite. *YZ* section. Sample T97-21. Plane-polarized light.

The microstructures and fabric geometries observed in *YZ* sections can be divided into two groups. Samples in the first group lack a preferred orientation of minerals parallel to the *Y*-axis and exhibit a predominantly linear fabric at the hand-sample scale. Samples in the second group contain a distinguishable planar fabric (S1), parallel to the *Y*-axis, at the hand-sample and thin-section scales, and microstructures which shorten that fabric. Shortening microstructures observed in samples of the second group include upright and close microfolds that have axial planes subperpendicular to the *Y*-axis (Fig. 5C). The enveloping surfaces of these microfolds define the S1 fabric at the hand-sample scale; at the thin-section scale, the microfolds shorten S1 parallel to the *Y*-axis. In other samples that contain a distinguishable S1, shortening was accommodated by kinking of the (001) planes of porphyroblastic biotites. The trace of the kink bands is generally subparallel to the *Z*-axis (Fig. 5D), and these microstructures are interpreted to have accommodated shortening along the *Y*-axis. No microstructures were observed that accommodated extension along the *Y*-axis. The microstructures that indicate *Y*-axis shortening are considered qualitative evidence for constrictional finite strain in the Chipan granitic gneiss.

Analysis of fold orientation and tightness. In the phyllitic rocks in domain IIa, microstructural indicators of the state of finite strain are not common. In thin sections cut parallel to foliation, chlorite and actinolite strain shadows around amphibole porphyroclasts and quartz and mica strain fringes around pyrite framboids are straight and subparallel to the mineral lineation. The strain shadows and strain fringes maintain a constant width from the core object to their distal ends, implying no shortening along the *Y*-axis and plane-strain deformation.

Folds with hinge lines parallel to the inferred direction of tectonic transport have been observed in many orogenic belts, e.g., the Caledonides (Fossen, 1993; Gillotti and Hull, 1993), the Appalachians (Bryant and Read, 1969), and the metamorphic core complexes in western North America (Fletcher and Bartley, 1994). In domains I, IIa, and IIb in the pre-Tertiary metamorphic basement, the hinge lines of mesoscale folds trend subparallel to the mean orientation of the L1 lineation (Fig. 6A), which is interpreted to be the direction of transport during noncoaxial flow. The orientation and tightness of folds were analyzed to determine whether the interlimb angle of individual folds is related to the separation angle between the hinge line and the mean orientation of L1. Folds that are initiated with their hinge lines at an angle to the transport direction will progressively rotate toward the transport direction and undergo progressive tightening of their interlimb angles. In contrast, folds that are initiated with their hinge lines parallel to the transport direction during a progressive noncoaxial deformation will maintain a constant interlimb angle regardless of the magnitude of the shear strain (Skjernaa, 1989). Because folds may be initiated with their hinge lines parallel to the transport direction as a result of shortening along the *Y*-axis, determination of the initial distribution of orientations of fold hinge lines may provide information on the state of strain.

In domain IIa, folds with hinge lines subparallel to the mean L1 lineation have interlimb angles that range from 0° to 105°, and no correlation is observed between the tightness of a fold and the angle separating its hinge line from the mean orientation of L1 (Fig. 6B). This result implies that the mesoscopic folds in the phyllites of the pre-Tertiary metamorphic basement have not undergone significant rotation during progressive deformation. The folds are, therefore, inferred to have been initiated with their hinge lines oriented subparallel to the mean orientation of L1.

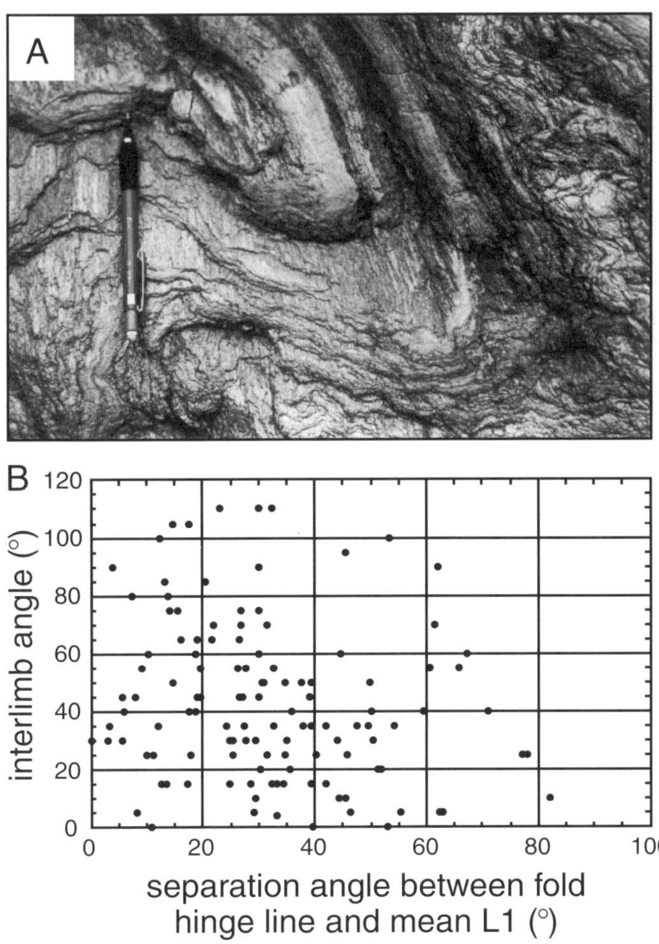

Figure 6. (A) Mesoscale fold in interlayered marbles and phyllites. Fold hinge line and L1 (shown by pencil) are subparallel. (B) Graph of interlimb angle vs. separation angle between fold hinge line and mean orientation of L1 for mesoscale folds in domain IIa.

Lateral-displacement gradients within a shear zone, termed "combined thrust and wrench shear" (Ridley, 1986; Ridley and Casey, 1989), and noncoaxial flow in a zone that is shortening laterally (parallel to the *Y*-axis) (Skjernaa, 1989; Fossen, 1993; Gillotti and Hull, 1993; Fletcher and Bartley, 1994) produce folds that are initiated with hinge lines parallel to the stretching lineation. Lateral-displacement gradients are gradients in the magnitude of shear strain perpendicular to the direction of transport in a shear zone, as might be expected along the lateral edges of a ductile thrust sheet (Ridley, 1986). The shear strain across the shear zone implied by the lateral gradient is termed a "wrench shear" and is treated as a simple shear on a plane orthogonal to the thrust but with the same displacement direction (see Fig. 2 of Ridley [1986]). In combined thrust and wrench shear, the separation angle between fold hinges and the stretching lineation depends on both the lateral displacement gradient (the thrust to wrench ratio) and the magnitude of finite strain. Given the regionally consistent orientation of hinge lines of mesoscopic folds across the pre-Tertiary metamorphic basement and the absence of evidence for lateral gradients in displacement, combined thrust and wrench shear does not provide an adequate explanation for the geometry and orientation of the folds. The parallelism of the fold hinge lines and the L1 lineation in domain IIa is interpreted, therefore, to be the result of noncoaxial flow combined with lateral shortening perpendicular to the transport direction. This interpretation is also consistent with the L > S fabrics and evidence for constrictional strain in the Chipan granitic gneiss.

INTERPRETATION OF STRUCTURAL DOMAINS

Four structural domains are recognized in the pre-Tertiary metamorphic basement along the Central Cross-Island Highway on the basis of the present orientation of the S1 foliation and L1 lineation. Central to any interpretation of the structural history of the pre-Tertiary metamorphic basement is whether this variation in orientation by domain of S1 and L1 is a primary feature of the orogen or the result of later rotations of one domain relative to another.

The potential number of configurations of primary structural domains is evaluated in terms of geometry and geologic likelihood. The number and configuration of primary structural domains are limited by the present orientation of S1 and L1 in each domain as well as the sense of shear. In essence, if two domains were originally a single domain, rotation of one domain relative to the other must bring the mean orientations of

S1 and L1 in each domain into coincidence and yield the same sense of shear. The geologic likelihood of the inferred rotation is assessed by noting if structures are present that could have accommodated rotation about the required axis.

Determination of primary structural domains and necessary block rotations

The simplest primary configuration is a single structural domain. Restoration of the mean S1 foliation and L1 lineation for one domain into coincidence with that of a second domain is possible by using a single, unique rotation axis; however, restoration of the four domains to a single domain requires rotations about two different axes. Rotation about a gently northwest-plunging axis, subparallel to the pole to S1, is required for domains I, IIa, and IIb. A second rotation about a subhorizontal, northeast-trending axis, subparallel to the strike of S1, is required for domain IIa. The same two rotation axes are required for restoration to a primary configuration consisting of two structural domains.

Although the two rotation axes just noted provide a geometrically feasible restoration of the domains, the geologic likelihood of these rotation axes must also be evaluated. Given the general northeast trend of structures in the pre-Tertiary metamorphic basement and the orogenic belt as a whole, rotations about a northwest-plunging axis, subparallel to the pole to S1, are not reasonable. Consequently, primary configurations with either one or two structural domains are considered unlikely.

In contrast, rotation about a subhorizontal, northeast-trending axis is consistent with the geology. Axes of outcrop-scale folds throughout the transect are subhorizontal and trend northeast or southwest. In addition, poles to the S1 foliation define a great circle with a subhorizontal, northeast-trending best-fit fold axis. Because geologic structures are commonly observed to be self-similar at different scales, we infer that these outcrop-scale structures reflect the orientation of the regional-scale rotation axis that accommodated block rotation of the structural fabrics. Rotations of structural fabrics about subhorizontal, northeast-trending axes have also been described for the pre-Tertiary metamorphic basement along the Southern Cross-Island Highway (Stanley et al., 1981; Clark et al., 1992; Crespi et al., 1996).

Rotation about a subhorizontal, northeast-trending axis, i.e., an axis subparallel to the strike of S1, can reduce the number of primary structural domains to three: domain I, restored domain II, and domain III. Domain II is reconstructed by a counterclockwise rotation (senses of rotation are described viewed to the northeast) of domain IIa about a subhorizontal, northeast-trending axis that is sufficient to bring the mean orientations of S1 and L1 into coincidence with those of domain IIb. This rotation results in a top-to-the-northeast sense of shear for domain IIa that is identical to that observed for domain IIb. It also restores S2 in domain IIa to an orientation that is consistent with its development within a zone of oblique, left-lateral thrusting. Restored domain II, which consists of domain IIb in its present orientation and rotated domain IIa, has a steeply southeast-dipping S1, an oblique, shallowly south-southwest-plunging L1, and oblique, left-lateral thrust kinematics (Fig. 7).

In contrast, clockwise rotation of domain IIa to bring the mean orientations of S1 and L1 into coincidence with those of domain I results in oblique, right-lateral thrust kinematics for domain IIa, which does not agree with the sense of shear observed in domain I. Domains I and IIa, therefore, cannot be restored to a single domain by using a single, geologically realistic rotation axis.

A primary configuration, consisting of domain I, restored domain II, and domain III, is reconstructed such that the S1 foliation dips southeast in all three domains. This interpretation is justified by several lines of evidence. First, no major east-vergent structures are recognized in the pre-Tertiary metamorphic basement along the Central Cross-Island Highway. Second, the slaty cleavage is southeast-dipping in the Slate Belt to the west along the Central Cross-Island Highway. Finally, a southeast-dipping S1 is consistent with the overall west vergence expected in an east-dipping subduction zone.

Timing constraints

Geochronologic data bearing on the absolute age of D1 structural fabrics in the pre-Tertiary metamorphic basement are not available. However, biostratigraphic data and geologic relationships in southern Taiwan imply that D1 structural fabrics in the pre-Tertiary metamorphic basement formed during the Pliocene–Holocene orogeny. Eocene slates on the eastern flank of the Central Range in southern Taiwan are characterized by a shallowly plunging lineation and, on the basis of the geometry

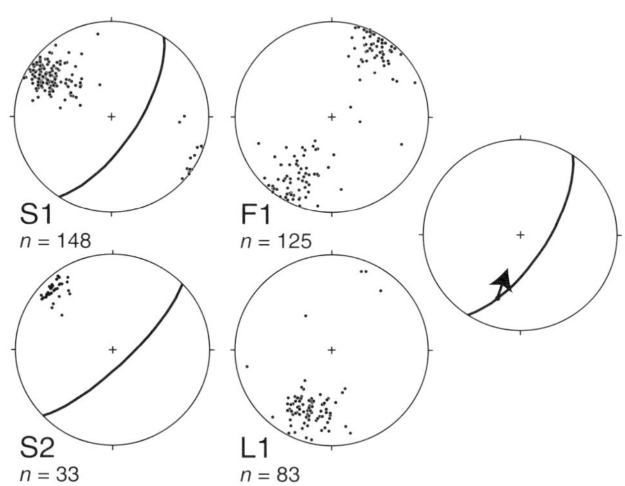

Figure 7. Orientation of structural elements in domain IIa after counterclockwise rotation of 67° about an axis oriented at 035, 04. Stereoplot at far right shows mean S1, mean L1, and sense of shear. Stereoplots are equal-area, lower-hemisphere projections.

of strain fringes around pyrite framboids, a constrictional state of strain (Fisher, 1997). Although no fossils have been found in the slates of the eastern Central Range in southern Taiwan, they have been assigned an Eocene age on the basis of lithologic similarities with the foraminifera-bearing Pilushan Formation in the Backbone Range (Ho, 1988). The similarity in orientation of the lineation and state of strain in the Eocene slates of the eastern Central Range in southern Taiwan and in the pre-Tertiary metamorphic basement along the Central Cross-Island Highway strongly suggests that the rocks were deformed during the same orogenic event. D1 structural fabrics are, therefore, inferred to be Eocene or younger. Because the geochronologic data support a single orogenic event in the Cenozoic (Lo and Onstott, 1995), D1 structural fabrics in the pre-Tertiary metamorphic basement are interpreted to be a result of the Pliocene–Holocene orogeny and not an earlier event.

The regional-scale block rotation of domain IIa postdates development of the D1 structural fabrics. Semibrittle to ductile normal-sense shear zones and brittle normal faults crosscut the S1 foliation in the pre-Tertiary metamorphic basement. In addition, kinematic analysis of brittle faults from the pre-Tertiary metamorphic basement and the Slate Belt along the Central Cross-Island Highway yields a consistent subhorizontal extension axis that trends west-northwest (Crespi et al., 1996). Because the kinematic axes do not vary in orientation, regional-scale rotation of structural domains is inferred to have occurred prior to the onset of normal faulting along the Central Cross-Island Highway.

Kinematic reconstruction

Detailed analysis of the geometric and geologic relationships between observed structural domains in the pre-Tertiary metamorphic basement along the Central Cross-Island Highway in east-central Taiwan has allowed reconstruction of the initial configuration of domains at deep levels of the orogenic belt (Fig. 8). In this reconstruction, the hinterland is divided into three primary structural domains: domain I, restored domain II, and domain III.

Although the three primary domains are assumed in the reconstruction to have a similarly oriented, southeast-dipping S1 foliation, they differ with respect to the orientation of the L1 lineation and kinematics. The orientation of the L1 lineation is slightly oblique to strike in domains I and II but approximately downdip in domain III. Strongly asymmetric kinematic indicators are observed in thin sections cut parallel to the L1 lineation and perpendicular to the S1 foliation. This finding suggests that the lineation provides a close approximation of the transport direction during noncoaxial flow. In domain I, the L1 lineation plunges shallowly to the northeast, and the sense of shear is described by oblique, left-lateral, normal-sense motion. In restored domain II, the L1 lineation plunges shallowly to the south-southwest, and the sense of shear is described by oblique, left-lateral, thrust-sense motion. In domain III, the L1 lineation is approximately downdip; the sense of shear is not known. The similarity in lineation orientation between domain III and the Slate Belt suggests that, in terms of structure, domain III belongs with the Slate Belt. We include domain III with the Slate Belt and refer to this composite domain as the slate-belt domain. The slate-belt domain is characterized by an approximately downdip L1 lineation and predominantly northwest-vergent thrusting (Clark et al., 1992; Clark, 1994; Tillman and Byrne, 1995; Tillman and Byrne, 1996).

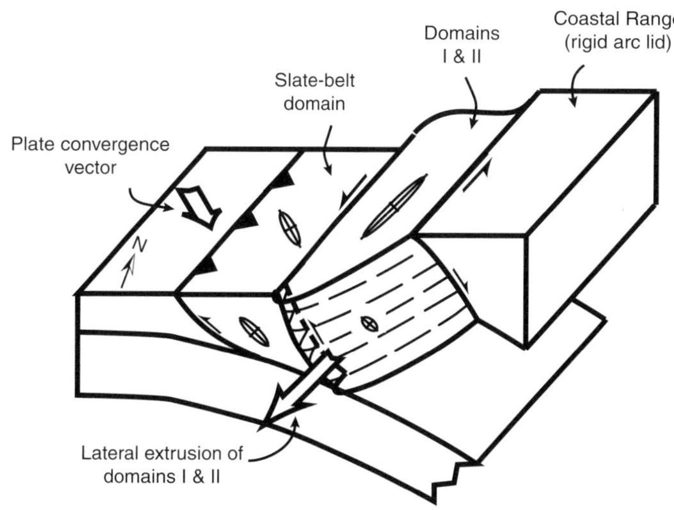

Figure 8. Schematic diagram depicting transpression and lateral extrusion in pre-Tertiary metamorphic basement in response to oblique convergence of Philippine Sea and Eurasian plates.

The change from predominantly oblique, left-lateral motion in domains I and II to predominantly dip-slip motion in the slate-belt domain coincides with a change in the state of strain. Microstructural evidence for shortening parallel to the *Y*-axis in the Chipan granitic gneiss in domain I and the presence of folds with hinge lines subparallel to the L1 lineation in the phyllites in domain II indicate that the state of strain is constrictional. In the Slate Belt along the Central Cross-Island Highway, synkinematic fibers in strain fringes around pyrite framboids are straight, parallel to the L1 lineation, and maintain a constant width in cleavage-parallel sections, indicating plane-strain deformation (Clark et al., 1992, 1993; Clark, 1994; Tillman and Byrne, 1995).

Faure et al. (1991) presented an interpretation for the structural evolution of the hinterland of Taiwan in which they argued that the pre-Tertiary metamorphic basement in the eastern Central Range is on the overturned limb of a regional-scale fold nappe. In this interpretation, unfolding of the nappe restores the L1 lineation to a northwest trend. The pre-Tertiary metamorphic basement and Slate Belt are, therefore, considered to be a single structural unit characterized by northwest-vergent folds and southeast-dipping thrust faults. We argue against this interpretation for several reasons. First, there are no data to show that the pre-Tertiary metamorphic basement is either upright or overturned. Second, the northeast-trending L1 lineation is subparallel to the mesoscale fold axes and the π-axis defined by the variation in orientation of S1; unfolding will, therefore, not rotate the L1 lineation to a downdip orientation. Third, the interpretation does not account for the difference in state of strain between the pre-Tertiary metamorphic basement and the Slate Belt. Finally, the pre-Tertiary metamorphic basement is treated as a single unit in the rotation scheme, which does not account for the variation in orientation by domain of structural fabrics.

DISCUSSION

Models of transpression describe deformation within zones of oblique convergence as a combination of an orogen-parallel, strike-slip simple-shear component and an orogen-perpendicular, pure-shear shortening component with or without volume change (Sanderson and Marchini, 1984; Robin and Cruden, 1994; Tikoff and Teyssier, 1994). Although these idealized models generally assume a subvertical zone of deformation, they can be used to provide a first approximation of deformation within an obliquely convergent orogenic wedge such as Taiwan, whose shape and taper are controlled by competing factors such as the relative amounts of underplating, accretion, erosion, and internal deformation (Suppe, 1981; Platt, 1986).

Early models (Sanderson and Marchini, 1984; Robin and Cruden, 1994; Tikoff and Teyssier, 1994) assumed that zones of transpression are laterally confined and have a fixed lower surface and a free upper surface. In these models, horizontal shortening across the zone results in vertical extension and flattening strains. In laterally confined zones of transpression with a fixed lower surface and constant-volume deformation, two factors control the shape of the finite strain ellipsoid and the orientation of structural fabrics: (1) the angle between the plate-convergence vector and the normal to the plate boundary and (2) the degree of partitioning, i.e., the amount of orogen-parallel motion taken up along discrete zones of noncoaxial flow (Tikoff and Teyssier, 1994; Teyssier et al., 1995).

Given the relatively small degree of obliquity (20°) between the plate-convergence vector and the normal to the plate boundary in Taiwan, the model of Tikoff and Teyssier (1994) predicts a zone of transpression dominated by orogen-perpendicular contraction. In a laterally confined zone with no partitioning, the rocks would be characterized by a vertical maximum finite extension direction and flattening strains, inconsistent with the shallowly plunging L1 lineation in domains I and II in the pre-Tertiary metamorphic basement along the Central Cross-Island Highway. This inconsistency led Clark (1994) to suggest that deformation in the Taiwan hinterland is strongly partitioned and approaches end-member simple shear in the pre-Tertiary metamorphic basement.

In the interpretation of Clark (1994), oblique convergence is accommodated by orogen-perpendicular contraction in the Slate Belt and orogen-parallel simple shear in the pre-Tertiary metamorphic basement. Partitioning of deformation between the two belts results in distinct lineation orientations and kinematics. In the pre-Tertiary metamorphic basement, orogen-parallel motion is manifested as a shallowly plunging L1 and left-lateral, strike-slip kinematics, whereas in the Slate Belt, orogen-perpendicular contraction is evidenced by an approximately downdip L1 and top-to-the-northwest thrust kinematics.

Although the shallowly plunging L1 lineation and distributed left-lateral, east-side-to-the-northeast kinematics are consistent with partitioning of orogen-parallel simple shear into the pre-Tertiary metamorphic basement, the evidence for constrictional strain in domains I and II is not. Simple shear, by definition, is a plane-strain deformation. Although deformation partitioning between the pre-Tertiary metamorphic basement and the Slate Belt is likely, it is not sufficient to explain the difference in the state of strain between the two belts.

More recent models (Dias and Ribeiro, 1994; Jones et al., 1997; Fossen and Tikoff, 1998; Teyssier and Tikoff, 1999) have considered transpression within zones that are not laterally confined, i.e., the zone can extend parallel to the trend of the orogen. In these models, both lateral extrusion and vertical thickening are allowed, resulting in a range of possible strain states from pure constriction to pure flattening. We follow Jones et al. (1997) in defining lateral extrusion (also termed "lateral escape"; Dias and Ribeiro, 1994) as a horizontal stretch that results in the lengthening of a deformation zone relative to the adjacent wall rocks of the zone margins.

We argue that the constrictional state of strain and the shallowly plunging L1 lineation in domains I and II of the pre-Tertiary metamorphic basement imply that orogen-perpendicular contraction was accommodated by lateral extrusion. In contrast, evidence for plane-strain deformation and an approximately downdip lineation suggest that vertical thickening accommodated the orogen-perpendicular contraction in the slate-belt domain. Lateral extrusion has been described by workers in several orogenic belts, e.g., the Scandinavian Caledonides (Gillotti and Hull, 1993), southwest Cyprus (Jones et al., 1997), the Ibero-Armorican arc (Dias and Ribeiro, 1994), and the Himalaya (Tapponnier et al., 1982).

The D1 structural fabrics in the pre-Tertiary metamorphic basement are interpreted to have developed within a zone of transpression in which oblique convergence was accommodated by orogen-perpendicular contraction, lateral extrusion, and orogen-parallel strike-slip motion. Deformation in the slate-belt domain, in contrast, is characterized by orogen-perpendicular contraction balanced by vertical thickening.

Two features of the pre-Tertiary metamorphic basement are not easily reconciled with the predictions of the transpression models: (1) the obliquity of the shallowly plunging L1 lineation to the trend of the orogen observed in both domains I and II and (2) the change in the sense of the dip-slip component from normal in domain I to thrust in domain II in our reconstruction. Although the consistent shallow plunge of L1 suggests some vertical thickening, the constrictional state of strain implies that the ratio of vertical to horizontal extension was relatively small.

In addition, the opposing senses of the dip-slip component in domains I and II required in the reconstruction reflect the exhumation of the pre-Tertiary metamorphic basement relative to both the Slate Belt and the Coastal Range. The differences between the observations and the predictions of the models may be related to the simplifying assumptions used to create the models. A vertical zone of deformation may not accurately approximate the orientation and kinematics of structures within wedge-shaped orogenic belts.

The change from oblique extension and constrictional strain in domains I and II of the pre-Tertiary metamorphic basement to downdip extension and plane strain in the Slate Belt may be a function of their different locations within the orogenic wedge. The pre-Tertiary metamorphic basement was deformed in the back of the wedge and may have been confined vertically by a southeast-dipping, rigid arc lid, resulting in lateral extrusion and a constrictional state of strain within a zone of left-lateral transpression. The Slate Belt, in contrast, is characterized by vertical thickening and plane-strain deformation, which may be characteristic of deformation in front of the toe of the arc lid. The distinct structural style of the pre-Tertiary metamorphic basement is inferred to be the result of different boundary conditions at different positions within the Taiwan orogenic wedge.

CONCLUSIONS

Four structural domains can be distinguished in the pre-Tertiary metamorphic basement along the Central Cross-Island Highway in east-central Taiwan on the basis of the systematic variation in the orientation of penetrative D1 structural fabrics. In domains I, IIa, and IIb, the S1 foliation generally strikes northeast and the L1 lineation plunges shallowly to the northeast or southwest, raking 30°–45° on S1. Domain III, which is structurally similar to the Slate Belt, has an approximately downdip L1. A consistent left-lateral component of motion is revealed for domains I, IIa, and IIb by microscale kinematic indicators; the sense of the dip-slip component changes from top-to-the-northeast, normal sense in domain I to top-to-the-southwest, normal sense in domain IIa to top-to-the-southwest, thrust sense in domain IIb. Analysis of microstructures in the Chipan granitic gneiss and the geometry of mesoscale folds in interlayered marbles and phyllites provide qualitative evidence for a constrictional state of strain in domains I and II.

An evaluation of the geometrically and geologically reasonable primary arrangements of structural domains suggests that the pre-Tertiary metamorphic basement and Slate Belt initially formed as a southeast-dipping package composed of domain I, domain II (rotated IIa and IIb), and the slate-belt domain (domain III and the Slate Belt); post-S1 block rotation resulted in the present configuration of domains.

D1 structural fabrics in domains I and II of the pre-Tertiary metamorphic basement formed within a transpressional shear zone, characterized by an oblique, left-lateral sense of shear and a constrictional state of strain. The orogen-perpendicular contractional component of oblique collision was accommodated by lateral extrusion, rather than vertical thickening, possibly due to the presence of a shallowly dipping, rigid arc lid.

ACKNOWLEDGMENTS

This research was supported by National Science Foundation grant EAR-9706801 (to Crespi and Byrne) and by a Harold T. Stearns Fellowship award from the Geological Society of America (to Pulver). Pulver is especially grateful to Ching-Hua Lo and Tung-Yu Zhou of National Taiwan University for their support during our field seasons and for introducing him to the geology of Taroko Gorge. We thank Ching-Hua Lo, Chia-Yu Lu, and Yu-Chang Chan for many informative discussions about the geology of Taiwan. Pulver thanks the staff of Taroko Gorge National Park in Taiwan, especially Nan-Tsong Yang and Mr. Morgan, who provided generous accommodations and logistical support during Pulver's research.

REFERENCES CITED

Bryant, B., and Read, J.C., 1969, Significance of lineations and minor folds near major thrust faults in the southern Appalachians and the British and Norwegian Caledonides: Geological Magazine, v. 106, p. 412–429.

Chai, B.H.T., 1972, Structure and tectonic evolution of Taiwan: American Journal of Science, v. 272, p. 389–422.

Chan, Y.-C., 1990, Ductile shear structures from the Hojen to Sanchan area on the eastern flank of the Central Range, Taiwan [M.S. thesis]: Taipei, National Taiwan University, 72 p. (in Chinese).

Chen, C.H., 1989, A preliminary study of the fossil Dinoflagellates from the Tananao Schist, Taiwan [M.S. thesis]: Taipei, National Taiwan University, 89 p. (in Chinese).

Chi, W.R., Namson, J., and Suppe, J., 1981, Stratigraphic record of plate interactions in the Coastal Range of eastern Taiwan: Memoir of the Geological Society of China, no. 4, p. 491–530.

Clark, M.B., 1994, Kinematics and structural evolution of the slate belt and metamorphic core of an active arc-continent collision, Taiwan [Ph.D. thesis]: University Park, Pennsylvania State University, 264 p.

Clark, M.B., Fisher, D.M., and Lu, C.-Y., 1992, Strain variations in the Eocene and older rocks exposed along the Central and Southern Cross-Island Highways, Taiwan: Acta Geologica Taiwanica, no. 30, p. 1–10.

Clark, M.B., Fisher, D.M., Lu, C.-Y., and Chen, C.-H., 1993, Kinematic analyses of the Hsuheshan Range, Taiwan: A large-scale pop-up structure: Tectonics, v. 12, p. 205–217.

Crespi, J.M., Chan, Y.-C., and Swaim, M.S., 1996, Synorogenic extension and exhumation of the Taiwan hinterland: Geology, v. 24, p. 247–250.

Dias, R., and Ribeiro, A., 1994, Constriction in a transpressive regime: An example in the Iberian branch of the Ibero-Armorican arc: Journal of Structural Geology, v. 16, p. 1543–1554.

Ernst, W.G., 1983, Mineral parageneses in metamorphic rocks exposed along Tailuko Gorge, Central Mountain Range, Taiwan: Journal of Metamorphic Petrology, v. 1, p. 305–329.

Faure, M., Lu, C.-Y., and Chu, H.-T., 1991, Ductile deformation and Miocene nappe-stacking in Taiwan related to motion of the Philippine Sea plate: Tectonophysics, v. 198, p. 95–105.

Fisher, D., 1997, Kinematics of ductile deformation within the eastern Central Range, *in* Proceedings, International Conference and Sino-American Symposium on Tectonics of East Asia: Chungli, Taiwan p. 66–67.

Fletcher, J.M., and Bartley, J.M., 1994, Constrictional strain in a non-coaxial shear zone: Implications for fold and rock fabric development, central Mojave metamorphic core complex, California: Journal of Structural Geology, v. 16, p. 555–570.

Fossen, H., 1993, Linear fabrics in the Bergsdalen Nappes, southwest Norway: Implications for deformation history and fold development: Norsk Geologisk Tidsskrift, v. 73, p. 95–108.

Fossen, H., and Tikoff, B., 1998, Extended models of transpression/transtension and application to tectonic settings, *in* Holdsworth, R.E., Strachan, R.A., and Dewey, J.F., eds., Continental transpressional and transtensional tectonics: Geological Society [London] Special Publication 135, p. 15–33.

Gillotti, J.A., and Hull, J.M., 1993, Kinematic stratification in the hinterland of the central Scandinavian Caledonides: Journal of Structural Geology, v. 15, p. 629–646.

Ho, C.S., 1986, A synthesis of the geologic evolution of Taiwan: Tectonophysics, v. 125, p. 1–16.

Ho, C.S., 1988, An introduction to the geology of Taiwan and explanatory text of the geologic map of Taiwan: Taipei, Taiwan, Republic of China, Central Geological Survey, Ministry of Economic Affairs, 192 p.

Jahn, B.M., 1988, Pb-Pb dating of young marbles from Taiwan: Nature, v. 332, p. 429–432.

Jahn, B.M., Martineau, F., Peucat, J.J., and Cornichet, J., 1986, Geochronology of the Tananao schist complex, Taiwan, and its regional tectonic significance: Tectonophysics, v. 125, p. 103–124.

Jones, R.R., Holdsworth, R.E., and Bailey, W., 1997, Lateral extrusion in transpression zones: The importance of boundary conditions: Journal of Structural Geology, v. 19, p. 1201–1217.

Liou, J.G., and Ernst, W.G., 1984, Summary of Phanerozoic metamorphism in Taiwan: Memoir of the Geological Society of China, no. 6, p. 133–152.

Lo, C.-H., and Onstott, T.C., 1995, Rejuvenation of K-Ar systems for minerals in the Taiwan Mountain Belt: Earth and Planetary Science Letters, v. 131, p. 71–98.

Lo, C.-H., and Wang Lee, C., 1981a, Marble enclave in the Chipan granitic gneiss, Hualien, eastern Taiwan: Proceedings of the Geological Society of China, no. 24, p. 137–140.

Lo, C.-H., and Wang Lee, C., 1981b, Mineral chemistry in some gneissic bodies, the Hoping-Chipan area, Hualien, eastern Taiwan: Proceedings of the Geological Society of China, no. 24, p. 40–55.

Lu, C.-Y., Lin, N.-T., and Chu, H.-T., 1997, Field trip guide to the eastern part of Central Cross Island Highway, *in* Chu, H.-T., and Lee, J.-C., eds., International Conference and Sino-American Symposium on Tectonics of East Asia, Field Trip Guidebook: Taipei, Taiwan, Republic of China, p. 113–118.

Lu, C.-Y., and Wang Lee, C., 1986, The sheath folds in the Tananao Group between Tienshiang and Tailuko, East-West Cross-Island Highway, Taiwan: Tectonophysics, v. 125, p. 125–131.

McCaffrey, R., 1992, Oblique plate convergence, slip vectors, and forearc deformation: Journal of Geophysical Research, v. 97, p. 8905–8915.

Mount, V.S., and Suppe, J., 1987, State of stress near the San Andreas fault: Implications for wrench tectonics: Geology, v. 15, p. 1143–1146.

Northrup, C.J., and Burchfiel, B.C., 1996, Orogen-parallel transport and vertical partitioning of strain during oblique collision, Efjorden, north Norway: Journal of Structural Geology, v. 18, p. 1231–1244.

Oldow, J.S., Bally, A.W., and Ave Lallemant, H.G., 1990, Transpression, orogenic float, and lithospheric balance: Geology, v. 18, p. 991–994.

Passchier, C.W., and Trouw, R.A.J., 1996, Microtectonics: Berlin, Springer-Verlag, 289 p.

Platt, J.P., 1986, Dynamics of orogenic wedges and the uplift of high pressure metamorphic rocks: Geological Society of America Bulletin, v. 97, p. 1037–1053.

Ramsay, J.G., and Huber, M.I., 1983, The techniques of modern structural geology 1: Strain analysis: London, Academic Press, 307 p.

Ridley, J., 1986, Parallel stretching lineations and fold axes oblique to shear displacement direction: A model and observations: Journal of Structural Geology, v. 8, p. 647–653.

Ridley, J., and Casey, M., 1989, Numerical modeling of folding in rotational strain histories: Strain regimes expected in thrust belts and shear zones: Geology, v. 17, p. 875–878.

Robin, P.-Y.F., and Cruden, A.R., 1994, Strain and vorticity patterns in ideally ductile transpressive zones: Journal of Structural Geology, v. 16, p. 447–466.

Sanderson, D.J., and Marchini, W.R.D., 1984, Transpression: Journal of Structural Geology, v. 6, p. 449–458.

Seno, T., 1977, The instantaneous rotation vector of the Philippine Sea plate relative to the Eurasian plate: Tectonophysics, v. 42, p. 209–226.

Seno, T., and Maruyama, S., 1984, Paleogeographic reconstruction and origin of the Philippine Sea: Tectonophysics, v. 102, p. 53–84.

Skjernaa, L., 1989, Tubular folds and sheath folds: Definitions and conceptual models for their development, with examples from the Grapesvare area, northern Sweden: Journal of Structural Geology, v. 11, p. 689–703.

Stanley, R.S., Hill, L.B., Chang, H.C., and Hu, H.N., 1981, A transect through the metamorphic core of the Central Mountains, southern Taiwan: Memoir of the Geological Society of China, no. 4, p. 443–473.

Suppe, J., 1981, Mechanics of mountain building and metamorphism in Taiwan: Memoir of the Geological Society of China, no. 4, p. 67–89.

Tapponnier, P., Peitzler, G., Le Dain, A.Y., and Armijo, R., 1982, Propagating extrusion tectonics in Asia: New insights from simple experiments with plasticine: Geology, v. 10, p. 611–616.

Teyssier, C., and Tikoff, B., 1999, Fabric stability in oblique convergence and divergence: Journal of Structural Geology, v. 21, p. 969–974.

Teyssier, C., Tikoff, B., and Markley, M., 1995, Oblique plate motions and continental tectonics: Geology, v. 23, p. 447–450.

Tikoff, B., and Teyssier, C., 1994, Strain modeling of displacement-field partitioning in transpressional orogens: Journal of Structural Geology, v. 16, p. 1575–1588.

Tillman, K.S., and Byrne, T.B., 1995, Kinematic analysis of the Taiwan Slate Belt: Tectonics, v. 14, p. 322–341.

Tillman, K.S., and Byrne, T.B., 1996, Out-of-sequence thrusting in the Taiwan Slate Belt: Journal of the Geological Society of China, v. 39, p. 189–208.

Wallis, S.R., 1995, Vorticity analysis and recognition of ductile extension in the Sanbagawa Belt, SW Japan: Journal of Structural Geology, v. 17, p. 1077–1093.

Wang Lee, C., Wang, Y., Yen, T.P., and Lo, C.-H., 1982, Polymetamorphism in some gneissic bodies, Hoping-Chipan area, Hualien, eastern Taiwan: Acta Geologica Taiwanica, no. 21, p. 122–139.

Yen, T.P., 1953, On the occurrence of late Paleozoic fossils in the metamorphic complex of Taiwan: Bulletin of the Geological Survey of Taiwan, no. 4, p. 23–29.

Yen, T.P., 1960, A stratigraphic study of the Tananao Schist in northern Taiwan: Bulletin of the Geological Survey of Taiwan, no. 12, p. 53–66.

Manuscript Accepted by the Society April 12, 2001

Printed in the U.S.A.

Geological Society of America
Special Paper 358
2002

Estimation of the thermal structure of a young orogenic belt according to a model of whole-crust thickening

Teh-Ru Alex Song*
Kuo-Fong Ma
Institute of Geophysics, National Central University, Chung-Li, 320-54, Taiwan, Republic of China

ABSTRACT

A comprehensive understanding of the thermal properties beneath the Central Range of Taiwan is essential to decipher the process of mountain building and interpret the observed geophysical features, such as seismic velocity and seismicity. A first-order-approximation thermal model based on crustal thickening followed by constant erosion processes is described for the Central Range. The effect of groundwater circulation on the observed surface heat flow has been excluded in our evaluation. Our results indicate that crustal thickening causes a decrease in the geothermal gradient and the temperature. Erosion causes an increase in the geothermal gradient and the temperature. Through modeling, we derived the optimum geotherm below the Central Range by considering a thickening factor of 2 for crustal thickening occurring at 10 Ma following constant erosion since 5 Ma. The preferred final geotherm estimated a moderate geothermal gradient of ~17 °C/km and a temperature of ~210–550 °C at a depth of 10–30 km. The uncertainties in the thickening factor, the time of crustal thickening, and the prethickening crustal thickness have a temperature-difference effect of only a few tens of degrees compared to the temperature yielded by the preferred final geotherm. Other geotherm parameters such as radiogenic heat flow, scale depth, and surface heat production are also tested in the final geotherm calculations. Seismicity cutoff depth (i.e., the brittle-to-ductile transition depth) and seismic Q_p values set limits on the middle- and lower-crust temperatures determined by our model. The resulting moderate geothermal gradient differs from the notion that the crust beneath the Central Range is "hot," although the temperature in the shallow crust needs further investigation.

INTRODUCTION

Low seismic velocity and a lack of seismicity have been observed in the crust of several Cenozoic orogenic areas, such as the Alps (Meissner, 1986), Rocky Mountains (Prodehl and Pakiser, 1980), Himalaya (Mishra, 1982; Volvovsky et al., 1983), and Tibet (Rodgers and Schwartz, 1997). Similar phenomena from seismic tomography and local seismicity observations have also been revealed in Taiwan, a Cenozoic orogenic belt characterized by the collision between the Eurasian and Philippine Sea plates (Teng, 1990) (Fig. 1). Results of seismic tomography studies of Taiwan (Rau and Wu, 1995; Ma et al., 1996) indicate a low velocity in middle to lower crust beneath the Central Range. Figure 2 shows the tomographic P-wave

*Present address: Seismological Laboratory, California Institute of Technology, Pasadena, California 91125, USA.

Song, T.A., and Ma, K.-F., 2002, Estimation of the thermal structure of a young orogenic belt according to a model of whole-crust thickening, *in* Byrne, T.B., and Liu, C.-S., eds., Geology and Geophysics of an Arc-Continent Collision, Taiwan: Boulder, Colorado, Geological Society of America Special Paper 358, p. 121–136.

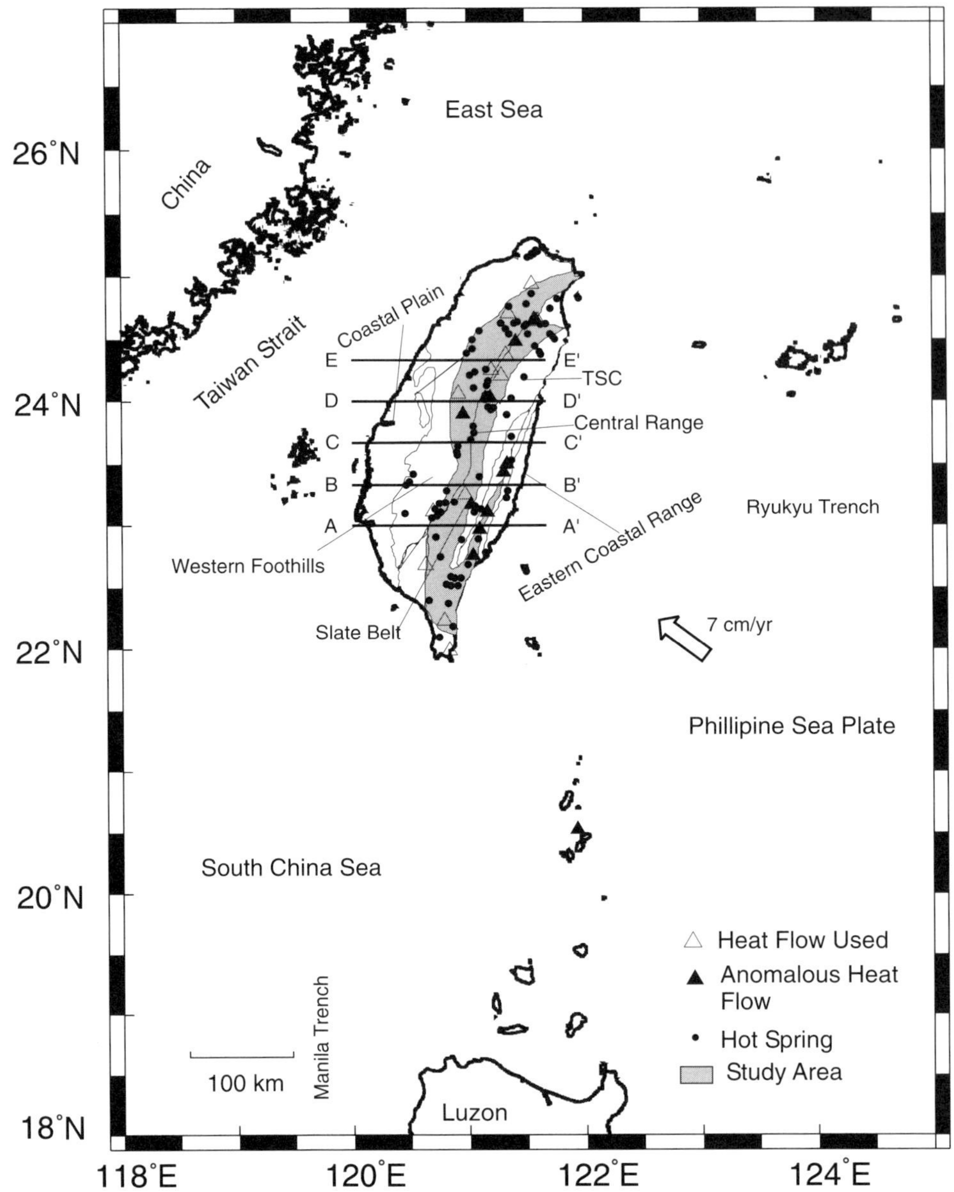

Figure 1. Geologic provinces, the metamorphic belt of Taiwan, and the distribution of heat flow (Lee and Cheng, 1986) used in the calculations (open triangles). The solid triangles indicate the abnormal heat flow of greater than 167 mW/m^2 or less than 33 mW/m^2. Solid circles represent the location of hot springs in Taiwan. Shaded area denotes the study area. A–A′ to E–E′ are the locations of profiles of seismicity and seismic P-wave velocity, as shown in Figure 2. TSC—Tananao Schist Complex.

velocity of Ma et al. (1996) with the background seismicity of M > 3.0 recorded from 1991 to 1998 by the Central Weather Bureau Seismic Network (CWBSN) of Taiwan along the profiles of A–A′ to E–E′ of Figure 1. Obvious low-velocity zones between the depths of 15 and 30 km were found in the profiles of A–A′ to E–E′; low rates of seismicity were also recognized in the profiles at these depths. The apparent low velocity and low seismicity in the middle and lower crust are commonly attributed to a high temperature and high geothermal gradient through those regions of the crust (Wu, 1978; Tsai, 1986; Wang et al. 1994; Wu et al. 1997). To decipher the possible significance of the seismic features, however, knowledge of the thermal structure is of fundamental importance.

The thermal structure in the Taiwan region had been investigated by Barr and Dahlen (1989) and Hwang and Wang (1993), respectively, to understand the mountain-building process in Taiwan. Barr and Dahlen (1989) considered a critical-wedge model to obtain the thermal structure of an actively deforming critically tapered fold-and-thrust belt (Suppe, 1980). Hwang and Wang (1993) analyzed the thermal evolution of Taiwan according to the model of discrete sequential thrusting. In these two studies, heat comes mainly from friction heating and shear-strain heating. Both groups used the observed heat-flow data in the Central Range (Lee and Cheng, 1986) to place constraints on their models. However, the resulting heat-flow values were anomalously high (>5 HFU) as the thermal evaluations did not account for the local groundwater-circulation effect. Lachenbruch and Sass (1977) and Lachenbruch et al. (1995) showed that the upward and downward circulation of groundwater could result in anomalously high and low surface

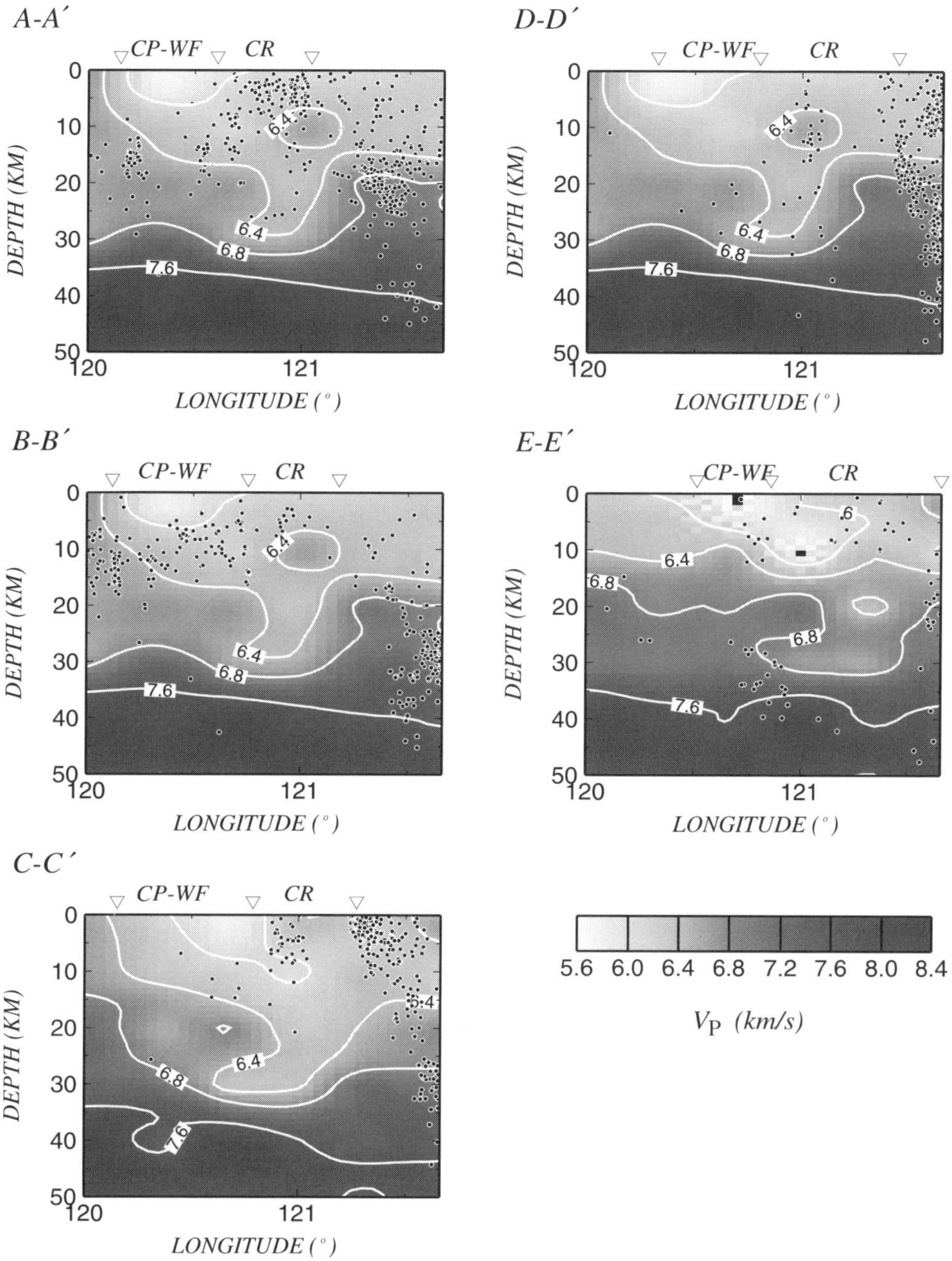

Figure 2. Seismicity and seismic P-wave velocity along profiles A–A′ to E–E′. Dots represent seismicity (M ≥ 3.0) recorded by the Central Weather Bureau Seismic Network in 1991–1998. Seismic velocity data are inferred from three-dimensional seismic tomography (Ma et al., 1996). CP-WF—Coastal Plain and Western Foothills, CR—Central Range.

heat-flow values. Lee and Cheng (1986) suggested that the anomalous surface heat flow observed in Taiwan (Fig. 1) is probably due to the groundwater circulation. Our groundwater-circulation simulation in this study supports this possibility. Thus, the use of the anomalously high surface heat flow as a constraint on the thermal-structure evaluation might not be appropriate.

In addition, both of these thermal models of Taiwan considered a contractional motion along a shallow décollement; this motion provided the frictional heating and internal strain heating for the deforming brittle wedge. However, several studies (England, 1978; Lachenbruch et al., 1995) have shown that the friction- and shear-strain heating might have only a small effect on the present-day heat flows within the possible lithospheric strength limitation of ~1 kbar (Ord and Hobbs, 1989; Lamontagne and Ranalli, 1996). For this reason, we propose a crust-thickening model, which is considered appropriate for thermal modeling of Taiwan because it is the site of active arc-continent collision. However, the type of thickening is debatable; thickening values vary with different tectonic concepts (e.g., Toksoz and Bird, 1977; England and Thompson, 1984; Batt and Braun, 1997). Our geotherm evaluation supports the model of whole-crust thickening, suggested by England and Thompson (1984).

Evaluation of geotherms is difficult in an orogenic belt, because the crustal deformation and the associated thermal evolution are not steady, but rather time dependent. In this study, we quantitatively evaluate the first-order thermal properties beneath the Central Range of Taiwan. A model involving whole-crust thickening is considered to satisfy the regional average surface heat flow, excluding the anomalously high and low values. Other minor heat sources, such as frictional heating, metamorphic heating, shear-strain heating, and heterogeneous thermal conductivity are neglected in our first-order approximation. It must be noted that our model is not sensitive to the geothermal gradient and temperature in the shallow crust (<10 km) because of the assumption of constant erosion rate. Horizontal heat advection caused by horizontal motion is also not considered in our calculations. Several synthetic tests on thermal parameters are conducted to justify our calculations.

GEOLOGIC SETTINGS

Taiwan is situated on the eastern margin of the Eurasian continent and is the consequence of an arc-continent collision between the Eurasian and Philippine Sea plates (Fig. 1). Radiometric dating indicates that the onset of the Cenozoic orogeny in Taiwan was at ca. 10–12 Ma (e.g., Lan et al., 1990; Lo and Yui, 1996). From basin sedimentary-sequence analysis (Teng, 1990), the relevant collision may have begun at 5 Ma. The degree of deformation increases eastward from the Coastal Plain through the Western Foothills, to the Central Range. These three areas represent, respectively, the undisturbed region and the fold-and-thrust belt and metamorphic core formed in the Taiwan orogeny. The Central Range, situated in the center of the Taiwan orogenic belt, has existed since middle Miocene time (Ho, 1982, 1988). The Central Range mainly consists of metamorphic terranes including the Slate Belt geologic province and the Tananao Schist Complex (Fig. 1). The Slate Belt consists of slates, argillites, and phyllites. The predominant constituents in the Tananao Schist Complex are green schists, black schists, and some minor gneisses. From west to east in the Central Range, the metamorphic facies increase from zeolite, prehnite-pumpellyite, to greenschist; amphibolite facies is finally reached in the northeastern part of the Central Range. High exhumation rates are not uncommon in the Central Range, especially in the region of the Tananao Schist Complex (Liu, 1982, 1988), according to analyses of fission-track data. Tsao (1996) showed that the exhumation rate has varied and has increased from the west to the east, on the basis of illite crystallinity, zircon fission-track ages, and K-Ar ages of metamorphic terranes in the Central Range, excluding the data related to faulting. Because of the abundance of faults and fractures, hot springs are well distributed in the Central Range (Chen, 1989) (Fig. 1). Owing to the complex structure in the Central Range, the thermal properties in the middle and lower crust are still under investigation. Because the thermal history of the Tananao Schist Complex in the Central Range is very complicated (Lan et al., 1990), in this study, we concentrate on resolving the thermal structure of the Slate Belt in the Central Range. The geotherm parameters used in our model therefore correspond to the Slate Belt region in the Central Range.

GEOTHERMAL ESTIMATIONS

In this study, we have tried to incorporate all of the available data related to the geothermal structure to obtain a first-order approximation of the regional average thermal structure beneath the Central Range of Taiwan. The present regional average thermal structure, in our model, is mainly controlled by three factors: (1) heat flow from the mantle, (2) the decay of radiogenic elements in the continental crust, and (3) the geotherm perturbation from crustal thickening, uplift, and erosion. The first two and the last factors represent the steady and transient factors, respectively.

The steady factor

As long as the tectonic environment is stable, only the radiogenic heat flow (Q_A) and mantle heat flow (Q_L) will contribute to the surface heat flow (Q_0) (England and Richardson, 1980), i.e., $Q_0 = Q_L + Q_A$ (all Q terms are in units of mW/m^2). The radiogenic heat flow is related to the amount of radiogenic heat produced and its depth distribution. The surface radiogenic heat production (A_0, in μW/m^3), which is related to the crustal content of potassium (K), thorium (Th), and uranium (U), can be evaluated from

$$A_0 = \rho \times 10^{-2} \times (3.48[\mathrm{K}] + 2.56[\mathrm{Th}] + 9.52[\mathrm{U}]), \quad (1)$$

where ρ is the density in kg/m^3, [K] is the K concentration in %, and [Th] and [U] are the Th and U concentrations in ppm (Rybach and Cermak, 1982). The heat production usually decays exponentially with depth in the form of $A(z) = A_0 \exp(-z/D)$, where $A(z)$ is the heat production at the depth z and D is the scale depth, which represents the depth at which the radiogenic heat production decays to $1/e$ of the surface radiogenic heat production. This exponential decay is due to a decrease in the abundance of the radiogenic elements with depth (Lachenbruch, 1970).

The nonradiogenic heat flow corresponds, in general, to the mantle heat flow in stable provinces (England and Richardson, 1980; Sclater et al., 1980; Morgan and Gosnold, 1989; Artemieva and Mooney, 2001). In this case, the geotherm in steady state could be calculated by

$$T(z) = \frac{Q_L}{k}z + \frac{D^2}{k}A_0\,(1 - e^{-z/D}), \quad (2)$$

for a one-dimensional regional average approximation (Lachenbruch, 1970), where $T(z)$ represents the temperature in °C at depth z, k is the thermal conductivity, D is the scale depth, and Q_L is the mantle heat flow from radioisotopes in the mantle and core.

The transient factors

If thickening, uplift, and erosion take place, time-dependent thermal perturbations will be recognized (e.g., Stuwe et al., 1994; Mancktelow and Graseman, 1997). In Taiwan, various orogenic models have been proposed by different groups (Teng, 1990; Lu and Hsu, 1992; Hsu and Sibuet, 1995; Wu et al., 1997; Hwang et al., 1997). However, which model is most appropriate is still under debate. In addition, the effect of the Taiwan orogeny on the lithospheric mantle is not well resolved. England and Thompson (1984) proposed a whole-crust homogeneous-thickening model for the active collisional environment of Taiwan. This model may be appropriate to use in investigating the regional thermal structure in the Taiwan orogenic belt. The model considers orogeny as due to a rapid phase of crustal thickening, during which little temperature change occurs in individual rocks, followed by erosion. The crustal thickening is assumed to occur by homogeneous horizontal shortening of the entire crust. The geothermal gradient below the crust is not affected by the thickening crust. In this case, the temperature perturbation (ΔT) due to crustal thickening can be estimated, for $z < f \times H$, as

$$\Delta \mathrm{T} = -\frac{\pi}{4}[1 + (l - fH)/l]\left[\frac{A_0De^{-1}(f^2 - 1)}{2k} + \frac{Q_LH(f - 1)}{k}\right], \tag{3a}$$

(modified from England and Thompson, 1984), where f is the thickening factor, l is the lithosphere thickness, and H is the crustal thickness before thickening. When $z > f \times H$, the temperature perturbation should be represented as

$$\Delta \mathrm{T} = -\frac{\pi}{4}(1 + (l - fH)/l)\left[\frac{A_0De^{-1}(f^2 - 1)}{2k} + \frac{Q_LH(f - 1)}{k}\right] \times (l - z)/(f \times H) \tag{3b}$$

ΔT represents the mean thermal perturbation between the surface and the base of the lithosphere when crustal thickening is complete. Equation 3 could express the thermal perturbation caused by the crustal thickening and the altering of heat production. The perturbed temperature was distributed in the lithosphere as a periodic cosine function that decays exponentially with time as

$$\Delta T(z, t) = \Delta T\cos\frac{\pi Z}{2l}\exp\left[\frac{-\pi^2\kappa t}{4l^2}\right], \tag{4}$$

where κ is the thermal diffusivity. $\Delta T(z,t)$ represents the geothermal perturbation at depth z after crustal thickening for time t. $\Delta T(z,t)$ is mainly dependent on the thickening factor (f), lithosphere thickness (l), and prethickening crust thickness (H).

When the crust undergoes tectonic uplift and erosion, these processes will lead to thermal advection and result in erosion-caused heat flow (Q_E), which contributes an increase of non-radiogenic heat flow. For a constant erosion rate, the erosion heat flow can be written as

$$Q_E(\lambda) = k\left.\frac{\partial T}{\partial z}\right|_{t=t_1}[1 + 4p^2 - 4i^2\mathrm{erfc}(p)]; \tag{5a}$$

where

$$p = \frac{H_E}{2\sqrt{\kappa\lambda}}, \tag{5b}$$

from England and Richardson (1980), where H_E is the erosion thickness, λ represents the duration from the initiation of erosion to the present time (called the "erosion time period" hereafter), and i^2 erfc represents the repeated integrals of the error function, and $\left.\frac{\partial T}{\partial z}\right|_{t=t_1}$ is the average thermal gradient above H_E at the time t_1 when the erosion had just been initiated. Erosion would cause not only heat-flow increase but also reduction of radiogenic heat production. The surface heat production A_0 and surface heat production before erosion $A_0(\lambda)$ has the relationship as

$$A_0 = A_0(\lambda)\exp(-H_E/D), \tag{6}$$

where H is the initial crust thickness before the onset of thickening. If we assumed that the crustal thickening occurred at time t_0 and that the erosion occurred at time t_1, the geotherm at time t_f due to crustal thickening and erosion can be written as

$$T(z, t_f) = \frac{Q_r}{k}z + \frac{D^2}{k}A_0(t_f - t_1)[1 - e^{-z/D}] + \Delta T(z,t_f - t_0), \tag{7}$$

where $T(z,t_f)$ represents the geotherm at time t_f, $A_0(t_f - t_1)$ is the surface heat production before erosion, and the reduced heat flow (Q_r) is now the combination of mantle heat flow (Q_L) and erosion heat flow (Q_E) for the erosion time period $\lambda = t_f - t_1$; therefore, $Q_r = Q_L + Q_E(\lambda)$.

Determination of geotherm parameters

Table 1 lists the geotherm parameters used in our calculation. However, several geotherm parameters are not available as they are difficult to obtain in the Central Range. Some arguments related to these geotherm parameters are addressed next. The sensitivities of the geotherm evaluation to these geotherm parameters are examined later.

Surface heat flow, Q_0. The distribution of the observed surface heat-flow data in the Central Range (Lee and Cheng, 1986) is shown in Figure 1. Lee and Cheng (1986) suggested that the observed anomalously high (>167 mW/m^2) and low heat-flow values (<33 mW/m^2) might be caused by groundwater circulation. Excluding the anomalously high and low heat-flow values, the average heat-flow value in the Central Range will be about $Q_0 = 97 \pm 32$ mW/m^2. To examine this hypothesis, we simulated the groundwater circulation to see its effect on the surface heat flows.

Lachenbruch and Sass (1977) showed that the surface heat flow at a local site will increase or decrease with the upward or downward movement of groundwater. The disturbance (ΔQ) in surface heat flow due to the moving groundwater can be represented by

$$\Delta Q = Q_0[e^{\pm(h/s)} - 1], \quad (8)$$

where h is the depth of circulating groundwater, $s = \alpha/V$ is the characteristic length scale for groundwater circulation, where α (~18.96 m^2/yr) is thermal diffusivity of water and V is velocity of the groundwater. The positive and negative signs of (h/s) indicate the upward and downward movement of groundwater, respectively. Chen (1989) suggested that the depth of groundwater circulation beneath the Central Range is ~4 km, which is comparable with the general observations (Elder, 1965; Lachenbruch and Sass, 1977). Because the velocity of groundwater is not well known in Taiwan, we used various V values in the calculations for a simple idealized one-dimensional model of groundwater circulation (see Appendix). After the calculations, we consider $V = 0.0075$ m/yr to be the velocity of groundwater circulation in the current study. For the circulated depth of 4 km, this velocity of groundwater circulation will require 130000 yr (the stabilization time) to achieve equilibrium, which is most comparable to the common observation of stabilization time of 10^5 yr for $h = 4$ km (Elder, 1965).

TABLE 1. PARAMETERS USED IN THE GEOTHERMAL CALCULATION

Thermal conductivity, k	2.5 mW/m^2
Present heat production, A_0	1.711 μW/m^3
Mantle heat flow, Q_L	27 mW/m^2, 38 mW/m^2
Scale depth, D	7 km, 11 km
Thermal diffusivity, κ	1.5×10^{-6} m^2/s
Crustal thickness, H	30 km
Lithospheric thickness, l	100 km
Surface heat flow, Q_0	97 ± 32 mW/m^2
Radiogenic heat flow, Q_A	39 ± 13 mW/m^2
Erosion thickness, H_E	8.2 km, 8 km
Heat production before erosion, $A_0(\lambda)$	5.52 μW/m^3, 3.56 μW/m^3
Thickening factor, f	2
Erosion heat flow, Q_E*	22.57 mW/m^2

*Q_E estimated for $Q_L = 38$ mW/m^2, $D = 7$ km, and $\lambda = 5$ Ma. Other estimations are listed in Table 3.

The corresponding characteristic length, s, of the groundwater circulation with the velocity of 0.0075 m/yr is ~2528 m. Considering $Q_0 = 97 \pm 32$ mW/m^2, $h = 4$ km, and $s = 2528$ m, from Equation 8, we obtain anomalously high heat-flow values of ~250 to 500 mW/m^2 and anomalously low heat-flow values of ~10 to 30 mW/m^2 due to the upward and downward movement of groundwater circulation, respectively. Figure 3 shows that the observed anomalous heat-flow values are compatible with our estimation. The plot suggests that the measured high and low heat-flow values in the Central Range mentioned by Lee and Cheng (1986) probably result from groundwater circulation. These anomalous heat-flow data, thus, are excluded from our thermal modeling. It would not be adequate to take these values as constraints in thermal-structure evaluation.

Radiogenic heat flow, Q_A. It is apparent that the radiogenic heat production should be reduced by the erosion process, while the augmentation caused by the crustal-thickening mechanism would yield an increase of crust materials. Thus, describing the radiogenic heat flow from orogenic crust is rather complicated. Huerta et al. (1996, 1998) attempted to explain the thermal structure as a response of accretion, erosion, and radiogenic heating. They mainly considered the trade-offs among erosion rate, accretion velocity, and the heat production of those accreted materials to the thermal structure. A higher accretion rate and a lower erosion rate accompanying high heat-production material would result in higher temperature (~500 °C) at the depths of 20–30 km. However, determination of such parameters and their mutual effects on the geotherm beneath Taiwan are rather difficult because of the lack of data. Thus, for the estimation of radiogenic heat flow, we adopt the global average value as ~40% of surface heat flow (i.e., $Q_A = 0.40Q_0$). This relationship has been shown to be suitable regardless of the age of geologic provinces (Pollack and Chapman, 1977; England and Richardson, 1980; Vitorello and Pollack, 1980). It implies that the radiogenic heat flow increases in proportion to the surface heat flow owing to the augmentation of heat production during the orogenic process. Considering this approximation, we estimated the radiogenic heat flow to be ~39 ± 13 mW/m^2 in the Central Range. However, other approximations of 35% and 45% of the surface heat flow as being due to radiogenic heat flow are also considered later to examine their effect on the final geotherm evaluation.

Mantle heat flow, Q_L. Sclater et al. (1980) considered that the reduced heat flow (Q_r) usually increases with transient heat flow and then decays over geologic time; Q_r reaches a constant heat flow of ~21–27 mW/m^2 as the background mantle heat flow. England and Richardson (1980) obtained the reduced heat flow of ~38 m/Wm^2 for Precambrian shields; this value is treated here as the background mantle heat flow. Because the mantle heat flow in Taiwan is not available, we use the value of 27 mW/m^2 and 38 mW/m^2 as the background mantle heat flow in our calculation. The effect of the uncertainty on the mantle heat flow is discussed in the next section.

Scale depth, D. Scale depth usually can be estimated from

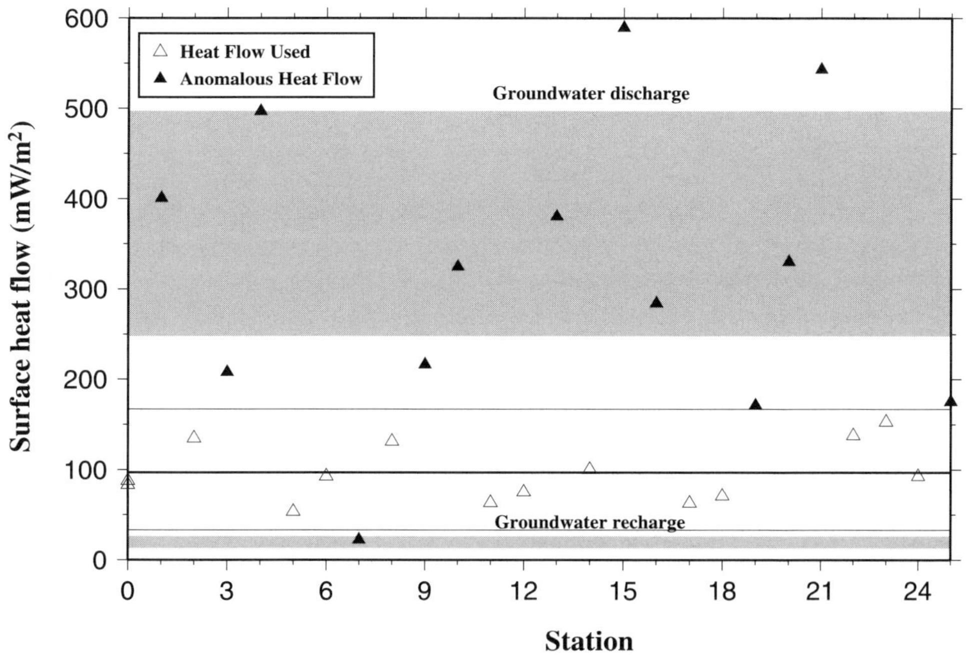

Figure 3. The range of calculated anomalous heat flows due to groundwater recharge (lower shaded region) and discharge (upper shaded region) compared with the observed anomalous heat flows (solid triangles) of greater than 167 mW/m^2 and less than 33 mW/m^2, suggested by Lee and Cheng (1986). Heat-flow values used in this study are shown by open triangles. Two thin solid lines are the heat flows of 167 mW/m^2 and 33 mW/m^2. Thick solid line represents the mean surface heat flow.

the observed surface heat flow (Q_0) and surface radiogenic heat production (A_0) as $Q_0 = Q_r + A_0D$ (Lachenbruch, 1970). After analyzing the heat-flow and production pairs reported in the global data, we concluded that the scale depth does not vary significantly among different geologic province (Vitorello and Pollack, 1980; Sclater et al., 1980; Haack, 1983). It is generally within 5–15 km. Because there is no observation of a pair of heat-flow and heat-production values in Taiwan, in this study, we adopt the scale depths of 7 and 11 km from Vitorello and Pollack (1980). The effect of the uncertainty of the scale depth is also discussed in the next section.

Erosion thickness, H_E, and heat production before erosion, $A_0(\lambda)$. The heat production usually decays exponentially with depth owing to the decrease in the radiogenic elements in the rocks, but the production also can be treated as constant within the crust. Chang (1989) surveyed the Central Range and compiled the amount of radiogenic elements in different lithologies. Table 2 shows the amount of radiogenic elements in different lithologies and their corresponding heat production. Radiogenic heat production varies with lithologies and decreases with the increase in the grade of metamorphism, in other words, it decreases with depth. Thus, the assumption that heat production decreases exponentially with depth might be more appropriate than constant heat production. Given that the metamorphic belts of Taiwan consist of about two-thirds slate and one-third phyllite, A_0 could be estimated at $\sim$1.711 $\mu W/m^3$. The present surficial radiogenic heat production (A_0) would have been buried at the depth z in the past before erosion and uplift (Vitorello and Pollack, 1980). If the crust is eroded by H_E during the erosion time period λ, the heat production before erosion, $A_0(\lambda)$, thus, can be estimated as $A_0(\lambda) = A_0\exp(H_E/D)$. Considering the reduced radiogenic heat flow due to the occurrence of erosion, the erosion thickness can be estimated as

$$H_E = D\ln\left[\frac{DA_0\exp(-H/D) - Q_A}{DA_0}\right], \qquad (9)$$

For crustal thickness H = 30 km and radiogenic heat flow = 39 mW/m^2, we estimate the erosion thickness H_E = $\sim$8.2 km and 8 km for the scale depths of 7 km and 11 km, respectively. According to the estimation of H_E, we can obtain $A_0(\lambda)$ values of 5.52 $\mu W/m^3$ and 3.56 $\mu W/m^3$ for H_E = 8.2 and 8 km, respectively. From sediment-accumulation analysis and radiometric dating complied by Teng (1990), the estimated erosion thickness is $\sim$8 km and the exhumation of rock is $\sim$10 km. Given the elevation of 1–4 km in the Central Range and H_E = 8 km, our calculations yield the exhumation depth of $\sim$9–12 km, which is compatible with the geologic evidence.

Thickening factor, f. The thickening factor indicates the ratio of the crustal thickness after and before crustal thickening. In the Taiwan orogenic belt, the prethickening crustal thickness could be reasonably postulated as 30 km, which is about the crustal thickness of the undisturbed Coastal Plain on the western side of Taiwan derived from P_n-wave study (Ma and Song, 1997), reflection experiments (Shih et al., 1998), and

TABLE 2. THE CONTENT OF RADIOGENIC ELEMENTS IN SURFACE METAMORPHIC ROCKS IN THE CENTRAL RANGE

Metamorphic Rocks	K (%)	U (ppm)	Th (ppm)	Density (kg/m³)	Radiogenic heat production (μW/m³)
Phyllite	3.13	2.81	16.3	2400	1.905
Slate	2.80	2.62	13.5	2400	1.662
Black schist	2.96	2.47	13.4	2400	1.637
Green schist	1.07	0.43	1.62	2400	0.287
Gneiss	1.90	1.85	9.70	2700	1.325

ray-tracing methods (Lee, 1999). The crustal thickness in the Central Range is ~45 km according to the P_n study of Ma and Song (1997). Therefore, it could be reasonable to consider the thickening factor as at least 1.5. Taking 9–12 km to be the amount of rock exhumation, as derived in the previous paragraph, the thickening factor becomes ~2. We take the thickening factor of 2 in our geotherm evaluation. Other thickening factors such as 1.5 and 2.5 are also tested later to justify the effects of the thickening factor on the geotherm estimation.

Time of thickening t_0 and time of erosion t_1 vs. erosion heat flow, Q_E. According to geologic and geochronologic data (Teng, 1990; Chen and Wang, 1995; Tsao, 1996), the collisional process in Taiwan is thought to have initiated at ca. 12–10 Ma. This time range is derived from the time of metamorphism, the direction change of plate motion, and sediment sequence analysis. We consider the time of deep burial metamorphism of 10 Ma (Tsao, 1996) as the initiation of crustal thickening in our thermal model. The time of initiation of erosion was estimated from the sediment-accumulation analysis in the Western Foothills and the Coastal Range (on the east side of Taiwan) (Teng, 1990). The relevant onset of erosion ought to be 5–3 Ma. Although there is no direct evidence to indicate when the erosion starts after crustal thickening, the delay between crustal thickening and erosion might be related to the density increase at the bottom of the crust from the granulite to eclogite transition (Richardson and England, 1979). Because the timing of initiation of erosion is sensitive to erosion heat flow (Q_E), we calculated the corresponding erosion heat flow for the erosion time period λ = 10 Ma, 5 Ma, and 3 Ma. By using the average surface heat flow as a constraint and considering the factor R,

$$R = \left[\frac{Q_A + Q_L + Q_E(\lambda)}{Q_0 - 1}\right] \times 100\%, \qquad (10)$$

we can estimate the optimized erosion time period λ in our calculation. The factor R indicates the compatibility between the sum of estimated Q_A, Q_L, and Q_E values and the observed surface heat flow. The smaller the absolute value of R ($|R|$), the more compatible the calculated results to the observed heat flow.

RESULTS AND DISCUSSION

Geotherm evolution

After detailed discussions about the determination of geotherm parameters, the estimated geotherms and their variation due to crustal thickening and erosion are shown in Figure 4. For convenience, we use letters to denote the calculations using pairs of Q_L and D values as follows: (A) Q_L = 27 mW/m² and D = 7 km, (B) Q_L = 27 mW/m² and D = 11 km, (C) Q_L = 38 mW/m² and D = 7 km, and (D) Q_L = 38 mW/m² and D = 11 km. Hereafter, we refer to these pairs of Q_L and D values as A, B, C, and D, respectively. Figure 4A shows the postulated initial geotherms before thickening. The geothermal gradient in the middle and lower crust (10–30 km) is ~15 ± 2 °C/km. The corresponding temperatures range from 250 ± 50 to 550 ± 100 °C. Figure 4B represents the corresponding geotherms accompanied by crustal thickening. The effect of crustal thickening reduces the temperature by ~150–200 °C at the depth of 25 km and reduces the average geothermal gradient to ~8 ± 2 °C/km in the depth range of 10–30 km.

Table 3 lists the factor R calculated for Q_L and D pairs A to D and for λ = 3 Ma, 5 Ma, and 10 Ma. The corresponding erosion heat flows and their contributions to the surface heat flow are also listed in Table 3. It is obvious that the effect due to the variations of scale depth on the estimation of erosion heat flow is less than that caused by variations in the mantle heat flow. The most important factor in evaluating the erosion heat flow is λ. Among these calculations, the one based on C for λ = 5 Ma shows results that are the most comparable to the observed values with R = 2.73%. The calculation based on C for λ = 5 Ma generates an erosion heat flow of ~22.57 mW/m², which contributes ~23.4% to the observed average heat flow. Other calculations using A (for λ = 3 Ma) and D (for λ = 10 Ma) also produce heat-flow values similar to the observed surface heat flow ($|R|$ = ~5%). Figure 4C depicts the geotherms for different time intervals of erosion for the optimum calculation using C (for λ = 5 Ma), A (for λ = 10 Ma), and D (for λ = 3 Ma), respectively. The geotherm estimated from the optimum calculation using C is thus considered as the final preferred geotherm. The geotherms of the latter two calculations can be treated as the deviated ranges of the geotherm. The final preferred geotherm shows that the temperature increases by ~200 °C at the depth of 25 km after thickening and erosion as shown in Figure 4C.

In general, the geotherm perturbations from crustal thickening and erosion are compensated for during the relative short relaxation time within 10 m.y. The final preferred geotherm has a geothermal gradient of ~17 °C/km in the depth range of 10–30 km. The geothermal gradient ranges deviated from 11 to 20

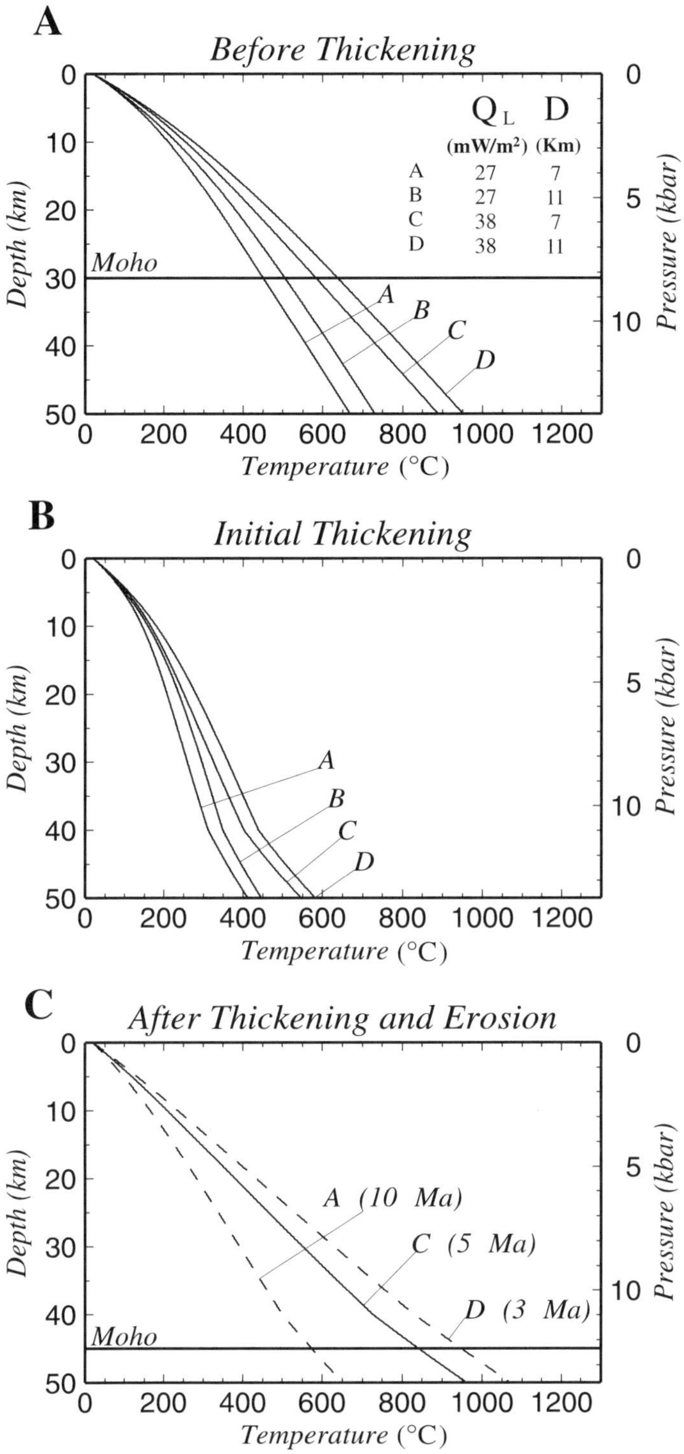

Figure 4. Average geotherm obtained considering three different parameters: (A) The radioactive heat production and mantle heat flow for different combinations of mantle heat flows (Q_L) and scale depths (D). (B) The crustal-thickening effect. (C) After thickening and erosion, the final preferred average geotherm was calculated by using the Q_L and D pair C (see text) for $\lambda = 5$ Ma. The results of calculations using Q_L and D pairs A (10 Ma) and D (3 Ma) are also shown for comparison (dashed lines).

°C/km at the depths of 10 km, 20 km, and 30 km are ~170–240 °C/km, 275–425 °C/km, and 390–630 °C/km, respectively. The effect of the uncertainty in the mantle heat flow (27 mW/m^2 or 38 mW/m^2) on the estimation of the geotherm, as compared in calculations using Q_L and D pairs A and C (Fig. 4C), results in an erosion time period uncertainty of ~2 m.y. and a temperature uncertainty of ~100 °C at the Moho (45 km).

Sensitivity test of geotherm parameters

Because the geotherm parameters used in our calculations are based on several assumptions, the uncertainties in our modeling caused by these assumptions are examined to justify our results.

First, we examine the effects of the thickening factor (f), the time of crustal thickening (t_0), and the prethickening crustal thickness (H) on the thermal perturbation due to crustal thickening because they are prior information for the calculation of the geotherm. Therefore, $f = 1.5$, 2, and 2.5, $t_0 = 8$ Ma, 10 Ma, and 12 Ma, and $H = 25$ km, 30 km, and 35 km are all simulated. On the basis of the equations 3 and 4, the thermal perturbations from the alternative 27 sets of thermal parameters are derived and shown in Figure 5. Comparing the thermal perturbations to the results of the preferred final geotherm calculation, where $f = 2.0$, $H = 30$ km, and $t_0 = 10$ Ma, the effect of changing the thickening factor (f) is seen to be the most significant. However, the effect of the ±0.5 difference in the thickening factor on the geotherm is only ~30 °C at the depth of 10 km and 60 °C at the depth of 30 km. The variation of crustal thickness yields a temperature difference of ~10 °C. As to the time of crustal thickening, it merely affects the temperature by ~5 °C for the relevant, short period since crustal thickening. Therefore, the uncertainties in the thickening parameter in our evaluation might not be significant to our result.

Second, we examine other geotherm parameters, such as present radiogenic heat flow (Q_A), scale depth (D), and present surface heat production (A_0). They are all closely related to the final geotherm. These parameters affect the estimation of the erosion thickness (H_E), initial surface radiogenic heat production ($A_0(\lambda)$), prethickening geotherm, and final geotherm. The estimation of erosion thickness would affect the evaluation of erosion heat flow and the final result as well. Besides using the radiogenic heat-flow value of 40% of surface heat flow (39 mW/m^2), radiogenic heat flows of 35% (34 mW/m^2) and 45% (44 mW/m^2) of surface heat flow are taken into consideration. In addition, D values of 5 km, 10 km, and 15 km and A_0 values of 1.0 μW/m^3, 1.5 μW/m^3, and 2.0 μW/m^3 are also considered in our calculations. Table 4 lists the estimations of erosion thickness (H_E) and initial surface heat production ($A_0(\lambda)$) from the various combinations of Q_A, D, and A_0. Smaller A_0 and larger D values would result in larger H_E values, but smaller $A_0(\lambda)$ values. Both $A_0(\lambda)$ and H_E increase with Q_A. For the same A_0 and D, Q_A will cause about 0.5–1.5 km and 0.3–1.0 μW/m^3 uncertainty in H_E and $A_0(\lambda)$, respectively. On the basis of the

TABLE 3. EROSION HEAT FLOWS (Q_E)

	Q_L (mW/m²)	D (km)	λ = 3 Ma			λ = 5 Ma			λ = 10 Ma		
			Q_E (mW/m²)	Q_E/Q_0 (%)	R (%)	Q_E (mW/m²)	Q_E/Q_0 (%)	R (%)	Q_E (mW/m²)	Q_E/Q_0 (%)	R (%)
A	27	7	35.24	36.50	+4.46	19.37	20.06	−11.97	11.90	12.33	−19.71
B	27	11	38.25	39.69	+7.58	21.57	22.34	−9.70	13.06	13.53	−18.51
C	38	7	41.12	42.59	+21.83	22.57	23.38	+2.73	13.80	14.29	−6.34
D	38	11	44.02	45.59	+24.95	24.78	25.67	+5.02	14.94	15.47	−5.17

Note: The erosion heat flows (Q_E) were obtained by considering different combinations of mantle heat flow (Q_L) and scale depths (D) for erosion time periods of λ = 3 Ma, λ = 5 Ma, and λ = 10 Ma, respectively.

parameters in Table 4, the estimation of the prethickening geotherm could be derived as depicted in Figure 6. Several estimations shown in this figure are comparable even with different combinations of these thermal parameters. Compared with the previous preferred geotherm calculation in which D = 7 km or 11 km and A_0 = 1.711 μW/m³, most estimations of the prethickening geotherm deviate from the results of previous calculations using Q_L and D pairs A–D by ~50 °C at 30 km. Overall, thermal parameters as surface heat production ($A_0(\lambda)$), scale depth (D), erosion thickness (H_E), radiogenic heat flow (Q_A), erosion time (λ), and mantle heat flow (Q_L) are included in the final geotherm calculation. We first exclude those estimations of H_E (<8 km) and $A_0(\lambda)$ (>6 μW/m³) in Table 4 that are incompatible with common observations (Teng, 1990; Rudnick, 1992); then we combine different values for mantle heat flow and erosion time period, and finally we obtain 72 sets of geotherm evaluations. By using the criterion $|R|$ as defined in the previous section, we obtain 12 possible geotherms with $|R| < 5\%$. The results are shown in Figure 7. The 12 possible final geotherms are almost within our previous calculations using Q_L and D pairs A–D. Although there are uncertainties regarding those critical thermal parameters, synthetic tests confirm that our results are robust. In the following, we discuss the correlation between our result and other independent constraints, such as seismic attenuation Q_p, the maximum depth of seismicity, and geologic evidence.

Seismic Q_p values set limits on deep-crustal temperature

For another independent seismic constraint, in this paper we consider the relationship between seismic anelasticity and temperature. The laboratory measurements on anelasticity of dry peridotite have shown that the seismic attenuation of compressional waves (Q_p^{-1}) depends on the ratio of temperature to the solidus temperature (T_m) as

$$Q_p/Q_{pm} = \exp[g(T_m/T - a)], \quad (11)$$

where Q_{pm} is the Q_p value at the solidus temperature T_m and T is the temperature (in kelvins). The constants g and a are 13.3 and 1.03 for $T_m/T > 1.08$. Q_{pm} is a linear function of pressure (Sato and Sacks, 1989):

$$Q_{pm} = Q_0 + P/P_0, \quad (12)$$

where P is the pressure and Q_0 and P_0 are 3.5 and 73 MPa, respectively. From the solidus of dry peridotite (Takahashi, 1986), the relationship between T and Q_p can be extrapolated to mantle depth. Once Q_p is derived, we can estimate temperature by equations 11 and 12. Chen et al. (1996) selected only high-quality data, employed a spectral decay method, and determined a three-dimensional Q_p structure of the Taiwan region to the depth of 100 km. According to the estimated average Q_p value in the Central Range, we infer its corresponding T_Q (temperature estimated from seismic Q_p) to compare with our final geotherm. Because the relationship of Q_p and T is limited to the upper mantle owing to the available laboratory measurements of peridotite, we only compare our result with T_Q at the depth of the upper mantle. The dry and wet solidus values for T_m had been determined by Sato (1994); we estimate the T_Q for dry and wet mantle, respectively. The estimated average Q_p at the depth of 35–50 km is ~610 from Chen et al. (1996). This Q_p value yields T_Q values of ~892 °C and 675 °C for dry and wet mantle, respectively. The temperature T_Q is approximately within the range of our geotherm evaluation, as depicted in Figure 8.

Seismicity cutoff depth sets limit on middle-crustal temperature

Actually, seismicity cutoff depth has been treated as a geothermometer in the crust (Doser and Kanamori, 1986; Bordi and Iizuka, 1993). It can be used to examine the temperature versus earthquake cutoff depth relationship (Sibson, 1982). If the seismicity cutoff depth (brittle-to-ductile transition) is determined, its corresponding temperature could be obtained. On the basis of this method, Song (1997) has shown that the temperatures related to the earthquake cutoff depths are ~270–400 °C and 350–510 °C, at depths of 15 km and 25 km, respectively. As depicted in Figure 8, these temperatures are also comparable with our estimated geotherm.

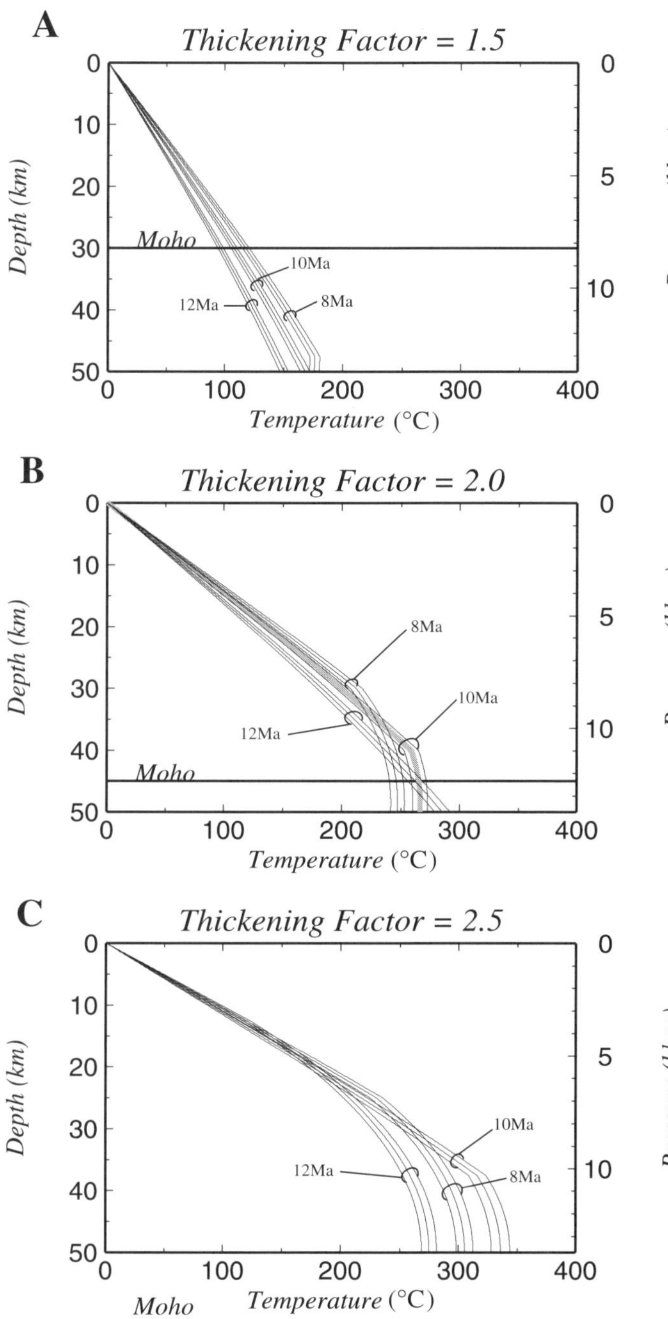

Figure 5. Results of thermal-parameter sensitivity test on whole-crust homogeneous thickening. Thickening times of 12 Ma, 10 Ma, and 8 Ma and prethickening crustal thicknesses of 25 km, 30 km, and 35 km are introduced in the test. The three lines in each thickening time correspond to the prethickening crustal thicknesses of 25 km, 30 km, and 35 km from left to right, respectively. (A) Temperature reduction with thickening factor of 1.5. (B) Temperature reduction with thickening factor of 2. (C) Temperature reduction with thickening factor of 2.5. The thick gray line in B represents the crustal-thickening effect considered in this study.

TABLE 4. THE EROSION THICKNESS

A_0 (µW/m³)	D	Q_A = 35% of Q_0		Q_A = 40% of Q_0		Q_A = 45% of Q_0	
		$A_0(\lambda)$ (µW/m³)	H_E (km)	$A_0(\lambda)$ (µW/m³)	H_E (µW/m³)	$A_0(\lambda)$ (µW/m³)	H_E (km)
1.0	5	6.68	9.5	7.69	10.2	8.67	10.8
1.0	10	3.42	12.3	3.90	13.6	4.35	14.7
1.0	15	2.38	13.0	2.68	14.8	3.00	16.5
1.5	5	6.77	7.5	7.58	8.1	8.55	8.7
1.5	10	3.44	8.3	3.92	9.6	4.37	10.7
1.5	15	2.44	7.3	2.77	9.2	3.08	10.8
1.71*	7*			5.52	8.1		
1.71*	11*			3.56	8.0		
2.0	5	6.77	6.1	7.63	6.7	8.61	7.3
2.0	10	3.46	5.5	3.95	6.8	4.41	7.9
2.0	15	2.51	3.4	2.83	5.2	3.15	6.8

Note: The erosion thickness (H_E) and $A_0(\lambda)$ were derived from a combination of different scale depths (D) and present surface heat production A_0. Q_A and Q_0 are in units of mW/m².
*The parameters used in our study.

TABLE 5. THE THERMAL PARAMETERS USED IN FINAL 12 POSSIBLE GEOTHERMS

No.	A (µW/m³)	D (km)	Q_A (mW/m²)	Q_L (mW/m²)	λ (Ma)
1	1.0	10	44	27	10
2	1.5	10	44	27	10
3	1.0	10	34	38	10
4	1.5	10	39	38	10
5	2.0	10	44	38	10
6	1.0	10	34	27	3
7	1.5	10	39	27	3
8	2.0	10	44	27	3
9	1.0	10	39	27	5
10	1.5	10	44	27	5
11	2.0	10	44	27	5
12	1.5	10	39	38	5

Geology and geochronology set limits on shallow-crustal temperature

Before comparing with the geologic evidence, it should be mentioned that the crustal heterogeneity is dramatic, especially in shallow crust. The exhumation rate, generally, is not constant though it is proportional to the metamorphic grade (Tsao, 1996). Therefore, our one-dimensional thermal model could not reflect this lateral variation, but only represent the average geotherm. Because of the complex thermal history in the Tananao Schist Complex (Lan et al., 1990; Lo and Yui, 1996), we do not attempt to discuss its thermal state. As mentioned previously, considering the amount of erosion and the topography of the Central Range, rocks beneath the Central Range might have been exhumed by 9–12 km. The metamorphic temperature and pressure are ~200–330 °C at 1–3 kbar in the Central Range (Chen and Wang, 1995), as depicted by the shaded box in Figure 8. After evaluating the possible geotherm in the Central Range, we calculate the corresponding pressure-temperature-time (*P-T-t*) path for rocks originally buried at 9 km (rock A) and 12 km (rock B) and exhumed at a constant rate, from our

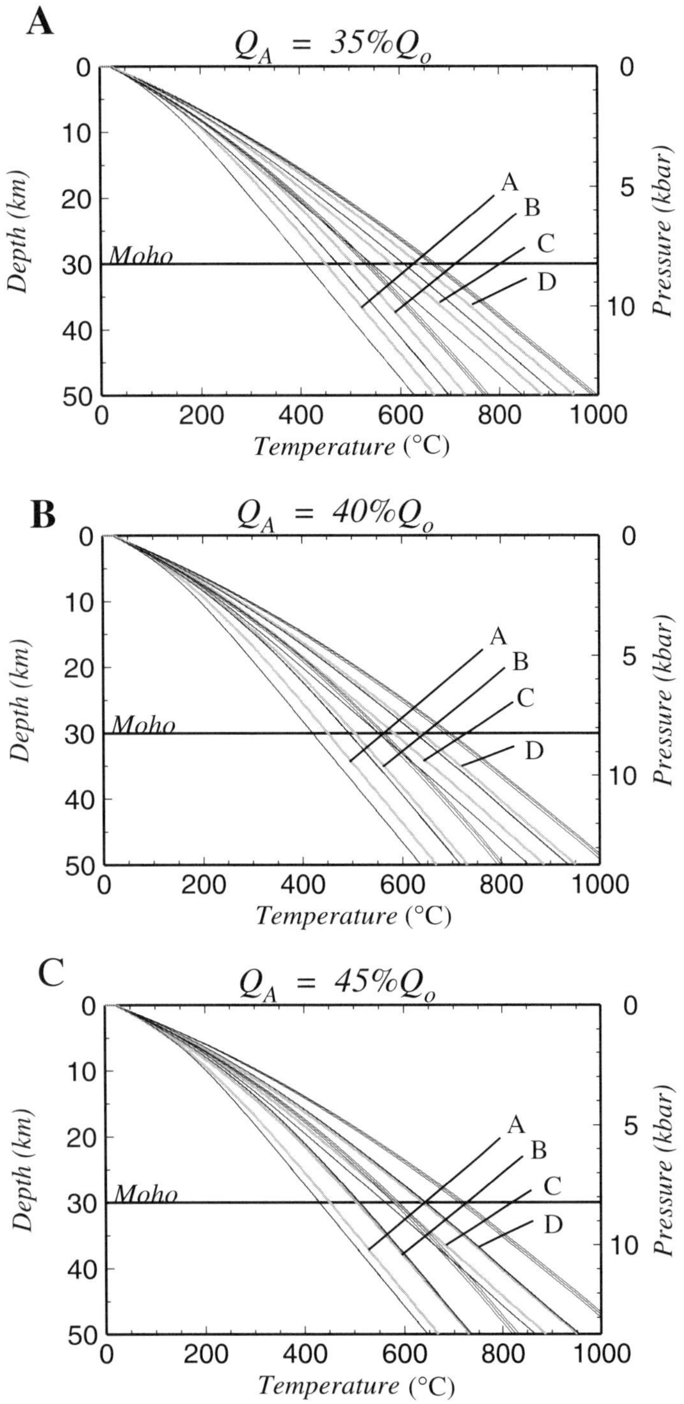

Figure 6. Prethickening geotherms (solid lines) calculated from the parameters used in Table 4 for the combination of A_0 values of 1.0 μW/m^3, 1.5 μW/m^3, 2.0 μW/m^3, and D values of 5 km, 10 km, and 15 km. Prethickening crustal thickness is 30 km. Results in (A), (B) and (C) represent the estimations with radiogenic heat flows of 35%Q_0, 40%Q_0, and 45%Q_0, respectively. Mantle heat flows of 27 mW/m^2 and 38 mW/m^2 are also involved in the calculation. The prethickening of geotherms Q_L and D pairs A–D used in our study are shown by gray lines for comparison.

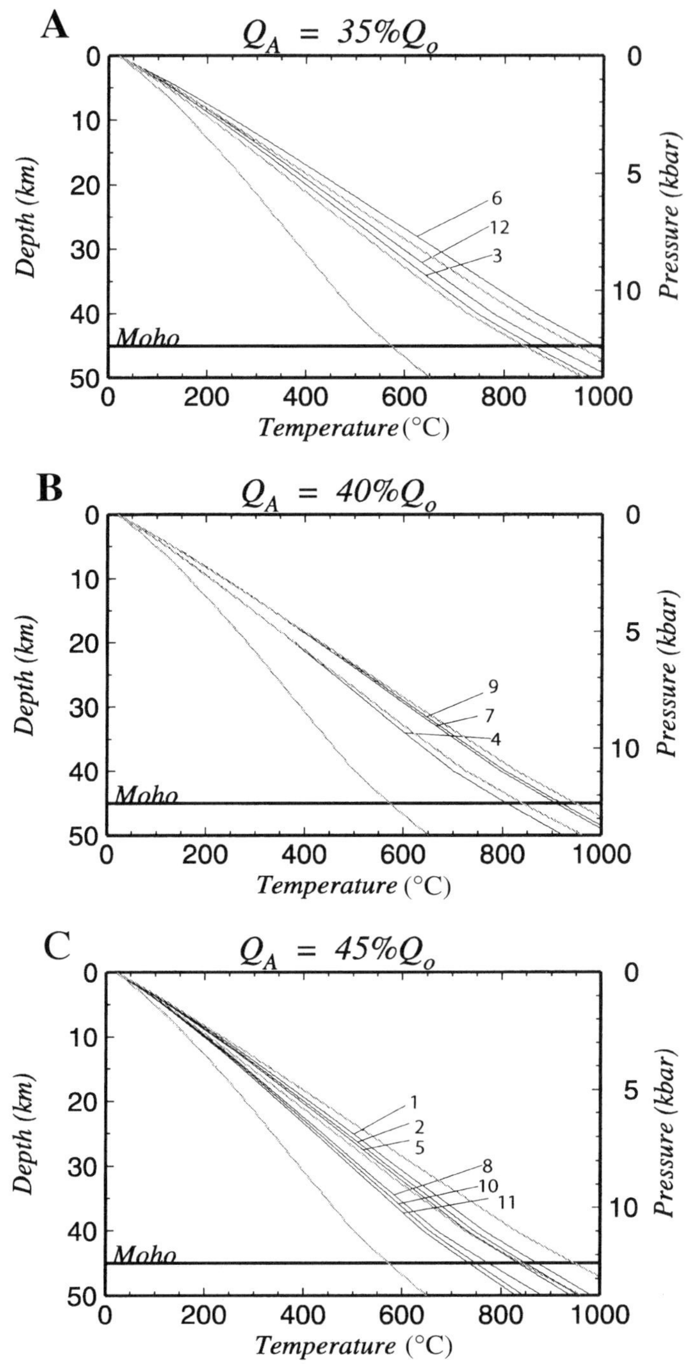

Figure 7. The final 12 geotherms obtained from 72 sets of geotherms in the thermal-parameter sensitivity test. The variables include radiogenic heat flow, scale depth, present surface heat production, mantle heat flow, and erosion time period as mentioned in the text. The solid lines are the final 12 geotherms. Table 5 shows the thermal parameters used in these 12 geotherms.

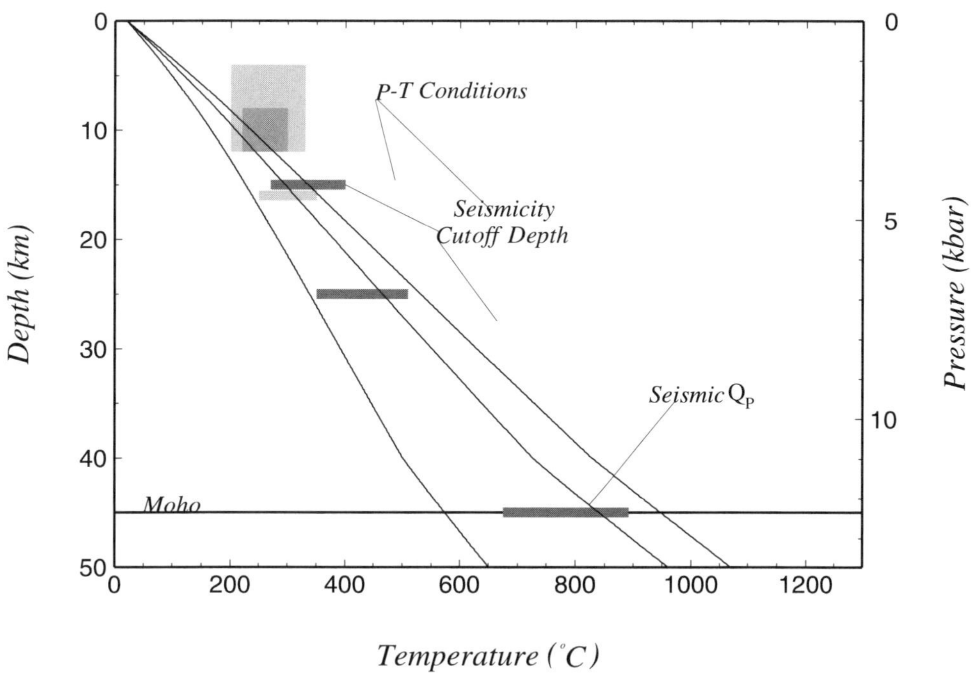

Figure 8. Comparison between the final preferred geotherm with *P-T* conditions determined by metamorphic facies (Chen and Wang, 1995), seismicity cutoff depth, and seismic Q_p (Song, 1997).

optimum results (calculations based on Q_L and D pairs A and C). The geotherm evolution since the initiation of erosion at 5 Ma and 3 Ma is shown in Figure 9, A and B, respectively. The blocking temperatures of apatite (~135 °C) and zircon (~235 °C) are represented by dotted lines. Figure 9A, for the erosion time period beginning at 5 Ma, shows that the exhumation rates corresponding to rock A and rock B are about 1.8 and 2.4 km/m.y., respectively. This exhumation rate is comparable with the exhumation rate of 0.3–2 km/m.y. in the Central Range obtained from the analysis of zircon fission-track data (Tsao, 1996). However, the maximum temperatures that rock A and rock B have undergone are merely ~110 °C and 130 °C, respectively, for the erosion time period beginning at 5 Ma. At their burial depths, neither of them is consistent with the blocking temperature of apatite and zircon and their cooling ages of 0.5 Ma and 2 Ma (Liu, 1982). A similar result could be seen in Figure 9B for the erosion time period of 3 Ma and exhumation rate of 3 and 4 km/m.y. for rock A and rock B, respectively. A deeper burial depth is necessary to satisfy the geochronologic evidence. The temperature in shallow crust seems a little underestimated, especially at a depth of less than 8 km. The principal reasons for the inconsistencies in the shallow crust are, probably, the assumption of constant erosion and exhumation rate and the whole-crust homogeneous thickening. Actually, the exhumation rate is getting faster in the Central Range (Liu, 1982, 1988; Teng, 1990; Tsao, 1996). If true, the real geotherm should slightly increase first and subsequently increase significantly. In addition, the whole-crust homogeneous-thickening model might not be adequate for shallow crust owing to its heterogeneity. Rocks usually deform brittlely in the shallow crust and ductilely in the middle and lower crust. Therefore, the whole-crust homogeneous thickening might be more adequate for the middle and lower crust because of their different rheological characteristics compared to the shallow crust. Instead of the whole-crust homogeneous-thickening model as in this study, for further study, a new model considering crustal thickening by thrusting above the brittle-to-ductile transition (Sibson, 1984) in the shallow crust, and homogeneous thickening in the middle and lower crust might be more appropriate. However, overall, the estimated average geotherm evaluation in the Central Range is considered satisfactory for the middle and lower crust.

In addition, as stated regarding the groundwater-circulation calculation, the anomalously high heat-flow values in the Central Range could be explained by the upward movement of groundwater. The purpose of the heat-flow measurements of Lee and Cheng (1986), in fact, was for hot-spring detection. The depths of those drilled holes are less than 150 m. For the shallow depth, local upward or downward movement of groundwater might have significant effects on the surface heat flow.

The moderate geothermal gradient inferred from our study beneath the Central Range suggests that temperature may not be the only factor responsible for the observed low seismic velocity and low seismicity in the middle and lower crust. Song (1997) has shown that a compositional difference could be responsible for the seismic velocity contrasts observed in the middle and lower crust between the Central Range and Coastal Plain–Western Foothills. Song (1997) adopted the composition and the thermal structure of the current study to construct the rheological structure and calculated the brittle-to-ductile behavior corresponding to the Central Range and Coastal Plain–Western Foothills. His results show that the seis-

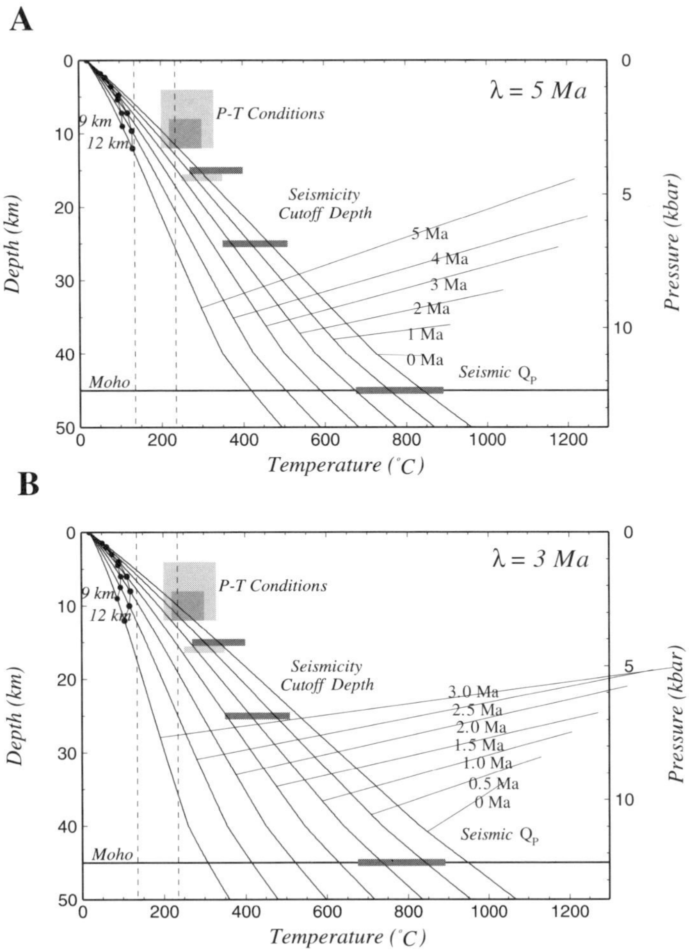

Figure 9. The geotherm evolution and *P-T-t* path derived in our study. (A) Geotherm evolution in 1 m.y. time intervals, considering a constant erosion rate since 5 Ma ($\lambda = 5$). (B) Geotherm evolution in 0.5 m.y. time intervals, considering a constant erosion rate since 3 Ma ($\lambda = 3$). Two vertical dashed lines represent the blocking temperature of 135 °C and 235 °C of apatite and zircon, respectively. Calculated *P-T-t* paths of rocks A and B initially buried at the depths of 9 km and 12 km are represented by dots. The depth of the Moho represents the present crustal thickness of ~45 km, denoted by horizontal thick lines.

micity of these two provinces can be satisfactorily explained by their brittle-to-ductile behavior. It suggests that the seismicity pattern and velocity beneath the Central Range might be controlled by not only temperature, but also by composition. A high geothermal gradient or temperature is not required for these apparent geophysical observations.

CONCLUSIONS

The thermal structure beneath the Central Range, Taiwan, has been estimated on the basis of the whole-crustal homogeneous-thickening model. Geotherms were calculated by considering as steady state the crustal radiogenic and mantle heat flow, whereas the ongoing tectonic collision provides a transient effect on the present geotherm. Groundwater circulation is considered to explain anomalously high and low heat flows. The transient factors of crustal thickening, uplift, and erosion have significant contributions to the geotherms. The crustal thickening results in a decrease of the temperature; the exhumation results in an increase of the temperature. The preferred final geotherm thus estimated has a geothermal gradient of ~17 °C/km in the middle and lower crust. Because most of the thermal parameters used in the geotherm calculation are subject to uncertainties owing to lack of available data, the sensitivity of our final geotherm calculation to the assumed geotherm parameters are examined. The sensitivity study shows that the parameters used in the final geotherm calculations are reasonable. The effects of the alternative geotherm parameters on the final geotherms are within the range of our estimated preferred geotherm. The preferred final geotherm is comparable with the temperature derived from the observed depth of seismic cutoff (i.e., the depth to the brittle-to-ductile transition depth) and seismic attenuation Q_p. However, the temperature at the shallow crust is not consistent with *P-T* conditions observed in the Central Range because of the assumption of constant erosion rate and exhumation rate in our model. Simulations of groundwater circulation show that the anomalous heat-flow values probably result from the upward and downward movement of groundwater. Crustal thickening plays an important role in the Taiwan orogeny. The moderate geothermal gradient and temperatures beneath the Central Range suggest to us that temperature might not be the only factor causing the observed lack of seismicity and seismic velocity in the Central Range.

ACKNOWLEDGMENTS

We thank reviewers B.-Y. Kuo, J.-H. Wang, and W.D. Mooney. Discussions with C.-H. Lo, F. Wu, and C.-Y. Wang were very helpful. This work was supported by NSC89-2119-M-008-001.

Appendix

For a circulation depth of ~4 km (Elder, 1965; Lachenbruch and Sass, 1977; Chen, 1989), from a simple idealized one-dimensional model for groundwater circulation, the temperature at the circulation depth due to the groundwater circulation can be estimated (Lachenbruch and Sass, 1977) as

$$T(t) = T(0) - T(0)[1 - (1/N)][1 - e^{-\beta^2}\,\mathrm{erfc}(\beta)], \quad \text{(A-1)}$$

where $T(0)$ is the temperature at the bottom of the circulated depth ($h = 4$ km) whereas the groundwater effect has not yet been taken into account. $\beta = [(N^2\alpha)/h^2]t$, and N has the form

$$N = \tfrac{1}{2}(e^{h/|s|} + e^{-h/|s|}), \quad \text{(A-2)}$$

by assuming a layer with uniform vertical flow, upward over half the area and downward over the other half. As $t \approx h^2/(N^2\alpha)$, termed the "stabilization time" t_W, $T(t = t_W)$ will approach the equilibrium temperature $T(0)/N$.

Table A-1 lists the corresponding stabilization times t_W and equilibrium temperatures for various groundwater velocities. For the cir-

TABLE A-1. THE EQUILIBRIUM TIME (t_w) AND TEMPERATURE (T) OF GROUNDWATER FOR VARIOUS GROUNDWATER VELOCITIES (V)

Velocity (10^{-2} m/yr)	Stabilization time, t_w (10^3 yr)	Temperature T (°C)
3.00	0.0107	6.22
2.00	0.730	8.7
1.50	6	14
1.00	48.126	28.75
0.75	131.239	43.61
0.50	325.601	65.31
0.25	646.678	89.64
0.20	709.828	93.63

culation depth of 4 km assumed in this study, Lachenbruch and Sass (1977) suggested the stabilization time should be on the order of about 10^5 yr. As listed in Table A-1, for the velocity of 0.0075 m/yr, the corresponding stabilization time t_W is about 130 000 yr, which is on the order of magnitude 10^5. We consider the groundwater velocity of 0.0075 m/yr as the appropriate velocity in our study.

REFERENCES CITED

Artemieva, I.M. and Mooney, W.D., 2001 Thermal thickness and evolution of Precambrian lithosphere: A global study: Journal of Geophysical Research, v. 106, p. 16 387–16 414.

Barr, T.D., and Dahlen, F.A., 1989, Brittle frictional mountain building. 2. Thermal structure and heat budget: Journal of Geophysical Research, v. 94, p. 3923–3947.

Batt, T.D., and Braun, J., 1997, On the thermomechanical evolution of compressional orogens: Geophysical Journal International, v. 128, p. 364–382.

Bordi, B., and Iizuka, S., 1993, Thermal regime, rheology, and seismicity in Central Range: Tectonophysics, v. 217, p. 1–9.

Chang, H.C., 1989, Radioactivity and heat generation of rocks in the Taiwan region (in Chinese) [Ph.D. thesis]: Chung-Li, National Central University, 159 p.

Chen, C.H., and Wang, C.H., 1995, Explanatory notes for the metamorphic facies map of Taiwan (in Chinese), (second edition): Central Geological Survey Special Publication, 51 p.

Chen, C.H., 1989, Hot springs and geotherm heat in Taiwan (in Chinese): Ti-Chih, v. 9, p. 327–340.

Chen, K.J., Yeh, Y.H., and Shyu, C.T., 1996, Qp structure in the Taiwan area and its correlation to seismicity: TAO, Terrestrial, Atmospheric and Oceanic Sciences, v. 7, p. 409–429.

Doser, D.I., and Kanamori, H., 1986, Depth of seismicity in the Imperial Valley region (1977–1983) and its relationship to heat flow, crustal temperature, and the October 15, 1979, earthquake: Journal of Geophysical Research, v. 91, p. 675–788.

Elder, J.W., 1965, Physical processes in geothermal areas, *in* Lee, W.H.K., ed., Terrestrial heat flow: American Geophysical Union Geophysical Monograph 8, p. 211–235.

England, P.C., 1978, Some thermal considerations of the Alpine metamorphism: Past, present and future: Tectonophysics, v. 46, p. 21–40.

England, P.C., and Thompson, A.B., 1984a, Pressure-temperature-time paths of regional metamorphism. 1. Heat transfer during the evolution of regions of thickened continental crust: Journal of Petrology, v. 25, p. 894–928.

England, P.C., and Richardson, S.W., 1980, Erosion and the age dependence of continental heat flow: Geophysical Journal of the Royal Astronomical Society, v. 62, p. 421–437.

Haack, U., 1983, On the content and vertical distribution of K, Th, and U in the continental crust: Earth and Planetary Science Letters, v. 62, p. 360–366.

Ho, C.S., 1982, Tectonic evolution of Taiwan and explanatory text of the tectonic map of Taiwan: Taipei, Taiwan, Republic of China, Central Geological Survey, Ministry Economic Affairs, 126 p.

Ho, C.S., 1988, An introduction to the geology of Taiwan and explanatory text of the geologic map of Taiwan: Taipei, Taiwan, Republic of China, Central Geological Survey, Ministry of Economic Affairs, 192 p.

Hsu, S.K., and Sibuet, J.C., 1995, Is Taiwan the result of arc-continent or arc-arc collision?: Earth and Planetary Science Letters, v. 136, p. 315–324.

Huerta, A.D., Royden, L.H., and Hodges, K.V., 1996, The interdependence of deformation and thermal process in mountain belts: Science, v. 273, p. 637–642.

Huerta, A.D., Royden, L.H., and Hodges, K.V., 1998, The thermal structure of collision orogens as a response to accretion, erosion, and radiogenic heating: Journal of Geophysical Research, v. 103, p. 15287–15302.

Hwang, C.Y., Wu, W.Y., Chang, C.P., Tsao, S., Yuan, P.B., Lin, C.W., and Yuan, X.K., 1997, Tectonic evolution of accretionary prism in the arc-continental collision terrane of Taiwan: Tectonophysics, v. 281, p. 31–51.

Hwang, W. T., and Wang, C.Y., 1993, Sequential thrusting model for mountain building: Constraints from geology and heat flow of Taiwan: Journal of Geophysical Research, v. 98, p. 9963–9973.

Lachenbruch, A.H., and Sass, J.H., 1977, Heat flow in the United States and the thermal regime of the crust, *in* Heacock, J.G., ed., The Earth's crust: Its nature and physical properties: American Geophysical Union Geophysical Monograph 20, p. 626–675.

Lachenbruch, A.H., 1970, Crustal temperature and heat production: Implications of the linear-flow relation: Journal of Geophysical Research, v. 75, p. 3291–3300.

Lachenbruch, A.H., Sass J.H., Clow, G.D., and Weldon, R., 1995, Heat flow at Cajon Pass, California, revisited: Journal of Geophysical Research, v. 100, p. 2005–2012.

Lamontagne, M., and Ranalli, G., 1996, Thermal and rheological constraints on the earthquake depth distribution in the Charlevoix, Canada, intraplate seismic zone: Tectonophysics, v. 257, p. 55–69.

Lan, C.Y., Lee, T., and Lee, C.W., 1990, The Rb-Sr isotope record in Taiwan gneiss and its tectonic implication: Tectonophysics, v. 183, p. 129–143.

Lee, S.J., 1999. The S-wave velocity structure under center-west Taiwan by string ground motion waveform modeling (in Chinese) [M. S. thesis]: Chung-Li, National Central University, 138 p.

Lee, C.R., and Cheng, W.T., 1986, Preliminary heat flow measurement in Taiwan: Proceedings, 4th Circum-Pacific Energy and Mineral Resources Conference, Singapore, p. 31–36.

Liu, T.K., 1982, Tectonic implication of fission track ages from the Central Range, Taiwan: Proceedings of the Geological Society of China, no. 25, p. 22–37.

Liu, T.K., 1988, Fission-track dating of Hsueshan Range: Thermal record due to arc-continent collision in Taiwan: Acta Geolologica Taiwanica, v. 26, p. 279–290.

Lo, C.H., and Yui, T.F., 1996, ^{40}Ar/^{39} Ar dating of high-pressure rocks in the Tananao basement complex, Taiwan: Journal of the Geological Society of China, v. 39, p. 13–30.

Lu, C.Y., and Hsu, K.J., 1992, Tectonic evolution of the Taiwan mountain belt: Petroleum Geology of Taiwan, v. 27, p. 21–46.

Ma, K.F., and Song, T.R., 1997, Pn velocity and Moho depth in Taiwan: Journal of the Geological Society of China, v. 40, p. 167–184.

Ma, K.F., Wang, J.H., and Zhao, D., 1996, Three-dimensional seismic velocity structure of the crust and uppermost mantle beneath Taiwan: Journal of Physics of the Earth, v. 44, p. 85–105.

Mancktelow, N.S., and Graseman, B., 1997, Time-dependent effects of heat advection and topography on cooling histories during erosion: Tectonophysics, v. 270, p. 167–195.

Meissner, R., 1986, The continental crust: A geophysical approach: Orlando, Florida, Academic Press, Inc., p. 361–362.

Mishra, D.C., 1982, Crustal structure and dynamics under Himalaya and Pamir ranges: Earth and Planetary Science Letters, v. 57, p. 415–420.

Morgan, P., and Gosnold, W.D., 1989, Heat flow and thermal regimes in the continental crust, *in* Pakiser, L.C., and Mooney, W.D., eds., Geophysical framework of the Conterminous United States: Geological Society of America Memoir 172, p. 493–522.

Ord, A., and Hobbs, B.E., 1989, The strength of the continental crust, detachment zones and the development of plastic instabilities: Tectonophysics, v. 158, p. 269–289.

Pollack, H.N., and Chapman, D.S., 1977, On the regional variation of heat flow, geotherms and lithospheric thickness: Tectonophysics, v. 38, p. 279–296.

Prodehl, C., and Pakiser, L.C., 1980, Crustal structure of the southern Rocky Mountains from seismic measurements: Geological Society of America Bulletin, v. 91, p. 147–155.

Rau, R.J., and Wu, F.T., 1995, Tomographic imaging of lithospheric structures under Taiwan: Earth and Planetary Science Letters, v. 133, p. 517–532.

Richardson, S.W., and England, P.C., 1979, Metamorphic consequences of crustal eclogite production in overthrust orogenic zones: Earth and Planetary Sciences Letters, v. 42, p. 183–190.

Rodgers, A.J., and Schwartz, S.Y., 1997, Low crustal velocities and mantle lithospheric variation in southern Tibet from regional Pnl waveforms: Geophysical Research Letters, v. 24, p. 9–12.

Rudnick, R.L., 1992, Restites, Eu anomalies, and the lower continental crust: Geochimica et Cosmochimica Acta, v. 56, p. 963–970.

Rybach, L., and Cermak, V., 1982, Radioactive heat generation in rocks, *in* Angenheister, G., ed., Physical Properties of Rocks, Numerical data and functional relationships in sciences and technology, New Series, Group 5, Volume 1a: Landolt-Bornstein, Berlin, Springer Verlag, p. 353–371.

Sato, H., 1994, H_2O and magmatism in island arc mantle inferred from seismic anelasticity and heat flow data: Journal of Physics of the Earth, v. 42, p. 439–453.

Sato, H., and Sacks, I.S., 1989, Anelasticity and thermal structure of the oceanic upper mantle: Temperature calibration with heat flow data: Journal of Geophysical Research, v. 94, p. 5705–5715.

Sclater, J., Jaupart, G.C., and Galson, D., 1980, The heat flow through oceanic and continental crust and the heat loss of the earth: Reviews of Geophysics, v. 18, no. 1, p. 269–311.

Shih, R.C., Lin, C.H., Lai, H.L., Yeh, Y.H., Huang, B.B., and Yen, H.Y., 1998, Preliminary crustal structures across central Taiwan from modeling of the onshore-offshore wide-angle seismic data: TAO, Terrestrial, Atmospheric, and Oceanic Sciences, v. 9, p. 317–328.

Sibson, R.H., 1982, Fault zone models, heat flow, and the depth distribution of earthquakes in the continental crust of the United States: Bulletin of the Seismological Society of America, v. 72, p. 151–163.

Sibson, R.H., 1984, Roughness at the base of the seismogenic zone: Contributing factors: Journal of Geophysical Research, v. 89, p. 5791–5799.

Song, T.-R.A., 1997, Characteristics of continental crust in Taiwan orogenic belts (in Chinese) [M.S. thesis]: Chung-Li, National Central University, 134 p.

Stuwe, K., White, L., and Brown, R., 1994, The influence of eroding topography on steady-state isotherms: Application to fission track analysis: Earth and Planetary Science Letters, v. 124, p. 63–74.

Suppe, J., 1980, A retrodeformable cross-section of northern Taiwan: Proceedings of the Geological Society of China, v. 23, p. 46–55.

Takahashi, E., 1986, Melting of a dry peridiotite KLB-a up to 14 GPa: Implications on the origin of peridotitic upper mantle: Journal of Geophysical Research, v. 91, p. 9367–9382.

Teng, L.S., 1990, Geotectonic evolution of late Cenozoic arc-continental collision in Taiwan: Tectonophysics, v. 183, p. 57–76.

Toksoz, M.N., and Bird, P., 1977, Modeling of temperature in continental convergence zones: Tectonophysics, v. 41, p. 181–193.

Tsai, Y.B., 1986, Seismotectonics in Taiwan: Tectonophysics, v. 125, p. 17–38.

Tsao, S., 1996, The geological significances of illite crystallinity, zircon fission-track ages, and K-Ar ages of metasedimentary rocks of the Central Range of Taiwan [Ph.D. thesis]: Taipei, National Taiwan University, 272 p.

Vitorello, I., and Pollack, H.N., 1980, On the variation of continental heat flow age and the thermal evolution of continents: Journal of Geophysical Research, v. 85, p. 983–995.

Volvovsky, B.S., Volvovsky, I.S., and Kambarov, N.S., 1983, Geodynamics and seismicity of the Pamir-Himalaya region: Physics of the Earth and Planetary Interiors, v. 31, p. 307–312.

Wang, J.H., and Chen, K.C., and Lee, T.Q., 1994, Depth distribution of shallow earthquakes in Taiwan: Journal of the Geological Society of China, v. 37, p. 125–142.

Wu, F.T., 1978, Recent tectonics of Taiwan: Journal of Physics of the Earth, v. 26, p. s265–s299.

Wu, F.T., Rau, R.J., and Salzberg, D., 1997, Taiwan orogeny: Thin-skinned or lithospheric collision?: Tectonophysics, v. 274, p. 191–220.

Manuscript Accepted by the Society April 12, 2001

Geological Society of America
Special Paper 358
2002

Synorogenic extension, Taiwan: Implications of physical and numerical modeling

Wei-Hau Wang
Wen-Jie Hung
Institute of Applied Geophysics, National Chung Cheng University, Minghsiung, Chia-Yi, Taiwan, Republic of China

ABSTRACT

We conducted two-dimensional physical and numerical modeling to simulate the synorogenic extension in the Taiwan mountain belt that is created by the collision between the Eurasian and the Philippine Sea plates at the Chinese continental margin. Our results show that crustal extension may occur around the crest of the Central Range because of gravitational spreading. Accompanying the extension, active thrust belts appear on either side of this extensional area. Overprinting of late-stage normal faults onto early-stage compressional structures on the east flank of the Central Range may result either from changes in the stress field near the mountain crest at various stages of the mountain-building event or from west to east compression at depth and tension near the surface. The concentration of normal faulting on the east flank of the Central Range suggests that the mantle suture between the Philippine Sea plate and the Eurasian plate might be farther west than the Longitudinal Valley. Unlike the hypothesis of mountain collapse, implying an overall reduction of elevation, our model shows that the synorogenic extension bounded by active thrust belts appears in the area undergoing fast surface uplift, as observed in Taiwan.

INTRODUCTION

The collision between the Eurasian and the Philippine Sea plates at the Chinese continental margin created the island of Taiwan, with a predominance of compressional structures. Increasing evidence, however, indicates that normal faults also appear in this growing mountain belt (e.g., Lee, 1995, Crespi et al., 1996). This phenomenon of crustal extension in an active collision belt is not uncommon, especially in large mountain belts such as the Himalayas, Alps, and Andes (e.g., Burchfiel and Royden, 1985; Selverstone, 1988; Dalmayrac and Molnar, 1981). Angelier et al. (1995) attributed the crustal extension in Taiwan to the following mechanisms: (1) backarc extension (the Okinawa Trough), (2) bending and downwarping of the lithosphere near subduction zones (southwestern Taiwan), (3) local effects related to folding and transfer faulting (Western Foothills), (4) increasing bending of the northern belt (northern Taiwan), (5) lateral extrusion related to collision (northeast of Taiwan), and (6) gravitational collapse of thickened lithosphere (mountain crest and the eastern border of the Central Range). In this paper, we focus our interest on the mechanism that drives crustal extension on the east flank of the Taiwan mountain belt. The interest of this area lies in the that even with prevailing normal faulting, this region has a higher exhumation rate than that observed near the crest of the mountain belt (e.g., Tsao et al., 1992; Liu, 1982; Lee, 1995) and is still in surface uplift at a rate of 4 cm/yr (Liu, 1995). This fact led Lee (1995) to attribute the extension to postcollision collapse of the Central Range. In other words, he suggested that the mountain range of Taiwan underwent two stages of deformation: an initial stage of growth, followed by a stage of collapse. This conceptual model, however, does not answer why the Central Range is gaining altitude.

Buck and Sokoutis (1994) found that both compression and extension could coexist in their scale model if rocks behaved

Wang, W.-H., and Hung, W.-J., 2002, Synorogenic extension, Taiwan: Implications of physical and numerical modeling, *in* Byrne, T.B., and Liu, C.-S., eds., Geology and Geophysics of an Arc-Continent Collision, Taiwan: Boulder, Colorado, Geological Society of America Special Paper 358, p. 137–146.

according to linear viscous rheology. They noticed that when linear viscous rheology is applied, surface extension is distinct; however, if power-law rheology is used, surface extension is reduced or completely disappears when the exponent become infinity. A recent study by Willett (1999) indicated that shallow-crustal extension occurs at high Argand numbers with a linear viscous rheology; power-law rheology suppresses extension, and no extension can be found in models with Coulomb plastic rheology. These studies provoke us to test the two-stage deformation hypothesis. In this study, both the velocity field and the strain field of the Taiwan mountain belt are simulated through conducting physical and numerical modeling. We would like to understand whether the strain field observed in the upper crust is similar to that in the lower crust of the Central Range. In addition, we attempt to explain how the ductile, compressional structures could have been overprinted by brittle, extensional structures in the Central Range (e.g., Lee and Yang, 1994), by a process other than mountain collapse.

TECTONIC SETTING OF TAIWAN

The Luzon arc riding on the Philippine Sea plate has been impinging on the Eurasian plate in Taiwan since the Pliocene–Pleistocene with a convergence velocity of 7 cm/yr (Seno et al., 1993, Fig. 1). The surface geology of Taiwan can be divided into six units: (1) the Coastal Plain, (2) the Western Foothills, (3) the Hsüehshan Range and the Backbone Range, (4) the Tananao Metamorphic Complex, (5) the Longitudinal Valley, and (6) the Coastal Range. The Coastal Plain is covered by mildly deformed Quaternary alluvial deposits with inactive east-trending normal faults cutting through the underlying Neogene strata. East of the Coastal Plain are the Western Foothills, a foreland fold-and-thrust belt with intensely shortened and nonmetamorphosed Neogene clastic sedimentary rocks. The Hsüehshan Range and the Backbone Range are composed mainly of argillite, slate, and phyllite. Thick-bedded sandstone also appears in the Hsüehshan Range. Although these two mountain ranges have similar lithologies, they are different in metamorphic grade and structural fabrics (e.g., Chen et al., 1983; Tillman and Byrne, 1995). The oldest stratigraphic member in Taiwan is the Tananao Metamorphic Complex. East of the Tananao Complex, the Longitudinal Valley separates the Coastal Range and the Central Range and is thought to be the suture zone between rocks of Eurasian affinity and rocks of Philippine Sea plate affinity (e.g., Ho, 1986). The Coastal Range, composed of Neogene volcanic rocks and flysch of the Luzon arc, has been thrust faulted and imbricated along a series of en echelon faults.

The deep structures in Taiwan are much less well understood. Suppe (1981, 1984) described the orogeny in Taiwan in terms of thin-skinned tectonics induced by eastward subduction of the Eurasian plate. Tsai (1986), however, indicated that most of the Taiwan island has no sign of a Benioff zone; rather, seismicity is characterized by shallow-crustal earthquakes except in the region south of 23°N. This fact, together with other

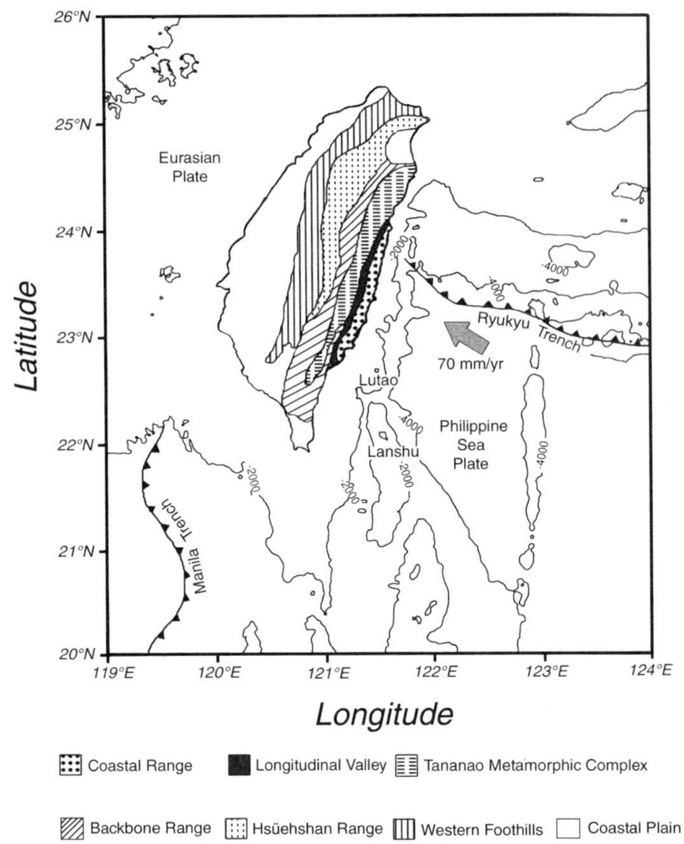

Figure 1. Tectonic setting of Taiwan and the surrounding area (modified from Ho, 1986). The Central Range referred to in the text consists of the Hsüehshan Range, the Backbone Range, and the Tananao Metamorphic Complex.

geophysical evidence, led Wu et al. (1997) to argue that the Taiwan orogeny was the result of lithospheric thickening instead of thin-skinned tectonics. Recent studies on wide-angle deep seismic profiling suggested significant crustal thickening in both the Eurasian plate and the Philippine Sea plate near the suture zone (e.g., Shih et al., 1998; Yeh et al., 1998), which is inconsistent with the thin-skinned model.

A schematic diagram (Fig. 2) illustrates the tectonic model used in this study according to the previous considerations. In this diagram, we interpreted the crustal thickening in the Philippine Sea plate as the result of underthrusting of the Philippine Sea plate beneath the Luzon arc. However, it could also result from deflection of the plate or from piling up crustal material during the arc-continent collision. Wang (1995) argued that thickening of the continental crust, as suggested by many authors (e.g., Lin, 1996; Wu et al., 1997), leads the adjacent Philippine Sea plate to act as a backstop. This is because the oceanic lithosphere has a deeper brittle-to-ductile transition than the adjacent continental crust. However, because both the Central Range and the Luzon arc are highly deformed (e.g., Chen and Liu, 1992; Yu et al., 1992, 1997; Yu and Chen, 1994), neither geologic domain should be regarded as a backstop, in the sense of a relatively rigid member embedded in much weaker Earth materials.

Although compressional features are dominant over the orogenic belt, recent studies indicate that the east flank of the Central Range underwent extension during the late stage of mountain building. Crespi et al. (1996) indicated that late-stage normal faults are concentrated in both the Backbone Range (near the mountain crest) and the Tananao Complex (east flank of the Central Range). This observation generally agrees with recent GPS measurements (e.g., Yu and Chen, 1994; Yu et al., 1997) even though only a few GPS stations are located on the Central Range. Additional evidence comes from a recent study on focal mechanisms by Lin et al. (1998) who found that most extensional earthquakes were located in the area of the Tananao Complex.

MODEL CRUSTAL RHEOLOGY AND TECTONIC FRAME

Two types of rheological models for continental crust are used here: one assumes that the continental crust is linear viscous and has a constant viscosity. The other assumes that the upper and the lower parts of the crust are linear viscous but with independent values of viscosities. In the following text, we refer these two models as the "one-layer" and the "two-layer" models. The one-layer model assumes that the whole continental crust in nature has an effective viscosity of $1 \sim 10^{21}$ Pa·s, which has been suggested by Royden (1996). On the other hand, the two-layer model suggests that the viscosity of the upper crust with a thickness of 15 km is determined by the power-law flow of quartz at depth of 7.5 km, whereas the lower crust with a thickness of 20 km is determined by the power-law flow of feldspar (albite) at a depth of 25 km. The effective viscosity can then be estimated by the following equation:

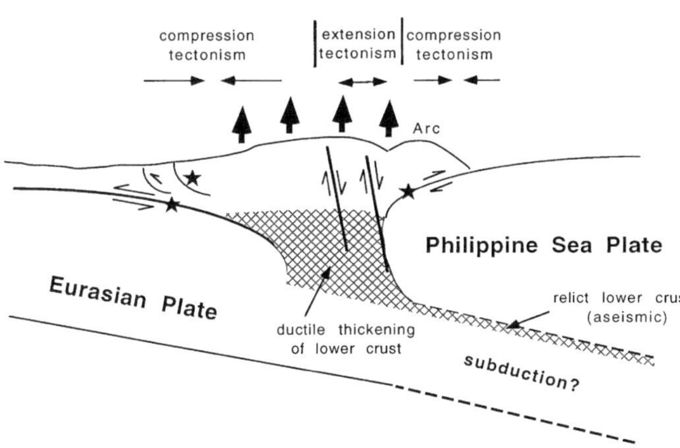

Figure 2. Schematic diagram illustrating a likely cross section of the Taiwan orogen. Note that this diagram is not properly scaled. Extension on the east flank of the mountain belt is bounded by two active thrust belts. The extension is associated with fast uplift in the Central Range that may be related to ductile thickening of the lower continental crust.

$$\eta_{\text{eff}} = (\tfrac{1}{2}A^{1/n})\ \dot{\varepsilon}^{1/(n-1)}\ \exp(Q/nRT),$$

where η is effective viscosity; A is a material constant; $\dot{\varepsilon}$ is strain rate, Q is an activation energy; R is the gas constant; T is absolute temperature; and n is the effective power exponent of the power-law flow such that $\dot{\varepsilon} = A\sigma^n \exp(-Q/RT)$ with σ denoting differential stress. The values of the material parameters, A, Q, and, n for these two minerals are listed in Table 1, which were obtained from experimental studies (Shelton and Tullis, 1981; Kronenberg and Tullis, 1984). We adopt a strain rate of 10^{-14} s^{-1} (e.g., Barr and Dahlen, 1989) and a geothermal gradient of 27° C/km (e.g., Suppe and Wittke, 1977) for the Taiwan orogen. The geothermal gradient we used is slightly smaller than the average geothermal gradient of 28.5–30 °C/km measured from drilling wells in the Cenozoic basins of western Taiwan (Huang, 1990). However, because Huang's result was obtained mostly from gas fields and some from hot-spring fields, the resulting average geothermal gradient might be overestimated (see also Song and Ma, this volume). Nevertheless, if the crustal extension that is driven by the motion of thermally activated lower crust appears under this "cool" condition, it should also appear under any "hotter" condition.

We regard the deep part (below 11 km where brittle-to-ductile transition occurs in the adjacent continental crust) of the Philippine Sea plate as a backstop according to the previous numerical study (Wang, 1995). Two geometries are tested in this study: one with a flat backstop at depth of 35 km and the other with the backstop dipping 20° west. In this paper, we focus on the deformation induced from plate collision at a crustal level because we are more interested in heterogeneous deformation, which essentially occurs at a crustal level due to the presence of the large strength contrast between the Eurasian and Philippine Sea plates (e.g., Wang, 1995). We do not mean to imply, however, a bias toward either the thin-skinned deformation model or the whole-lithospheric deformation model for the generation of the Taiwan orogen.

PHYSICAL MODELING

Model setup and similarity criterion

For the physical models, a piece of hardboard (either flat or tilting 20° to the left) was used to simulate the strong Philippine Sea plate (the backstop), which moved to the left analogous the collision between the Eurasian plate and the Philippine Sea plate (Fig. 3). The materials in front of and above the

TABLE 1. MATERIAL PROPERTIES FOR QUARTZ AND ALBITE

Parameters	Quartz	Albite
A ($GPa^{-n}\cdot s^{-1}$)	1.0×10^2	1.6×10^6
n	2.4	3.9
Q ($KJ\cdot mol^{-1}$)	1.56×10^2	2.34×10^2

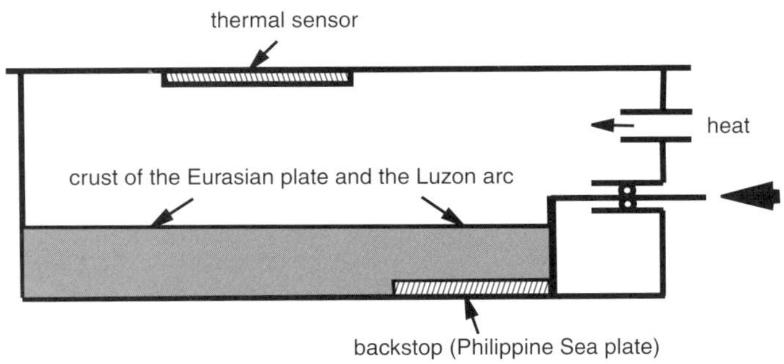

Figure 3. Schematic cross section illustrating the operation of the scale model. A backstop consisting of wood or a hardboard either is flat as shown in the diagram or dips 20° to the left. Temperature was kept constant during each experimental run.

TABLE 2. PARAMETER VALUES FOR THE SCALE MODEL AND FOR NATURE

Parameters	Scale Model	Nature
Convergence velocity v(m/s)	1.6×10^{-3}	2.2×10^{-9}
Gravitational constant g(m/s^2)	9.8	9.8
Density of continental crust ρ(kg/m^3)	8.45×10^{2}	2.7×10^{3}
Thickness of continental crust* h(m)	1.4×10^{-2}	3.5×10^{4}
Thickness of upper crust† h_u(m)	6.0×10^{-3}	1.5×10^{4}
Thickness of lower crust† h_l(m)	8.0×10^{-3}	2.0×10^{4}
Viscosity of continental crust* η(Pa·s)	6.92×10^{1}	1.0×10^{21}
Viscosity of upper crust† η_u(Pa·s)	7.96×10^{2}	1.16×10^{22}
Viscosity of lower crust† η_l(Pa·s)	4.13×10^{1}	6.0×10^{20}

*Values used for the one-layer model only.
†Values used for the two-layer model only.

backstop that simulate the crusts of the Eurasian plate and the Luzon arc are mixtures of hydrocarbons, which exhibit either viscoplastic or brittle behavior dependent on their viscosity and strain rate. Note that the model setup in Figure 3 shows the mechanical members instead of geologic members. The backstop, which includes the part of oceanic crust at a depth more than 11 km plus the underlying mantle, represents the strong part of the Philippine Sea plate compared to the adjacent continental plate. Although the Luzon arc is genetically related to the Philippine Sea plate, mechanically it is closer to continental crust as both have been intensely deformed and have similar lithology.

This scale modeling meets the following similarity criterion (Shemenda, 1994): $\eta v/\rho g h^2$ = constant, where η is the viscosity of the continental crust, v is the convergence velocity between the Eurasian plate and the Philippine Sea plate, ρ is the density of the continental crust, h is the thickness of the continental crust, and g is the acceleration due to gravity. The parameter values used for our scale model and those for nature are listed in Table 2.

Simulation results

One-layer model. Initially, only ductile, contractional shear zones (under compression) appeared in our model as shown in Figure 4A. At the later stage, as the model grew, extension appeared around the crest of the model mountain belt (Fig. 4, B and C). However, compressional shear zones still developed on either side of the mountain crest. During the experiment (or collision period), we did not observe any reduction in height of the mountain belt. A similar deformation history is observed in a model with an inclined backstop (Fig. 5).

Two-layer model. At the initial stage of the two-layer model, brittle faulting appeared in the upper crust and created a pop-up structure while ductile shearing occurred in the lower crust (Fig. 6A). This distribution of deformation reflects the fact that deformation in the upper crust is controlled by its brittle strength instead of ductile creep. Although we were not able to measure the exact brittle strength of the model upper crust, it must be smaller than its ductile strength, analogous to the rheology of upper crusts in nature. At the later stage (Fig. 6, B and C), the fault blocks near the crest of the deforming wedge that were once thrust over each other were pulled apart by the extensional flow in the model lower crust. A similar deformation history is also shown in the model with an inclined backstop (Fig. 7).

NUMERICAL MODELING

We conducted finite-element modeling to better understand and quantify the results from the scale modeling. Boundary conditions for the numerical models are similar to those in the physical models (Fig. 3): a free surface on the top with a rigid backstop moving at a speed of 7 cm/yr toward the left. No mass flux is allowed across the left and bottom boundaries. In addition, a no-slip boundary is used as the bottom boundary in front of the backstop.

We adapted the values of parameters for "nature" listed in Table 2. Although the model width is 1400 km, in the following section we will only display cross sections of 200 km around the toe of the backstop where deformation is most intense. Assuming that crustal material deforms as the Navier-Stokes flow, we solved for the velocity and the strain fields in the model mountain belt by using the commercial software package "PDEase."

Simulation results

One-layer model. At the onset of the one-layer model, vertical velocities increased around the toe of the backstop (Fig. 8A). As the mountain belt grew (Fig. 8B), horizontal velocities increased in the region in front of the backstop and decreased in the area behind the toe of the backstop, because of the tendency of gravitational sliding along the surface slope. At the later stage (Fig. 8C), horizontal velocity was almost vanished near the surface at a distance of about 50 km behind the toe of the backstop where the force induced by gravitational sliding along the east (right) flank of the model mountain belt is bal-

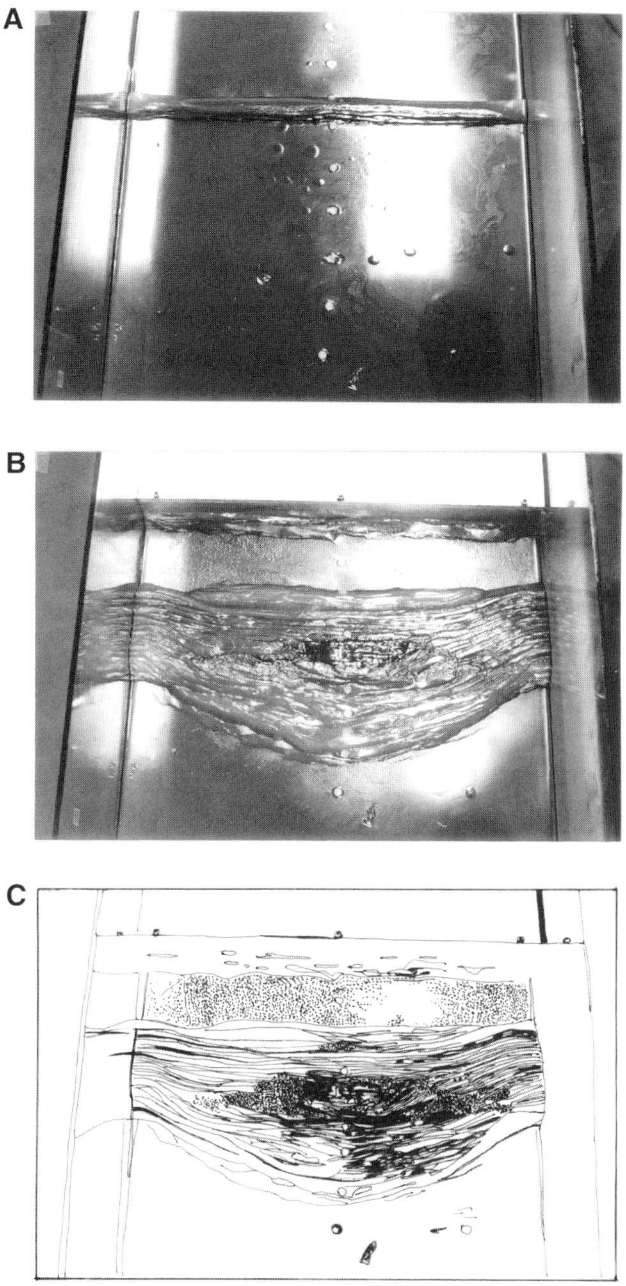

Figure 4. Top ("map") views showing deformation in a "one-layer" model with a flat backstop. (A) An initial "pop-up" bounded by two oppositely dipping shear zones developed in front of the backstop. Note that equally spaced markers (small dots) were placed on the surface of our model, and the lower half of the model crust was stained black for later comparison. (B) Extension around the crest of the deformed wedge developed at the final stage. (C) Schematic diagram corresponding to B. Note that the lower crust (in black) was brought up to the place near the crest of the contractional wedge and pulled the overlying thrust sheets apart. Adjacent to this area, compressional shearing is still active.

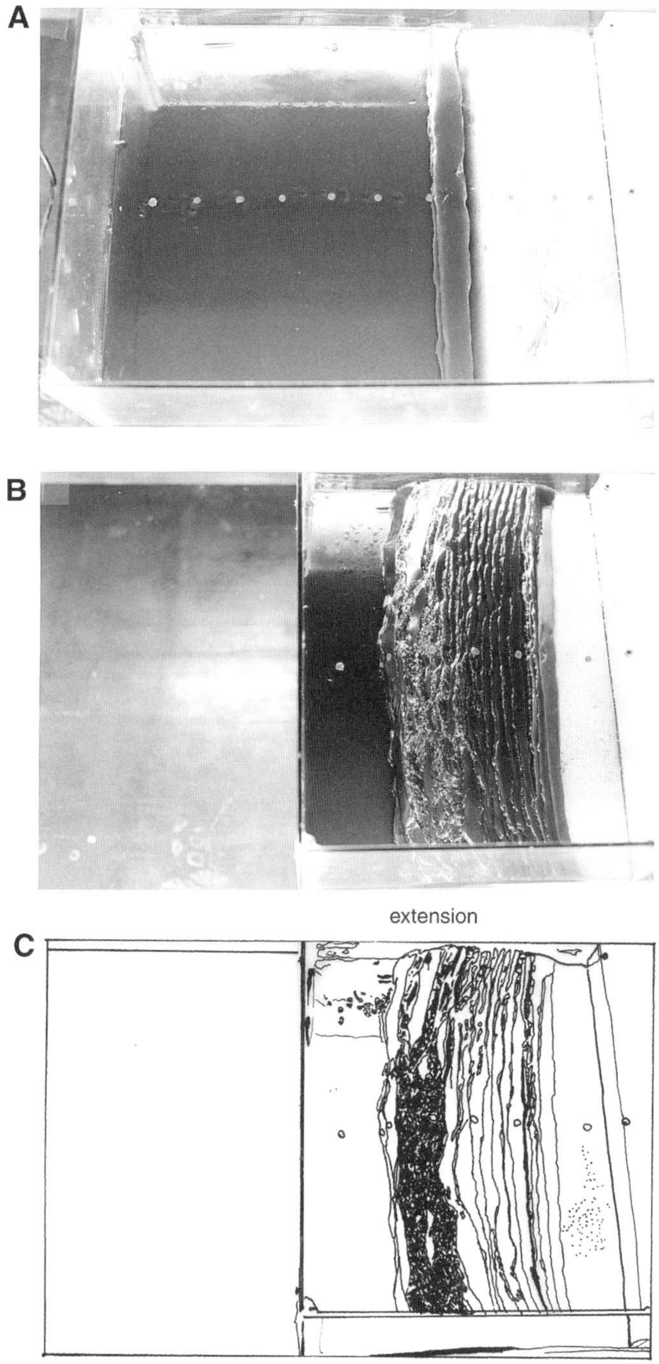

Figure 5. Top ("map") views showing deformation in a "one-layer" model with an inclined backstop. The resulting structures are quite similar to those in the model with a flat backstop shown in Figure 4.

anced by the plate-collision force. Note that during these stages, the model mountain belt has been continuously gaining height. In other words, no sign of mountain collapse was found.

In terms of strain, intense compression initially occurred around the toe of the backstop (Fig. 9A); then, with larger displacements, extension appeared around the mountain crest with

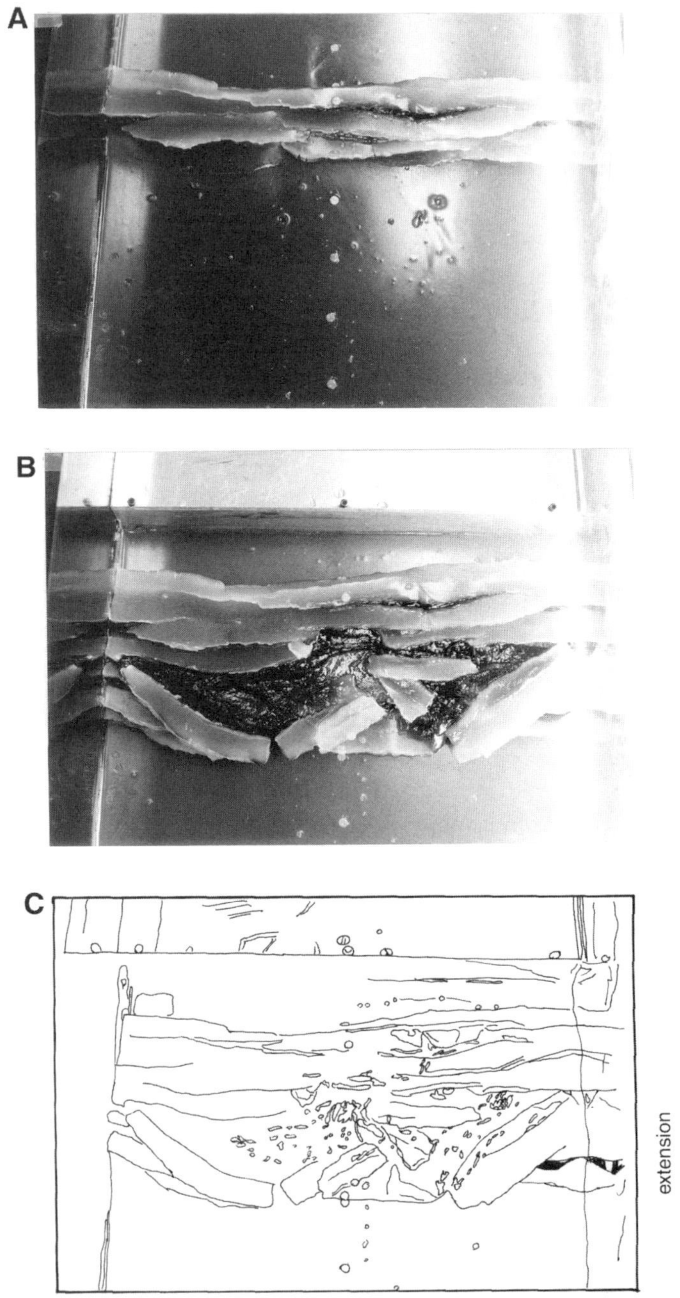

Figure 6. Top ("map") views showing deformation in a "two-layer" model with a flat backstop. The lower half of the model crust was stained black. (A) An initial "pop-up" appears in front of the backstop. The upper crust deforms by brittle thrusting instead of ductile shearing. (B) Extension around the crest of the deforming wedge develops at the final stage. (C) Schematic diagram corresponding to Figure 6B. Note that the lower crust (in black in part C) was brought up to near the crest of the contractional wedge, pulling the overlying thrust sheets apart.

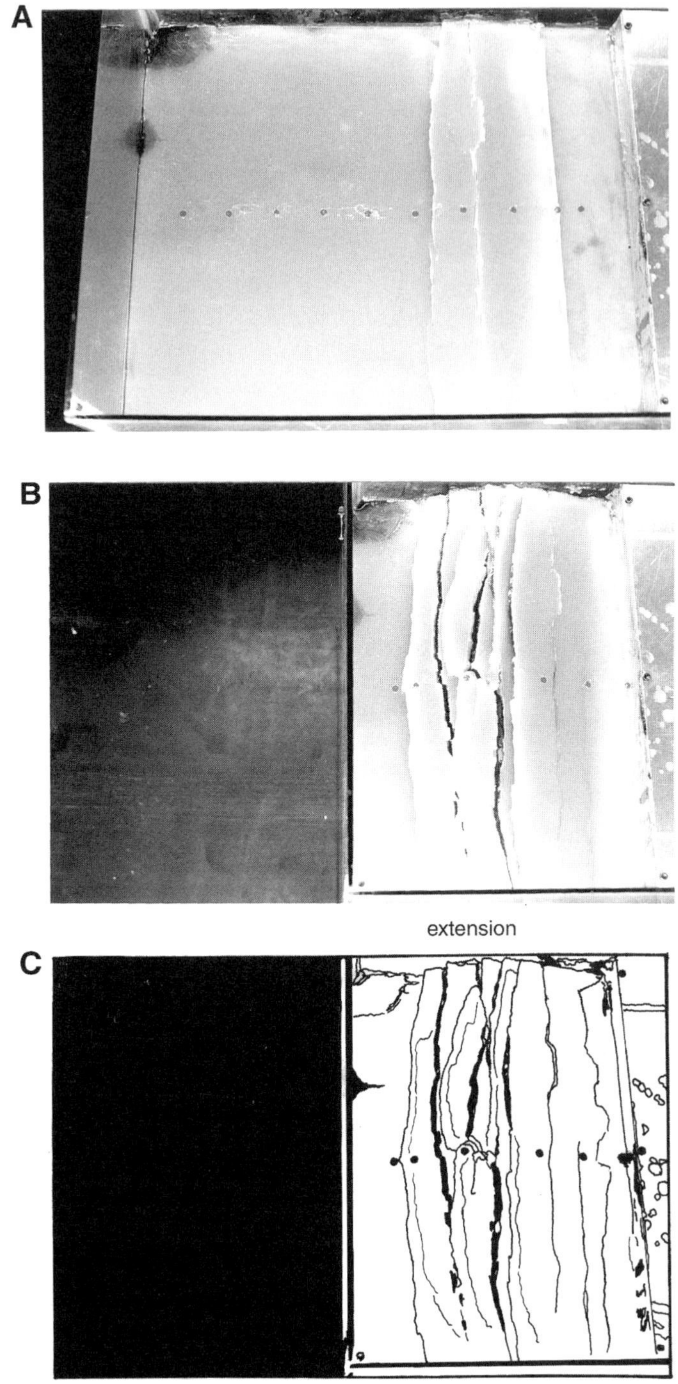

Figure 7. Top ("map") views showing deformation in a "two-layer" model with an inclined backstop. The resulting structure is quite similar to the two-layer model with a flat backstop shown in Figure 6.

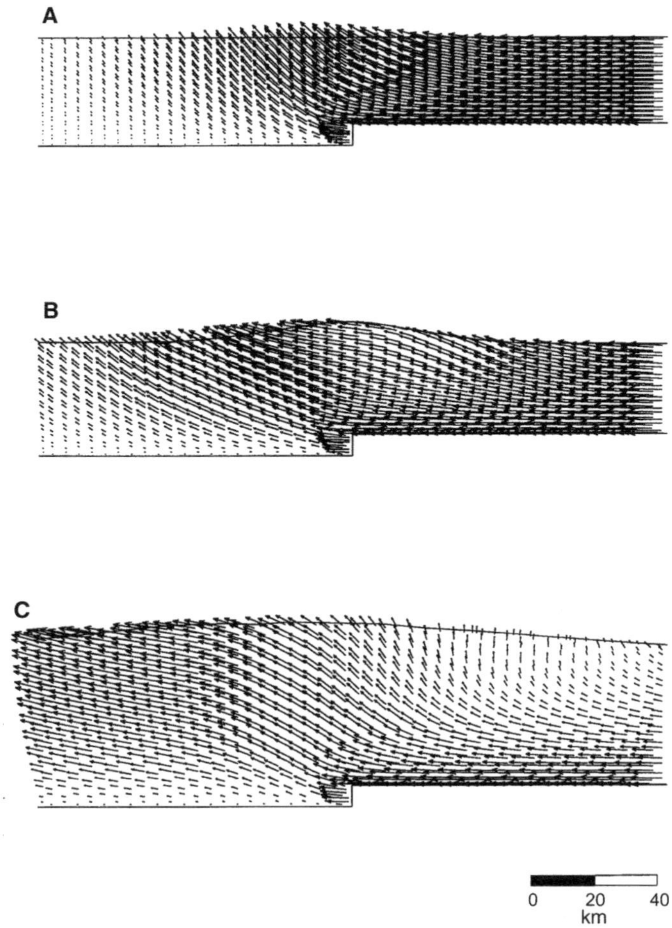

Figure 8. Velocity variations in a cross-sectional "one-layer" model with a flat backstop. The length of a vector is proportional to its velocity. (A) Initially, vertical velocity increases near the toe of the backstop. (B) Later (0.25 m.y. after the onset of collision), velocity increases in front of the backstop and decreases behind the toe of the backstop. (C) Finally (1.6 m.y. after the onset of collision), horizontal flow was greatly reduced at the area about 50 km behind the toe of the backstop.

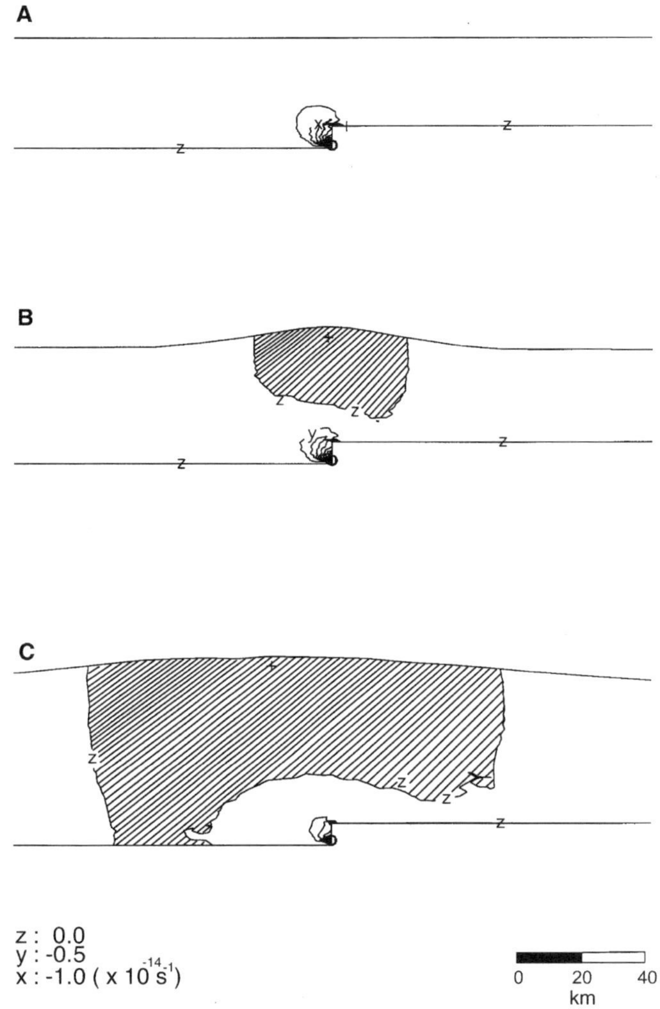

Figure 9. Horizontal strain rate corresponding to the velocity field in Figure 8. Extension occurred in the area marked with diagonal lines. (A) At the very beginning of convergence, compression was concentrated on the area near the toe of the backstop. (B) Later, extension developed above the toe of the backstop. (C) Finally, extension penetrated the deeply seated crust except the area near the toe of the backstop.

compression at depth (Fig. 9B). At the final stage, extension penetrated the whole crust; however, the area where the plates contact (the toe of the backstop) was still under compression.

Two-layer model. As observed in the one-layer model, velocity progressively increases in the region in front of the backstop in the two-layer model and decreases in the area behind the toe of the backstop as the mountain belt grows (Fig. 10). Because the upper layer has a higher viscosity than that used in the one-layer model, changes in velocity along the two flanks of the model mountain belt are less distinct in this case. Nevertheless, no sign of mountain collapse was observed.

In terms of strain rate, intense compression occurred around the toe of the backstop at the initial stages (Fig. 11A). However, extension near the surface did not immediately appear as the mountain grew (Fig. 11B). Only when a large area was elevated did we find extension near the crest of the mountain belt (Fig. 11C). At the final stage, extension penetrated the whole crust, but the area in front of the toe of the backstop was still under compression.

We also tested a two-layer model with a dipping backstop (Fig. 12). During the initial stages, the whole mountain belt was under compression (Fig. 12A). As in the two-layer model with a flat backstop, surface extension did not immediately show up as the mountain grew (Fig. 12B). Later, extension was initiated at the top of the lower crust instead of at the surface of the mountain belt (Fig. 12C). At the final stage (Fig. 12D), extension appeared behind the toe of the backstop instead of right above it.

DISCUSSION

Both our numerical and physical models demonstrate that surface extension is almost an inevitable consequence in the

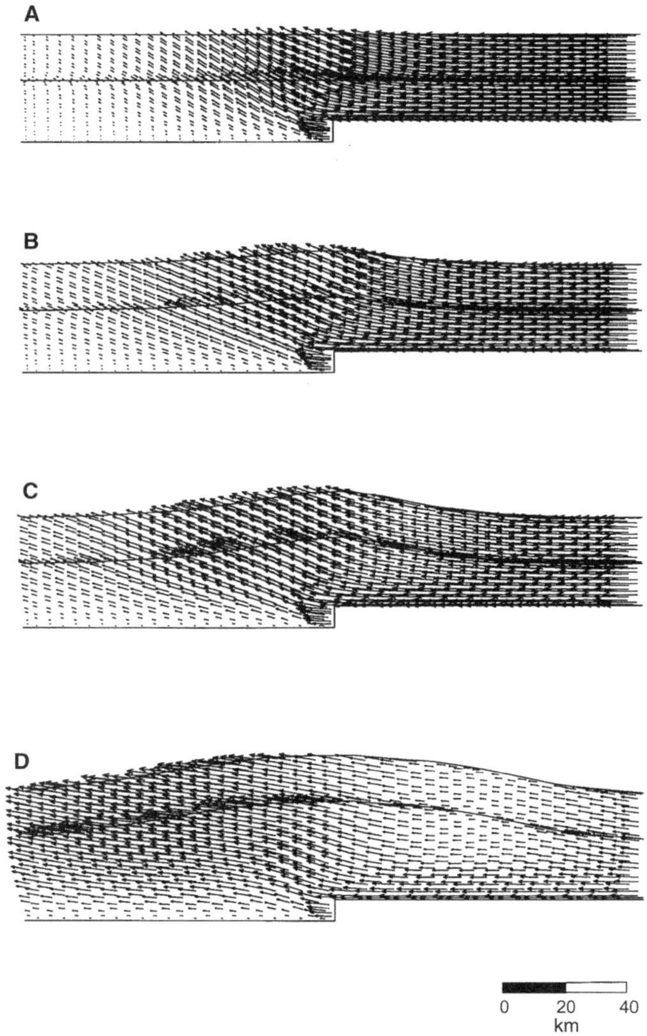

Figure 10. Velocity variations in a cross-sectional "two-layer" model with a flat backstop. The length of a vector is proportional to its velocity. The progressive increase in velocity in front of the backstop and decrease in velocity behind the toe of the backstop are shown. Elapsing time relative to the onset of Taiwan orogeny: (A) 0 m.y. (B) 0.25 m.y. (C) 0.5 m.y. (D) 1.6 m.y.

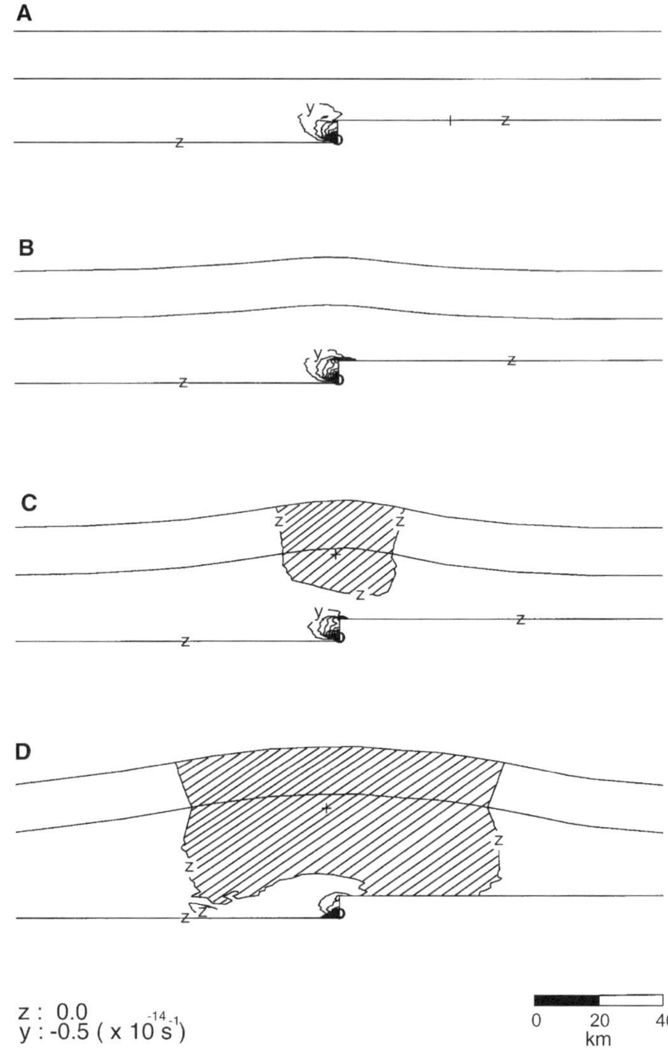

Figure 11. Horizontal strain rate corresponding to the velocity field in Figure 10. Extension occurred in the area marked with diagonal lines. (A) At the very beginning, only compressional features were observed. (B) Extension did not appear at this stage (elapsing time of 0.25 m.y.), although noticeable relief had built up. (C) Later (elapsing time of 0.5 m.y.), extension developed above the toe of the backstop. (D) Finally (elapsing time of 1.6 m.y.), extension penetrated the deeply seated crust except the area near the toe of the backstop.

growth of a mountain belt with a linear viscous rheology. This extensional behavior, however, did not show up in previous studies on convergent belts when the model crust is brittle or frictional (e.g., Mulugeta, 1988; Wang and Davis, 1996). Dahlen (1984) analyzed the internal structures of a critically tapering Coulomb wedge with given internal brittle strength and basal-friction resistance. Gravity spreading by normal faulting, based on his study, will appear in a critically tapered wedge only when the basal dip of the wedge is quite small or negative, or when the surface slope of the wedge is high and the basal friction of the wedge is negligible. It is questionable whether the mountain belt of Taiwan meets these strict requirements. Willett (1999) tested several reasonable combinations of internal and basal-friction coefficients for a contractional Coulomb-plastic wedge and found that no extension was observed except when the internal friction coefficient was equal to the basal-friction coefficient. In that unique case, only very near-surface extension was observed. Therefore, we infer that the extension found near the crest of the Central Range is more likely caused by the induced flow of the ductile lower crust above the toe of the Philippine Sea plate, as illustrated in our scale and numerical models.

As noted earlier, the late-stage normal faulting overprints the early-stage compressional, ductile structures on the east flank of the Central Range (e.g., Lee and Yang, 1994; Crespi et al., 1996). The tectonic alteration could have resulted from changes in the stress and strain fields from compression to extension near

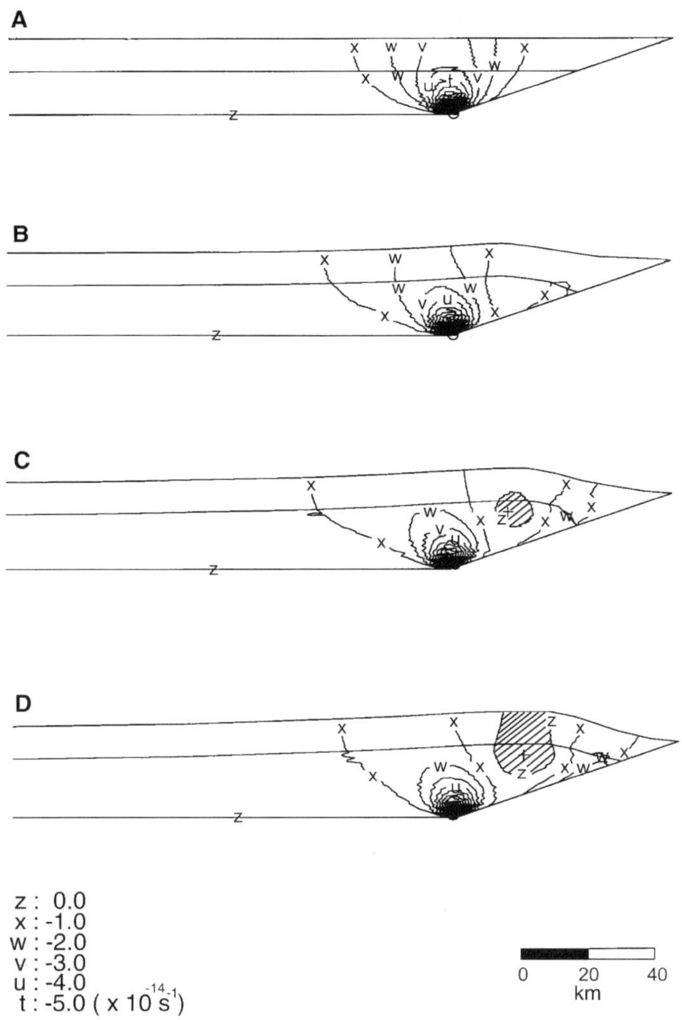

Figure 12. Horizontal strain rate in a cross-sectional two-layer model with an inclined backstop. Extension occurred in the area marked with diagonal lines. (A and B) No extension was observed during first two stages. (C) Extension was initiated at the top of the lower crust behind the toe of the backstop and (D) eventually extended to the upper crust behind the toe of the backstop.

the mountain crest at various stages of the mountain building. Alternatively, even for a steady-state orogen, such a deformation history could apply to rocks that underwent a compressional regime near the toe of the Philippine Sea plate or within the fold-and-thrust belt at an early stage and then migrated eastward and were uplifted, entering the extensional regime.

The velocity fields in our numerical models are quite different from those in a steady-state, Coulomb wedge given by Dahlen and Barr (1989), who forced particles to move upward along a nearly vertical boundary. In addition, the importance of the frictional heating along a shallowly seated décollement suggested by Barr and Dahlen (1989) needs to be reexamined when the lower crust is also involved. Limited rock strength at depths below the brittle-to-ductile transition cannot generate the large frictional heating that was predicted by Barr and Dahlen (1989). Both issues should be addressed by future study.

All our models predict crustal extension to almost symmetrically occur around the crest of the mountain belt. In other words, we fail to simulate the crust extension that concentrates on the east flank of the mountain belt. This consequence may result from ignoring the effect of erosion because it may bring rocks originally seated below the mountain crest, a regime of extension, to the east flank of the Central Range. Lin et al. (1998) pointed out that according to recent leveling surveys, the east flank of the Central Range has a much higher surface uplift rate than the area around its crest (argillite-slate belt). If so, then erosion must be very intense in this area and cannot be ignored. In fact, fission-track data also indicate that the east flank has a higher exhumation rate than the areas around mountain crest (Liu, 1982). Kinematic movement of rocks in the Central Range is therefore important to understand the distribution of extensional features in the Taiwan orogen.

If the Philippine Sea plate tilts toward the west as shown in Figure 2, we should expect that the mantle contact between the Eurasian and the Philippine Sea plates to be located west of the Longitudinal Valley (see, e.g., Fig. 12). The exact location of this suture, however, strongly depends on the geometry of the Philippine Sea plate and the strength contrast between the upper and the lower crusts.

Although the extension observed in our models is induced by gravity, we feel that the usage of the term "mountain collapse," which implies an overall reduction in elevation, is inappropriate. In neither of our physical or numerical models have we found large-scale elevation reduction near the mountain crest. These results are also consistent with the lack of correlation between extension and the topographic crest in the Central Range. The fact that the eastern Central Range has a high surface uplift and exhumation rate and is undergoing extension agrees with our prediction that extension correlates with active surface uplift of a mountain belt. The word "collapse" does not really describe this phenomenon and can be misleading.

CONCLUSIONS

We have conducted both scale and numerical modeling to simulate the orogenic processes in Taiwan. Our results indicate that viscous deformation of the crust during contraction can produce spatially limited extension as observed in Taiwan. This extension is associated with surface uplift and hence is not correctly referred to as gravitational collapse.

Our numerical simulation also indicates that an inclined backstop is more likely to induce asymmetric deformation in a contractional belt: compression in front of (landward of) the toe of the backstop and extension behind (seaward) the toe of the backstop. The westward deflection of the Philippine Sea plate, therefore, implies that the mantle suture between the Eurasian plate and the Philippine Sea plate is located farther west than the Longitudinal Valley.

There are two reasonable explanations for the overprinting of the late-stage normal faults onto the early-stage compressional fabrics on the east flank of the Central Range. The overprinting could result from change in the stress or strain field from compression to extension around the mountain crest at various stages of the mountain building. Alternatively, erosion will lead to west to east motion of rocks to replace eroded mass. As a result, rocks having undergone the compression west of the mountain crest now are brought up into the regime of extension.

ACKNOWLEDGMENTS

We thank Tim Byrne and Sean Willett for constructive reviews that improved the paper. We are grateful to Daniel M. Davis for helpful comments on the original manuscript. This work was supported in part by National Science Council grant 852111M194004.

REFERENCES CITED

Angelier, J., and others, 1995, Crustal extension in an active orogen: Taiwan: International Conference and 3rd Sino-French Symposium on "Active Collision in Taiwan", Taipei, Taiwan, Republic of China, Abstracts, p. 25–32.

Barr, T.D., and Dahlen, F.A., 1989, Brittle frictional mountain building. 2. Thermal structure and heat budget: Journal of Geophysical Research, v. 94, p. 3923–3947.

Burchfiel, B.C., and Royden, L., 1985, North-south extension within the convergent Himalayan region: Geology, v. 13, p. 679–682.

Buck, R., and Sokoutis, D., 1994, Analogue model of gravitational collapse and surface extension during continental convergence: Nature, v. 369, p. 737–740.

Chen, C.H., Chu, H.T., Liou, J.G., and Ernst, W.G., 1983, Explanatory notes for the metamorphic facies map of Taiwan: Special Publication of the Central Geological Survey of Taiwan, Republic of China, 32 p.

Chen, Y.G., and Liu, T.K., 1992, Vertical crustal movement of tectonic uplifting volcanic island—Lutao: Journal of the Geological Society of China, v. 35, p. 231–246.

Crespi, J., Chan, Y.C., and Swaim, M., 1996, Synorogenic extension and exhumation of the Taiwan hinterland: Geology, v. 24, p. 247–250.

Dahlen, F.A., 1984, Noncohesive critical Coulomb wedges: An exact solution: Journal of Geophysical Research, v. 89, p. 10 125–10 133.

Dahlen, F.A., and Barr, T.D., 1989, Brittle frictional mountain building. 1. Deformation and mechanical budget: Journal of Geophysical Research, v. 94, p. 3906–3922.

Dalmayrac, B., and Molnar, P., 1981, Parallel thrust and normal faulting in Peru and constraints on the state of stress: Earth and Planetary Science Letters, v. 55, p. 473–481.

Ho, C.S., 1986, A synthesis of the geologic evolution of Taiwan: Tectonophysics, v. 125, p. 1–16.

Huang, L.S., 1990, Preliminary study on the subsurface temperature and geothermal gradients of the Late Cenozoic basin in western Taiwan: Bulletin of the Central Geological Survey, no. 6, p. 117–144.

Lee, Y.-H., 1995, A study of stress field system evolution of the Taiwan mountain belt, Central Cross-Island Highway and surrounding regions: Bulletin of the Central Geological Survey, no. 10, p. 51–90.

Lee, Y.-H., and Yang, C.N., 1994, Structural evolution in the Tayulin area of the Central Range: Bulletin of the Central Geological Survey, no. 9, p. 77–106.

Kronenberg, N.J., and Tullis, J., 1984, Flow strengths of quartz aggregates: Grain size and pressure effects due to hydrolytic weakening: Journal of Geophysical Research, v. 89, p. 4281–4297.

Lin, C.-H., 1996, Crustal structures estimated from arrival differences of the first P-waves in Taiwan: Journal of the Geological Society of China, v. 39, p. 1–12.

Lin, C.-H., Yeh, Y.-H., Yen, H.-Y., Chen, K.-C., Huang, B.-S., Roecker, S.W., and Chiu, J.-M., 1998, Three-dimensional elastic wave velocity structure of the Hualien region of Taiwan: Evidence of active crustal exhumation: Tectonics, v. 17, p. 89–103.

Liu, C.H., 1995, Geodetic monitoring of mountain building in Taiwan: Eos (Transactions, American Geophysical Union), v. 76, p. 636.

Liu, T.K., 1982, Tectonic implication of fission track ages from the Central Range, Taiwan: Proceedings of the Geological Society of China, v. 25, p. 22–37.

Mulugeta, G., 1988, Modeling of the geometry of the Coulomb thrust wedges: Journal of Structural Geology, v. 10, p. 847–859.

Royden, L., 1996, Coupling and decoupling of crust and mantle in convergent orogens: Implications for strain partitioning in the crust: Journal of Geophysical Research, v. 101, p. 17 679–17 705.

Selverston, J., 1988, Evidence for east-west extension in the eastern Alps: Implications for the unroofing history of the Tauern window: Tectonics, v. 7, p. 87–105.

Seno, T., Stein, S., and Gripp, A.E., 1993, A model for the motion of the Philippine Sea plate consistent with NUVEL-1 and geological data: Journal of Geophysical Research, v. 98, p. 17941–17948.

Shelton, G., and Tullis, J., 1981, Experimental flow laws for crustal rock: Eos (Transactions, American Geophysical Union), v. 62, p. 396.

Shemenda, A.I., 1994, Subduction: Insights from physical modeling: Dordrecht, The Netherlands, Kluwer Academic Publishers, 215 p.

Shih, R.C., Lin, C.H., Lai, H.L., Yeh, Y.H., Huang, B.S., and Yen, H.Y., 1998, Preliminary crustal structures across central Taiwan from modeling of the onshore-offshore wide-angle seismic data: TAO, Terrestrial, Atmospheric and Oceanic Sciences, v. 9, p. 317–328.

Suppe, J., 1981, Mechanics of mountain building and metamorphism in Taiwan: Memoir of Geological Society of China, v. 4, p. 67–89.

Suppe, J., 1984, Kinematics of arc-continent collision, flipping of subduction and back-arc spreading near Taiwan: Memoir of Geological Society of China, v. 6, p. 21–33.

Suppe, J., and Wittke, J.H., 1977, Abnormal pore-fluid pressures in relation to stratigraphy and structure in the active fold-and-thrust belt of northwestern Taiwan: Petroleum Geology of Taiwan, no. 14, p. 11–24.

Tillman, K.S., and Byrne, T.B., 1995, Kinematic analysis of the Taiwan Slate Belt: Tectonics, v. 14, p. 322–341.

Tsai, Y.B., 1986, Seismotectonics of Taiwan: Tectonophysics, v. 125, p. 17–37.

Tsao, S., Li, T.C., Tien, J.L., Chen, C.H., Liu, T.K., and Chen, C.H., 1992, Illite crystallinity and fission-track ages along the east Central Cross-Island Highway of Taiwan: Acta Geologica Taiwanica, v. 30, p. 45–64.

Wang, W.-H., 1995, Numerical simulation of arc-continent collision: Geophysical Society of China, 1995 Annual Meeting, Taipei, Taiwan, Abstracts, p. 30.

Wang, W.-H., and Davis, M.D., 1996, Sandbox model simulation of forearc evolution and noncritical wedges: Journal of Geophysical Research, v. 101, p. 11329–11339.

Willett, S., 1999, Rheological dependence of extension in wedge models of convergent orogens: Tectonophysics, v. 305, p. 419–435.

Wu, F.T., Rau, R.-J., and Salzberg, D., 1997, Taiwan orogeny: Thin-skinned or lithospheric collision?: Tectonophysics, v. 274, p. 191–220.

Yeh, Y.H., Shih, R.C., Lin, C.H., Liu, C.C., Yen, H.Y., Huang, B.S., Liu, C.S., and Wu, F.T., 1998, Onshore/offshore wide-angle deep seismic profiling in Taiwan: Terrestrial, Atmospheric and Oceanic Sciences, v. 9, p. 301–316.

Yu, S.B., and Chen, H.Y., 1994, Global Positioning System measurements of crustal deformation in the Taiwan arc-continent collision zone: Terrestrial, Atmospheric and Oceanic Sciences, v. 5, p. 477–498.

Yu, S.B., and Chen, H.Y., and Kuo, L.C., 1997, Velocity field of GPS stations in the Taiwan area: Tectonophysics, v. 274, p. 41–59.

Yu, S.B., Yu, G.K., Kuo, L.C., and Lee, C., 1992, Crustal deformation in the southern Longitudinal Valley area, eastern Taiwan: Journal of the Geological Society of China, v. 35, p. 219–230.

Manuscript Accepted by the Society April 12, 2001

Printed in the U.S.A.

Geological Society of America
Special Paper 358
2002

Transition tectonics of northern Taiwan induced by convergence and trench retreat

Jyr-Ching Hu
Department of Geosciences, National Taiwan University, No. 1, Sec. 4, Roosevelt Road, Taipei 106, Taiwan
Shui-Beih Yu
Institute of Earth Sciences, Academia Sinica, P.O. Box 1-55, Nankang, Taipei, Taiwan, Republic of China
Hao-Tsu Chu
Central Geological Survey, P.O. Box 968, Taipei, Taiwan, Republic of China
Jacques Angelier
*Tectonique Quantitative, URA 1315, CNRS, Université Pierre et Marie Curie,
4 place Jussieu, T26, E-1, 75252 Paris Cedex 05, France*

ABSTRACT

As an active collision zone between the Luzon arc and the China continental margin, the Taiwan mountain belt is undergoing strong crustal shortening and rapid uplift, especially in its south-central part. In contrast, northeast Taiwan, located near the junction between the Taiwan orogen and the Ryukyu arc, is subject to both the compressional force exerted by the indentation of Philippine Sea plate and the tensile force induced by trench retreat related to the suction force at the Ryukyu Trench. As a function of these two forces, the stress regime rapidly changes from south to north, from compression in the Taiwan collision zone to extension in the Ryukyu subduction zone. By using a two-dimensional plane-stress finite-element model with elasto-plastic rheologies, we analyze the relationship between the kinematics of convergence, the crustal deformation, and the stress distribution in the present Taiwan collision zone between the Ryukyu and Luzon subduction zones. Our model predicts that in northeastern Taiwan, a transition zone may develop between pure compression (which characterizes most of the collision zone of Taiwan) and pure extension (which prevails in the northeastern part of the model). The transition in the stress field is characterized by two intermediate regimes, one mainly compressional and one mainly extensional, where significant strike-slip deformation is expected. We show that the interaction between the opening of the Okinawa Trough and the collision at the northwestern boundary of the Philippine Sea plate plays a crucial role in controlling the stress and strain patterns. The location of the transition zone in the north is a balance between these two factors, which may explain the variability of tectonic regimes in the region.

INTRODUCTION

The island of Taiwan is located at the junction of the Ryukyu arc and the Luzon arc at the western edge of the Philippine Sea plate. The plate interaction around this segment of the convergent plate boundary between the Eurasian and Philippine Sea plates is complex (Fig. 1). Seismicity northeast of Taiwan indicates the presence of a Wadati-Benioff zone, where the Philippine Sea plate subducts northward along the Ryukyu Trench (e.g., Tsai, 1986). In contrast, south of Taiwan, the Eurasian

Hu, J.-C., Yu, S.-B., Chu, H.-T., and Angelier, J., 2002, Transition tectonics of northern Taiwan induced by convergence and trench retreat, *in* Byrne, T.B., and Liu, C.-S., eds., Geology and Geophysics of an Arc-Continent Collision, Taiwan: Boulder, Colorado, Geological Society of America Special Paper 358, p. 147–160.

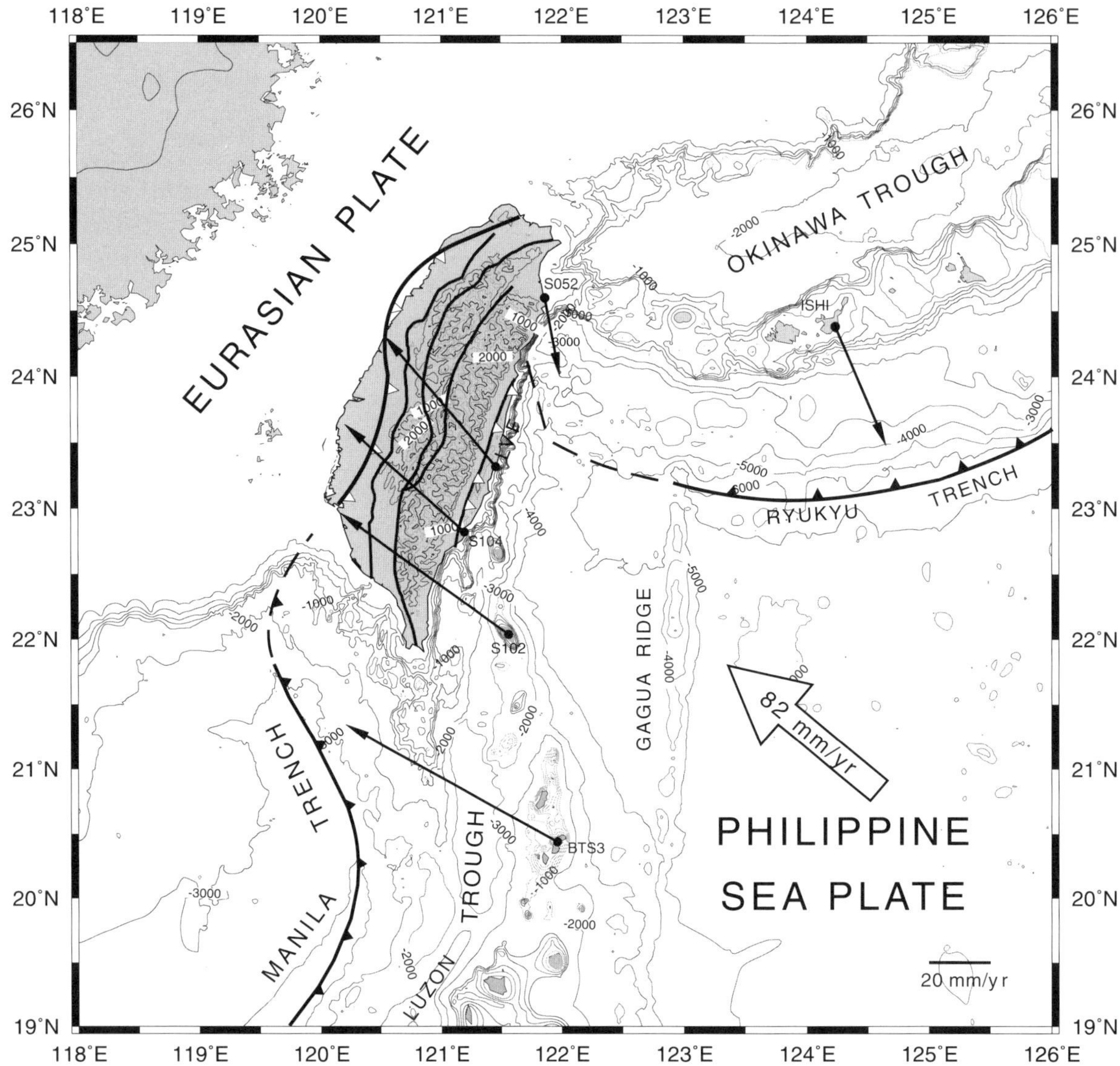

Figure 1. Bathymetric and geodynamic map of the vicinity of Taiwan. Subduction along the Ryukyu and Manila Trenches is represented with closed barbs on the overriding plate. The Longitudinal Valley fault (LVF) strikes northward along the East Coast and shows both oblique left-lateral and reverse (open barbs on hanging wall) movement (e.g., Barrier, 1985). Large open arrow and number indicate the relative-plate-motion vector between the Philippine Sea and Eurasian plates as determined by Yu et al. (1997). GPS velocity vectors relative to Paisha station (Penghu) appear as thin arrows. The data of ISHI station at Ishigaki in the Ryukyu Islands are provided by ERI (Earthquake Research Institute).

plate subducts southeastward beneath the Philippine Sea plate along the Manila Trench (e.g., Wu, 1978; Tsai, 1986; Pezzopane and Wesnousky, 1989). The Taiwan collision zone thus marks a segment of plate boundary where subduction undergoes a polarity reversal. Furthermore, Taiwan is where the Luzon arc initially collided with the Eurasian continent at ca. 5 Ma (e.g., Ho, 1986; Angelier, 1990; Teng, 1990, 1996). The fold-and-thrust belt of Taiwan advanced northwestward, while the orogen growth was propagating southwestward along the passive continental margin of the Eurasian plate (Suppe, 1984). As an active collision zone between the Luzon arc and the China continental margin, the Taiwan mountain belt, particularly its south-central part, is undergoing strong crustal shortening (e.g., Yu and Chen, 1994) and rapid uplift (e.g., Liew et al., 1990; Liu and Yu, 1990; Wang and Burnett, 1990; Pirazzoli et al., 1993).

Extensive paleostress analyses have been made throughout the island of Taiwan since 1984; thus the evolution of tectonic paleostresses is well known in terms of regional distribution and major trends (Barrier, 1985, 1986; Barrier and Angelier, 1986; Angelier et al., 1986, 1990; Lee, 1986; Lee and Wang, 1988; Chu, 1990; Rocher et al., 1996; Lacombe et al., 1997). These paleostress determinations allowed not only description of local mechanisms, but also understanding of the regional tectonic patterns at the scale of the plates. The regional knowledge of the late Cenozoic paleostress field thus provided strong constraints in the geodynamic interpretation of the Taiwan collision belt. In order to address the problem of consistency between paleostress and deformation, and taking advantage of these constraints, several numerical experiments considering strain-stress relationships were carried out (e.g., Huchon et al., 1986; Hu et al., 1996). These experiments aimed to evaluate the general compatibility of the kinematic boundary conditions, the geologic structure, rheology, and paleostress patterns (Hu et al., 1996, 1997).

During the past few years, the Global Positioning System (GPS) became one of the most important geodetic tools for studying crustal deformation (e.g., Dixon, 1991). In Taiwan, accurate relative positioning with the GPS in a dense network allowed description of crustal deformation at the scale of the collision belt and surrounding island (Yu and Chen, 1994). To better quantify both the rate and style of deformation currently taking place along the complex convergent plate boundary of Taiwan, the 131 stations of the "Taiwan GPS Network" were surveyed four to six times from 1990 to 1996 with dual-frequency geodetic receivers (Yu et al., 1997). Such accurate geodetic data on present-day deformation thus provided new opportunities to study the relationships between the crustal deformation and the plate kinematics. It became possible to describe not only the type and orientation, but also the magnitudes of tectonic effects, by using GPS-based displacement data rather than just the paleostress orientation data as constraints in modeling (Hu et al., 1997).

In this paper, we aim to characterize the crustal deformation in northern Taiwan induced by the Taiwan collision and surrounding subduction zones. In the westernmost Ryukyu-Taiwan region, the rapid change from oblique subduction to collision is most intriguing because it involves all major seismogenic structures and is accomplished in a very short distance. The relationship among structures showing different patterns of crustal strain provides fundamental constraints on the accommodation of deformation along the subduction-collision transition. We combine numerical simulation, deformation data from GPS studies, and geologic constraints in order to better quantify the rate, style, and distribution of deformation currently taking place along the complex oblique subduction to collision boundary of northern Taiwan.

TECTONIC SETTING

The island of Taiwan is located along a segment of the convergent boundary between the Philippine Sea plate and Eurasia plate, where collision and subduction processes occur (Fig. 1). The Philippine Sea plate moves in a northwest direction and subducts beneath Eurasia along the Ryukyu Trench, but flips subduction polarity and overrides the Eurasia plate along the Manila Trench (Fig. 1). The southeast-facing Ryukyu arc-and-trench system extends from southern Kyushu (Japan) to east of Taiwan. The morphology of the Ryukyu Trench disappears west of 123°E where it is intercepted by the Gagua Ridge (Fig. 1). This subduction zone is associated with the backarc spreading in the Okinawa Trough, which ends in the Ilan Plain of northeastern Taiwan (Fig. 1) (Tsai, 1986). In contrast, the west-facing Luzon-Manila arc-and-trench system extends from the Philippines to about 22°N (Fig. 1). North of 22°N, the bathymetric signature of the Manila Trench disappears and the Manila Trench–Luzon arc system gradually loses its identity as it gets close to the continental margin of southeast China. The Manila Trench system extends onshore in southeastern Taiwan and forms the Quaternary fold-and-thrust belt of southern Taiwan (Liu et al., 1997). At present, the collisional process is still active in southern and central Taiwan as demonstrated by the crustal shortening and intense seismicity (e.g., Wu, 1978; Tsai, 1986; Yu and Chen, 1994).

A characteristic feature of the Taiwan mountain belt is the major change in the general structural trends between central Taiwan and northern Taiwan. On average, the mountain ranges in central Taiwan trend north-northeast south of 24.5°N, whereas in northeastern Taiwan they trend east-northeast (Suppe, 1984; Angelier et al., 1990). The existence of major change in structural trends of about 55° between 24°N and 25°N (Fig. 2) has been interpreted to be the result of subduction flipping and opening of the Okinawa Trough (Suppe, 1984).

In contrast with south-central Taiwan, northeast Taiwan has undergone postcollisional processes and has been incorporated in the southern Ryukyu arc system. Evidence for this late evolution is provided by the presence of Quaternary extensional structures (Lee and Wang, 1988; Lu et al. 1995) and extensional earthquake focal mechanisms (Tsai et al., 1977; Tsai, 1986; Yeh et al., 1991; Kao et al., 1998). All of these features indicate that the polarity of subduction has flipped in northern Taiwan; accordingly, the compressional tectonism that resulted in the growth of the Taiwan orogen has been replaced by the extensional tectonism of the Okinawa Trough (Suppe, 1984; Lee and Wang, 1988; Teng, 1996). On the basis of integration of available radiometric age data, Teng et al. (1992) pointed out a westward-younging character for the volcanism in northern Taiwan. It appears that both the onset and the cessation of the calc-alkalic volcanic activity migrated from east to west near the western tip of the Ryukyu arc in northern Taiwan. Teng et al. (1992) proposed that the arc magmatism was initiated by the westward encroachment of the Ryukyu subduction system and ceased subsequently as a result of the southwestward propagation of the opening of the Okinawa Trough. Therefore, the emplacement of the ultrapotassic rock may have resulted from the onset of postcollision extension, as recently claimed on the basis of geochemical studies (Chung et al., 1995; Chen et al., 1995).

The east-northeast-trending belt segment of northeastern Taiwan is composed of several stratigraphic units, which belong to the northern Hsüehshan Range, the Western Foothills, the Coastal Plain, and the Pleistocene volcanic rocks (Fig. 3a). In

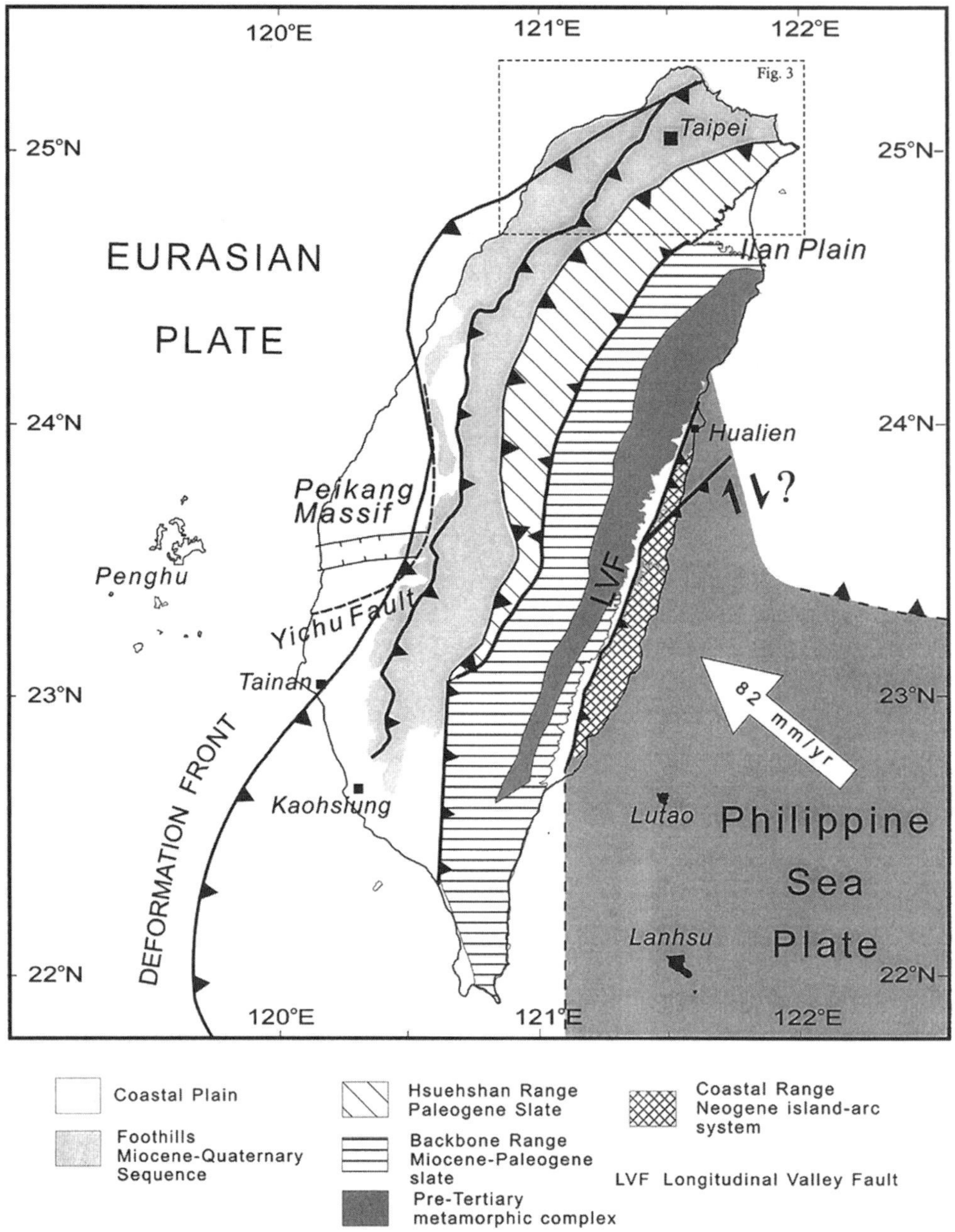

Figure 2. Tectonic framework and main structural units in Taiwan. Main tectonic units follow Ho (1986) and Biq (1989). Major thrust faults have triangles on the hanging wall.

the Coastal Plain and Western Foothills, clastic sedimentary rocks crop out, with ages ranging from Oligocene to Quaternary. The Hsüehshan Range (Fig. 2) includes a passive-margin shallow-marine sequence of Eocene to Oligocene age that is folded and metamorphosed. Metamorphic grades increase from northwest to southeast, from unmetamorphosed or slightly metamorphosed rocks in the Coastal Plain area to lower-greenschist facies in the inner Hsüehshan Range. Studies in the northern Hsüehshan Range (Lu et al., 1995) show that it was subject to complex deformation processes including contraction, regional horizontal bending, transcurrent faulting, block rotation, and extension. The structural analyses in central and northern Taiwan indicate that during the late Cenozoic, the Hsüehshan Range underwent a pop-up style of deformation (Clark et al., 1993).

The late Cenozoic structure of northern Taiwan (Fig. 3) is characterized by numerous northeast-trending fold-and-thrust faults, which are deformed by broadly spaced strike-slip faults and some normal-fault systems. There is a slight obliquity between the regional trends of the major thrust units and the fold axes, revealing a poorly marked en echelon arrangement of several folds and thrusts between Taipei and Keelung. Pliocene–Quaternary normal faults are present in several basins and depression structures within the compressional mountain chain,

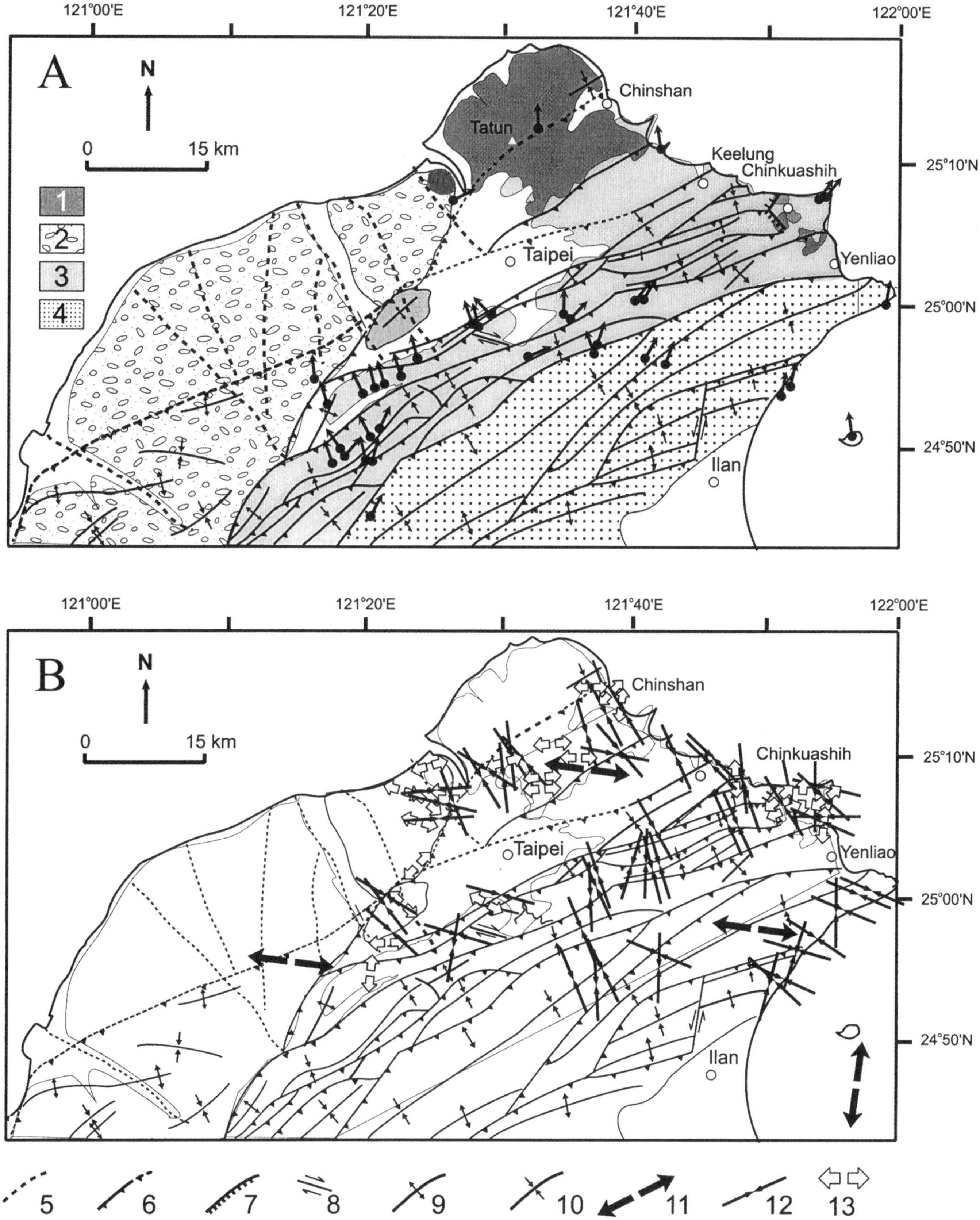

Figure 3. Location map showing general geologic structures interpreted from the geologic maps (Ho, 1986) and SLAR (side-looking airborne radar) image of northern Taiwan (Mars Inc., 1981). Keys: 1—Pleistocene volcanic rocks, 2—Pleistocene conglomerate, 3—Western Foothills, 4—Hsüehshan Range, 5—major fracture, 6—thrust fault, 7—strike-slip fault, 8—normal fault, 9—anticline axis, 10—syncline axis, 11—extension based on focal mechanism, 12—compression, 13—extension based on normal faults (modified after Lu et al., 1995). (A) Small dots with arrows describe the declinations of magnetic fabrics in the rocks of northeastern Taiwan (after Lue et al., 1991). (B) Summary of the results of the paleostress analyses. For extension, each pair of divergent arrows (local trend of σ_3) corresponds to the analysis of normal faults and focal mechanism. For compression, each pair of convergent solid arrows (local trend of σ_1) corresponds to the analysis of a site with many compressional structures, including thrusts, strike-slip faults, fracture cleavages, folds, and joints (modified after Lu et al., 1995).

like the Taipei Basin, the Ilan Basin, and the Yenliao depression together with the Chinkuashih gold-copper field (Fig. 3A). Extensional structures widely cut across the old contractional structures and are thought to be associated with downfaulting of Quaternary intramontane basins (Lee and Wang, 1988).

Recent studies of fault-slip data sets (e.g., Lu et al., 1995), borehole breakouts (Suppe et al., 1985), and focal mechanisms of earthquakes (Yeh et al., 1991) have shown that north-south and east-west extensional movements are common throughout northernmost Taiwan and the Ilan Plain area. According to most analyses, north-south extension dominates near the western tip of the Okinawa Trough (e.g., Ilan Plain and adjacent offshore areas), whereas east-west extension prevails in northernmost Taiwan (e.g., Taipei basin and the Tatung volcano field, see Fig. 3). To the east, the north-south extension is widespread in the Ryukyu arc and corresponds to backarc spreading activity of the Okinawa Trough. The Okinawa Trough shows various extensional features including grabens identified from regional bathymetry and seismic reflection profiles (e.g., Kimura, 1985; Sibuet et al., 1987, 1998; Sato et al., 1994) as well as earthquake focal mechanisms (e.g., Kao and Chen, 1991). In addition, geodetic and tomographic studies suggest that the extension associated with the opening of the Okinawa Trough extends westward beneath northeast Taiwan (e.g., Yeh et al., 1989; Liu, 1995).

STRESS AND STRAIN PATTERNS IN NORTHERN TAIWAN

Present-day stress and paleostresses related to Pliocene–Quaternary collision

Although it is not described in detail hereafter, the distribution of recent and present-day tectonic stress patterns provides an important key for understanding the recent deformation in northern Taiwan. On the basis of the results of several studies, the paleostresses are summarized in Figure 3B in terms of trends of the maximum horizontal compressional stress axes (σ_1) corresponding to different compressional structures, including thrusts, strike-slip faults, fracture cleavages, folds, and joints (Lu et al., 1995). Considering the dispersion due to uncertainties of measurements and paleostress determinations, the variations of σ_1 trends cannot be simply accounted for by the natural irregularities and stress-field inhomogeneity related to a single event affecting a rigid mass. In addition, these Pliocene–Quaternary patterns of compressional stress related to collision cannot be considered as a whole because it has been demonstrated that significant changes have occurred over time. For example, the late Pliocene–Pleistocene records of compressional paleostresses could be separated into two principal tectonic events, as proposed by Angelier et al. (1986). Further detailed analyses, especially in the Hsüehshan Range in northern Taiwan (Angelier et al., 1990) have revealed a complex set of the paleostress regimes that have been successively recorded since the late Miocene (Chu, 1990).

The present-day distribution of stress in Taiwan was obtained from the analysis of 1972 to 1986 earthquake focal mechanisms (Fig. 3B) (Yeh et al., 1991). Although a compressional stress regime (with reverse- and strike-slip faulting) dominates in most of Taiwan, results of focal mechanism analyses from northern Taiwan (Fig. 3B) indicate that the present state of stress of this part of the island is compatible with strike-slip and extensional regimes (Yeh et al., 1991).

In summary, all determinations of present-day stress indicated that the regional stress patterns in Taiwan are dominated by (1) northeast-southwest compression in the orogenic belt, due to collision between Eurasia and the Philippine Sea plate, and (2) active north-south and east-west extension in northeasternmost Taiwan and the western tip of the Okinawa Trough (Fig. 3B).

Spatial variation of crustal strain of northern Taiwan

The changes in baseline lengths determined from the repeated GPS surveys in the northern region of Taiwan are used to assess the spatial variation of the horizontal crustal strain over the region. On the basis of a consideration of the geologic structures and the distribution of geodetic stations, northern Taiwan is divided into 17 subnetworks denoted NT1 to NT17 (Fig. 4). Each of these subnetworks includes 4 to 13 GPS stations (Fig. 4). It is assumed that, spatially, the crustal strain accumulates uniformly over each subnetwork (after tensor determination, this assumption is checked by misfit review), and that the rate of strain accumulation is constant over the time period considered (1990–1997). Note that no very large earthquakes occurred during this period. The maximum magnitude (M_L) was 6.20 so that these data principally reflect the interseismic deformation. Following the procedures of Prescott et al. (1979), the average annual strain rate for each net was calculated for each subnetwork. The strain rate for each baseline, $\dot{e}$, is expressed as $\dot{e} = L^{-1}(dL/dt)$, where L is the baseline length and dL/dt is the average annual strain-rate change. It is related to the surface strain-rate tensor, $\dot{E}_{ij}$, by

$$\dot{e} = \dot{E}_{11}\sin^2\theta + \dot{E}_{12}\sin 2\theta + \dot{E}_{22}\cos^2\theta, \quad (1)$$

where θ is the azimuth (measured clockwise from north) of the baseline and the strain-rate tensor, $\dot{E}_{ij}$, is referred to a geographic coordinate system with axis 1 directed to the east and axis 2 directed to the north. Three components of the average strain-rate tensor, $\dot{E}_{11}$, $\dot{E}_{12}$, and $\dot{E}_{22}$, were thus calculated by using all of the surveys for each of the 17 subnetworks or tectonic blocks (Fig. 4).

The average principal strain rates are computed from three components of the surface strain-rate tensor using the following formula:

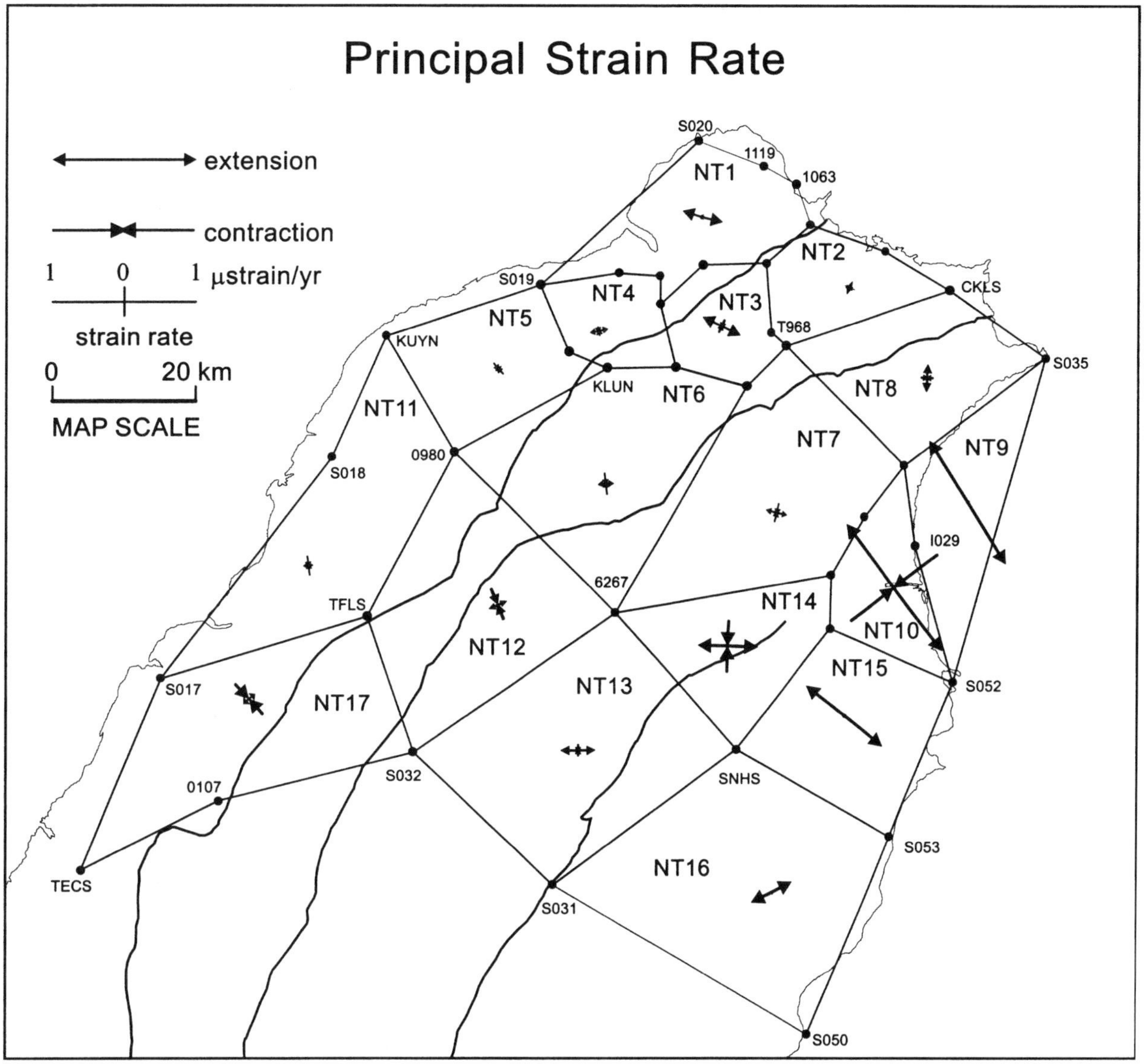

Figure 4. Principal strain rates for 17 subnetworks in northern Taiwan. Thick line denotes contraction, whereas thin line represents extension. Solid circles indicate GPS stations.

$$\dot{\varepsilon}_{1,2} = \tfrac{1}{2}(\dot{E}_{11} + \dot{E}_{22}) \pm [\dot{E}_{12}^{2} + \tfrac{1}{4}(E_{11} - \dot{E}_{22})^{2}]^{1/2}, \quad (2)$$

where $\dot{\varepsilon}_1$ and $\dot{\varepsilon}_2$ are the algebraically larger and smaller principal strain rates, respectively. The azimuth of $\dot{\varepsilon}_1$, φ, is given by

$$\varphi = \tfrac{1}{2}\tan^{-1}[2\dot{E}_{12}/(\dot{E}_{22} - \dot{E}_{11})]. \quad (3)$$

The average principal strain rates for each of the 17 subnetworks are given in Table 1. All uncertainties quoted are ± 1 standard deviation. The strain rates are given in units of microstrain (µstrain) per year (1 µstrain $= 10^{-6}$). Positive values denote extension, whereas negative values represent shortening or contraction.

The most prominent feature of the strain distribution patterns in northern Taiwan certainly corresponds to the Ilan Plain. Significant extension and shortening rates of 0.19 and 0.07/yr in azimuths of 2° and 92°, respectively, are observed in northeastern Taiwan (NT8). The Ilan Plain (NT10) and its eastern offshore area (NT9) show remarkable extension rates of 1.03–1.11 µstrain/yr on an azimuth of 143°–148°. Also in the Ilan Plain (NT10), the largest shortening rate is found (0.77 µstrain/yr with an azimuth of 143°) with a large extension rate in the perpendicular direction (1.13 µstrain/yr with an azimuth of 53°), indicating strike-slip deformation mode. Because this area is considered as a western extension of the Okinawa Trough (Yu and Tsai, 1979; Yeh et al., 1989; Liu, 1995), the extensional deformation is assumed to be related to extension in the Okinawa Trough (Kimura, 1985; Letouzey and Kimura, 1985; Sibuet et al., 1987).

TABLE 1. GPS-DERIVED PRINCIPAL STRAIN RATES OF 17 SUBNETS IN NORTHERN TAIWAN (1990–1997)

Net	$\dot{\varepsilon}_1$ (μ strain/yr)	$\dot{\varepsilon}_2$ (μ strain/yr)	(°)
NT1	0.15 ± 0.02	−0.03 ± 0.02	−74.2 ± 5.5
NT2	0.11 ± 0.04	−0.06 ± 0.04	77.9 ± 9.9
NT3	0.07 ± 0.03	−0.08 ± 0.03	23.0 ± 11
NT4	0.15 ± 0.03	−0.09 ± 0.03	−64.7 ± 5.4
NT5	−0.05 ± 0.03	−0.09 ± 0.03	47.8 ± 33.2
NT6	0.05 ± 0.04	−0.15 ± 0.04	83.7 ± 8.7
NT7	0.09 ± 0.05	−0.12 ± 0.05	−76.9 ± 9.1
NT8	0.19 ± 0.08	−0.07 ± 0.04	1.7 ± 6.5
NT9	1.03 ± 0.06	0.02 ± 0.05	−31.9 ± 3.8
NT10	1.13 ± 0.03	−0.77 ± 0.04	−37.2 ± 0.9
NT11	0.05 ± 0.04	−0.13 ± 0.03	82.2 ± 9.9
NT12	0.09 ± 0.04	−0.21 ± 0.04	67.4 ± 5.8
NT13	0.24 ± 0.06	−0.07 ± 0.06	89.4 ± 5.0
NT14	0.43 ± 0.04	−0.35 ± 0.08	−88.0 ± 3.3
NT15	0.69 ± 0.04	−0.01 ± 0.05	−51.2 ± 3.1
NT16	0.15 ± 0.03	0.01 ± 0.03	64.4 ± 10.7
NT17	0.05 ± 0.02	−0.28 ± 0.04	48.4 ± 4.2

In the northwestern part of Taiwan, insignificant or minor deformations are observed. Overall the shortening rates in this area clearly increase from south to north, which is consistent with the southward propagation of collision and GPS data from the rest of Taiwan (Yu et al., 1997). In the Taipei area, subnetworks NT1–NT4 have small extension rates (0.07–0.15 μstrain/year) and small contraction rates (0.03–0.09 μstrain/year). The extension deformations in these networks mostly trend east-west to east-northeast. Overall, the shortening rates in these areas clearly increase from north to south, which is consistent with the southward propagation of collision and with GPS data of Taiwan (Yu et al., 1997).

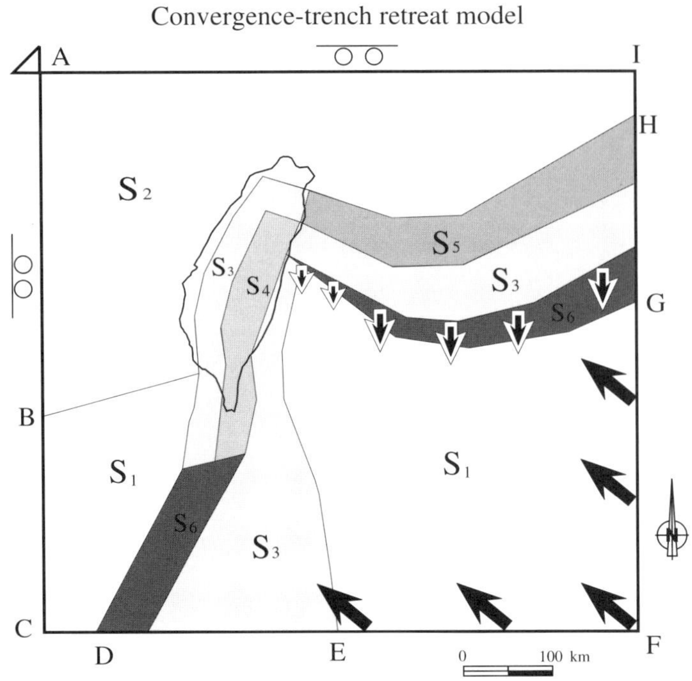

Figure 5. Geometry and boundary conditions of the finite-element model of Taiwan collision. For geographic coordinates, see Figure 1; coast of Taiwan shown as thin line. S_1, S_2, S_3, S_4, S_5, and S_6 are subdomains with different rheologies (Young's modulus, Poisson's ratio, and yield stress listed in Table 2). Uppercase letters A–I indicate segments of frame with different boundary conditions: open triangle—fixed boundary (point A), open circles—roller in one direction (north-south for AB, east-west for AI), no symbol—free segments (BC, CD, DE, IH, and GH). Large black arrows, common direction of displacement based on plate kinematics (EF and FG).

TABLE 2. RHEOLOGY AND PARAMETERS USED IN THE FINITE-ELEMENT MODELING

Subdomain	Young's modulus (GPa)	Poisson's ratio	Yield stress (MPa)
S1	60	0.25	300
S2	40	0.25	260
S3	30	0.25	240
S4	10	0.25	240
S5	5	0.25	120
S6	1	0.20	60

Note: Compare with Figure 5.

FINITE-ELEMENT MODELING

The finite-element technique

The computer code MODULEF was used for the numerical simulation. This finite-element technique was developed by the INRIA (Institut National de Recherche en Informatique et en Automatique) (George et al., 1986). The model used here (Fig. 5) is a two-dimensional model with a plane-stress condition that includes a variety of subdomains with different material properties (Table 2), in order to represent the mechanical behavior of different regions (for details, see Hu et al., 1996). This configuration was chosen according to the regional structural framework of the area under investigation (Fig. 1). As expected for an elastic model that fails to take into account the stress-buffering effects of yielding, purely elastic models produced stresses that were progressively larger and that become unrealistic with increasing amounts of displacements. Owing to the obvious limitations of the elastic model, we adopted an elasto-plastic model with the von Mises criterion, which is more realistic and better accounts for the actual deformation of the mountain belt. With this elasto-plastic model, relatively large deformation can be considered. The elasto-plastic yield limit obviously varies with depth, so that the values adopted represent an average effective yield strength for the model. The depth considered for the brittle deformation is about 20–30 km. Ranalli and Murphy (1987) showed that the depth-dependence of lithospheric rheology varies with tectonic provinces; in many cases the continental lithosphere includes one or two soft ductile layers sandwiched between brittle layers. As a result, brittle behavior may occur either in the upper crust or in the uppermost mantle. In our model, we aim at discussing the brittle deformation in the upper crust through a simplified two-dimensional analysis.

We used plane-stress approximations, which correspond to a two-dimensional model of an "infinitely thin" plate. Under this condition, the vertical component of stress is zero, or is small relative to the horizontal component of stress. The two-dimensional models used may be regarded as the first-order approximation of the problem under consideration; i.e., it does not consider the effects of topography, buoyancy, basal drag, coupling at depth, and thermomechanical deformation within the lithosphere. Another important omission is the possible presence of a mechanically weak layer at mid-crust depth in the continental lithosphere, which could result in mechanical decoupling of the crustal layers. Furthermore, mechanical coupling that may be present in the convergence zone, due to underthrusting oblique to the plate margin, is not considered in our two-dimensional modeling experiments discussed herein.

Geometric configuration of the model and boundary condition

The model covers a rectangular area between latitudes of 19°N and 27°N and between longitudes of 118°E and 127°E (Fig. 5). The 10 subunits of the model represent major structural units according to the regional tectonic setting. Because a rigorous introduction of physically defined parameters for all features is not feasible, the boundaries of the major structural units should be regarded as a first approximation. The model contains 519 triangular linear elements, with three nodes in the summits.

We have used displacement-boundary conditions for the model, because the velocity of plate movement between the Philippine Sea plate and the Eurasian plate is given by the NUVEL-1 global-kinematics results (e.g., Seno et al., 1993) and GPS data (Yu et al., 1997). The northwestern corner of the model is fixed (point A in Fig. 5); displacement is left free along the east-west direction for the northern boundary (AI in Fig. 5) and along the north-south direction for the northwestern boundary (AB in Fig. 5). For the southeastern boundaries of the model (EF and FG in Fig. 5), we imposed an initial horizontal displacement of the plate motion of the Philippine Sea plate relative to Eurasia on the basis of the global plate-kinematic model. The imposed displacement in the models of Figure 5 is 7 cm/yr along the azimuth 310° (Seno et al., 1993). The GPS observed velocity of Lanhsu relative to Paisha is 81.5 ± 1.3 mm/yr, along an azimuth of 306.3° ± 1.2° (Yu et al., 1997), which is slightly larger than the velocity estimated by Seno et al. (1993). We also tried to evaluate the rate of opening of the Okinawa Trough, which influences the deformation field in the northeastern Taiwan area. Letouzey and Kimura (1985) suggested a spreading rate of about 46 mm/yr for the southern Okinawa Trough, based on the study of magnetic anomalies. This rate is consistent with GPS results that reveal a velocity of 56.2 mm/yr along an azimuth of 156° (Fig. 1). It is reasonable to assume that the retreating rate of the Ryukyu Trench is smaller than the opening rate of Okinawa Basin, owing to shortening in the Ryukyu accretionary prism between the Ryukyu

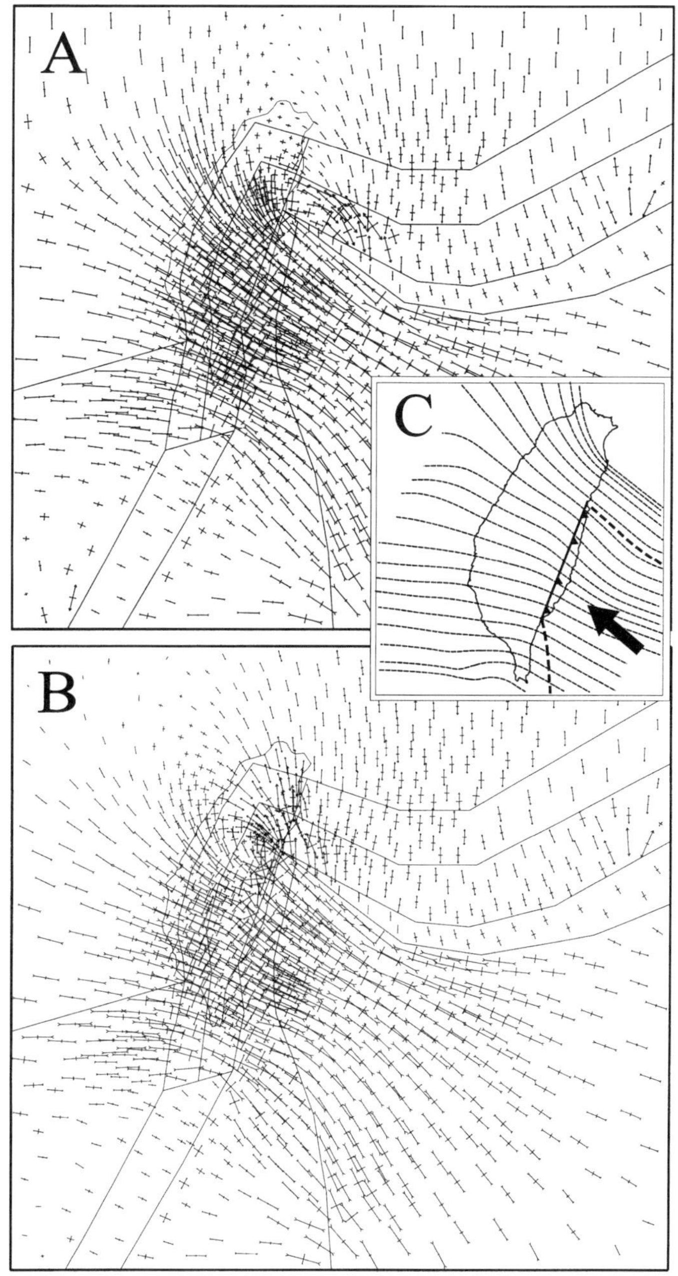

Figure 6. (A) Stress distribution calculated in the convergence and trench-retreat elasto-plastic model. Structure, rheology, and boundary conditions defined in Figure 5. Principal stresses in the horizontal plane shown as small couples of arrows. Pairs of convergent arrows represent maximum compressive stress. Note the stress concentration in the Taiwan collision zone, the fan-shaped distribution throughout the island, and the pronounced stress trajectory deviation near the northwestern corner of the Philippine Sea plate. Pairs of divergent arrows represent minimum extensional stress, especially northeast of Taiwan, in the Okinawa Trough area. (B) Influence of southward trench-retreat velocity. Same explanation as for A except for an increase of southward trench-retreat velocity (see text). Note the westward shift of the transition zone, relative to A. Note also the local occurrence of the east-northeast extensional stress trends in the northeastern Taiwan. (C) Distribution of Quaternary and present-day stresses throughout Taiwan. Main compressional stress field associated with the Quaternary collision from various sources (focal mechanisms, borehole breakouts, and Quaternary fault-slip data). Plate boundary added as thick line with triangles on the hanging wall. Vector of relative plate motion shown by black arrow.

arc and the Ryukyu Trench. We consequently adopted a retreating rate of about 15 mm/yr for the model of Figure 6. We set the retreating rate to zero near Taiwan (because the Okinawa Trough extension obviously disappears west of the Ilan Plain), with a linear increase from west to east between 122°E and 123°E (Fig. 5). In order to evaluate the influence of this additional southward displacement of the Ryukyu Trench, we carried out experiments with various velocities. The results obtained with a rate of 15 mm/yr are shown in Figure 6A; the results with a higher rate of 30 mm/yr along the Ryukyu Trench front and 15 km/m.y. near Taiwan are summarized in Figure 6B.

Rheology and material properties

We divided the region studied into 10 subdomains, shown in Figure 5, with six different properties listed in Table 2. The main criteria to define these major rheological properties deal with geologic composition (oceanic crust, continental crust, island-arc crust, foreland belt, metamorphic belt, and accretionary prism), crustal thickness, and thermal state (e.g., the Okinawa Trough area), etc. The materials themselves are considered homogeneous and isotropic. The elastic properties are described by Young's modulus and Poisson's ratio. For the elasto-plastic deformation, a third parameter, the yield stress, is added. The actual values of these parameters are not well determined; as a consequence, we tested models with a variety of values in order to explore the effects of changes in parameters and the sensitivity of our results to various choices. The Young's modulus thus ranges from the highest value in the oceanic lithosphere of the Philippine Sea plate and the South China Sea (S_1 in Fig. 5) to the lowest value in the thinned lithosphere of the Okinawa Trough backarc basin (S_5 in Fig. 5). Intermediate values were assigned to the continental shelf and margin (S_2 in Fig. 5), island arcs (S_3 in Fig. 5), outer and inner mountain belt (S_3 and S_4, respectively, in Fig. 5), and the Okinawa backarc basin (S_5 in Fig. 5). All these values are listed in Table 2. Average Poisson's ratios for the continental and oceanic crusts are taken to be 0.265 and 0.30, respectively (Christensen, 1996), although a Poisson's ratio of 0.25 may be acceptable for common rock types. Abnormally low values of Young's modulus and yield stress were assigned along the subduction zones of the Ryukyu Trench and the Manila Trench (S_6 in Fig. 5). These low values (Table 2) reflect the presence of mechanical decoupling at the trenches. The final values retained in Table 2 may be regarded as somewhat arbitrary. However, numerical experiments revealed that the calculated patterns of stress trajectories are not sensitive to the values of the assigned elastic constants within a reasonable range of choices.

THE CONVERGENCE AND TRENCH-RETREAT MODEL

The results of the modeling experiments defined in Figure 5 and the previous section are shown in Figures 6 and 7. The fan-shaped distribution of compressional stress trajectories was thoroughly discussed in an earlier paper (Hu et al., 1996). The fan-shaped stress pattern is in agreement with measurements of Quaternary and present-day stresses (Suppe et al., 1985; Barrier and Angelier, 1986; Angelier et al., 1986, 1990; Lee, 1986; Chu, 1990; Yeh et al., 1991), as summarized in Figure 6C. Figures 6 and 7 also show the importance of north-south extensional stress in the Okinawa Trough–Ryukyu arc region, which is consistent with the seismotectonic data available in the western Ryukyu arc and the Ilan Plain (Yeh et al., 1991; Kao et al., 1998). Our model predicts that a transitional regime in northeastern Taiwan exists between the pure compressional tectonic regime in the collision zone of Taiwan and the extensional tectonic regime in the northeastern part of the Taiwan. Furthermore, the transition in the stress field, where significant strike-slip deformation can be expected, is characterized by two intermediate subregimes: one is mostly compressional, the other one is mostly extensional, as indicated by patterns of crosses and triangles, respectively, in Figure 7. From west to east across the transition zone, the relative magnitudes of stress vary from compression to extension (Fig. 7).

An interesting conclusion of the experiments lies in the significant migration of the transition zone as boundary conditions vary. A simple increase in the velocity of the southward retreat of the subduction zone (30 mm/yr in Fig. 7B in stead of 15 mm/yr in Fig. 7A) results in a southwestward shift of more than 50 km of the transition zone (compare Fig. 7, A and B). Thus the transition between regional compression (Taiwan collision belt) and extension (Okinawa-Ryukyu system) is extremely sensitive to variations in boundary conditions. The general influence of these changes is shown in Figure 6, A and B, with same boundary conditions as for Figure 7, A and B, respectively, showing that the pattern of compressional stress in the collision belt of central and southern Taiwan undergoes little perturbation. In other words, the high sensitivity of the stress regimes to the variations in trench retreat is restricted to northern Taiwan. To the south, these effects sharply diminish and practically vanish in the southern half of Taiwan (Fig. 6). We conclude that the effects of the suction force related to the Ryukyu subduction zone play a very important role in northern Taiwan, but have very little influence farther south, in most of the collision zone.

DISCUSSION

The finite-element models presented herein are significantly simplified with respect to the actual structure and tectonism in and around Taiwan. However, they provide a mechanically consistent framework (Fig. 5) for interpreting the observed stress and strain patterns (Figs. 3 and 4), in relation to the major features and kinematics of this collision-subduction area. The interaction between opening of the Okinawa Trough and the collision of eastern Taiwan plays a crucial role in determining the regional stress patterns (e.g., Hu et al., 1996; Teng

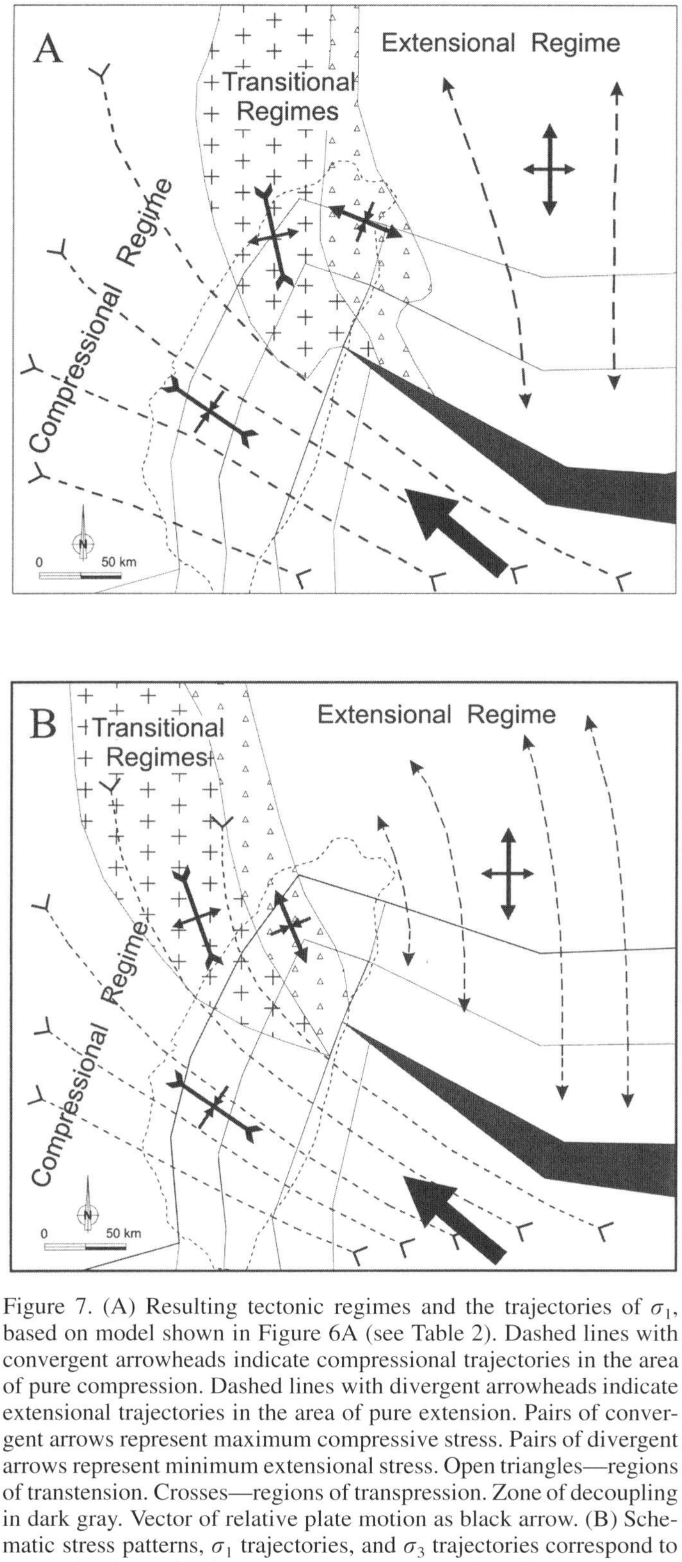

Figure 7. (A) Resulting tectonic regimes and the trajectories of σ_1, based on model shown in Figure 6A (see Table 2). Dashed lines with convergent arrowheads indicate compressional trajectories in the area of pure compression. Dashed lines with divergent arrowheads indicate extensional trajectories in the area of pure extension. Pairs of convergent arrows represent maximum compressive stress. Pairs of divergent arrows represent minimum extensional stress. Open triangles—regions of transtension. Crosses—regions of transpression. Zone of decoupling in dark gray. Vector of relative plate motion as black arrow. (B) Schematic stress patterns, σ_1 trajectories, and σ_3 trajectories correspond to the model shown in Figure 6B. Note that the main difference, relative to Figure 6A, is a westward shift of the transition from compression (Taiwan) to extension (Ryukyu).

and Lee, 1996). Teng (1996) suggested that the orogenic collapse of northern Taiwan is mainly triggered by the change in the boundary forces induced by the change in vergence of subduction. According to this interpretation, the continental margin of southern Ryukyu and northern Taiwan first underwent a lithospheric thickening, as induced by collision, which was followed by lithospheric stretching, as induced by trench suction. This general idea was proposed by Viallon et al. (1986) based on numerical modeling and by Teng (1996) based on qualitative approach.

Our numerical simulation indicates that the crustal deformation in the Ryukyu-Taiwan area may be strongly affected by the interaction between the indentation of the Philippine Sea plate and the suction at the Ryukyu Trench (Fig. 5). This interaction results in a varying stress-strain field. Three subregimes may be distinguished: a compressional regime in south-central Taiwan, an extensional regime in northeast Taiwan and south Ryukyu, and, between these domains, a transition zone with transtension and transpression subdomains. The transition zone migrates significantly as a function of the variable ratios between velocities of northwest-southeast convergence and north-south trench retreat.

Our numerical models also predict that the maximum tension in the Okinawa Trough area rotates counterclockwise from north-south to northwest-southeast, with magnitudes that decrease as one approaches northeast Taiwan (Figs. 6 and 7). Similarly, the trends of the maximum compression in central Taiwan gradually turn clockwise from northwest-southeast to north-south and even to northeast-southwest as one approaches northeast Taiwan, while the magnitudes decrease rapidly.

The transition zone between these different tectonic regimes coincides with the area where the Okinawa seismic zone, the interface seismic zone, the collision seismic zone, and the lateral-compression seismic zone merge (Kao et al., 1998). The observation of sharp changes in the trends of minimum stress, from approximately north-south in the Ilan Plain and Okinawa Trough to approximately east-west in northernmost Taiwan supports our numerical simulation. The model is also supported by the distribution of earthquake focal mechanisms in this area (Kao et al., 1998). The S waveform's splitting observed in northern Taiwan, interpreted to be caused by stress-aligned cracks in the upper crust (Kuo et al., 1994), is consistent with the present conclusion that northeastern Taiwan is currently under the combined actions of extension due to the Okinawa Trough opening and compression due to the collision between the Philippine Sea plate and Eurasia.

Lu and Malavieille (1994) used three-dimensional sandbox-modeling experiments to illustrate the kinematic processes of the Taiwan thrust wedge. The results showed a major consequence of oblique collision with an asymmetric indenter in the development of an asymmetric thrust wedge with a characteristic distribution of tectonic domains. The deformation within the belt is a combination of shortening, rotations, and stretching, which locally results in a partitioning among thrusting, strike-slip faulting, and normal faulting. On the basis of these results, Lu et al. (1995) suggested that the northeastern segment of the Taiwan belt (Fig. 8) first underwent contractional deformation (such as shortening and oblique thrust-sheet stacking), followed by right-lateral transcurrent faulting and normal faulting. Lu et al. (1995) pointed out that the indentation results not only in contractional tectonics across the belt (Fig. 8A), but also in divergent strike-slip components and block rotation as well as local occurrence of extension roughly parallel to the trends of the major units of the belt (Fig. 8A). These authors concluded that extensional structures such as the Taipei Basin developed in response to increasing belt curvature and related block rotations (Fig. 3A), and, due to the asymmetry of the collision (Fig. 8B), the amounts of right-lateral strike slip and clockwise rotation east of the Tanshui River axis (Fig. 8A) are larger than those of the left-lateral strike slip and counterclockwise rotation west of it (Fig. 8B). These interpretations were supported not only by tectonic analyses but also by paleomagnetic reconstruction of rotation in northern Taiwan (Lee et al., 1991; Lue et al., 1991). The presence of nearly east-west extension in northeasternmost Taiwan (Fig. 3) is also in good agreement with the reconstruction in Figure 8.

The complex deformation of northern Taiwan, however, cannot be accounted for solely by the effects of wedge convergence. Northeast off Taiwan, the Okinawa Trough area is dominated by extensional tectonics (Kimura, 1985; Letouzey and Kimura, 1985, 1986; Sibuet et al., 1987) as a result of the suction force related to the Ryukyu Trench retreat. Part of the extensional regime present in northeastern Taiwan, especially for the north-south extension near the Ilan Plain (Tsai, 1986; Yeh et al., 1989, 1991), is the westernmost expression of the slab-related backarc extension. Hu et al. (1996) confirmed that the extension in the Ryukyu arc and the western Okinawa Trough cannot be explained solely by lateral extrusion, as a simple consequence of the Taiwan collision. Here we point out that the suction force related to the southward retreat of the Ryukyu subduction zone may play a major role.

CONCLUSIONS

The study of two-dimensional elasto-plastic finite-element modeling of the subduction-collision in and around Taiwan allows us to estimate the influences of different parameters in producing the stress and strain patterns. The main constraints

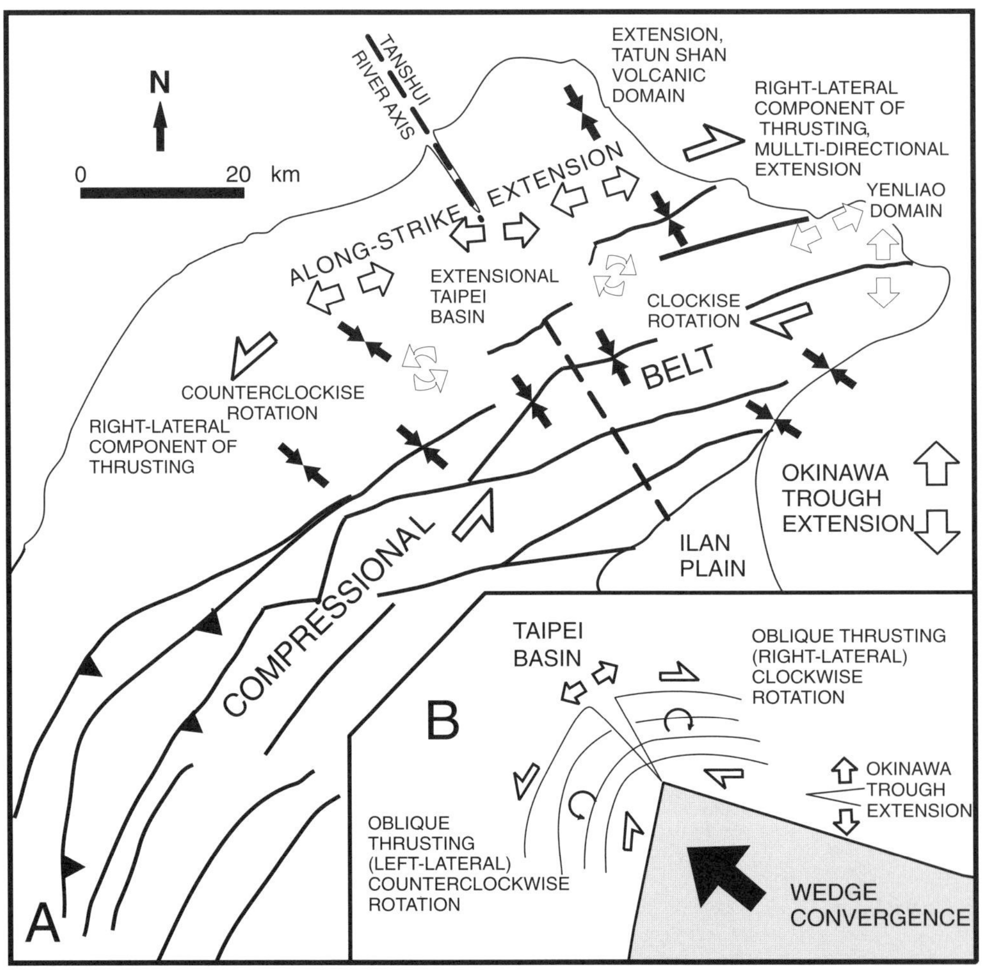

Figure 8. (A) Interpretative map and (B) explanatory sketch. Model of Pliocene–Quaternary tectonic deformation in the foreland thrust belt of northern Taiwan (modified after Lu et al., 1995). Note the existence of three major domains: (1) the eastern domain with right-lateral oblique-slip thrusting, right-lateral strike-slip, and clockwise rotations; (2) the central domain characterized by indentation in the inner belt and extensional deformation in the outer belt; and (3) the western domain with left-lateral oblique thrusting, right-lateral strike-slip, and counterclockwise rotations. Explanation of symbols in A.

for testing the model experiments are derived from both the tectonic and seismic studies and the GPS analyses of Taiwan deformation. Additional constraints are provided by consideration of structural geology and paleomagnetic results. According to our two-dimensional numerical simulations, the crustal deformation in the Ryukyu-Taiwan area is controlled by two major factors: the indentation of the Philippine Sea plate into Eurasia and the trench retreat due to suction at the Ryukyu Trench (Fig. 6). The interaction of these two forces results in a complex stress-strain field that includes a compressional regime in south-central Taiwan, an extensional regime in northeast Taiwan and south Ryukyu, and a narrow transitional regime in between. The interaction between the opening of the Okinawa Trough and the collision along eastern Taiwan may play a crucial role in determining the regional stress and strain patterns, and its variations resulted in significant migration of the transition zone (Fig. 7).

The westward migration of the Okinawa Trough seems to be the reason for increasing intensity of extension with time in northern Taiwan. According to recent studies (Chung et al., 1995; Chen et al., 1995), the eruption of Quaternary andesitic volcanic rocks in this area is probably a result of the change in stress field from pure compression to extension. The interaction already discussed can explain the spatial clockwise rotation of stress axes from northwest-southeast to nearly north-south compression, then progressively to east-west extension and finally to north-south extension. The migration of the Okinawa Trough system and the variations in time of the two sources can explain the local successions of stress regimes, although the ages of the tectonic events are poorly determined (Figs. 6 and 7). Our numerical models predicted that the maximum tension in the Okinawa Trough rotates significantly from north-south to northwest-southeast, and magnitudes decrease rapidly toward northeast Taiwan (Fig. 6). The trajectory of the maximum compression follows the relative convergence direction along the Longitudinal Valley, but gradually turns to northeast-southwest with decreased magnitude in northeast Taiwan. The transition zone between these two stress regimes coincides with the place where the major seismogenic structures merge.

ACKNOWLEDGMENTS

We are grateful to L. Royden, C. Wang, and T. Byrne for constructive and helpful reviews of the manuscript. This research was financially supported by Academia Sinica and the National Science Council (NSC) of the Republic of China under grant NSC 86-2116-M-001-025. Thanks are due to our colleagues C.-H. Yen and H.-H. Su for their assistance in preparing this manuscript and the illustrations. This work is also part of the Taiwan-France cooperation framework supported by the NSC and the French Institute in Taipei. Some figures are generated by the GMT software written by Paul Wessel and Walter H.F. Smith.

REFERENCES CITED

Angelier, J., 1990, Foreword, Geodynamics of the eastern Eurasian Margin: Tectonophysics, Special Issue, v. 183, p. vii-x.

Angelier, J., Barrier, E. and Chu, H.-T., 1986, Plate collision and paleostress trajectories in a fold-and thrust belt: The Foothills of Taiwan: Tectonophysics, v. 125, p. 161–178.

Angelier, J., Bergerat, F., Chu, H.-T. and Lee, T.-Q., 1990, Tectonic analysis and the evolution of a curved collision belt: The Hsüehshan Range, northern Taiwan: Tectonophysics, v. 183, p. 77–96.

Barrier, E., 1985, Tectonique d'une chaîne de collision active: Taiwan: Paris, Université de Pierre et Marie Curie, Mém. Sci. Terre, no. 85–29, 492 p.

Barrier, E., 1986, The double collision of Taiwan: An active orogen: Tectonophysics, v. 125, p. 39–72.

Barrier, E. and Angelier, J., 1986, Active collision in eastern Taiwan: The Coastal Range: Tectonophysics, v. 125, p. 39–72.

Biq, C., 1989, The Yushan-Hsüehshan megashear zone in Taiwan: Proceedings of the Geological Society China, v. 32, p. 7–20.

Chen, C.-H., Lee T., Shieh Y.-N., Chen, C.-H., and Hsu, W.-Y., 1995, Magmatism at the onset of back-arc basin spreading in the Okinawa Trough: Journal of Volcanology and Geothermal Research, v. 69, p. 313–322.

Christensen, N.I., 1996, Poisson's ratio and crustal seismology: Journal of Geophysical Research, v. 101, p. 3139–3156.

Chu, H.-T., 1990, Néotectonique cassante et collision plio-quaternaire à Taiwan: Paris, Université Pierre et Marie Curie, Mém. Sci. Terre, no. 90–28, 292 p.

Chung, S.-L., Yang, T.F., Lee, C., and Chen, C.-H., 1995, The igneous provinciality in Taiwan: Consequence of continental rifting superimposed by Luzon and Ryukyu subduction systems: Journal of Southeast Asian Earth Sciences, v. 11, p. 73–80.

Clark, M.B., Fisher, D.M., Lu, C.-Y., and Chen, C.-H., 1993, Kinematic analysis of the Hsüehshan Range, Taiwan: A large-scale pop-up structure: Tectonics, v. 1, p. 1153–1172.

Dixon, T.H., 1991, An introduction to the Global Positioning System and some geological applications: Review of Geophysics, v. 29, p. 249–276.

George, P.L., Laug, P., Muller, B., and Vidrascu, M., 1986, Guide d'utilisation et normes de programmation: Le Chesnay, France, Institute de National Recherche Informatique Automatique, 113 p.

Ho, C.-S., 1986, A synthesis of the geologic evolution of Taiwan: Tectonophysics, v. 125, p. 1–16.

Hu, J.-C., Angelier, J., Lee, J.-C., Chu, H.-T., and Byrne, D., 1996, Kinematics of convergence, deformation and stress distribution in the Taiwan collision area: 2-D finite-element numerical modeling: Tectonophysics, v. 255, p. 243–268.

Hu, J.-C., Angelier J., and Yu, S.-B., 1997, An interpretation of the active deformation of southern Taiwan based on numerical simulation and GPS studies: Tectonophysics, v. 274, p. 145–170.

Huchon, P., Barrier, E., De Bremaecker, J.C., and Angelier, J., 1986, Collision and stress trajectories in Taiwan: A finite element model: Tectonophysics, v. 125, p. 179–191.

Kao, H., and Chen, W.-P., 1991, Earthquakes along the Ryukyu-Kyushu arc: Strain segmentation, lateral compression, and the thermomechanical state of the plate interface: Journal of Geophysical Research, v. 96, p. 21443–21485.

Kao, H., Shen, S.-J., and Ma, K.-F., 1998, Transition from oblique subduction to collision: Earthquakes in the southernmost Ryukyu arc–Taiwan Region: Journal of Geophysical Research, v. 94, p. 17561–17579.

Kimura, M., 1985, Back-arc rifting in the Okinawa Trough: Marine Petroleum Geology, v. 2, p. 222–240.

Kuo, B.-Y., Chen, C.-C., and Shin, T.-C., 1994, Split S waveforms observed in northern Taiwan: Implication for crustal anisotropy: Geophysical Research Letters, v. 21, p. 1491–1494.

Lacombe, O., Angelier, J., Chen H.-W., Deffontaines, B., Chu, H.-T., and Rocher, M., 1997, Syndepositional tectonics and extension-compression relationships at the front of the Taiwan collision belt: A case study in the Pleistocene reefal limestones near Kaohsiung, SW Taiwan: Tectonophysics, v. 274, p. 109–120.

Lee, C.-T., 1986, Methods of stress analysis and paleostress changes in northern Taiwan due to arc-continent collision [Ph.D. thesis]: Taipei, National Taiwan University, 370 p.

Lee C.-T., and Wang Y., 1988, Quaternary stress changes in northern Taiwan and their tectonic significance: Proceedings of the Geological Society of China, no. 31, p.154–168.

Lee, T.-Q., Angelier, J., Chu, H.-T., and Bergerat, F., 1991, Rotations in the northern eastern collision belt of Taiwan: Preliminary results from paleomagnetism: Tectonophysics, v. 199, p. 109–120.

Letouzey, J., and Kimura, M., 1985, The Okinawa Trough genesis, structure and evolution of a back-arc basin developed in a continent: Marine Petroleum Geology, v. 2, p. 111–130.

Letouzey, J., and Kimura, M., 1986, The Okinawa Trough: Genesis of a back-arc basin developing along a continental margin: Tectonophysics, v. 125, p. 209–230.

Liew, P.-M., Hsieh, M.-L., and Lai, C.-K, 1990, Tectonic significance of Holocene marine terraces in the Coastal Range, eastern Taiwan: Tectonophysics, v. 183, p. 121–127.

Liu C.-C., and Yu, S.-B., 1990, Vertical crustal movements in eastern Taiwan and their tectonic implications: Tectonophysics, v. 183, p. 111–119.

Liu, C.-C., 1995, The Ilan Plain and the southwestward extending Okinawa Trough: Journal of Geological Society of China, v. 38, p. 229–242.

Liu, C.-S., Lundberg, N., Huang, I.L., and Teng, L.S., 1997, Structure features off southwestern Taiwan: Marine Geology, v. 137, p. 305–319.

Lu, C.-Y., and Malavieille, J., 1994, Oblique convergence indentation and rotation tectonics in the Taiwan Mountain Belt: Insights from experimental modeling: Earth and Planetary Science Letters, v. 121, p. 474–494.

Lu, C.-Y., Angelier, J., Chu, H.-T., and Lee, J.-C., 1995, Contractional, transcurrent, rotational and extensional tectonics: Examples from Northern Taiwan: Tectonophysics, v. 125, p. 129–146.

Lue, Y.-T., Lee, T.-Q., Horng, C.-S., and Wang, Y., 1991, Magnetic fabric in the non-metamorphosed terrain of the northwestern Foothills-Hsüehshan belts of Taiwan: Proceedings of the Geological Society of China, no. 34, p. 131–146.

Mars Inc., 1981, Radar mosaic image of northern Taiwan (sheet 1, south looking direction): Taipei, Taiwan, MRSO, Industrial Technology Research Institute, Mars Contract, 78086.

Pezzopane, S.K., and Wesnousky, S.G., 1989, Large earthquakes and crustal deformation near Taiwan: Journal of Geophysical Research, v. 94, p. 7250–7264.

Pirazzoli, P.A., Arnold, M., Giresse, P., Hsieh, M.-L., and Liew, P.-M., 1993, Marine deposits of late glacial times exposed by tectonic uplift on the east coast of Taiwan: Marine Geology, v. 110, p. 1–6.

Prescott, W.H., Savage, J.C., and Kinoshita, W.T., 1979, Strain accumulation rates in the western United States between 1970 and 1978: Journal of Geophysical Research, v. 86, p. 6067–6072.

Ranalli, G., and Murphy, D., 1987, Rheological stratification of the lithosphere: Tectonophysics, v. 132, p. 281–295.

Rocher, M., Lacombe, O., Angelier, J., and Chen, H.-W., 1996, Mechanical twin sets in calcite as markers of recent collisional events in a fold-and-thrust belt: Evidence from the reefal limestones of southwestern Taiwan: Tectonics, v. 15, p. 984–996.

Sato, T., Koresawa, S., Shiozu, Y., Kusano, F., Uechi, S., Nagaoka, O., and Kasahara, J., 1994, Microseismicity of back-arc rifting in the middle Okinawa Trough: Geophysical Research Letters, v. 21, p. 13–16.

Seno, T., Stein, S., and Gripp, A.E., 1993, A model for the motion of the Philippine Sea plate consistent with NUVEL-1 and geological data: Journal of Geophysical Research, v. 98, p. 17941–17948.

Sibuet, J.-C., Letouzey, J., Barbier, F., Charvet, J., Foucher, J.P., Hilde, T.W.C., Thomas, W.C., Kimura, M., Chiao, L.-Y., Marsset, B., Muller, C., and Stephan, J.F., 1987, Back arc extension in the Okinawa trough: Journal of Geophysical Research, v. 92, p. 14041–14063.

Sibuet, J.-C., Deffontaines B., Hsu, S.-K., Thareau, N., Formal, J.-P., Liu, C.-S., and ACT party, 1998, Okinawa trough backarc basin: Early tectonic and magmatic evolution: Journal of Geophysical Research, v. 103, p. 30245–30267.

Suppe, J., 1984, Kinematics of arc-continent collision, flipping of subduction and back-arc spreading near Taiwan: Memoir of the Geological Society of China, v. 6, p. 21–33.

Suppe, J., Hu, C.-T., and Chen, Y.-J., 1985, Present-day stress direction in western Taiwan inferred from borehole elongation: Petroleum Geology of Taiwan, n. 21, p. 1–12.

Teng, L.S., 1990, Geotectonic evolution of late Cenozoic arc-continent collision in Taiwan: Tectonophysics, v. 183, p. 57–76.

Teng, L.S., Chen, C.-H., Wang, W.-S., Liu, T.-K., Jung, W.-S., and Chen, J.-C., 1992, Plate kinematic model for Late Cenozoic arc magmatism in northern Taiwan: Journal of Geological Society of China, v. 35, p. 1–18.

Teng, L.S., 1996, Extensional collapse of the northern Taiwan mountain belt: Geology, v. 24, p. 949–952.

Teng, L.S., and Lee, C.-T., 1996, Geomechanical appraisal of seismogenic faults in northeast Taiwan: Extensional collapse of the northern Taiwan mountain belt: Journal of the Geological Society of China, v. 39, p. 125–142.

Tsai, Y.-B., Teng, T.L., Chiu, J.-M., and Liu, H.-L., 1977, Tectonic implications of the seismicity in the Taiwan region: Memoir of the Geological Society of China, v. 2, p. 13–41.

Tsai, Y.-B., 1986, Seismotectonics of Taiwan: Tectonophysics, v. 125, p. 17–37.

Viallon, C., Huchon, P., and Barrier, E., 1986, Opening of the Okinawa basin and collision in Taiwan: A retreating trench model with lateral anchoring: Earth and Planetary Science Letters, v. 80, p. 145–155.

Wang, C.-H., and Burnett, W.C., 1990, Holocene mean uplift rates across an active plate collision boundary in Taiwan: Science, v. 248, p. 204–206.

Wu, F.T., 1978, Recent tectonics in Taiwan: Journal of Physical Earth, v. 26, p. s265–s299.

Yeh, Y.-H., Lin, C.-H., and Roecker, S.W., 1989, A study of upper crustal structures beneath northeastern Taiwan: Possible evidence of the western extension of Okinawa Trough: Proceedings of the Geological Society of China, no. 32, p. 139–156.

Yeh, Y.-H., Barrier, E., Lin, C.-H., and Angelier, J., 1991, Stress tensor analysis in the Taiwan area from focal mechanisms of earthquakes: Tectonophysics, v. 200, p. 276–280.

Yu, S.-B., and Tsai, Y.-B., 1979, Geomagnetic anomalies of the Ilan Plain, Taiwan: Petroleum Geology of Taiwan, no. 16, p. 19–27.

Yu, S.-B., and Chen, H.-Y., 1994, Global positioning system measurement of crustal deformation in the Taiwan arc-continent collision zone: TAO, Terrestrial, Atmospheric and Oceanic Sciences, v. 5, p. 477–498.

Yu, S.-B., Chen, H.-Y., and Kuo, L.-C., 1997, Velocity field of GPS Stations in the Taiwan area: Tectonophysics, v. 274, p. 41–59.

Manuscript Accepted by the Society April 12, 2001

Printed in the U.S.A.

Geological Society of America
Special Paper 358
2002

Crustal structure of the convergent plate-boundary zone, eastern Taiwan, assessed by seismic tomography

Win-Bin Cheng
General Education Center, Jin-Wen Institute of Technology, 99 Ann-Chung Road, Hsin-Tein, Taipei County, Taiwan, Republic of China
Chengsung Wang
Institute of Applied Geophysics, National Taiwan Ocean University, Keelung, Taiwan, Republic of China
Chuen-Tien Shyu
Institute of Oceanography, National Taiwan University, Taipei, Taiwan, Republic of China
Tzay-Chyn Shin
Seismological Observation Center, Central Weather Bureau, 64 Kung-Yuan Road, Taipei, Taiwan, Republic of China

ABSTRACT

The three-dimensional P-wave velocity structure of the obliquely convergent zone in the eastern Taiwan area has been determined by using traveltimes of seismic waves from 1826 local earthquakes and air-gun shots recorded by the Central Weather Bureau Seismographic Network, and 8334 earthquakes have been relocated for better understanding of the current tectonics. The possible location of the plate boundary between the Eurasian and Philippine Sea plates, characterized by a sharp gradient in the velocity structure, is found beneath the eastern flank of the Central Range to the north of 23.5°N, eastern Taiwan. To the south of 23.5°N, this boundary is generally along the Longitudinal Valley fault and its southern projection. The distribution of the relocated earthquakes also shows a spatial pattern closely related to the boundary and the state of plate collision. To the east of the boundary, a prominent high-velocity anomaly in the middle to lower crust is found beneath the Longitudinal Valley and the Coastal Range in eastern Taiwan; this anomaly could be interpreted as the oceanic crust of the Luzon forearc. To the west of the boundary, the Central Range has a relatively low velocity at the same depth. The velocity structure and relocated seismicity have led to the recognition of interaction between the materials on opposite sides of the boundary. The relatively high P-wave velocity of the Luzon forearc suggests that it can accumulate strain energy and then release it as brittle failure. However, the relatively low P-wave velocity of the Central Range implies that it responds to the convergence by silent or ductile deformation.

INTRODUCTION

East of Taiwan, the Philippine Sea plate is subducting beneath the Eurasian plate along the Ryukyu Trench (Fig. 1). South of Taiwan, the South China Sea lithosphere is subducting eastward under the Philippine Sea plate. Connecting these two subduction zones, the active convergent margin in the eastern Taiwan area is characterized by rapid crustal movement (Yu et al. 1997), regional-scale crustal faulting (e.g., Barrier and Angelier, 1986), high seismicity (e.g., Tsai, 1986), and the Luzon volcanic arc. The oblique collision of the Luzon volcanic arc against the edge of the Asian continent resulted in the uplift of

Cheng, W.-B., Wang, C., Shyu, C.-T., and Shin, T.-C., 2002, Crustal structure of the convergent plate-boundary zone, eastern Taiwan, assessed by seismic tomography, *in* Byrne, T.B., and Liu, C.-S., eds., Geology and Geophysics of an Arc-Continent collision, Taiwan, Republic of China: Boulder, Colorado, Geological Society of America Special Paper 358, p. 161–175.

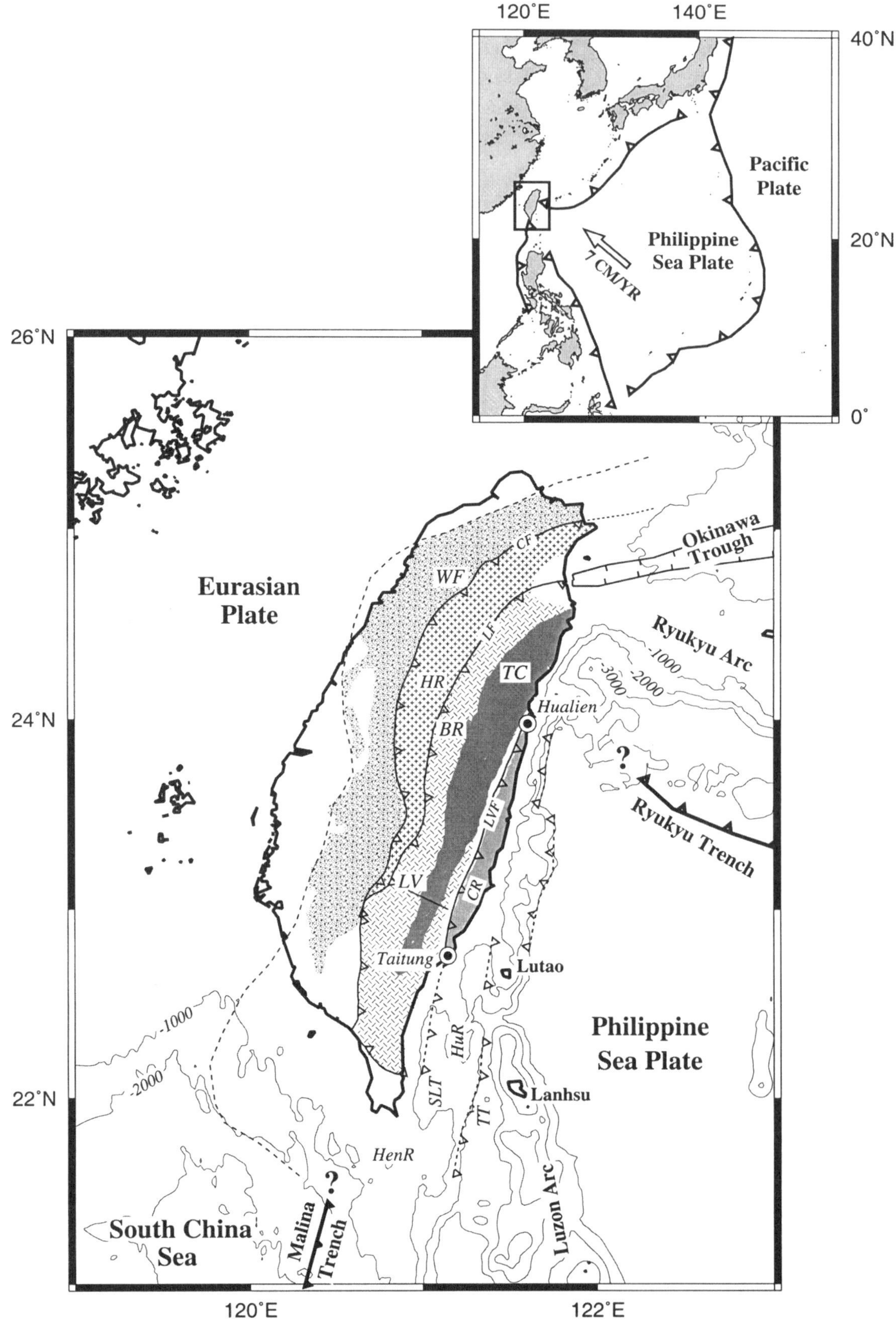

Figure 1. Tectonic framework and main structural units in Taiwan (geodynamic setting in the upper right). Bathymetry is in meters. Major thrust faults at sea and on land are also shown with open triangles on the upper side. Thick lines indicate subduction with solid triangles on the overriding plate. Several thrust faults located in the eastern and southeastern Taiwan offshore area were proposed by Malavieille et al. (this volume). BR—Backbone Range, CF—Chuchih fault, CR—Coastal Range, HR—Hsüehshan Range, HuR—Huatung Ridge, HenR—Hengchun Ridge, LF—Lisan fault, LV—Longitudinal Valley, LVF—Longitudinal Valley fault, SLT—Southern Longitudinal Trough, TC—Tananao Complex, TT—Taitung Trough, WF—Western Foothills.

the Taiwan mountain belt to more than 3 km above sea level in central Taiwan. Continuous GPS (Global Position System) monitoring between 1990 and 1995 shows that, in the area of eastern Taiwan, shortening rates of about 3 cm/yr are measured across the Longitudinal Valley (Yu et al., 1997). Such oblique convergence often creates arc-parallel migration of the forearc (e.g., McCaffrey, 1994) and produces extreme seismic hazard as a consequence of the relative motions of forearc blocks (e.g., Kanamori, 1995). Three-dimensional velocity models using natural earthquake data from the Taiwan area have been constructed and discussed (Roecker et al., 1987; Chen, 1995; Rau and Wu, 1995; Ma et al., 1996). However, their resolution is not good enough to reveal the crustal structure of the Luzon forearc, which is likely to have played a significant role in both the collision that built the mountain belt and the seismogenic zone of the convergent boundary.

In this study, we determined the three-dimensional P-wave velocity structure of the convergent zone in eastern Taiwan by using traveltimes of seismic waves from local earthquakes and air-gun shots recorded by the Central Weather Bureau Seismographic Network (CWBSN). We also relocated the earthquakes for more reliable seismicity data based on the obtained P-wave velocity structure. Our goal was to answer the following questions: (1) In the velocity structure, what is the characteristic velocity feature of the Luzon forearc? (2) How does the forearc behave during collision? (3) What is the role of the forearc in the plate tectonics?

GEOLOGIC SETTING

Taiwan is the result of active, oblique collision between the Eurasian and Philippine Sea plates. It has two main tectonic provinces divided by the Longitudinal Valley (Fig. 1). To the east of the Longitudinal Valley, the Coastal Range, a part of the Luzon arc, is mainly composed of Miocene to Pliocene andesitic volcanic units and associated flyschoid and turbiditic sedimentary deposits (Ho, 1986). To the west of the Longitudinal Valley, the Central Range and the Hsüehshan Range are composed of a pre-Tertiary metamorphic basement overlain by Paleogene sedimentary rocks metamorphosed at low grade, Neogene folded and thrust-faulted sedimentary rock layers, and Quaternary alluvial deposits. The Hsüehshan Range is separated from the Central Range by the Lisan fault, which branches off from the Chuchih fault. The Chuchih fault is a major upthrust fault that follows the contact between the Slate Belt and the fold-and-thrust belt of the Western Foothills region (Ho, 1982). The Western Foothills is a province of typical cover tectonics in which only upper Tertiary and Quaternary rocks are involved (Fig. 1). In addition to the exposed part in Taiwan, deformed rock sequences of the orogenic structural features are thought to extend northeastward and southward to offshore Taiwan (e.g., Liu et al., 1997).

The southeastern Taiwan area is a transitional zone between an intensive mountain building due to the collision of the Luzon arc against the Eurasian plate and the typical eastward subduction of the South China Sea lithosphere. On the basis of a morphologic study, Chen and Juang (1986) divided the southeastern offshore area of Taiwan into five physiographic zones: the Hengchun Ridge, the Southern Longitudinal Trough, the Huatung Ridge, the Taitung Trough, and the Lanhsu Ridge, from west to east, respectively (Fig. 1). By using magnetic and gravity data, Liu et al. (1992) determined that the material of the basement of the Taitung Trough belongs to a volcanic arc, whereas the basements of the Southern Longitudinal Trough and Huatung Ridge do not. The Taitung Trough, the northern part of the North Luzon Trough, narrows and shallows toward the north and ends at the southern Coastal Range on eastern Taiwan (Page and Suppe, 1981). Lutao and Lanhsu are the two northernmost islands of the Lutao-Babuyan Ridge of the Luzon volcanic arc (Bowin et al., 1978). The Luzon arc presumably continues northward to include the Chimei Igneous Complex in the middle of the Coastal Range of Taiwan (Ho, 1986). Radiometric dating shows a southward younging of volcanic ages in this ridge belt (Richard et al., 1986). In this part of the Luzon arc, the main volcanic activity ceased in the middle Miocene (Ho, 1982, 1986).

SEISMIC DATA AND ANALYSIS

The purpose of this study is to obtain a high-resolution wave-velocity structure by using earthquake and air-gun data and a three-dimensional (3-D) tomographic technique (Thurber, 1983, 1993; Eberhart-Phillips, 1990) for delineating the plate-boundary structure between the Eurasian and Philippine Sea plates beneath eastern Taiwan. In eastern Taiwan, the CWBSN stations cover a long and narrow area with high seismicity. We thus divide the study area into four subareas from north to south (Fig. 2), in order to increase the model resolution. For an earthquake with magnitude larger than 4.5 that occurs at a distance of less than 50 km, the amplitude of the P-coda always is saturated and makes the S-wave reading difficult and unreliable owing to the limited dynamic range of the velocity-type sensor used by the CWBSN. For this reason, we consider only the P-wave arrivals (and only those for earthquakes with magnitude greater than 4.5) to be reliable. Without an S-wave reading, we are doubtful about the accuracy of earthquake location, especially the hypocentral depth. Although the wave-velocity modeling of a broad area would decrease the resolution because of omitting most minor earthquakes or a number of moderate earthquakes, the small-area analysis used in this study can include minor to moderate earthquakes with well-defined P- and S-wave arrivals and increase model resolution.

It must be emphasized that we want to provide the highest resolution in the middle crust instead of either the deeper crust or upper mantle. Therefore, most of the earthquakes selected in this study are crustal events with focal depth generally shallower than 40 km in depth. Because the deeper events occur offshore in areas east of Taiwan that are outside the CWBSN and far from the study area, the model solution would be con-

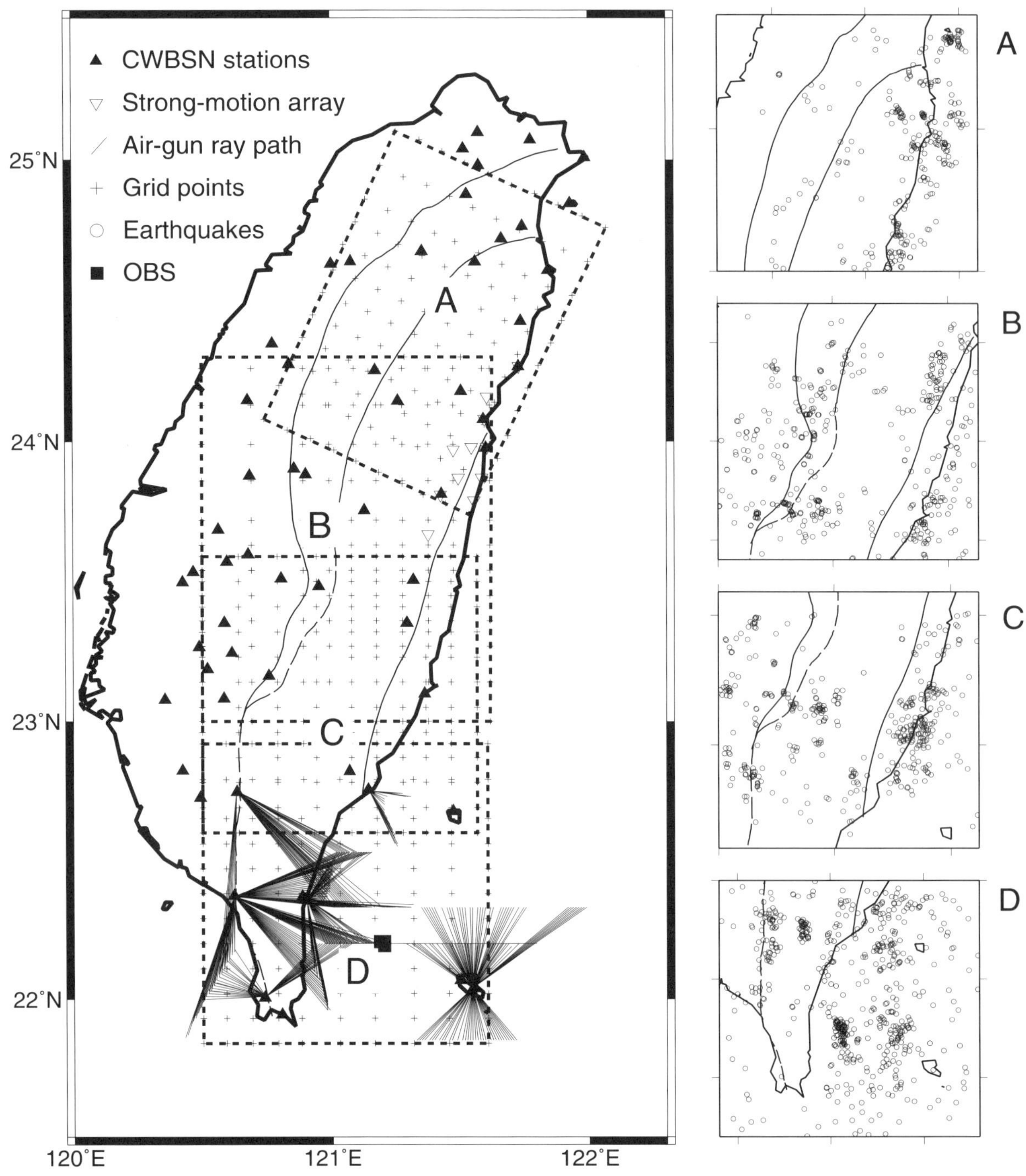

Figure 2. Locations of receivers (left) and sources (right) are superimposed on the four subareas' base maps; x is positive toward the east, and y is positive toward the north except for subarea A.

taminated owing to the location uncertainty and long travel paths outside the study area if the traveltimes of the deeper earthquakes were used. On the other hand, the large quantity of local crustal events can be better located by the CWBSN, which permits selection of high-quality data sets. Moreover, the Lanhsu and Lutao stations situated on the Luzon arc (Fig. 2) keep a large number of the selected crustal events inside the CWBSN network, yielding a good opportunity to investigate the lateral crustal variation across a volcanic island, forearc basin, and subduction complex.

Traveltime data

Two types of traveltime data were used in this study: (1) first-arrival times from air-gun shots for subarea D (Fig. 2) and (2) the arrival times of P- and S-waves generated by earthquakes occurring in the vicinity of Taiwan. Air-gun sources were released by the R/V *Maurice Ewing* on the southeastern offshore area of Taiwan in September 1995 (e.g., Yeh et al., 1998) and recorded by the CWBSN stations in subarea D. In addition, the air-gun signals recorded by an ocean bottom seismometer (OBS) (Chen and Nakamura, 1998) were also included in the data set for the 3-D inversion. A study of the crustal structure using air-gun and earthquake data in subarea D can provide us with a detailed velocity model from shallow to deeper crust that permits us to study the crustal deformation associated with the transition from subduction to collision.

The earthquake data used in this study are P- and S-wave arrivals of local earthquakes recorded in real time by the CWBSN. The CWBSN uses three-component short-period digital seismographs having velocity-type sensors with an adjustable natural frequency of from 0.75 to 1.1 Hz (Shin, 1993). The earthquakes were selected to provide a set of hypocenters with the best available high-quality ray coverage of the subareas. They are digital data recorded by the CWBSN during 1991 to 1997. Events were restricted to those with at least 12 arrival times (both P and S waves) and with an azimuthal gap of less than 180° between stations. In order to represent the full spatial distribution of the arrival data, offshore events with at least six high-quality P arrivals were included that improve the spatial distribution of the data set. The Lutao station on the Luzon arc was shut down at the end of 1991. Thus, the earthquake data recorded at the Lutao station during 1986–1991 were also used. Furthermore, 110 events recorded in 1995–1996 by a temporary strong-motion array deployed in the Hualien area are included in the data to provide a denser coverage of stations for better resolution in modeling of the northern Coastal Range (Fig. 2). In addition, those strong-motion seismometers also provide well-defined S-wave arrivals near the Hualien area even for a moderate earthquake. The traveltime data were weighted according to their accuracy inferred from the signal-to-noise ratio of the seismic signals (weights from 1 to 0.25 for reading errors ranging from 0.01 s to 0.1 s for the P-wave). In this study, the data set includes 641 air-gun arrivals and 1826 earthquakes that provide 24 745 P- and 13 297 S-wave arrivals (Table 1).

Method and inversion procedure

Local earthquake and air-gun data were used in a simultaneous inversion for 3-D P-wave velocity and hypocenters (Thurber, 1983, 1993; Eberhart-Phillips, 1990). The velocity of the medium is parameterized by assigning velocity values at the 3-D grid of points. We compiled traveltimes for each receiver as a function of the 3-D spatial location of source and inverted them for the 3-D velocity structure. The initial hypocenter locations and origin times of earthquakes input were those determined by the CWBSN. The ray-tracing is done by using an approximate 3-D algorithm with curved nonplanar ray paths (Um and Thurber, 1987). The velocity for a point along a ray path and the velocity partial derivatives are computed by linearly interpolating between the surrounding eight grid points. Parameter separation (Pavlis and Booker, 1980) is operated on the matrix of hypocentral and velocity partial derivatives so that the hypocentral calculation is separated from the velocity calculation. The damped least-squares approach is used to solve for velocity in the inversion (Thurber, 1983). To refine the hypocenter of an earthquake and determine a more reliable one-dimensional (1-D) velocity model, the VELEST program (Kissling et al., 1994) was run for each subarea. The initial 1-D velocity model for subareas was taken from the model in Chen (1995), who inverted earthquake traveltimes for the 3-D velocity model and applied a station correction for the Taiwan area.

The inversion procedure was carried out for each subarea with independent data sets and velocity parameterizations (Fig. 2). The grid spacing was decreased from the coarse grid to a fine one in the study subareas during inversion. The model calculated at each step was used as the input model for the following step of inversion. We performed a trade-off analysis (Eberhart-Phillips, 1986) of the traveltimes and the model variances in order to choose the most suitable damping parameter to be used in the damped least-squares inversion. Finally, a damping value that reduced the data variance with only a moderate increase in solution variance was selected for each subarea as shown in Table 1.

We measured the solution quality by computing a spread function (Michelini and McEvilly, 1991). As explained by Toomey and Foulger (1989) in detail, the spread function can describe the resolution in a better way than by solely examining the diagonal element. Because it combines the amount of information and the smearing in evaluating resolution for a node, the spread function does not have physical units. The higher

TABLE 1. SUMMARY OF INVERSION PARAMETERS

	Total stations	Total events	P-wave arrivals	S-wave arrivals	Air-gun arrivals	Damping value
Area A	43	294	4910	2607	0	100
Area B	49	537	10069	6009	0	30
Area C	19	400	5365	2764	0	150
Area D	14	595	4401	1917	641	30

the value of the spread function, the smaller the quantity of information coming from the data.

RESULTS

The 3-D velocity structure in the middle crust of eastern Taiwan is displayed by map views at 13, 17, and 25 km depths in Figure 3 and by a series of east-trending cross sections, arranged from north to south in Figure 4. Both figures reveal particularly strong velocity variations in the middle crust where the spread function indicates that the model resolution is good. The S-wave velocity model obtained (but not shown) reveals the same interesting features as those in the P-wave velocity model. We discuss only the latter in this paper.

Velocity anomalies and crustal blocks

The tomographic model in Figures 3 and 4 defines two main crustal units. The eastern unit extends from the eastern flank of the Central Range to the offshore area in eastern Taiwan. It can be further separated in the two subunits, the Luzon arc subunit in the east and the forearc subunit in the west, seen most clearly in Figure 4F. The forearc subunit shows a high P-wave velocity (V_p) feature in the shape of a north-south–elongated body in the middle crust beneath the Longitudinal Valley, the Coastal Range, and the Southern Longitudinal Trough (shown in Fig. 3, subareas B, C, and D). The velocity of the high-V_p volume varies from about 6.4 to 7.3 km/s at 17 km depth. The resolution is good, so the existence of the forearc high-velocity zone is reliable. The hypocenters of the events used in the inversion within a 10 km distance from each cross section are also plotted in Figure 4. The distribution of the hypocenters shows that inside the forearc high-V_p volume there are fewer earthquakes. The low-velocity material, lying above the forearc high-V_p volume, is about 10 km in thickness.

The two CWBSN stations (Lutao and Lanhsu stations) deployed on the northern Luzon arc (Fig. 1) provide us with two important sites to resolve the velocity structure of the Luzon arc subunit. The typical structure of the Luzon arc is the lenticular body beneath the Lutao islet as shown in Figure 4F. An interesting feature shown in Figure 4 is that the Coastal Range and Huatung Ridge are underlain by the forearc high-V_p crust in eastern Taiwan. Although the Coastal Range is thought to be the remnant of a westward-facing Neogene island arc on the leading edge of the Philippine Sea plate (Biq, 1972; Bowin et al., 1978; Chi et al., 1981), our observation suggests that the core of the Coastal Range is probably the uplifted forearc basin sediment, which filled in after the formation of the forearc basin; the forearc basin was then shortened and diminished. Our profiles of V_p show that the upper part of the Coastal Range above 10 km in depth corresponds to the part of the sediment filled in (Figs. 4D to 4F).

The second main unit, the Central Range, bounded on the west by the Chuchih fault and on the east by the Longitudinal Valley fault (Fig. 4), has a large middle-crustal region with low V_p. The velocity of the Central Range unit is about 0.1 to 0.6 km/s lower than average velocities in subareas B, C, and D. The spatial distribution of the low-V_p volumes beneath the Central Range indicates that they are continuous from south to north. In the area between 23.5°N and 24°N, there is a high-velocity zone beneath the western flank of the Central Range (Figs. 4B and 4C). Wu et al. (1997) also reported the high-velocity zone, which they speculated represents the extrusion of high-velocity materials from the middle or lower crust to shallow depth. North of 24°N, the velocity contours are relatively flattened beneath the Central Range (Fig. 4A).

Relocation of earthquakes

The 3-D velocity model was used to relocate 8334 earthquakes recorded by the CWBSN during 1991 to 1997. The epicenters of these events are plotted in Figure 5. When the relocated hypocenters are compared with the positions of the same earthquakes determined by the CWBSN, significant systematic changes are apparent. The horizontal changes vary from 0 to 15 km and move landward in a direction of northwest to southeast for subareas B, C, and D, as shown by the rose diagrams in Figure 5. The reason for this systematic variation in the forearc high-V_p structure has not been taken into account in the CWBSN 1-D velocity model. The vertical changes are mostly in the range of 0 to 5 km.

COMPARISON WITH GRAVITY ANOMALIES

In order to check the tomographic inversion results and enable estimation of crustal densities associated with the forearc high-V_p volume in eastern Taiwan, we perform 3-D gravity modeling directly from the 3-D velocity structure (Figs. 3 and 4). Gravity anomalies can be calculated and compared with observed gravity values, if an assumption is made about the relationship between velocity and density. We assume that lateral perturbations to velocity and to density are linearly related. In addition, because the digital elevation model offshore Taiwan is available (Liu et al., 1998), it might be more correct to compare the results with Bouguer than free-air anomalies at sea. Therefore, we determined the Bouguer gravity anomaly in eastern Taiwan from the 3-D V_p model by finding the velocity-contrast layers and then calculating the anomalies for each layer by using the formula of Parker (1972). In spite of some discrepancies between the observed (Yen et al., 1995) and model-derived gravity anomalies in the area of the Luzon arc (Fig. 6), they show many similarities in the spatial pattern, in particular the location of the on-land gravity high in eastern Taiwan. The implication of Figure 6 is that the source of the gravity high in eastern Taiwan may be the forearc high-V_p volume. In addition, although the gravity high is continuous along the strike of the Longitudinal Valley, its amplitude decreased from the 40–80 mgal values east of Longitudinal Valley to about 10 mgal to

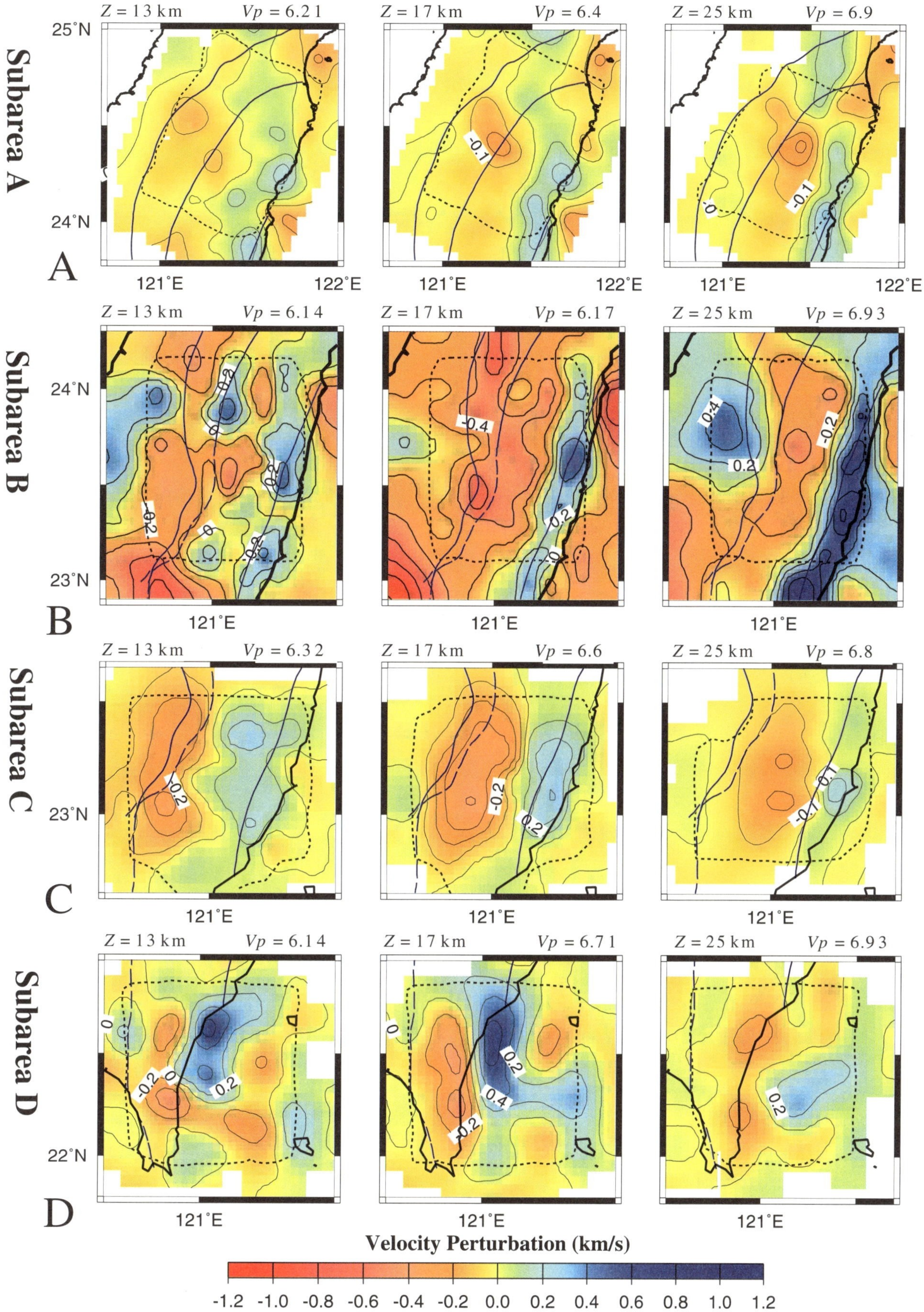

Figure 3. Results of three-dimensional inversion for P-wave velocities. Velocities are shown in map view at three slices at 13, 17, and 25 km for each subarea. Major faults are also shown. Average velocity for each depth slice is shown in upper-right corner. See Figure 2 for reference map. The area for which the spread function is 4.0 or smaller (inside the dashed lines) is also shown. Spread function smaller than 4.0 (i.e., inside the contour lines) indicates that the velocity value is well resolved and spatially well determined.

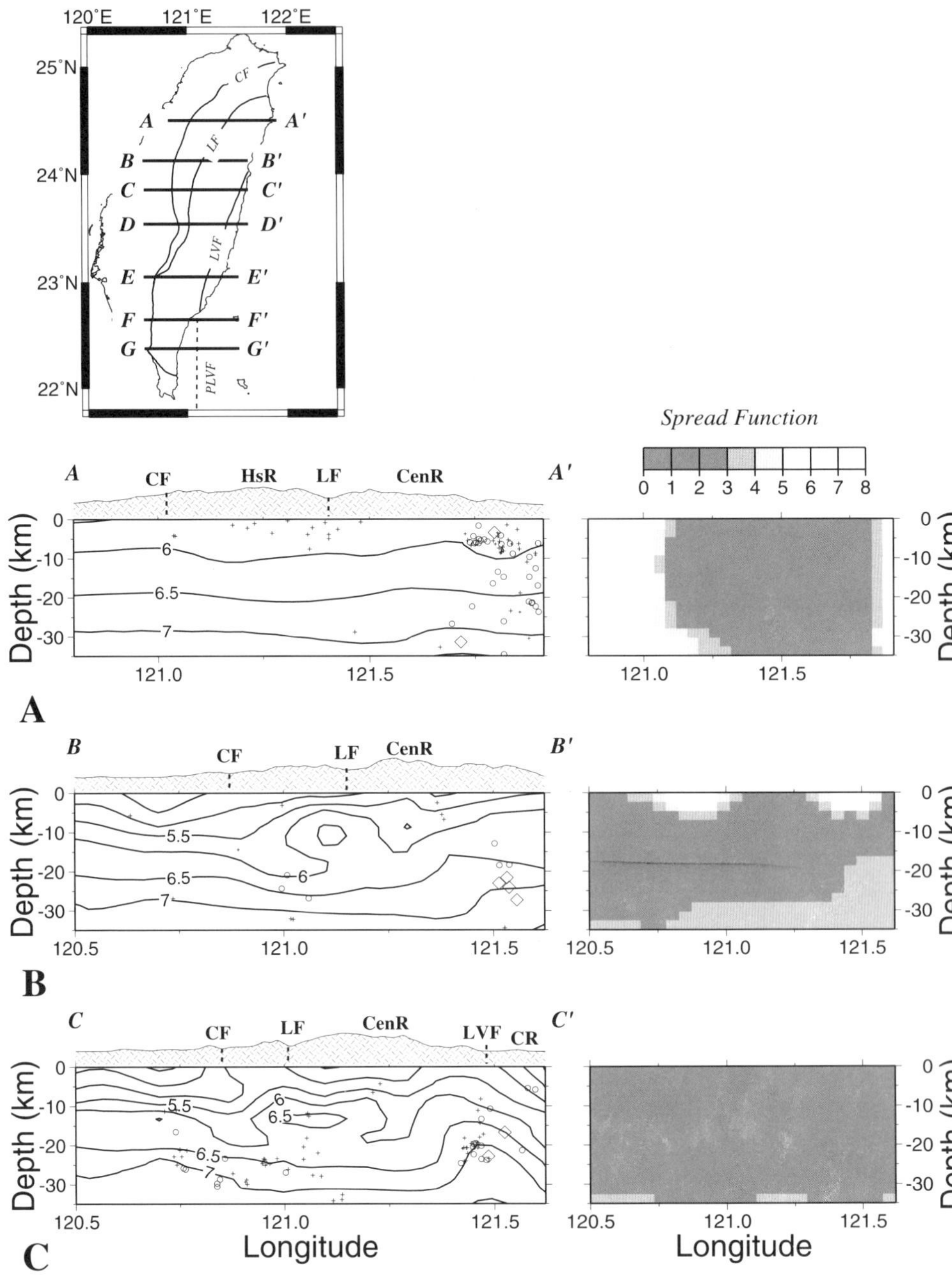

Figure 4 (on this and facing page). Cross sections A–A′ to G–G′ (locations shown in the top reference map) of 3-D velocity structure contoured at 0.5 km/s intervals. Relocated hypocenters of events used in velocity inversion to 35 km depth are also shown. Diamonds—events of magnitude ≥ 5.0. Open circles—events of magnitude ≥ 4.0. Crosses—events of magnitude ≥ 2.0. Topography of each cross section is shown on top. Plots to the right of each cross section show the spread function. CenR—Central Range, HsR—Hsüehshan Range, CF—Chuchih fault, LF—Lisan fault, CR—Coastal Range, HR—Huatung Ridge, LVF—Longitudinal Valley fault, PLVF—southern prolongation of Longitudinal Valley fault.

the west (Fig. 6). This amplitude change could correspond to the transition from the forearc (oceanic crust) to the Central Range (continental crust).

LUZON FOREARC SLIVER

The most prominent velocity feature of the 3-D velocity model is the elongated forearc high-velocity volume along the strike of the Longitudinal Valley beneath eastern Taiwan (Figs. 4 and 5). It is evident that the forearc high-velocity volume narrows northward and extends at least 200 km in eastern Taiwan. The head waves generated from 10 events that occurred in eastern Taiwan (Fig. 7) confirm the existence of the narrow forearc high-V_p volume in the middle crust. The high-V_p volume is also the possible mass that generates the gravity high in that area.

In a convergent zone, the existence of such a high-V_p volume is not unusual; examples include the imbricated oceanic crust near the Contact fault zone, southern Alaska (Wolf et al., 1991; Fuis et al., 1991), the shattered oceanic crust near the Santa Lucia Escarpment in central California (Howie et al., 1993), and the remnant oceanic crust beneath Vancouver Island, western Canada (Spence et al., 1985). By using a temperature versus depth model derived from thermal-gradient data (Lee

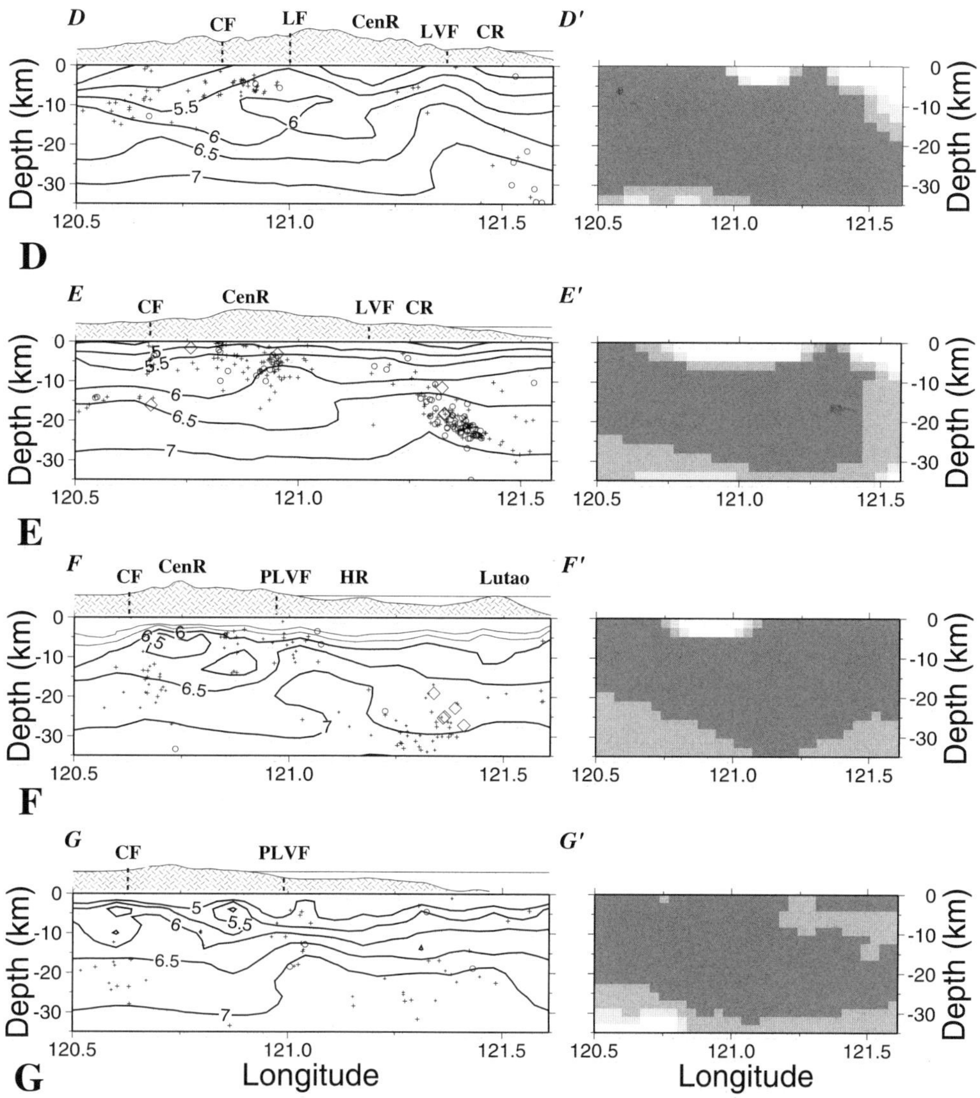

and Chang, 1986) and velocity-depth functions for various rocks (Christensen and Mooney, 1995), Cheng et al. (1998) concluded that the high-velocity volume beneath the Longitudinal Valley corresponds to a trapped oceanic crustal sliver of the Luzon forearc. In the forearc sliver, the oceanic crust was covered with deformed ocean-floor sediments, uplifted trench-floor and trench-slope sediments, and the depositional fills of subsiding forearc basins during the convergence of Eurasian and Philippine Sea plates. Because of the various lithologies represented, the crustal velocity structure of the forearc sliver shows a signature differing from that of a typical oceanic crustal section.

The dense clusters of background seismicity in eastern Taiwan show that the forearc sliver is almost coincident with a less seismic area delineated by two subparallel seismic zones (Fig. 5). The locations of high-moment release events appear to correlate well with the high-V_p volume, which is consistent with the positive correlation between increasing velocity and increasing ability of the materials to accumulate strain energy and release it as brittle failure (Michael and Eberhart-Phillips, 1991). In contrast, the low-V_p zone beneath the central and northern Central Range, which is also a relatively quiescent zone of seismicity, might imply that the crustal material of the Central Range is ductile or weak and deforms and uplifts silently in response to the plate convergence. We thus conclude that the forearc sliver plays an important role in the arc-continent collision in the Taiwan area and in the tectonic evolution of the Central Range uplift.

A velocity cross section along the strike of the Longitudinal Valley is plotted in Figure 8. The plate interface and dip angle of the subducted Philippine Sea lithosphere are taken from a local seismicity and source-parameter study in the southernmost Ryukyu-Taiwan area (Kao et al., 1998), by approximately delineating the upper envelope of the dipping seismic

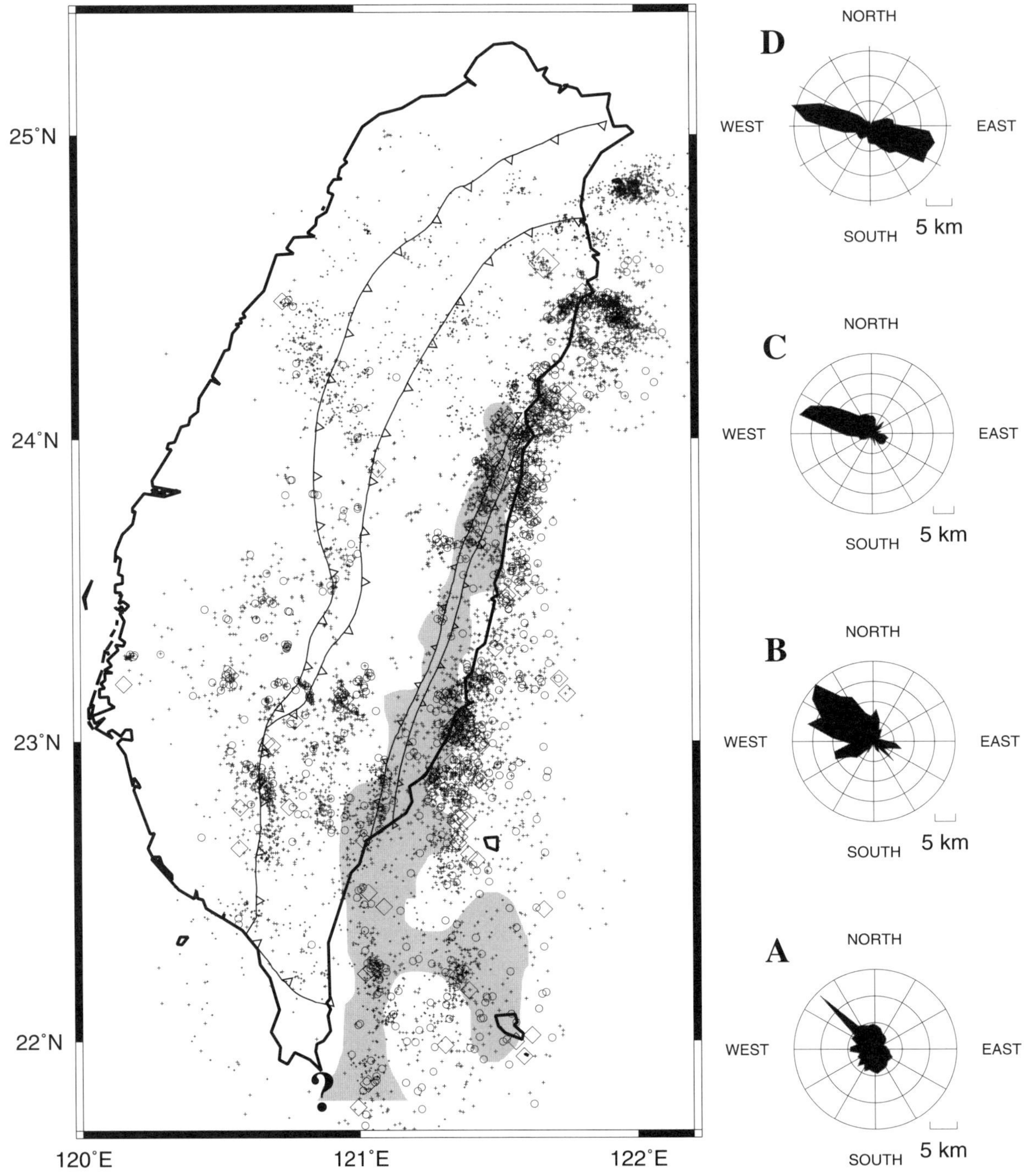

Figure 5. Map of relocated epicenters of earthquakes that occurred during 1991 to 1997. Hypocenters are relocated with the 3-D velocity model. Diamonds—events of magnitude ≥ 5.0. Open circles—events of magnitude ≥ 4.0. Crosses—events of magnitude ≥ 2.0. Major faults are also shown. Shaded area outlines the region with V_p generally greater than 6.5 km/s. The southern extension of the high-velocity zone cannot be resolved in this study. Rose diagrams shown on right indicate horizontal variations and azimuths between initial (located by Central Weather Bureau Seismographic Network) and relocated positions of events.

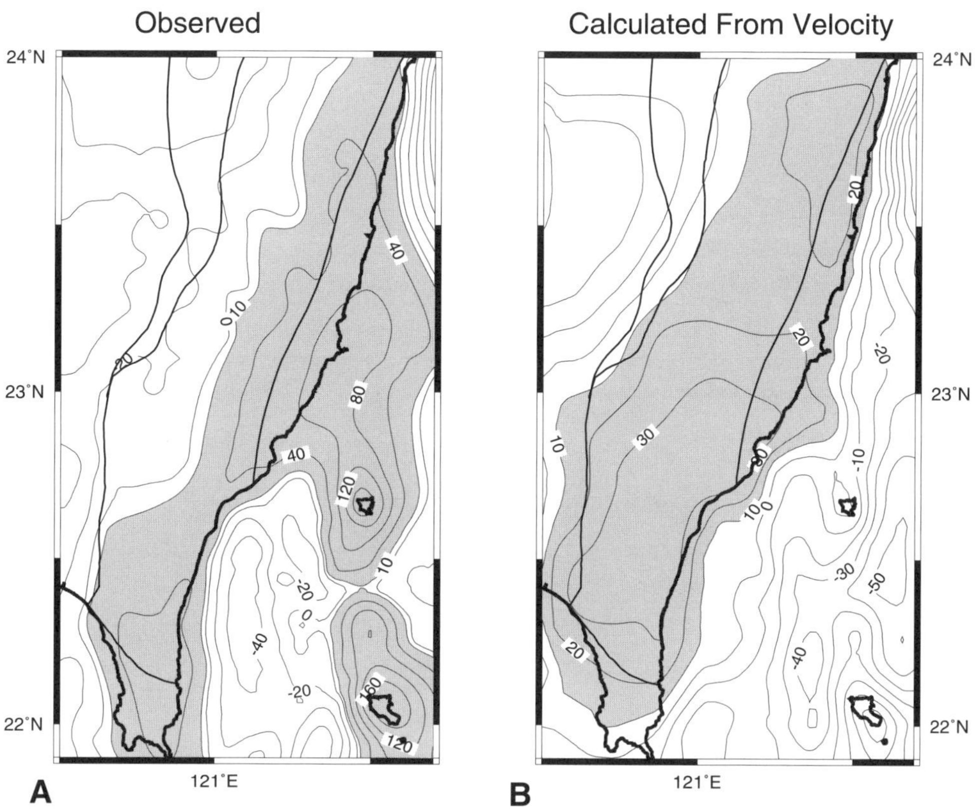

Figure 6. Map of (A) observed and (B) calculated gravity. Observed is from Yen et al. (1995) and shows Bouguer values at sea and on land. Right shows Bouguer values calculated from the 3-D velocity model. Shaded areas indicate Bouguer values greater than +10 mgal.

zone (e.g., Isacks and Barazangi, 1977). The forearc high-V_p volume above the plate interface is intriguing because of its location and the density of earthquakes occurring around it. A possible interpretation for the existence of the forearc high-V_p volume above the subducted plate is that it is the Luzon forearc oceanic crust that was torn off and separated from the Philippine Sea plate after the Luzon arc had been formed; the crust has been shortened in the east-west direction during the collision of the Eurasia and Philippine Sea plates.

The oblique collision between the Luzon arc and the Eurasian continental shelf started near the Hualien area at ca. 4 Ma and moved progressively southward to reach the Taitung area in southeastern Taiwan by ca. 1 Ma (Lee et al., 1991). As shown in Figure 1, the spacing between the Luzon arc and the Longitudinal Valley is about 50 km in the Taitung area (measured from Lutao to Taitung city) and decreases to less than 20 km in the Hualien area. The northward narrowing of the spacing between the Luzon arc and the Longitudinal Valley is similar to that of the forearc high-V_p volume. A feasible interpretation for the northward-narrowing feature is that most of the forearc sliver has been shortened or plunged into the mantle. Chemenda et al. (1997) had justified this interpretation as one of the possible evolutions for the Taiwan collision on the basis of physical modeling.

The low-V_p zone beneath the Central Range coincides with the region of most intense active deformation (e.g. Yu et al., 1997), suggesting that it is relatively weak. In contrast, the V_p of the Luzon forearc sliver is relatively high in the middle to lower crust (Fig. 3). Thus, when the Luzon forearc sliver collides with the crust of the Central Range, the former may play a role as a backstop in the dynamics of the Taiwan orogeny. Such an occurrence may in turn result in a rapid rising of crustal materials to shallow depths, seen most clearly in Figure 4, B to E. The rapid uplift of the crustal material beneath the Central Range should have been accompanied by normal faulting along the eastern flank of the Central Range. The normal faulting observed in two transects along the Central and Southern Cross-Island Highways (Crespi et al., 1996) can be ascribed to the push of the forearc sliver.

PLATE BOUNDARIES

The plate boundaries between oceanic regions are narrow, such as the Mariana convergent plate margin (e.g., Fryer, 1996). Plate boundaries increase to hundreds of kilometers when they are between oceanic and continental lithospheres, such as the New Zealand subduction zone (e.g., Walcott, 1998) and the Cascadia subduction zone of the western United States (e.g., Goldfinger et al., 1997). However, seismicity and active faulting in continental regions are commonly dispersed over thousands of kilometers, such as Central Asia (Molnar and Tapponnier, 1988). The tectonic setting and width of the seismogenic zone in eastern Taiwan (Fig. 5) is generally consistent with the boundary between the oceanic and continental lithospheres. Al-

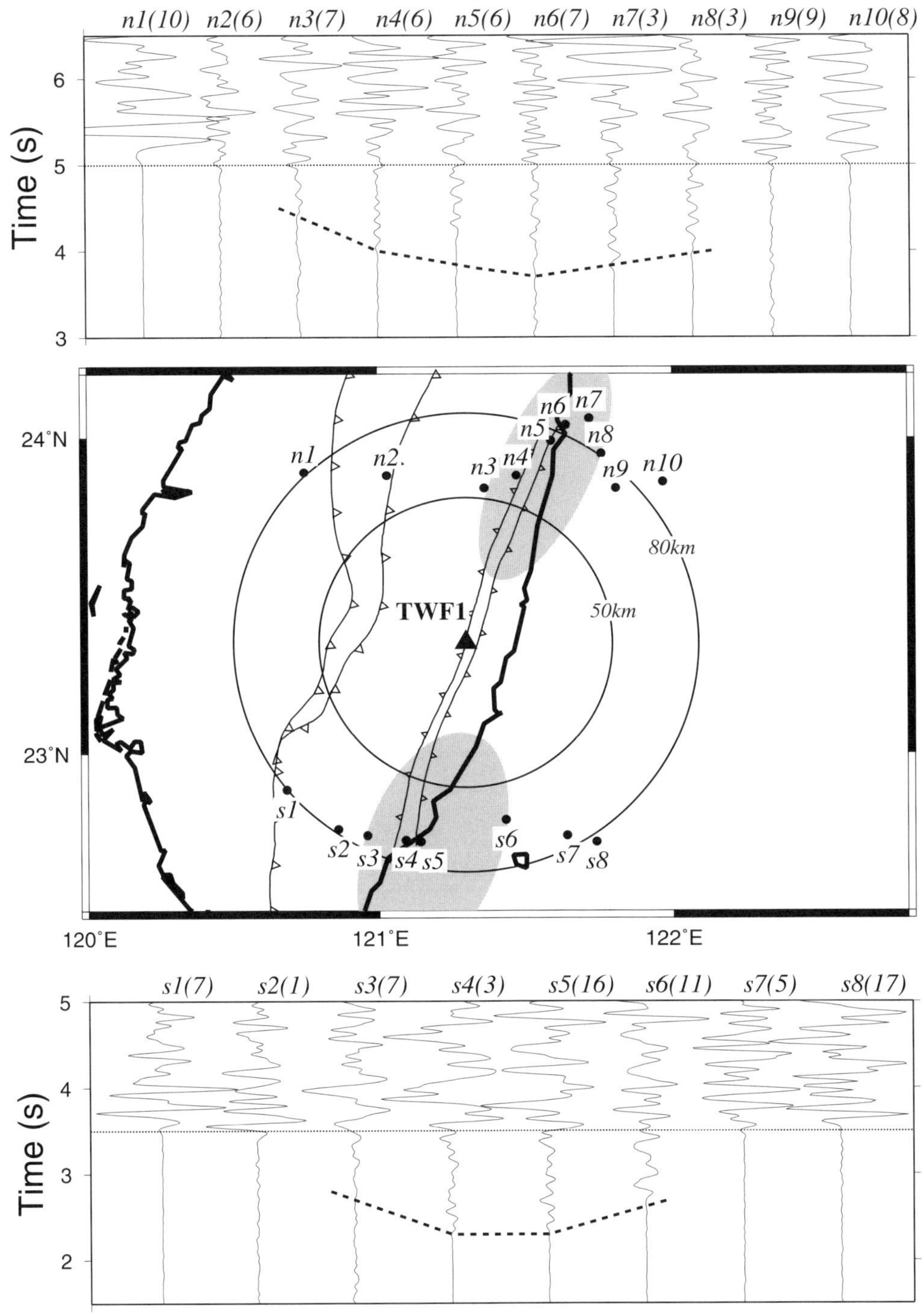

Figure 7. Seismograms of shallow events (solid circles) that occurred in eastern Taiwan and that were recorded by station TWF1 (solid triangle) of the Central Weather Bureau Seismographic Network. Dashed lines envelop the arrival times of the head wave. As shown, head waves are only generated from 10 events distributed within narrow zones (inside the shaded ellipsoids). Numbers in brackets indicate the hypocentral depths (in km).

though the Longitudinal Valley fault is generally accepted as the plate boundary between the Eurasian and Philippine Sea plates (Biq, 1972; Chai, 1972; Karig, 1973; Wu and Lu, 1976; Tsai et al., 1981; Chi et al., 1981; Hsu and Sibuet, 1995), the detailed geometry and nature of the plate boundary are difficult to evaluate from the highly deformed crust in the Taiwan region. Controversial opinions on the plate boundary in Taiwan naturally exist (e.g., Biq, 1972; Bowin et al., 1978).

In this study, based on the relocated seismicity and the obtained 3-D velocity structure, the possible location of the plate boundary could be generally outlined. It is characterized by a sharp across-fault velocity variation, which is a primary feature in the 3-D velocity structure (Figs. 3 and 4). It generally marks a boundary between the oceanic and continental crusts. To the south of 23.5°N, this boundary is generally along the mapped Longitudinal Valley fault trace and its southern pro-

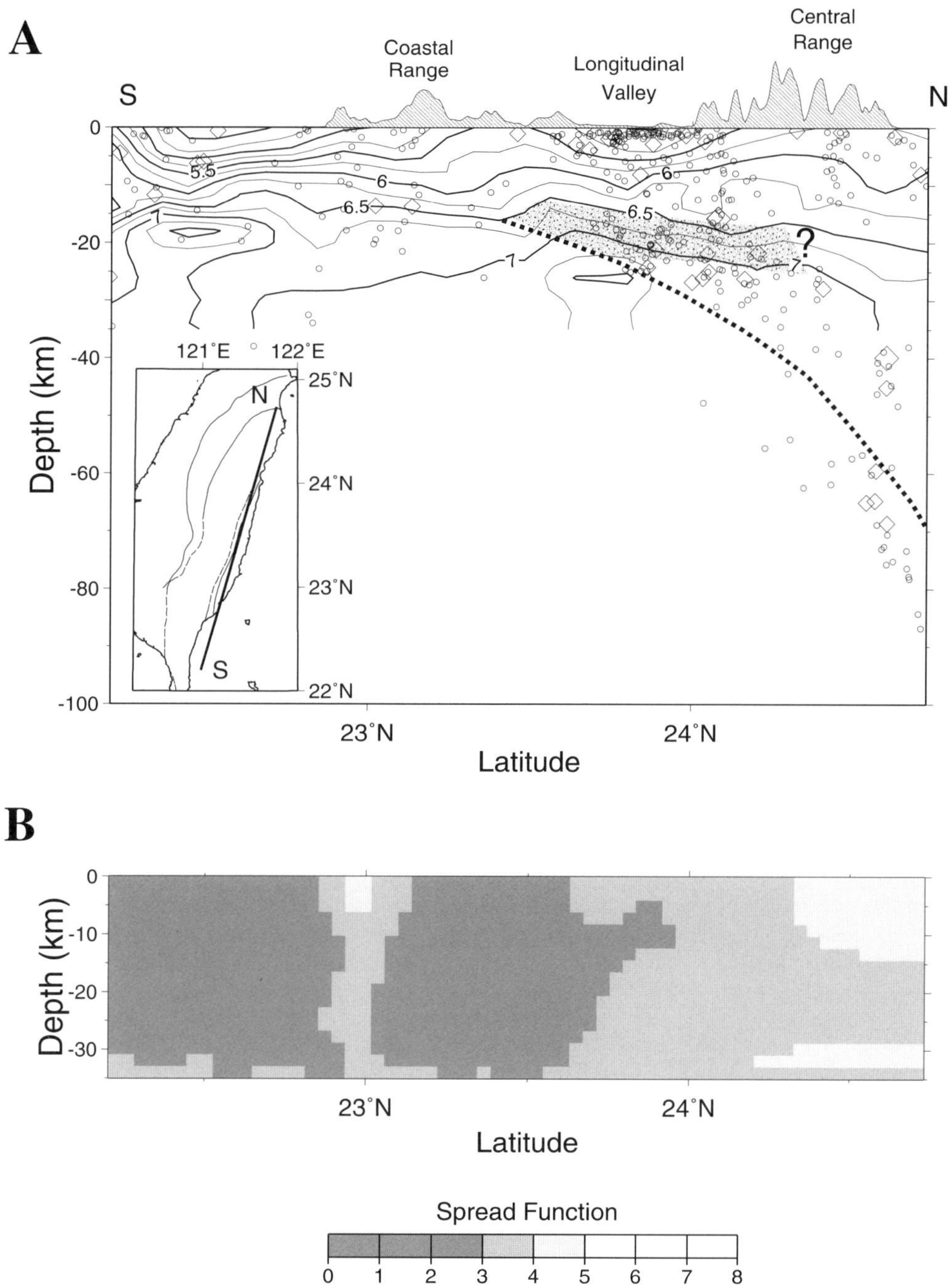

Figure 8. (A) A schematic interpretation of the tectonics based on the 3-D velocity model along the north-south section in inset. Thick dashed line represents the plate interface. Shaded area outlines part of the region of the forearc high-velocity volume. See text for explanation. Topography is shown on top. (B) Spread function is shown for the cross section to 35 km depth.

longation (Fig. 4, E to G). To the north of 23.5°N, it seems to move westward from the mapped trace of the Longitudinal Valley fault to the eastern flank of Central Range (Fig. 4, C and D). The east dip in the shape of the forearc high-V_p volume implies that the boundary may dip to the east along with the Luzon forearc sliver.

CONCLUSIONS

The three-dimensional P-wave velocity structure of the convergent zone in eastern Taiwan has been obtained by inversion of local earthquake and air-gun traveltimes. A large-volume high-velocity zone is observed in the middle to lower crust between the Central Range and Luzon arc in eastern Taiwan (i.e., beneath the Coastal Range and Longitudinal Valley). The high-velocity zone narrows northward and extends to at least 24°N; it can be interpreted as the oceanic crust of the Luzon forearc sliver. The Luzon forearc sliver is almost coincident with a less seismic area delineated by two subparallel seismic zones revealed by the relocated epicenters. The locations of high-moment release events appear to correlate well with the high-velocity volume, which is consistent with the positive correlation between increasing velocity and increasing ability of the material to store strain energy and release it as brittle failure.

West of the forearc high-velocity volume, the overall shape and location of the low P-wave velocity zone beneath the Central Range corresponds well to the mapped surface geology where deformation has been the most active. This low-velocity zone also corresponds to the relatively quiescent zone of seismicity in the central and northern Central Range. These results imply that the crustal material of the Central Range is weak or ductile and that when it collides with the Luzon forearc sliver, it might respond to the collision by silent or ductile deformation and crustal thickening.

Between the Luzon forearc sliver and the Central Range is the boundary between the oceanic and continental plates, which can be represented by a sharp velocity gradient in the three-dimensional crustal-velocity structure. This plate boundary is situated beneath the eastern flank of the Central Range to the north of 23.5°N and along the Longitudinal Valley to the south of 23.5°N and its southern prolongation in the offshore area. It may dip to the east along with the Luzon forearc sliver.

ACKNOWLEDGMENTS

This research was supported financially by the National Science Council of Taiwan. We are grateful to the Seismological Observation Center, Central Weather Bureau, for providing data of earthquakes and air-gun shots and to the TAICRUST group for providing the location of air-gun sources. We thank the members of the Institute of Earth Sciences, Academia Sinica of Taiwan, for their efforts in recording the air-gun signals. We also thank L.Y. Chiao, S.K. Hsu, S. Lallemant, and J. Malavieille for their valuable discussion.

REFERENCES CITED

Barrier, E., and Angelier, J., 1986, Active collision in eastern Taiwan: The Coastal Range: Tectonophysics, v. 125, p. 39–72.

Biq, C., 1972, Dual-trench structure in the Taiwan-Luzon region: Proceedings of Geological Society of China, no. 15, p. 65–75.

Bowin, C., Lu, R.-S., Lee, C-.S., and Schouten, H., 1978, Plate convergence and accretion in the Taiwan-Luzon region: American Association of Petroleum Geologists Bulletin, v. 62, p. 1645–1672.

Chai, B.H.T., 1972, Structure and tectonic evolution of Taiwan: American Journal of Science, v. 272, p. 289–442.

Chemenda, A.I., Yang, R.K., Hsieh, H.H., and Groholsky, A.L., 1997, Evolutionary model for the Taiwan collision based on physical modeling: Tectonophysics, v. 274, p. 253–274.

Chen, A.T., and Nakamura, Y., 1998, Velocity structure beneath the eastern offshore region of southern Taiwan based on OBS data: Terrestrial, Atmospheric and Ocean Sciences, v. 9, p. 409–424.

Chen, M.P., and Juang, W.S., 1986, Seafloor physiography off southeastern Taiwan: Acta Oceanographica Taiwanica, v. 16, p. 1–7.

Chen, Y.-L., 1995, Three dimensional velocity structure and kinematics analysis in Taiwan area [M.S. thesis]: Chung-Li, Taiwan, National Central University, 172 p. (in Chinese).

Cheng, W.B., Wang, C., Shyu, C.T., and Shin, T.C., 1998, A three-dimensional Vp model of the southeastern Taiwan area and its tectonic implications: Terrestrial, Atmospheric and Ocean Sciences, v. 9, p. 425–452.

Chi, W.R., Namson, J., and Suppe, J., 1981, Stratigraphic record of plate interactions in the Coastal Range of eastern Taiwan: Memoir of the Geological Society of China, v. 4, p. 155–194.

Christensen, N.I. and Mooney, W.D., 1995, Seismic velocity structure and composition of the continental crust: A global view: Journal of Geophysical Research, v. 100, p. 9761–9788.

Crespi, J.M., Chan, Y., and Swaim, M.S., 1996, Synorogenic extension and exhumation of the Taiwan hinterland: Geology, v. 24, p. 247–250.

Eberhart-Phillips, D., 1986, Three-dimensional velocity structure in northern California Coast Ranges from inversion of local earthquakes arrival times: Seismological Society of America Bulletin, v. 76, p. 1025–1052.

Eberhart-Phillips, D., 1990, Three-dimensional P and S velocity structure in the Coalinga region, California: Journal of Geophysical Research, v. 95, p. 15343–15363.

Fryer, P., 1996, Evolution of the Mariana convergent plate margin system: Review of Geophysics, v. 34, p. 89–125.

Fuis, G.S., Ambos, E.L., Mooney, W.D., Christensen, N.I., and Geist, E., 1991, Crustal structure of accreted terranes in southern Alaska, Chugach Mountains and Copper River Basin, from seismic refraction results: Journal of Geophysical Research, v. 96, p. 4187–4227.

Goldfinger, C., Kulm, L.D., Yeats, R.S., McNeill, L., and Hummon, C., 1997, Oblique strike-slip faulting of the central Cascadia submarine forearc: Journal of Geophysical Research, v. 102, p. 8217–8243.

Ho, C., 1982, Tectonic evolution of Taiwan and explanatory text of the tectonic map of Taiwan: Taipei, Taiwan, Republic of China, Central Geological Survey, Ministry Economic Affairs, Taipei, 126 p.

Ho, C., 1986, A synthesis of the geologic evolution of Taiwan: Tectonophysics, v. 125, p. 1–16.

Howie, J.M., Miller, K.C., and Savage, W.U., 1993, Integrated crustal structure across the south central California margin: Santa Lucia escarpment to the San Andreas Fault: Journal of Geophysical Research, v. 98, p. 8173–8196.

Hsu, S.K., and Sibuet, J.C., 1995, Is Taiwan the result of arc-continent or arc-arc collision?: Earth and Planetary Science Letters, v. 136, p. 315–324.

Isacks, B.L., and Barazangi, M., 1977, Geometric of Benioff zones: Lateral segmentation and downwards bending of subducted lithosphere, *in* Talwani, M., and Pittman, W.C., eds., Island arcs, deep sea trenches and back-arc basins: American Geophysical Union Maurice Ewing Series, v. 1, p. 99–114.

Kanamori, H., 1995, The Kobe (Hyogo-ken Nanbo) earthquake of January 16, 1995: Seismological Research Letters, v. 66, p. 6–10.

Kao, H., Shen, S.J., and Ma, K.F., 1998, Transition from oblique subduction to collision: Earthquakes in the southernmost Ryukyu-arc-Taiwan region: Journal of Geophysical Research, v. 103, p. 7211–7229.

Karig, D.E., 1973, Plate convergence between Philippines and the Ryukyu Islands: Marine Geology, v. 14, p. 153–168.

Kissling, E., Ellsworth, W.L., Eberhart-Phillips, D., and Kradolfer, U., 1994, Initial reference models in local earthquake tomography: Journal of Geophysical Research, v. 99, p. 19635–19646.

Lee, C.-R., and Chang, W.-T., 1986, Preliminary heat flow measurements in Taiwan, *in* Proceedings, Circum-Pacific Energy and Mineral Resources Conference, 4th, Singapore.

Lee, T.Q., Kissel, C., Barrier, E., Laj, C., and Chi, W.R., 1991, Paleomagnetic evidence for a diachronous clockwise rotation of the Coastal Range, eastern Taiwan: Earth and Planetary Science Letters, v. 104, p. 245–257.

Liu, C.S., Huang, I.L., and Teng, L.S., 1997, Structural features off southwestern Taiwan: Marine Geology, v. 137, p. 305–319.

Liu, C.S., Liu, S.Y., Lallemand, S.E., Lundberg, N., and Reed, D.L., 1998, Digital elevation model offshore Taiwan and its tectonic implications: Terrestrial, Atmospheric and Ocean Sciences, v. 9, p. 705–738.

Liu, S.-Y., Liu, C.-S., Reed, D., and Lundberg, N., 1992, Analysis of the tectonic structures in the region off southeastern Taiwan based on gravity-magnetic data, *in* Proceedings, Taiwan Symposium on Geophysics, 4th: Taipei, Taiwan, p. 575–585.

Ma, K.-F., Wang, J.-H. and Zhao, D., 1996, 3-D seismic structure of the crust and uppermost mantle beneath Taiwan: Journal of the Physics of the Earth, v. 44, p. 85–105.

Malavieille, J., Lallemand, S.E., Dominguez, S., Deschamps, A., Lu, C.-Y., Liu, C.-S., Schnürle, P., and the ACT Scientific Crew, 2002, Arc continent collision in Taiwan: New marine observations and tectonic evolution, *in* Byrne, T.B., and Liu, C.-S., eds., Geology and geophysics of an arc-continent collision, Taiwan, Republic of China (this volume).

McCaffrey, R., 1994, Global variability in subduction thrust zone-forearc systems: Pure and Applied Geophysics, v. 142, p. 173–224.

Michael, A.J., and Eberhart-Phillips, D., 1991, Relations among fault behavior, subsurface geology, and three-dimensional velocity models: Science, v. 253, p. 651–654.

Michelini, A., and McEvilly, T.V., 1991, Seismological studies at Parkfield. 1. Simultaneous inversion for velocity structure and hypocenters using cubic b-splines parameterization: Seismological Society of America Bulletin, v. 81, p. 524–552.

Molnar, P., and Tapponnier, P., 1988, Cenozoic tectonics of Asia: Effect of a continental collision: Science, v. 189, p. 419–426.

Page, B.M., and Suppe, J., 1981, The Pliocene Lichi Melange of Taiwan: Its plate tectonic and olistostromal origin: American Journal of Science, v. 281, p. 193–277.

Parker, R.L., 1972, The rapid calculation of potential anomalies: Geophysical Journal of the Royal Astronomical Society, v. 31, p. 447–455.

Pavlis, G.L., and Booker, J.R., 1980, The mixed discrete-continuous inverse problem: Application to the simultaneous determination of earthquake hypocenters and velocity structure: Journal of Geophysical Research, v. 85, p. 4801–4810.

Rau, R.J., and Wu, F.T., 1995, Tomographic imaging of lithospheric structures under Taiwan: Earth and Planetary Science Letters, v. 133, p. 517–532.

Richard, M., Bellon, H., Maury, R.C., Barrier, E., and Juang, W.S., 1986, Miocene to recent calc-alkalic volcanism in eastern Taiwan: K-Ar ages and petrography: Tectonophysics, v. 125, p. 87–102.

Roecker, S.W., Yeh, Y.H., and Tsai, Y.B., 1987, Three-dimensional P and S wave velocity structures beneath Taiwan: Deep structure beneath an arc-continent collision: Journal of Geophysical Research, v. 92, p. 10547–10570.

Shin, T.C., 1993, The calculation of local magnitude from the simulated Wood-Anderson seismograms of the short-period seismograms in the Taiwan area: Terrestrial, Atmospheric and Ocean Sciences, v. 4, p. 155–170.

Spence, G.D., Clowes, R.M., Ellis, R.M., 1985, Seismic structure across the active subduction zone of western Canada: Journal of Geophysical Research, v. 90, p. 6754–6772.

Thurber, C.H., 1983, Earthquake locations and three-dimensional crustal structure in the Coyote Lake area, central California: Journal of Geophysical Research, v. 88, p. 8226–8236.

Thurber, C.H., 1993, Local earthquake tomography: Velocities and V_p/V_s theory, *in* Iyer, H.M., and Hirahara, K., eds., Seismic tomography: Theory and practice: London, Chapman and Hall, p. 563–583.

Toomey, D.R., and Foulger, G.R., 1989, Tomographic inversion of local earthquake data from the Hengill-Grensdalur central volcano complex, Iceland: Journal of Geophysical Research, v. 94, p. 17497–17510.

Tsai, Y.B., Liaw, Z.S., Lee, T.Q., Lin, M.T., and Yeh, Y.H., 1981, Seismological evidence of an active plate boundary in the eastern Taiwan area: Memoir of the Geological Society of China, v. 4, p. 143–154.

Tsai, Y.B., 1986, Seismotectonics of Taiwan: Tectonophysics, v. 125, p. 17–37.

Um, L., and Thurber, C.H., 1987, A fast algorithm for two-point seismic ray tracing: Seismological Society of America Bulletin, v. 77, p. 972–986.

Walcott, R.I., 1998, Modes of oblique compression: Late Cenozoic tectonics of the South Island of New Zealand: Review of Geophysics, v. 36, p. 1–26.

Wolf, L.W., Stone, D.B., and Davies J.N., 1991, Crustal structure of the active margin, south central Alaska: An interpretation of seismic refraction data from the Trans-Alaska Crustal Transect: Journal of Geophysical Research, v. 96, p. 16455–16469.

Wu, F.T., and Lu, C.-P., 1976, Recent tectonics of Taiwan: Bulletin of Geological Survey Taiwan, v. 25, p. 97–111.

Wu, F.T., Rau, R.J., and Salzberg, D., 1997, Taiwan orogeny: Thin-skinned or lithospheric collision?: Tectonophysics, v. 274, p. 191–220.

Yeh, Y.H., and 10 others, 1998, Onshore/offshore wide-angle deep seismic profiling in Taiwan: Terrestrial, Atmospheric and Ocean Sciences, v. 9, p. 301–316.

Yen, H.Y., Yeh, Y.H., Lin, C.H., Chen, K.J., and Tsai, Y.B., 1995, Gravity survey of Taiwan: Journal of the Physics of the Earth, v. 43, p. 683–696.

Yu, S.B., Chen, H.Y., and Kuo, L.C., 1997, Velocity field of GPS stations in the Taiwan area: Tectonophysics, v. 274, p. 41–59.

Manuscript Accepted by the Society April 12, 2001

Printed in the U.S.A.

Geological Society of America
Special Paper 358
2002

Compressional subduction regime and initial arc-continent collision: Numerical modeling

J.-C. Tang
A.I. Chemenda
Géosciences Azur, UMR 6526, Université de Nice–Sophia Antipolis et CNRS, 250 Rue Albert Einstein–Sophia Antipolis, 06560 Valbonne, France
J. Chéry
S. Lallemand
Laboratoire de Géophysique et Tectonique, UMR 5573, Université Montpellier II, Case 060, Pl. E. Bataillon, 34095, Montpellier, Cedex 05, France
R. Hassani
Laboratoire d'Instrumentation Géophysique, Université de Savoire, 73376 Le Bourget du Lac Cedex, France

ABSTRACT

We performed two-dimensional finite-element numerical modeling of subduction to study the deformation and failure pattern of an overriding plate. This plate has elasto-plastic rheology with strain weakening and contains a "volcanic arc" represented in the model as a thinned and rheologically weakened zone. The subducting plate is not modeled directly. Its effect on the overriding plate is simulated by the normal and tangential stresses along the interplate surface, which correspond to the interplate pressure and interplate friction, respectively. Both stresses cause nonhydrostatic (tectonic) compression on this plate. The first approximation for the interplate pressure is derived from previous experimental modeling of subduction of oceanic lithosphere. Numerical modeling shows that the overriding plate fails under the arc along a trenchward-dipping fault. We then vary both the interplate pressure and the interplate friction and show that the failure direction does not change. Subduction of the continental margin modifies (increases) the interplate pressure because of the buoyancy of thick continental crust. The compressive stress in the overriding plate increases. Failure of this plate can occur in the same way as indicated or in the opposite direction along a fault that dips under the arc (away from the trench and toward the ocean. The failure mode depends in particular on the depth of the margin's subduction prior to lithosphere rupture. Failure along a fault dipping toward the continent results in a subduction reversal, whereas failure along a fault dipping toward the ocean is followed by subduction of a forearc block beneath the arc. Both scenarios seem to have natural analogues. We consider one analogue, the ongoing arc-continent collision in Taiwan, and argue that this process occurs according to the second scenario, which involves forearc subduction.

Tang, J.-C., Chemenda, A.-I., Chéry, J., Lallemand, S., and Hassani, R., 2002, Compressional subduction regime and initial arc-continent collision: Numerical modeling, *in* Byrne, T.B., and Liu, C.-S., eds., Geology and Geophysics of an Arc-Continent Collision, Taiwan: Boulder, Colorado, Geological Society of America Special Paper 358, p. 177–186.

INTRODUCTION: MECHANICAL ESSENTIALS OF SUBDUCTION

If one neglects the dynamic interaction between the overriding plate in a subduction zone and the underlying low-viscosity asthenosphere, then the stress-strain state of this plate will be defined only by its mechanical structure and the boundary conditions along the interplate zone. These conditions are the normal (interplate pressure) and tangential (frictional) stresses. Because interplate friction appears to be very small (Tichelaar and Ruff, 1993; Hassani et al., 1997) at least in some subduction zones, the major parameter controlling the stress-strain state of the overriding plate should be the interplate pressure. If the rigidity of the subducting plate were zero, the overriding plate would be in hydrostatic equilibrium, which means that the interplate pressure is equal to the hydrostatic pressure $P_h = -\rho_o g z$ (ρ_o is the density of the overriding plate; g is the acceleration due to gravity; z is the vertical coordinate) and there is no tectonic stress in the overriding plate. As the flexural rigidity of the subducting lithosphere normally is not negligible and the subducted segment is not in hydrostatic equilibrium, the interplate pressure P_n differs from P_h, which results in a tectonic pressure force F_p acting on the overriding plate (Fig. 1). This force can be either compressional (Fig. 1A) or extensional (Fig. 1B). The horizontal component of this force is equal to (Fig. 1)

$$F_h = \int_0^H \sigma_n dz, \quad (1)$$

where $\sigma_n = P_n - P_h$ is the nonhydrostatic interplate pressure, and H is the thickness of the overriding plate (Fig. 3). Experimental modeling of subduction (Shemenda, 1993, 1994) shows that the interplate pressure depends on the density contrast $\Delta\rho$ between the subducting plate and the asthenosphere as well as on the relative motion of the subducted plate and the surrounding mantle. If $\Delta\rho = 0$ and there is no dynamic interaction between the subducting plate and the mantle, then the overriding lithosphere is under compression during subduction (Fig. 1), which is due to the flexural rigidity of the subducting plate. When the subducting plate is denser then the underlying mantle ($\Delta\rho > 0$), then the pull force caused by the density contrast tends to separate the plates in the subduction zone along the lower part of the interplate surface. This separation reduces the interplate pressure (Fig. 1B), resulting in an extensional tectonic force F_h.

The subduction of a continental margin can follow subduction of oceanic lithosphere in either extensional or compressional regimes, but as progressively thicker and more buoyant continental crust is underthrust, the interplate pressure increases (Shemenda, 1994; Chemenda et al., 1997). Both the compression of the overriding plate and the uplift of its frontal wedge increase during this process. Roughly, the interplate pressure can be presented as having two components: the nonisostatic component due to the flexural rigidity of the subducting plate and its interaction with the mantle and the isostatic component due only to the buoyancy force caused by the subducted crustal wedge of the continental margin.

In this paper we apply numerical modeling to compute tectonic stresses within the overriding plate corresponding to the compressional regime of oceanic subduction and the subduction of a continental margin. We investigate the conditions and the

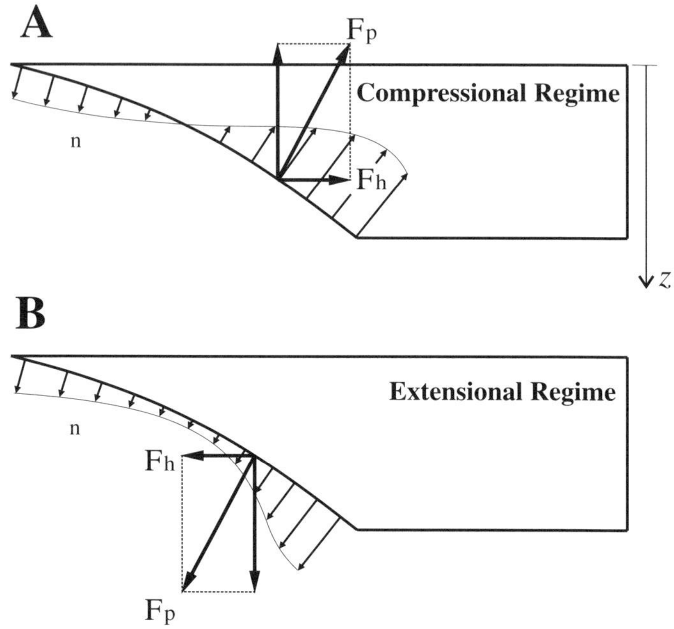

Figure 1. Distribution of the nonhydrostatic interplate pressure σ_n for (A) compressional and (B) extensional regimes of oceanic subduction (Shemenda, 1993, 1994). $\sigma_n = P_n - P_h$, where P_n is the total interplate pressure and P_h is the hydrostatic pressure. F_p is the nonhydrostatic interplate pressure force caused by σ_n, and F_h is the horizontal component of F_p.

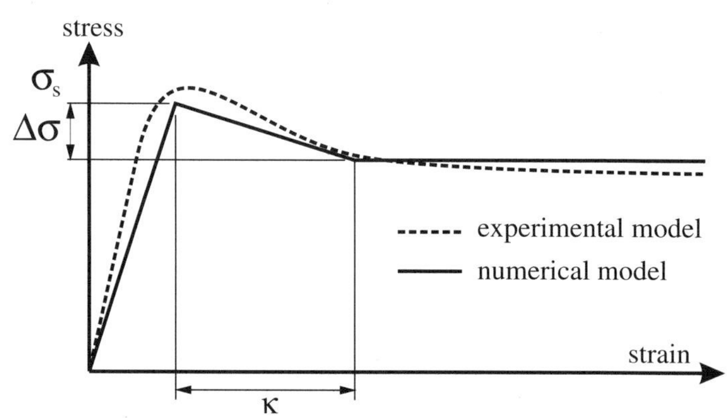

Figure 2. Stress-strain diagrams corresponding to the experimental and numerical models. In laboratory experiments (Shemenda, 1994), the yield limit of the model lithosphere for the normal load is 28 Pa, which corresponds to $\sigma_s = 210$ MPa in nature. In the numerical model presented here, $\sigma_s = 180$ MPa, Young's modulus $E = 20$ GPa; $\Delta\sigma = 36$ MPa; $\kappa = 0.3$; Poisson's ratio $\nu = 0.25$.

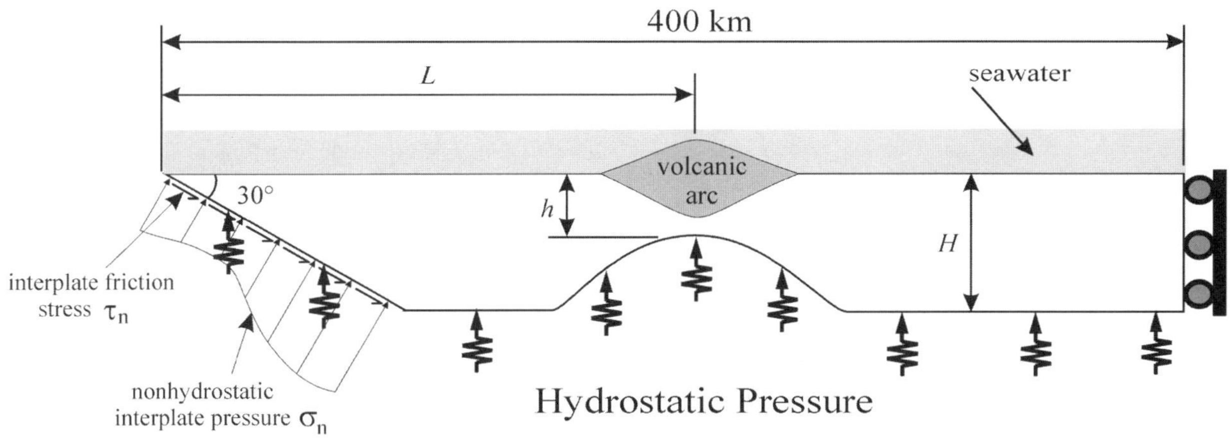

Figure 3. Setup of the numerical model. Thickness of the overriding plate away from volcanic arc $H = 60$ km; h is the thickness of the plate in the arc area; the density of the plate is $\rho_o = 3.3 \times 10^3$ kg/m^3, which is equal to the density of the asthenosphere. The volcanic arc has the same yield limit as the plate, and its density is 2.8×10^3 kg/m^3. The arc is completely isostatically compensated; its height is 4.2 km, and the maximum Moho depth is 15 km. Thickness of the ocean water is 4.5 km. L is the distance between the arc axis and the trench.

modes of possible failure of the overriding plate during subduction. Failure occurs in the volcanic arc area, which is a zone of weakness caused by such factors as induced convection in the mantle, magmatic activity (presence of the magma chambers and channels), and increasing water content. We then test the effect of the interplate friction. Finally, we compare the results to the compressional oceanic subduction zones of the Kuril and Japan as well as to the zone of the continent-margin subduction to the south of Taiwan.

OCEANIC LITHOSPHERE SUBDUCTION

Setting of numerical experiments

Following the conditions of the experimental modeling just described, we consider a one-layer model of oceanic overriding lithosphere. Its deformation, caused by the interplate normal and tangential stresses, is computed by using the two-dimensional finite-element code ADELI (Hassani, 1994). In the numerical experiments we apply the same elasto-plastic rheology of the lithosphere with strain softening as in the laboratory models (Shemenda, 1994) (Fig. 2). We use Von-Misses yield criterion for time-independent plastic deformation. The geometry and boundary conditions in the numerical models are shown in Figure 3. The overriding plate is underlined by the Winkler base with isostatic restoring forces (Williams, 1990; Wallace and Melosh, 1994) and zero viscosity. The normal or tangential (or both) stresses are applied along the interplate surface (left edge of the model). The kinematic boundary condition at the right edge allows no displacement in the horizontal direction. The density of the plate is the same as that of the asthenosphere and is equal to 3.3×10^3 kg/m^3. The lithosphere is covered by 4.5 km of ocean water.

The weak zone associated with a volcanic arc is introduced in three ways in different tests. In one set of experiments we simply make a notch at the base of an initially homogeneous plate. In another set of tests we add a "volcanic arc" with a density of 2.8×10^3 kg/m^3 (Fig. 4A). The arc is isostatically compensated and is 19.2 km thick, which yields 4.2 km of underwater relief. Such average relief is representative of real subduction zones, although this parameter has little influence on the modeling results. The arc is also associated with a thermal anomaly; the lithosphere here is characterized by higher temperature and hence lower strength. The thermal structure of the lithosphere in the arc area, however, depends on many poorly known factors and is the subject of much debate (e.g. Furukawa, 1993; Schmidt and Poli, 1998. The problem of rheology of the lithosphere in this area is even more controversial, but it seems to be obvious that the lithospheric strength decreases toward the arc axis. This decrease is taken into account in the setup in Figure 4B. The model in Figure 4C is designed to test the effect of the magma channels on the overriding plate failure.

We first investigate oceanic plate subduction in a compressional regime (Fig. 1A). The distribution of the interplate pressure σ_n for this case is obtained from the experimental model of Shemenda (1994), which shows that the σ_n amplitude is comparable with the yield limit (i.e., the plate strength) σ_s of the lithosphere under normal loading. It was not possible, however, to quantify exactly the σ_n value from the laboratory experiments. Therefore we have tested numerically different σ_n functions following the general form defined in the laboratory models. In each experiment presented here (except experiment 1), the overriding plate has been weakened enough for the lithosphere to fail. The results presented show that the way the plate is weakened has little effect on the failure mode.

Experimental results

Experiment 1. Both the setup and the result of this experiment are shown in Figure 5A. The deformation of the model with an

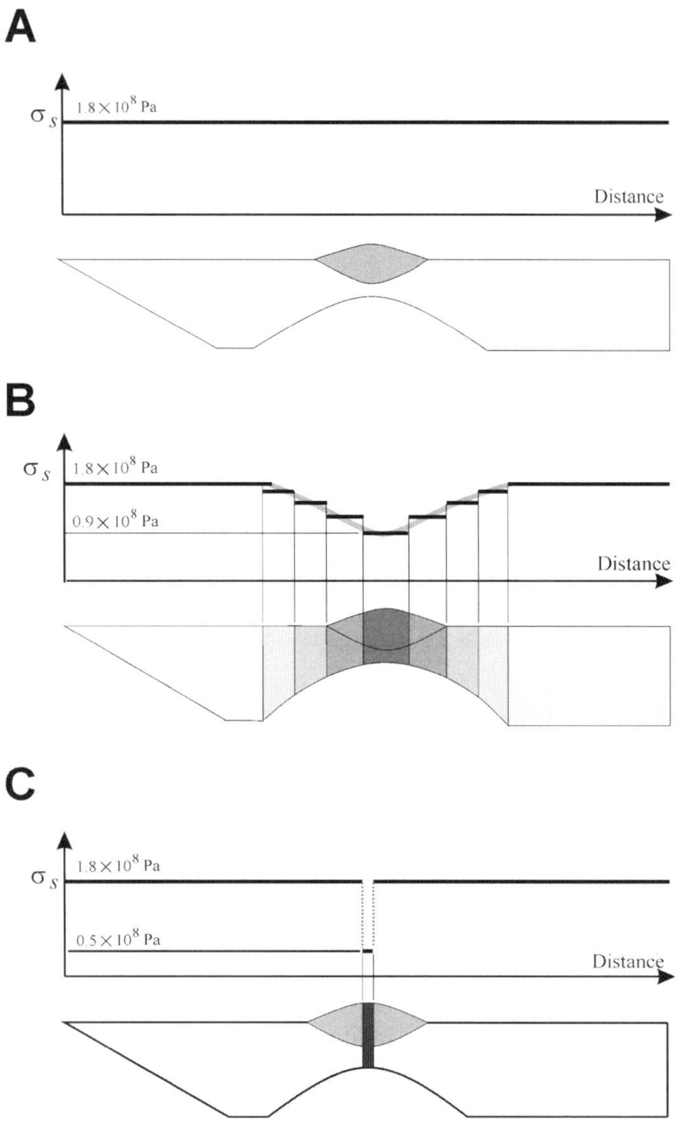

Figure 4. Three tested mechanical structures of the arc area: (A) model with a notch and a constant yield limit, (B) plate with a notch and a yield limit that steps down toward the arc axis, and (C) plate with the notch and a constant σ_s except a 5-km-wide axial zone in the arc, which is very weak (see text for explanations).

initially horizontal surface results in the formation of a 7.5-km-deep trench and of an ~2-km-high frontal nonisostatic arc. The lithosphere is under compression, and the tectonic normal horizontal stress, σ_h, at the left edge of the model is equal to 44 MPa. To ensure that the chosen model parameters are reasonable, we calculated free-air gravity anomaly Δg generated by the deformed model; in the trench $\Delta g \approx -250$ mgal and in the forearc $\Delta g \approx 230$ mgal. Such anomalies correspond well to the Kuril arc (Sandwell and Smith, 1997), which is an example of a compressional subduction zone. The σ_h value in this experiment is only about 24% of the plate strength σ_s. Therefore the model does not fail, whereas in the real compressional subduction zones, such a failure seems to be possible. There is evidence that the overriding plate is failing under compressive stress in the Kuril arc (Baranov and Lobkovsky, 1980; Savostin et al., 1983) and Japan (Nakamura, 1983; Tamaki, 1984; Lallemand et al., 1985) subduction zones. The failure occurs in the arc area, which is a weak zone.

Experiment 2. In this experiment we have weakened the overriding plate in the arc by making a notch deep enough for the plate to fail. The distance between the trench and the notch (arc) axis is 200 km, which corresponds to the average distance between these structures observed in nature. The other conditions are the same as in the previous experiment. The overriding plate fails along a shear zone that dips trenchward (Fig. 5B). Several experiments have been conducted under the same conditions but with a different number of finite elements. The results were always the same as in Figure 5B. We also performed a set of tests varying the arc-trench distance L. For $L > 180$ km and $L < 230$ km, the failure occurred as in Figure 5B. For $L > 230$ km, the shear zone dips in the opposite direction.

Experiment 3. Setup of this experiment corresponds to Figure 4A and is shown in Figure 5C. The failure occurs in the same direction as in the previous experiment. Variation of the L value yielded the following results: for $L > 180$ km and $L <$ 230 km, the failure occurs along the trenchward-dipping zone. For $L > 230$ km, the failure occurs along the zone dipping in the opposite direction.

Experiment 4. The setup corresponds to Figure 4B. The result in Figure 5D shows the same failure direction as in the previous experiments.

Experiment 5. The only difference with the previous experiment is that we increased L to 250 km. Figure 5E shows that this modification causes change in the failure direction. By further varying the L value, we have obtained almost the same results as in the models without an arc and with a constant yield limit of the lithosphere. The only difference is that the critical L value, corresponding to the change in failure mode, is now 250 km instead of 230 km in the previous experiments, so that when 180 km $< L <$ 250 km, the failure occurs as in Figure 5D and when $L \geq 250$ km, the failure occurs as in Figure 5E.

Experiment 6. In this experiment we increased the interplate pressure (Fig. 5F), which resulted in increase in the compressive horizontal stress: $\sigma_h = 70$ MPa. Such a compression is sufficient to cause failure of $h = 35$-km-thick lithosphere in this experiment. Variation of L yields the following: for 180 km $< L <$ 230 km, failure occurs as in Figure 5F, and for $L > 230$ km, the plate fails in the opposite direction.

Experiment 7. In reality, the lithosphere in the arc is not only thinned and weakened, but is also cut by magma channels. Therefore we also conducted experiments with a vertical cut (fault) along the arc axis. The introduction of this cut does not affect the plate deformation and failure. In this experiment we introduce the vertical weak zone according to Figure 4C. The result presented in Figure 6A shows that the plate still fails along an inclined fault. The critical L value is the same (230

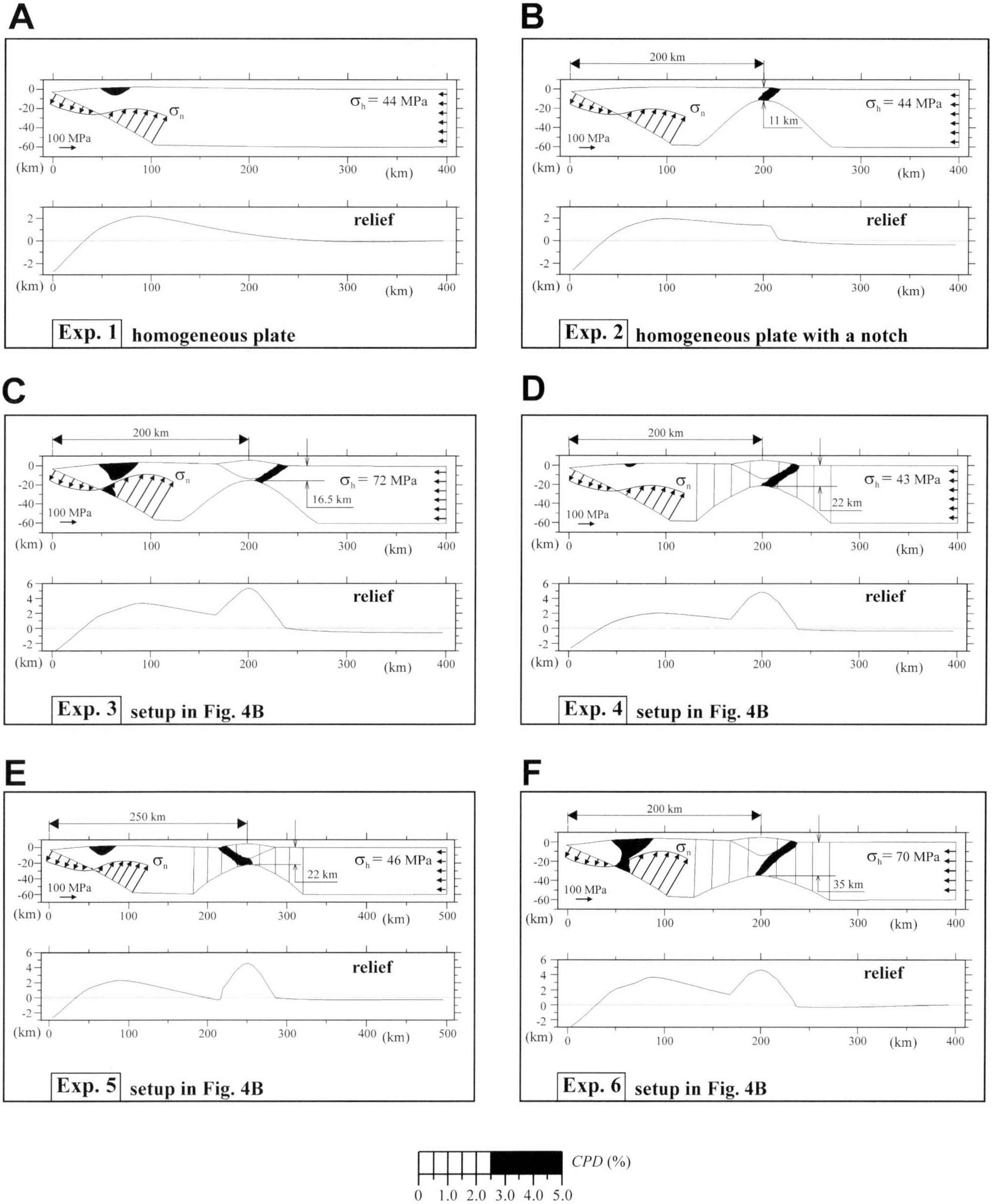

Figure 5. Results of the experiments (from experiment 1 to experiment 6 [plots A to F, respectively]) on oceanic subduction. Each experiment is represented by two of the plots. Cross sections A and B explain the model setup and show the general view on the deformed plate with zones of plastic strain localization. Cross sections C to F show the geometry (relief) of the initially horizontal plate surface. σ_n is the nonhydrostatic interplate pressure caused by the flexural rigidity of the subducting plate (see Fig. 1A); *CPD* is the cumulated plastic deformation shown in A and B. See text for more detailed description of the experiments.

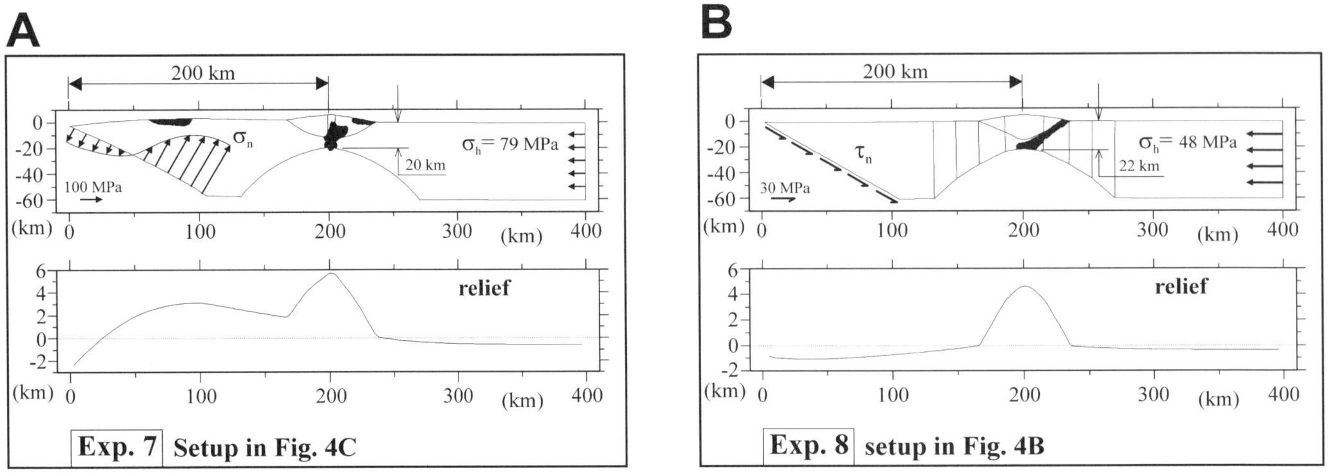

Figure 6. Two more experiments (experiments 7 and 8) on oceanic subduction (see caption to Fig. 5). τ_n is the interplate friction stress. See caption to Figure 5 for more explanations.

km) as in the other similar experiments (experiments 4 and 6) without the vertical cut.

Experiment 8. This experiment is designed to test the effect of the interplate friction. The interplate pressure σ_n is zero and a constant tangential (friction) stress $\tau_n = 30$ MPa is applied along the interplate surface. The value of the friction stress is chosen to be large enough to cause the failure of the model thinned as in experiments 4 and 5 to 22 km. This value corresponds to the available estimations of this parameter in nature, of the order of 10 MPa (Peacock, 1996). The overriding plate fails as shown in Figure 6B, i.e., in the same direction as in experiment 4 (Fig. 5D) where failure is caused by the interplate pressure. The orientation of the fault caused by the friction force does not depend on the L value (we tested 180 km $< L <$ 300 km).

CONTINENTAL-MARGIN SUBDUCTION

Model setup

Let us now suppose that the flexural rigidity of the subducting plate (continental margin) is negligible and the interplate friction is zero. In this case the interplate nonisostatic pressure σ_n can be due only to the buoyancy of the subducted continental crust. To quantify this pressure we have run the numerical test shown in Figure 7. The subducting plate is represented only by a very weak continental crust. The thickness of the crust h_c near the trench is assumed to be 15 km and is reduced linearly to zero at the base of the overriding plate. The computed interplate pressure σ_b caused by the subducted margin is shown in Figure 8A and is used in the next three experiments. In the first experiment (experiment 9), we apply only this pressure, and we assume that both the flexural rigidity of the subducting plate and the interplate friction are negligible. In the next experiment (experiment 10), the interplate friction τ_n is added to σ_b. In experiment 11, the applied interplate pressure is the sum of σ_b and the "nonisostatic" pressure σ_n (Fig.

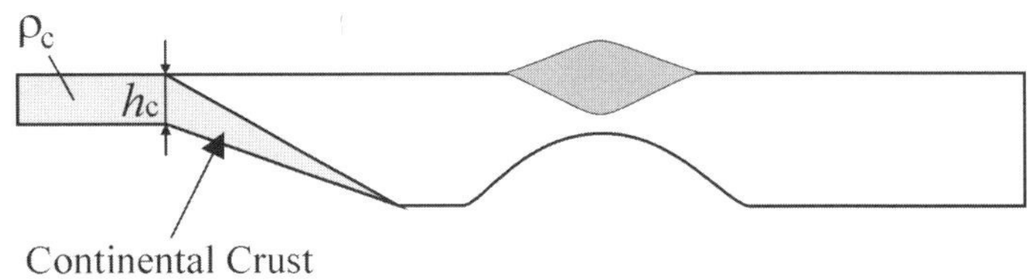

Figure 7. Model setup corresponding to subduction of the continental margin. Only the buoyancy effect of the subducted continental crust of thickness h_c is taken into account; the rigidity of this crust is negligible. ρ_c is the density of the continental crust.

5A) caused by flexural rigidity of the subducting plate, $\tau_n = 0$. In experiment 12, we apply the both pressures and also add the interplate friction.

Results

Experiment 9. The conditions are the same as in experiment 4, except that the interplate pressure is equal to σ_b. The overriding plate fails along the oceanward-dipping fault (Fig. 8A). Unlike the experiments on oceanic subduction, the failure direction remains the same at different L values (we tested 180 km $< L$ $<$ 300 km). The horizontal compressive stress σ_h is proportional, as expected, to the thickness h_c of the subducted crust; in experiment 8, $\sigma_h = 43$ MPa, which is comparable with the σ_h values corresponding to the oceanic subduction.

Experiment 10. In this experiment, the interplate pressure is the same as in the previous test, but we add the interplate friction $\tau_n = 15$ MPa. The introduction of the friction results in a higher compressive stress τ_h, which is enough now to cause a failure of a thicker lithosphere (Fig. 8B). Failure occurs along an oceanward-dipping fault as in experiment 9 regardless of the L value used

Experiment 11. In this experiment we test the combined effect of the interplate pressures due to the subducting plate's flexural rigidity σ_n and the subducted margin's buoyancy σ_b:

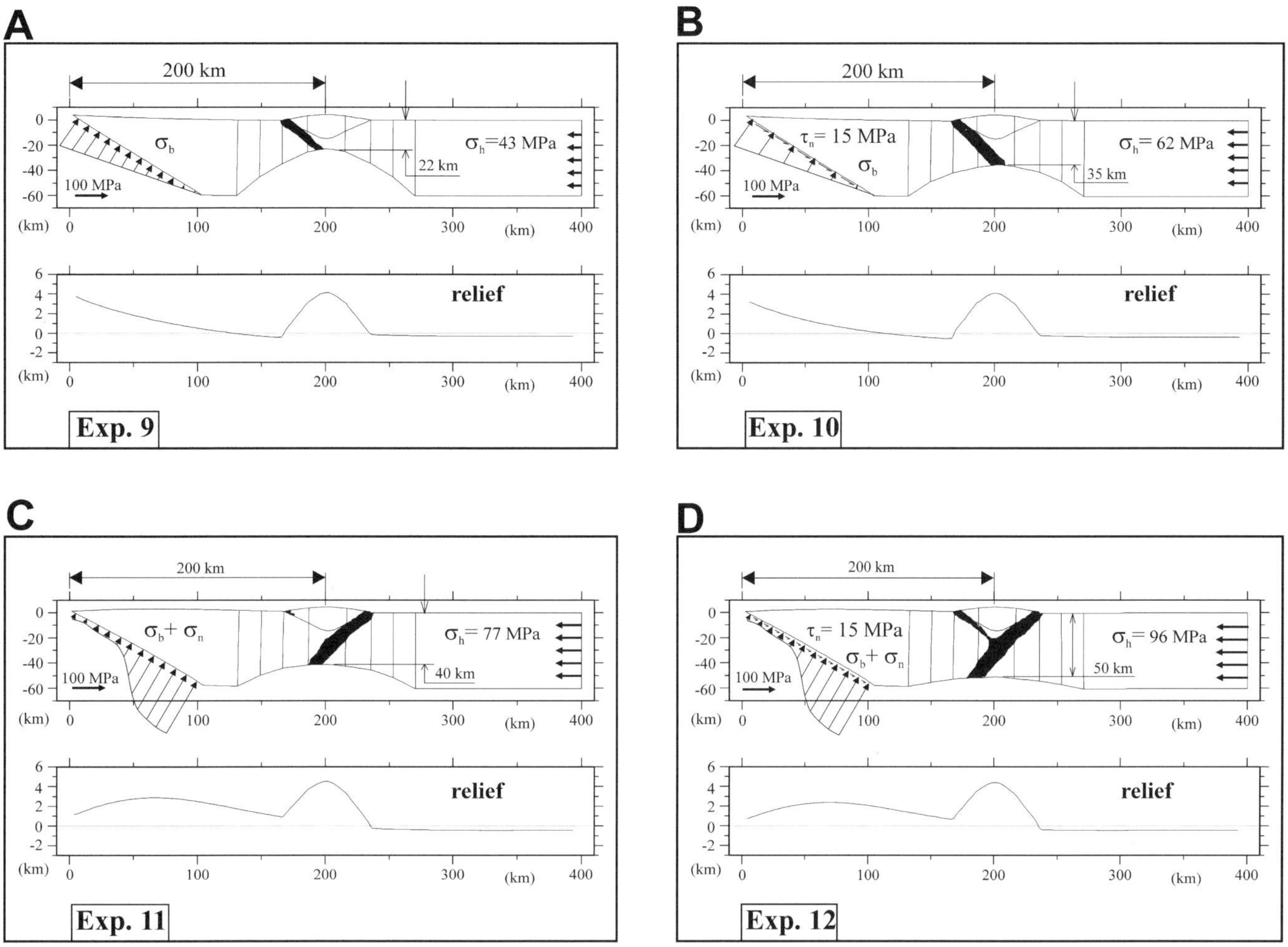

Figure 8. Four experiments (experiments 9 to 12) of continental-margin subduction. The overriding plate's structure in all experiments corresponds to that in Figure 4B. σ_b is the interplate pressure caused by the buoyancy effect of the subducted continental crust and has been calculated for $h_c = 15$ km and $\rho_c = 2.8 \times 10^3$ kg/m^3 (see Fig. 7). σ_n in experiments 11 and 12 is the same as in experiment 1 (Fig. 5A). The interplate friction stress in experiments 10 and 12 is $\tau_n = 15$ MPa. See the caption of Figure 5 for more explanations.

the total interplate pressure represents a sum of both pressures shown in Figures 5A and 8A. The compressive stress is higher, $\sigma_h = 77$ MPa, and the failure occurs along a continentward-dipping fault, which is the opposite to the experiment 10 direction (Fig. 8C). This direction of failure applies for 180 km $< L <$ 210 km. At $L \geq 210$ km, the failure occurs along an oceanward-dipping fault (as in Fig. 8B).

Experiment 12. The interplate pressure is the same as in the previous test, but we add the interplate friction (frictional stress) $\tau_n = 15$ MPa. Compressive stress σ_h increases more in this experiment and is enough to cause failure of a thicker plate ($h = 50$ km). As in the previous test, a failure occurs along the continentward-dipping fault (Fig. 8D).

DISCUSSION AND CONCLUSIONS

The general problem of geodynamic modeling (both experimental and numerical) is either the uncertainty of the rheology of the lithosphere (Kohlstedt et al. 1995) or the boundary conditions (e.g., the interplate pressure and interplate friction stress). The results of the modeling are frequently very sensitive to these factors. This sensitivity leads at times to an ambiguity in the modeling results and in their geodynamic interpretation. For example, it is certain that the lithosphere in the arc area is weakened, but we do not know to what extent. Weakening depends on many factors such as the duration of subduction, the nature of the overriding plate, the magmatic activity of the arc, the presence of magmatic chambers and channels, etc. All these factors are inadequately defined. Therefore we do not know a priori whether the stress generated within the overriding plate in compressional subduction zones is large enough to cause failure of this plate or not. On the other hand, natural situations show that such failure does occur. This evidence leads to the conclusion that the lithosphere in the arc is weak enough. The purpose of this work is to study how this failure occurs under different conditions. The results are as follows: there are two potential directions (surfaces) of failure that dip under the arc from both sides. In some tested cases, the failure (localized deformation) starts along both directions at once (e.g., Fig. 8D), but then one zone dies out and the main failure occurs along the remaining one. The failure direction is controlled by three factors: the distribution and the value of the pressure along the interplate surface, the interplate friction, and the distance between the trench and the arc (the weak zone) axis. Different combinations of these parameters result in different failure directions. We have shown, however, that under the conditions corresponding to oceanic plate subduction (experiments 1 to 6), failure always occurs along the trenchward-dipping fault whatever the combination of interplate pressure and friction. This fault orientation does not depend on the tested mechanical structures of the overriding plate at the realistic distances L between the arc and the trench ($L < 230$ km). Thus if the overriding plate in an oceanic subduction zone fails, the failure preferably occurs along a trenchward-dipping fault that results in subduction reversal. This conclusion is in agreement with the Kuril (Baranov and Lobkovsky, 1980; Nakamura, 1983; Savostin et al., 1983) and Japan subduction zones (Tamaki, 1984; Lallemand et al., 1985).

As far as subduction of a continental margin is concerned, the modeling results show (1) that the compression of the overriding plate is generally higher than for oceanic plate subduction and (2) that the plate failure can occur along either of the two possible directions. The failure mode depends largely on the flexural rigidity of the continental margin and on the thickness of the subducting continental crust (h_c in Fig. 7), which should be proportional to the amount of the continental margin subducted before the overriding plate's failure. Because compression is higher, failure is more probable during subduction of a continental margin. These results are in a good agreement with experimental modeling of the same process by Shemenda (1994) and Chemenda et al. (1997). The numerical experiments are based on the results from the laboratory experiments that provided boundary conditions for the mathematical model and allowed us to test the validity of the numerical code. On the other hand, numerical modeling allowed us to advance further in studying the influence of different parameters (the interplate friction in particular) on the style of lithospheric deformation and failure. Both modeling techniques show that the principal element of continental subduction is the failure of the overriding plate in the arc area. Failure along the continentward-dipping fault results in a subduction reversal, whereas failure in the opposite direction is followed by subduction of a forearc block beneath the arc. Both scenarios seem to have natural analogues. For example, structural (Snyder et al., 1996) and GPS (Global Positioning System) (Genrich et al., 1996) data clearly show that the subduction of the Australian continental margin under the Banda arc near Timor resulted in a subduction reversal. Next we consider another example of the ongoing arc-continent collision in Taiwan and argue that in this area the overriding plate failed along the oceanward-dipping fault.

Taiwan collision

Along the Luzon-Taiwan segment of the convergent plate boundary, the Eurasian plate is being subducted under the Philippine Sea plate, which is carrying the Luzon arc (Fig. 9A). The Philippine Sea plate moves toward the Eurasian plate at a rate of more than 7 mm/yr (Seno et al., 1993; Yu et al., 1997). South of about 21°N, subduction of the South China Sea plate beneath the active Luzon volcanic arc is taking place along the Manila Trench (Taylor and Hayes, 1983; Hayes and Lewis, 1984). This is a typical intraoceanic subduction zone (Liu et al., 1998). Near 21.25°N, the Manila Trench disappears and the Luzon arc becomes inactive, although rapid plate convergence continues in this region and results in the formation of the Taiwan collisional prism. At this latitude, the Asian margin has already begun to be underthrust below the Philippine Sea plate (Reed et al., 1991; Liu et al., 1992). Farther to the north, the continental subduction is more advanced. The arc-continent collision is thus propagating to the south, with the collision front being somewhere between 21.25°N and 22.65°N (Lundberg et al., 1992, 1997; Huang et al., 1997). North of 22.65°N, the Luzon arc overrides the collisional prism along a lithospheric-scale

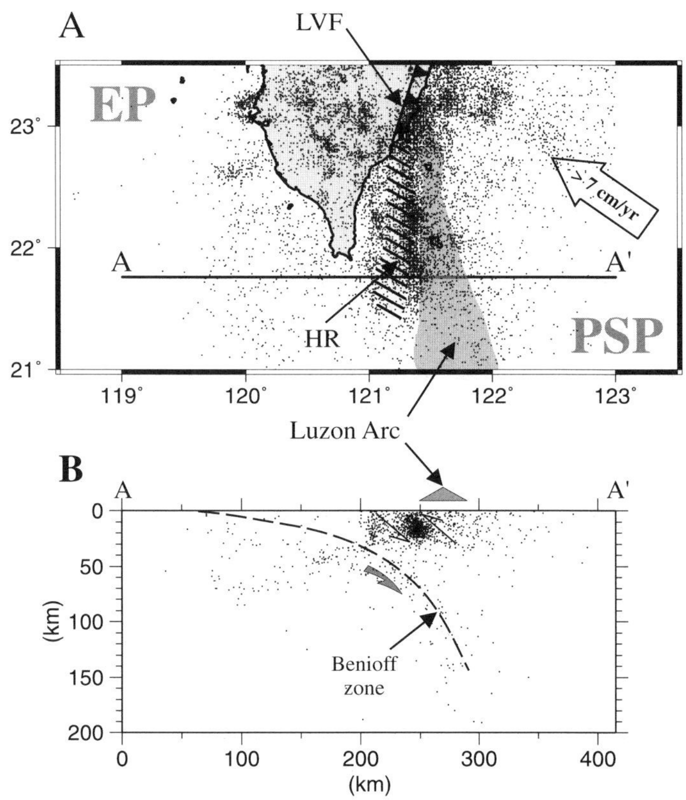

Figure 9. Seismicity around southern Taiwan (earthquakes of magnitude M > 3 between 0 and 30 km depth located by the Taiwan Telemeter Seismographe Network [TTSN] and the Central Weather Bureau Seismographe Network [CWBSN] during 1974–1997). (A) Seismicity map. (B) Seismicity section. EP—Eurasian plate, PSP—Philippine Sea plate, HR—Huatung Ridge, LVF—Longitudinal Valley fault.

east-dipping fault, the Longitudinal Valley fault (Fig. 9A). This fault is not mapped offshore to the south where the Philippine Sea plate appears to be deforming in a more complex manner (Huang et al., 1992; Lundberg et al., 1997). GPS data (Yu et al., 1997) show rapid shortening between the Luzon arc and southern Taiwan. This shortening is occurring along a west-dipping backthrust, causing the formation of the Huatung Ridge (Fig. 9A) made up of the deformed sediments of the forearc basin (Lundberg et al. 1997). The deep lithospheric deformation in this area can be inferred from the seismicity distribution. The most prominent feature of the seismicity pattern around southern Taiwan is a narrow, nearly linear seismicity concentration zone along the western border of the Luzon arc (Fig. 9A). The cross section in Figure 9B shows the deep geometry of this zone, which appears as a east-dipping swarm beneath the arc. We interpret this zone as corresponding to the Philippine Sea plate failure and underthrusting of the forearc block under the Luzon arc. According to Figure 9A, the failure front is located near 21.5°N (the southern termination of the seismicity concentration zone). The direction of the thrusting at depth (at lithospheric scale) inferred from the seismicity distribution in Figure 9B is thus inconsistent with the already-mentioned backthrusting revealed by the seismic profiling in the same area. The most likely explanation of this discrepancy between the model and the already-mentioned observations is that the superficial manifestation of the plate failure does not mimic directly the deformation at depth. To illustrate this point we use the well-known result of laboratory modeling of sedimentary-layer deformation at the convergent plate boundary, presented in Figure 10A (Malavieille, 1984; Byrne et al., 1988). When combined with our results, this picture leads to the model in Figure 10B.

ACKNOWLEDGMENTS

This research was accomplished in the framework of a Sino-French cooperative project. J.-C. Tang benefited from a grant from the Centre National des Oeuvres Universitaires et Scolaires, France, and the Ministry of Education at Taiwan, Republic of China. We thank T. Byrne for review and valuable suggestions that allowed us to significantly improve the manuscript. The seismicity data were provided by the Central Weather Bureau of Taiwan.

REFERENCES CITED

Baranov, B.V., and Lobkovsky L.I., 1980, Shallow seismicity behind the Kuril arc and its relation to the Benioff zone (in Russian): Doklady Akademii Nauk SSSR, v. 255, p. 67–71.

Byrne, D.E., Davis, D.M., and Sykes, L.R., 1988, Loci and maximum size of thrust earthquakes and the mechanics of the shallow region of subduction zones: Tectonics, v. 7, p. 833–857.

Chemenda, A.I., Yang, R.K., Hsieh, C.-H., and Groholsky, A.L., 1997, Evolutionary model for the Taiwan collision based on physical modeling: Tectonophysics, v. 274, p. 253–274.

Furukawa, Y., 1993, Magmatic processes under arcs and formation of the volcanic front: Journal of Geophysical Research, v. 98, p. 8309–8319.

Genrich, J.F., Bock, Y., McCaffrey, R., Calais, E., Stevens, C.W., and Subarya, C., 1996, Accretion of the southern Banda arc to the Australian plate margin determined by Global Positioning System measurements: Tectonics, v. 15, p. 288–295.

Hassani, R., 1994, Modélisation numérique de la déformation des systemes géologiques [Ph.D. thesis]: Montpellier, Université de Montpellier II, 139 p.

Hassani, R., Jongmans, D., and Chéry, J., 1997, Study of plate deformation and stress in subduction process using two-dimensional numerical models: Journal of Geophysical Research, v. 102, no. B8, p. 17951–17965.

Hayes, D.E., and Lewis, S., 1984, A geophysical study of the Manila Trench,

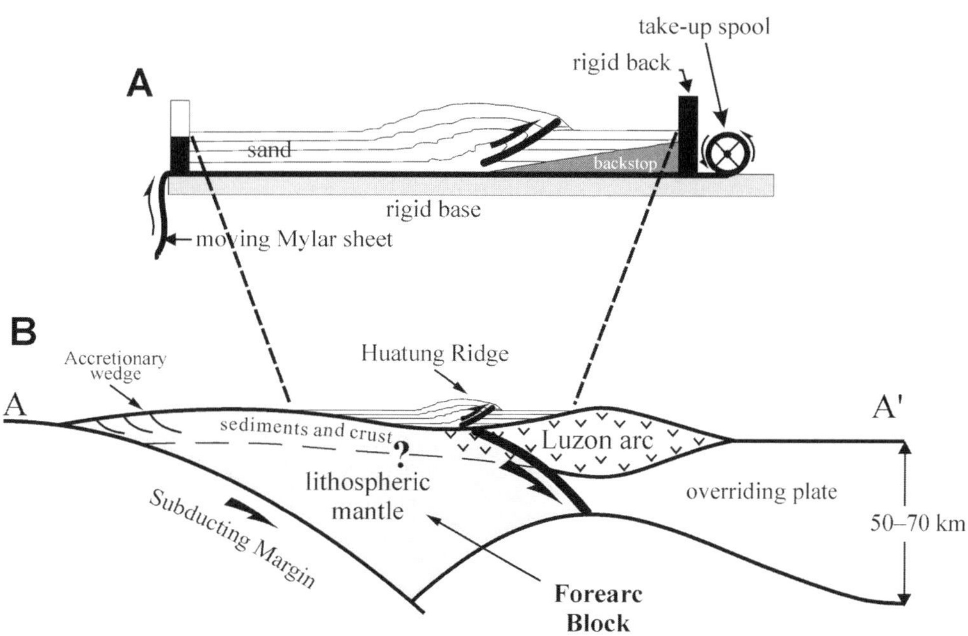

Figure 10. Scheme for the initial subduction of a forearc block (initial arc-continent collision) and a superficial manifestation of this lithospheric-scale process. (A) Sand-box modeling result from Malavieille (1984) and Byrne et al. (1988). (B) Superposition of cross section A with the lithospheric-scale failure model obtained in this work by numerical modeling.

Luzon, Philippines. 1. Crustal structure, gravity, and regional tectonic evolution: Journal of Geophysical Research, v. 89, p. 9171–9195.

Huang, C.-Y., Shyu, C.T., Liu, S.B., Lee, T.Q., and Sheu, D.D., 1992, Marine geology in the arc-continent collision zone off southeastern Taiwan: Implications for late Neogene evolution of the Coastal Range: Marine Geology, v. 107, p. 183–212.

Huang, C.-Y., Wu, W.Y., Chang, C.P., Tsao, S., Yuan, P.B., Lin, C.W., and Xia, K.Y., 1997, Tectonic evolution of accretionary prism in the arc-continent collision terrane of Taiwan: Tectonophysics, v. 281, p. 31–51.

Kohlstedt, D.L., Evans, B., and Mackwell, S.J., 1995, Strength of the lithosphere: Constraints imposed by the laboratory experiments: Journal of Geophysical Research, v. 100, p. 17587–17602.

Lallemand, S., Okada, H., Otsuka, K., and Labeyrie, L., 1985, Compressive tectonics along the eastern margin of the Japan sea: Evidence for eastward thrust: Comptes Rendus de l'Académie des Sciences, ser. 2, no. 3, v. 301, p. 201–206.

Liu, C.-S., Liu, S.Y., Kuo, B.Y., Lundberg, N., and Reed, D.L., 1992, Characteristics of the gravity and magnetic anomalies off southern Taiwan: Acta Geologica Taiwanica, n. 33, p. 121–130.

Liu, C.-S., Liu, S.Y., Lallemand, S.E., Lundberg, N., and Reed, D.L., 1998, Digital elevation model offshore Taiwan and its tectonic implications: TAO, Terrestrial, Atmospheric, and Oceanic Sciences, v. 9, p. 705–738.

Lundberg, N., Reed, D.L., Liu, C.-S., and Lieske, J., Jr., 1992, Structural controls on orogenic sedimentation, submarine Taiwan collision: Acta Geologica Taiwanica, n. 30, p. 131–140.

Lundberg, N., Reed, D.L., Liu, C.-S., and Lieske, J., Jr., 1997, Forearc-basin closure and arc accretion in the submarine suture zone south of Taiwan: Tectonophysics, v. 274, p. 5–23.

Malavieille, J., 1984, Modélisation expérimentale des chevauchements imbriqués: Application aux chaînes de montagne: Bulletin Société Géologique de France, v. 7, p. 129–138.

Nakamura, K., 1983, Possible nascent trench along the eastern Japan Sea as the convergent boundary between Eurasian and North American plates: University of Tokyo Publications in Bulletin of the Earthquake Research Institute, v. 58, p. 711–722 (in Japanese with English abstract).

Peacock, S.M., 1996, Thermal and petrologic structure of subduction zones, *in* Bebout, G.E., Scholl D.W., Kirby, S.H., and Platt, J.P., eds., Subduction top to bottom: American Geophysical Union Geophysical Monograph 96, p. 119–133.

Reed, D.L., Lundberg, N., Liu, C.-S., and McIntosh, K.D., 1991, Evidence of front thrust propagation and fluid migration in an offscraped sedimentary basin sequence: Offshore Taiwan, *in* Taicrust Workshop Proceedings, Institute of Oceanography: Taipei, Taiwan, Republic of China, National Taiwan University, p. 103–106.

Savostin, L., Zonenshain, L.P., and Baranov, B., 1983, Geology and plate tectonics of the Sea of Okhotsk, *in* Hilde, T.W.C., and Uyeda, S., eds., Geodynamics of the western Pacific-Indonesian Region: American Geological Union Geodynamics Series, v. 11, p. 189–222.

Sandwell, D.T., and Smith, W.H.F., 1997, Marine gravity anomaly from Geosat and ERS-1 satellite altimetry: Journal of Geophysical Research, v. 102, p. 10039–10054.

Schmidt, M., and Poli S., 1998, Experimentally based water budgets for dehydrating slabs and consequences for arc magma generation: Earth and Planetary Science Letters, v. 163, p. 361–379.

Seno, T., Stein, S., Gripp, A.E., 1993, A model for the motion of the Philippine Sea plate consistent with NUVEL-1 and geological data: Journal of Geophysical Research, v. 98, p. 17941–17948.

Shemenda, A.I., 1994, Subduction: Insights from physical modeling: Dordrecht, the Netherlands, Kluwer Academic Publishers, Modern Approaches in Geophysics, 215 p.

Shemenda, A.I., 1993, Subduction of the lithosphere and back-arc dynamics: Insights from physical modeling: Journal of Geophysical Research, v. 98, p. 16167–16185.

Snyder, D.B., Prasetyo, H., Blundell, D.J., Pigram, C.J., Barber, A.J., Richardson, A., and Tjokosaproetro, S., 1996, A dual doubly vergent orogen in the Banda Arc continent-arc collision zone as observed on deep seismic reflection profiles: Tectonics, v. 15, p. 34–53.

Tamaki, K., 1984, Active features and tectonics along the eastern margin of the Japan Sea: The Earth Monthly, v. 6, p. 38–48 (in Japanese).

Taylor, B., Hayes, D.E., 1983, Origin and history of the South China Sea Basin, *in* Hayes, D.E., ed., The tectonic and geological evolution of Southeast Asian seas and islands, Part 2: American Geophysical Union Geophysical Monograph 20, p. 129–155.

Tichelaar, B.W., and Ruff, L.J., 1993, Depth of seismic coupling along subduction zones: Journal of Geophysical Research, v. 98, p. 2017–2037.

Wallace, H.M., and Melosh, H.J., 1994, Buckling of a pervasively faulted lithosphere: PAGEOPH (Pure and Applied Geophysics), v. 142, p. 239–261.

Williams, C.A., 1990, Numerical modeling of fault formation and the dynamics of existing faults [Ph.D. thesis]: Tucson, University of Arizona, 297 p.

Yu, S.B., Chen, H.Y., and Kuo, L.C., 1997, Velocity field of GPS stations in Taiwan area: Tectonophysics, v. 274, p. 41–59.

Manuscript Accepted by the Society April 12, 2001

Printed in the U.S.A.

Geological Society of America
Special Paper 358
2002

Arc-continent collision in Taiwan: New marine observations and tectonic evolution

Jacques Malavieille*
Serge E. Lallemand
Stephane Dominguez
Anne Deschamps
Laboratoire de Géophysique, Tectonique et Sédimentologie
UMR 5573, UM2-CNRS, CC. 60, 34095 Montpellier, France, 33-467-14-36-58;
Chia-Yu Lu*
Department of Geology, National Taiwan University, 245 Chou-Shan Road, Taipei, Taiwan, Republic of China;
Char-Shine Liu
Philippe Schnürle
Institute of Oceanography, National Taiwan University, P.O. Box 23-13, Taipei, Taiwan, Republic of China
and the ACT (Active Collision in Taiwan) Scientific Crew[†]

ABSTRACT

Marine observations offshore of Taiwan indicate intense deformation of the Luzon arc-forearc complex, with episodic eastward migration of the active deformation front across the complex. This active tectonic domain absorbs a significant amount of shortening between the Eurasia margin and the Philippine Sea plate, which is moving toward N310°E at about 8 cm/yr relative to Eurasia. Swath bathymetry and backscattering data, together with seismic reflection and geopotential data obtained during the ACT (Active Collision in Taiwan) cruise onboard the R/V *L'Atalante*, showed major north to south changes in the tectonic style in both the indenting arc and the host margin.

Structural observations show that the forearc basement of the Luzon arc no longer exists north of 22°30′N. To the south, only a small part of the forearc domain may remain beneath the Huatung Ridge (rear portion of the former Manila Trench oceanic accretionary wedge including forearc and intra-arc sequences) and rear of the thrust wedge. A tectonic model involving the progressive underthrusting of large slices of the forearc basement may account for the contrasting styles of deformation encountered from south to north across the collisional orogen and apparent nonexistence of the forearc region. The progressive subduction of the continental margin of China induces: (1) to the south, major eastward backthrusting and shortening of the forearc domain between the former oceanic accretionary wedge and the Luzon arc volcanic edifice, (2) to the north, accretion of parts of the arc domain to the collisional belt associated with westward thrusting, eastward backthrusting at the base of the slope and block rotation.

*E-mails: Malavieille—malavie@dstu.univ-montp2.fr; Lu—lu@sun03.gl.ntu.edu.tw
[†]ACT Scientific Crew: J. Angelier, J.-Y. Collot, B. Deffontaines, M. Fournier, S.-K. Hsu, J.-P. Le Formal, S.-Y. Liu, J.-C. Sibuet, N. Thareau, and F. Wang.

Malavieille, J., Lallemand, S.E., Dominguez, S., Deschamps, A., Lu, C.-Y., Liu, C.-S., Schnürle, P., and the ACT Scientific Crew, , 2002, Arc-continent collision in Taiwan: New marine observations and tectonic evolution, *in* Byrne, T.B., and Liu, C.-S., eds., Geology and Geophysics of an Arc-Continent collision, Taiwan, Republic of China: Boulder, Colorado, Geological Society of America Special Paper 358, p. 187–211.

INTRODUCTION

Arc-continent collision and arc accretion along most continental margins mark the early episodes of mountain building. In the collision belts that can be studied in continental domains, the finite deformation exposed is very complex as a result of a long geologic history, generally involving numerous superimposed tectonic events. Thus, although ophiolite belts and slices of arcs, related to old oceanic suture zones, can be found in many mountain belts, it is often difficult to reconstruct the geodynamic setting and tectonic history prior to collision. To better understand the mechanics of mountain building, it is therefore fundamental to analyze the deformation processes acting currently in growing orogens.

The young Taiwan orogen has been the subject of study for many years, mainly on land (e.g., Biq, 1972; Chai, 1972; Jahn, 1972; Yen, 1973; Wu, 1978; Suppe, 1981, 1987; Barrier and Angelier, 1986; Ho, 1986; Lu and Hsü, 1992), but also offshore (e.g., Chen and Juang, 1986; Reed et al., 1992; Lundberg et al., 1992; Huang et al., 1992, 1997; Lallemand and Tsien, 1997). The geodynamic setting of the arc-continent collision is well defined, and the general kinematics are well determined. The island itself represents the emerged part of a southward-propagating collisional orogen located between the Eurasian and Philippine Sea plates. South of the island, initial stages of the collision between the Luzon volcanic arc and the Chinese continental platform can be observed (Lundberg et al., 1997; Huang et al., 1997). To the north, the island of Taiwan corresponds to a mature stage of collision and arc accretion, resulting in a 4-km-high mountain belt. A part of the evolving collision complex is under water to the east of the island (Fig. 1). This paper focuses on the active deformation occurring south and along the Coastal Range of Taiwan and its connection with the southern Ryukyu subduction zone (Lallemand et al., 1999).

Several geophysical cruises have been conducted in the offshore area of Taiwan since the early 1990s, providing multichannel and OBS (ocean-bottom seismograph) refraction seismic data (Liu et al., 1995; Wang et al., 1996; Lallemand et al., 1997a; Schnürle et al., 1998), seafloor mapping data (Reed et al., 1992; Lundberg et al., 1997; Liu et al., 1998), and gravity and magnetic data (e.g., Hsu et al., 1998). In 1996, EM12-Dual multibeam bathymetry, side-scan sonar imagery, six-channel seismic reflection data, and geopotential data were acquired during the 24-day ACT (Active Collision in Taiwan) cruise of the R/V *L'Atalante* around southern and eastern Taiwan (Lallemand et al., 1997a). The southernmost lines were acquired during a transit cruise between Kaoshiung and Nouméa after the acquisition of the ACT data set. The survey covers most of the offshore area that was, and is currently, deforming owing to the subduction of the Chinese continental margin beneath the Luzon arc. These newly acquired data greatly contribute to setting limits on the timing and mode of deformation through detailed mapping of the offshore structural elements, including major active faults, ridges, morphology, drainage, and collisional basins.

We address in this study several important questions: How does the Philippine Sea plate and Luzon volcanic arc deform offshore during increasing subduction of the Chinese continental margin? What is the fate of the deformed Luzon arc in advanced collisional stages (does it accrete to the Eurasian margin or does it disappear by subducting below the Philippine Sea plate)? Does the Central Range belong to the Chinese continental margin or to the Philippine Sea plate forearc domain (representing former parts of the China margin rifted apart during South China Sea opening)? How does exhumation of the metamorphic belt occur during convergence? Is the collisional process continuous, or does it occur in several distinct stages? Does the Longitudinal Valley, which was often considered as the on-land active plate boundary, extend offshore? What is the offshore equivalent of the Lichi mélange, currently described as a suture formation?

Data acquired during the ACT cruise build on a large body of previous work and provide for new interpretations of major structures and tectonic processes involved in the collision. As a result, an evolutionary model is proposed for the tectonic evolution of the Taiwan orogen for since 4 Ma.

GENERAL SETTING

Two plates are involved in the complex tectonic setting that surrounds the Taiwan orogen, the Philippine Sea plate, and the Eurasian plate (Fig. 1). Northeast of the Taiwan area, the Philippine Sea plate is being subducted beneath the Ryukyu arc, which belongs to the Chinese continental margin. The Okinawa backarc basin is developed to the north of the Ryukyu arc (Letouzey and Kimura, 1985) and is associated with active extension and recent volcanic activity (e.g., Sibuet et al., 1999). South of the Taiwan island, the oceanic lithosphere of the South China Sea subducts beneath the Philippine Sea plate, inducing the volcanism of the Luzon arc. The relative motion between the Philippine Sea plate and the Eurasian plate has resulted in the progressive subduction of the Chinese continental margin and the development of the Taiwan mountain belt.

The Philippine Sea plate

Global kinematics indicate that the Philippine Sea plate is currently moving northwestward, with respect to the Eurasian plate (Seno, 1977; Seno et al., 1993). GPS (Global Positioning System) measurements on Lanyu Island over the 1990–1995 period provide a precise N306° ± 1° azimuth for the convergence at a rate of 80–83 mm/yr relative to the Penghu Islands, which lie on the southern Chinese continental margin (Yu et al., 1997; Lallemand and Liu, 1998). The Philippine Sea plate near Taiwan consists of Eocene oceanic crust (Karig et al., 1975), and its western border is marked by the Luzon island arc. This arc and the associated Manila Trench are probably early Miocene in age, according to the oldest volcanic rocks (andesites) sampled in the Coastal Range of Taiwan by Juang

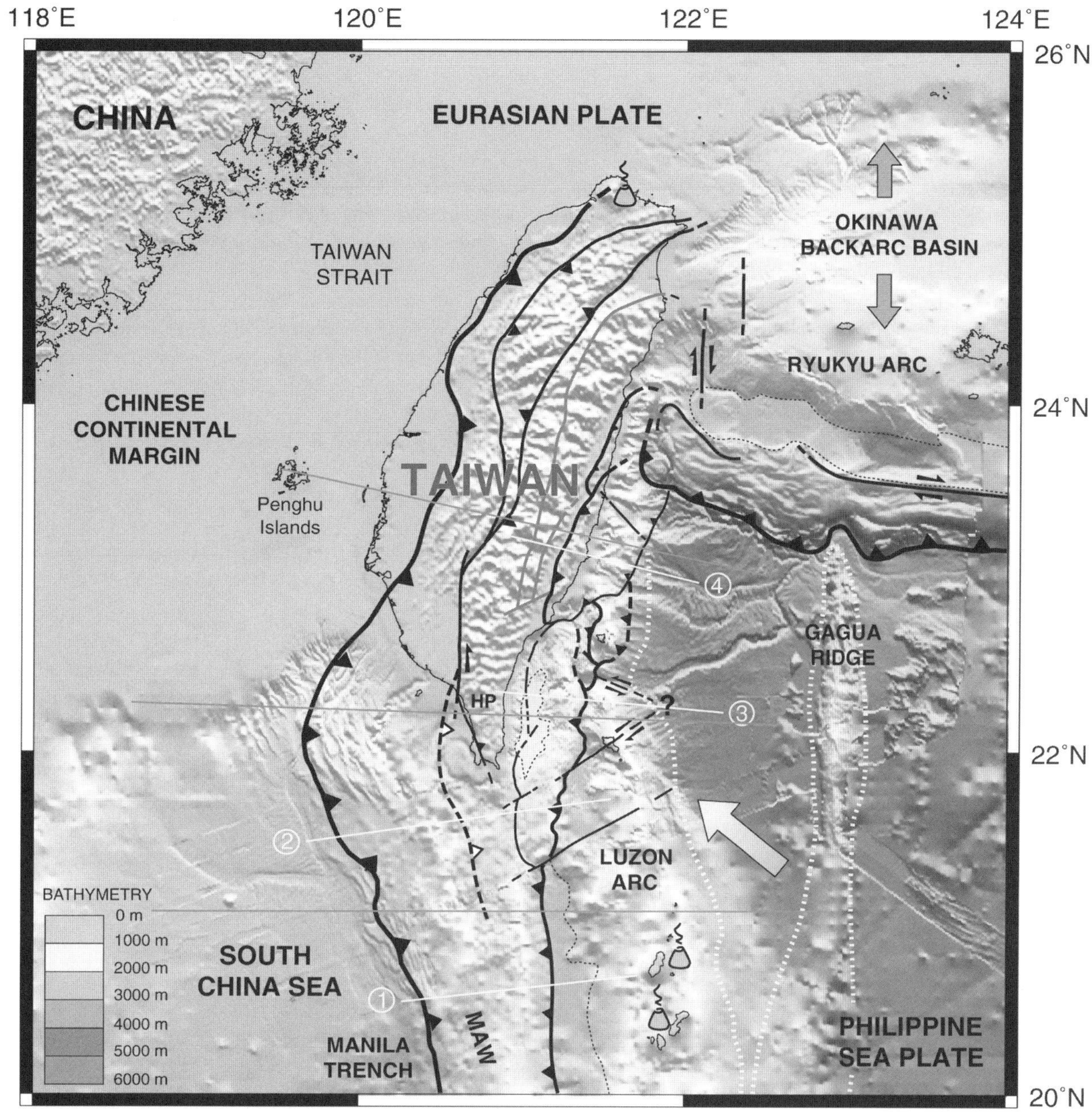

Figure 1. Geodynamic setting of the Taiwan orogen and location of the sections. The main tectonic units are shown on the shaded topographic view. MAW—Manila accretionary wedge, HP—Hengchun Peninsula.

and Bellon (1984). The northernmost present-day activity of the arc is observed at the latitudes of the Batan Islands. Younger ages, as recent as 30 ka, were also found by using K-Ar dating on Lutao and Hsiaolanyu (a small island located south of Lanyu) islands (Yang et al., 1996). The activity of the northern segment thus probably ceased during the Quaternary at the latitude of the islands of Lutao and Lanyu (Yang et al., 1996), or even farther north if younger volcanic rocks have been eroded from the uplifted Coastal Range (Richard et al., 1986).

A progressive increase in continental contamination of andesitic magmas correlates with progressively younger ages in Coastal Range volcanic rocks. This progression could reflect the systematic increase through time of continentally derived sediment entering the Manila Trench (Dorsey, 1992) or simply the transition from oceanic to continental subduction, shortly before the cessation of island-arc magmatism.

South China Sea, Manila Trench, and Chinese margin

The South China Sea opened during Oligocene–Miocene time (32 to 15 Ma). Magnetic lineations indicate that roughly north-northwest–south-southeast spreading migrated southwestward (Briais and Pautot, 1992). The fossil spreading ridge is currently subducting west of Luzon island (Pautot and Rangin, 1989). The oceanic lithosphere of the South China Sea was subsequently subducted beneath the Philippine Sea plate along the Manila Trench in the early Miocene, as suggested by the Neogene age of associated volcanism along the Luzon arc (Ho, 1986). The Manila accretionary wedge has developed between the two plates (Fig. 1) and is still growing today south of 21°20′N (Hayes and Lewis, 1984; Lewis and Hayes, 1984; 1989). North of 21°20′N, the wedge becomes incorporated in the arc-continent collision domain, involving progressively lower to middle Miocene slope and trench sediments (Reed et al., 1992). The southern tip of Taiwan Island, called the Hengchun Peninsula, extends offshore southward in the form of the Hengchun Ridge as far as 20°30′N (Fig. 1). This ridge is generally considered as the uplifted internal domain of the oceanic accretionary wedge currently undergoing the effects of the collision (Lundberg and Dorsey, 1990; Reed et al., 1992). A scenario for the tectonic evolution of the submarine accretionary wedge has been proposed by Huang et al. (1997) with the Kaoping slope west of the subduction wedge of the Hengchun Ridge representing the part of the prism associated with the active collision zone (e.g., Huang et al., 1997; Liu et al., 1997).

The continental shelf of the Eurasian plate bounds the South China Sea to the northwest. The shelf in the Taiwan Strait rarely exceeds 100 m depth and represents the flexural foreland basin of the Taiwan belt; up to 5 km of recent orogenic sediments has been deposited in former Cenozoic basins of the margin (e.g., Sibuet and Hsu, 1997). Indeed, in the early Tertiary, normal faulting resulted in the formation of Paleogene grabens along the Asian margin. Horst and graben structures, bounding these local basins filled with various thicknesses of sedimentary sequences, have played an important role in the development of the Taiwan thrust wedge (Suppe, 1988; Lu et al., 1998). The strata of these basins are currently exposed in the Western Foothills of Taiwan (Suppe, 1980; Davis et al. 1983), and some of the history and tectonic events of the arc-continent collision are registered in the young and nonmetamorphosed strata of the thrust wedge (Teng et al., 1991; Déramond et al., 1996). The transition between oceanic and continental crust in the Eurasian plate is marked approximately by the 2500 m isobath (Letouzey et al., 1988). Although being oriented approximately N70°E to the southwest of Taiwan, the present trend of this boundary is not regular, which suggests that before collision, the margin could have presented a step-like geometry in the Taiwan area (Pelletier, 1985). If such a situation has occurred, the southward propagation of collision would not have been as continuous a process as is commonly believed.

The Taiwan orogen

The geologic setting of the Taiwan collision belt can be summarized as follows. The sedimentary cover of the Chinese continental margin (see Taiwan Strait in Fig. 1) was accreted in the northwestern part of the island, and it is still accreting in the southwestern part, against a backstop formed by the pre-Tertiary rocks of the Central Range (e.g., Lu and Malavieille, 1994). Shallow-marine sequences of the passive continental margin as well as foreland sequences constitute the deformed units of the Coastal Plain, Western Foothills, and Hsüehshan Range (Ho, 1982, 1988). During convergence, they were progressively accreted to the collision prism along a series of east-dipping thrusts.

The Central Range includes the Eocene and Miocene (but not Oligocene) metamorphic Backbone Range. The Lishan-Laonung-Hengchun fault, west of the subduction wedge, is a high-angle west-dipping reverse fault (Lee et al., 1997), which separates the western Backbone Range from the Eocene–Oligocene units of the Hsüehshan Range, both of them forming the slate belt (Ho, 1986; Angelier et al., 1990; Teng et al., 1991; Clark et al., 1993; Tillman and Byrne, 1995). The Western Foothills correspond to a fold-and-thrust belt affecting Oligocene–Miocene strata overlain by a 4-km-thick sequence of Pliocene–Quaternary molasse (Lu and Hsü, 1992). This foreland thrust belt is inactive north of 24°N, but it is still growing south of this latitude as shown by GPS measurements (Yu et al., 1997). The Central Range is bounded to the east by the Longitudinal Valley, which separates the Central Range from the Coastal Range (i.e., northernmost segment of the Luzon volcanic arc). Deformation of the Philippine Sea plate lithosphere within the Coastal Range is suggested by the 50-km-thick seismogenic zone beneath the arc domain (Wu et al., 1997). The Coastal Range is still well coupled to the Philippine Sea plate, as attested to by the GPS velocities of 63 ± 9 mm/yr in the direction of N314° ± 8°, except north of 23°40′N where rates dramatically drop to 8–43 mm/yr along the azimuths N292°–352° (Yu et al., 1997). About 8 cm/yr of plate convergence is taken up in the offshore area north of 24°N, whereas south of 24°N, the convergence is almost entirely distributed across the island, with 3 cm/yr accommodated in the Longitudinal Valley area (Angelier et al., 1997) and across the other thrusts of the Coastal Range shown in Figure 1 (Yu et al., 1997). Lundberg and Dorsey (1990) estimated the uplift rate of Quaternary marine terraces along the Coastal Range to be in excess of 6 mm/yr.

MARINE OBSERVATIONS AND DATA ACQUISITION

The ACT cruise and the transit cruise between Kaoshiung and Nouméa aboard the R/V *L'Atalante* mapped an area of about 67 500 km^2 at 10 knots during 15 days with 100% bathymetric and backscattering coverage (Fig. 2A). Tracks were generally oriented parallel to the structures (Fig. 2B) in order to keep the swath width constant and because most of the previously acquired seismic lines were shot normal to the structural fabric. The ship is equipped with SIMRAD EM12-Dual and EM950 (for depths shallower than 300 m) multibeam systems that enable swath mapping and side-scan imagery over a max-

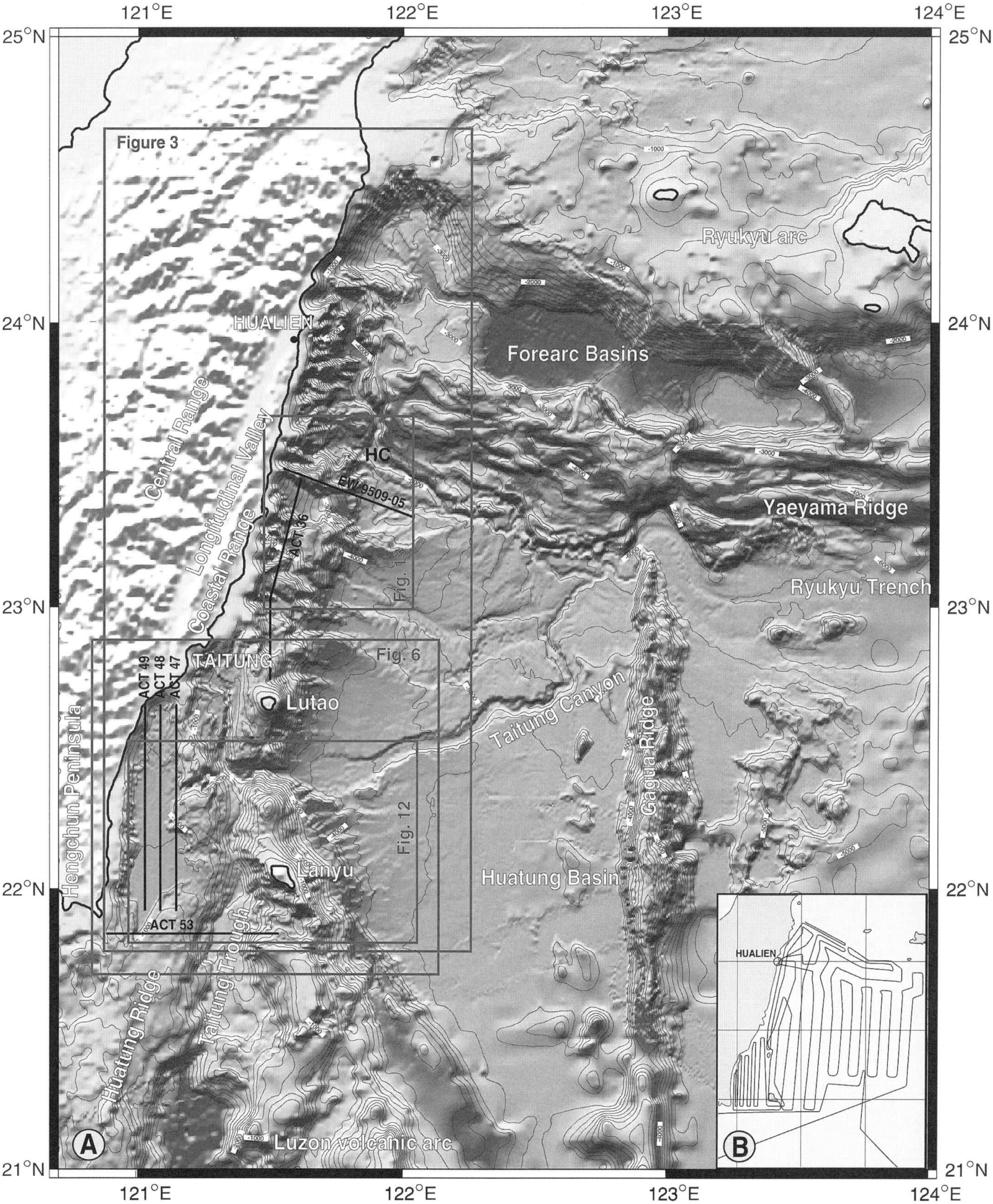

Figure 2. (A) Bathymetry of the collision zone offshore east of Taiwan. Location of the main morphotectonic units and the seismic lines described in the paper are shown. HC—Hualien Canyon. (B) Ship tracks.

imum 20-km-wide strip (151 simultaneous soundings) of seabed in a single pass. The side-scan imagery associated with the EM12-Dual system gives detailed information on the acoustic reflectivity, which is related to small-scale bathymetric features and to variations in the nature of the seafloor. Subbottom (3.5 kHz) and reflection seismic profiling and magnetic and gravity data were also recorded along the 20-km-spaced ship tracks (for great depths) and along more closely spaced tracks at shallower depths. Six-channel streamer was deployed with two 75 cubic inch GI guns at a pressure of 160 bars. The guns were fired in harmonic mode to generate a source signature centered on 20 Hz to enhance subbottom penetration of energy. Shot intervals were ~50 m. Seismic data were processed with ProMAX software to obtain poststack time-migrated profiles.

FROM INCIPIENT TO MATURE ARC-CONTINENT COLLISION OFFSHORE EAST TAIWAN

Arc-continent collision is classically considered to be exemplified by the subduction of the continental margin of China under the Luzon island arc (Fig. 1). The transition from oceanic to continental subduction along the Manila Trench occurs near 22°N (Reed et al., 1992), but the impact on deformation of this major change in the nature of the two plates is observed as far south as 21°N.

Main structural elements

The main tectonic features are defined on the tectonic maps (Figs. 1 and 3) and shaded bathymetric map (Fig. 4). Two structural domains constitute the collisional area:

A southern domain is composed of four morphotectonic structures that characterize the domain of incipient collision: (1) the Southern Longitudinal Trough south of the on-land Longitudinal Valley, (2) the elongated Huatung Ridge forming the eastward dam of the Southern Longitudinal Trough, (3) the Taitung Trough, marked by a sinuous V-shaped valley, and (4) the Luzon volcanic arc including the islands of Lanyu and Lutao.

A northern domain represents the area of well-developed collision; it includes (1) the Coastal Range and its offshore eastern flank, (2) the western termination of the Ryukyu accretionary wedge, and (3) the complex area that marks the junction between the collision zone and the Ryukyu Trench in which the Hoping Basin has developed.

The boundary between the two structural domains is characterized by a transfer zone of deformation that accommodates their different mechanical evolution. Figure 3 summarizes the major tectonic features observed and the kinematics of the main units given by the GPS data of Yu et al. (1997).

Four east-west geologic cross sections (Fig. 5; drawn with no vertical exaggeration; locations in Fig. 1) are presented; they are based on the analysis of data from the ACT cruise. These sections illustrate the major structural changes occurring from the south to the north because of the increasing involvement of the Chinese continental margin in the subduction and collision. The interpretation of the structure at depth proposed for the four sections and the deformation mechanisms involved are explained and discussed subsequently.

Section 1, based on interpretations in Reed et al. (1992), is located around 21°N and shows the typical geometry of the Manila accretionary wedge, with backthrusting at the rear that controls the development of the North Luzon Trough forearc basin. North of 21°20′N, the size, morphology, and structure of the accretionary wedge changes significantly owing to the first effects of the collision. The forearc basin is shortened in between the rear of the wedge and the volcanic edifice of the arc.

Section 2, redrawn from Reed et al. (1992), shows the structure of the wedge. Three main morphotectonic units appear: (1) a lower-slope wedge with a very low slope angle, (2) an uplifted domain of the wedge (the Hengchun Ridge) bounded by a slope break and characterized by a steeply dipping upper slope, (3) a smaller ridge, composed of deformed strata of part of the former wedge and forearc sedimentary deposits, separated from the rear of the wedge by a narrow north-south-oriented valley (see also in Fig. 1). Reed et al. (1992) suggested significant shortening of the central domain of the wedge, with out-of-sequence thrusting at the boundary between upper and lower slopes and major backthrusting in the forearc domain. North of 22°N, the Hengchun Peninsula represents the emerged continuation of the submarine Hengchun Ridge. At this latitude, a large part of the slope sediments from the continental margin is involved in the accretionary wedge.

Section 3 shows the development of a Quaternary collisional basin (the Southern Longitudinal Trough) between the eastern flank of the Hengchun Peninsula and the Huatung Ridge overthrusting onto the Luzon arc (Lundberg and Dorsey, 1988) (Fig. 1). Along this section again, major shortening occurs today in the forearc domain (Lundberg et al., 1997). The Southern Longitudinal Trough is partly filled with sediments coming from the erosion of the emerged Taiwan mountains (Lundberg and Dorsey, 1988). It is not the northern equivalent of the North Luzon Trough forearc basin, which is filled with arc-derived volcaniclastic sedimentary deposits and which ends south of the Huatung Ridge at about 21°20′N. From 22°40′N to about 24°N, the collision is completed and the forearc domain is drastically reduced.

Section 4 shows the active westward thrusting of the Coastal Range on the formerly exhumed metamorphic rocks of the Central Range. The on-land Coastal Range and its offshore slope continuation are composed of offscraped parts of the former volcanic arc edifice, intra-arc basins, and strongly deformed forearc strata. This tectonic unit may be decoupled from the Philippine Sea plate oceanic basement by an west-dipping thrust located close to the base of the slope (Fig. 1).

Deformation in the southern structural domain

The Southern Longitudinal Trough and surroundings. The Southern Longitudinal Trough (Fig. 4) is a proximal orogenic

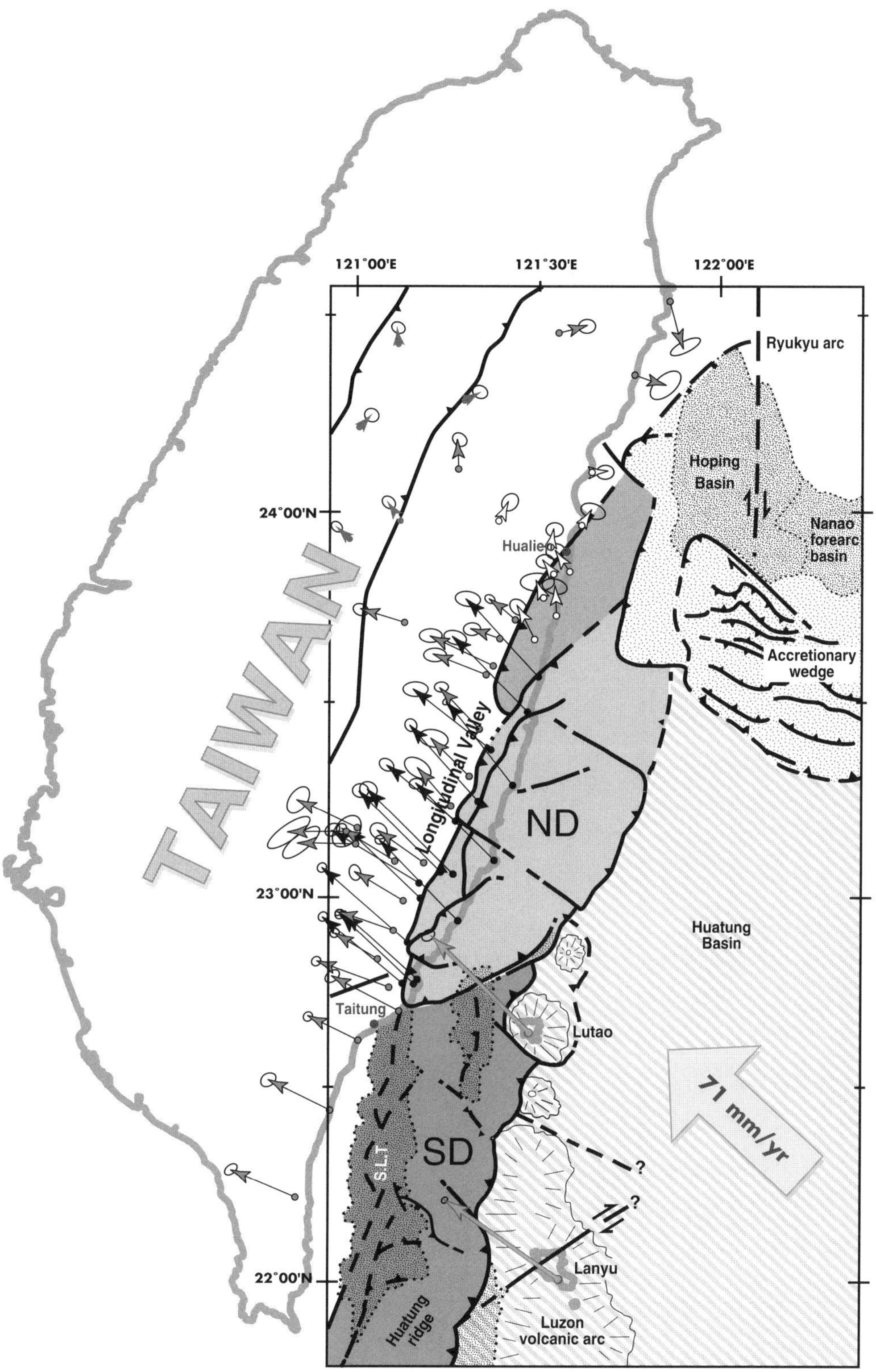

Figure 3. Schematic tectonic map of the study area showing the kinematics of the main tectonic units (GPS data after Yu et al. [1997]). The 95% confidence ellipse is shown at the tip of each velocity vector. ND—northern tectonic domain, SD—southern tectonic domain.

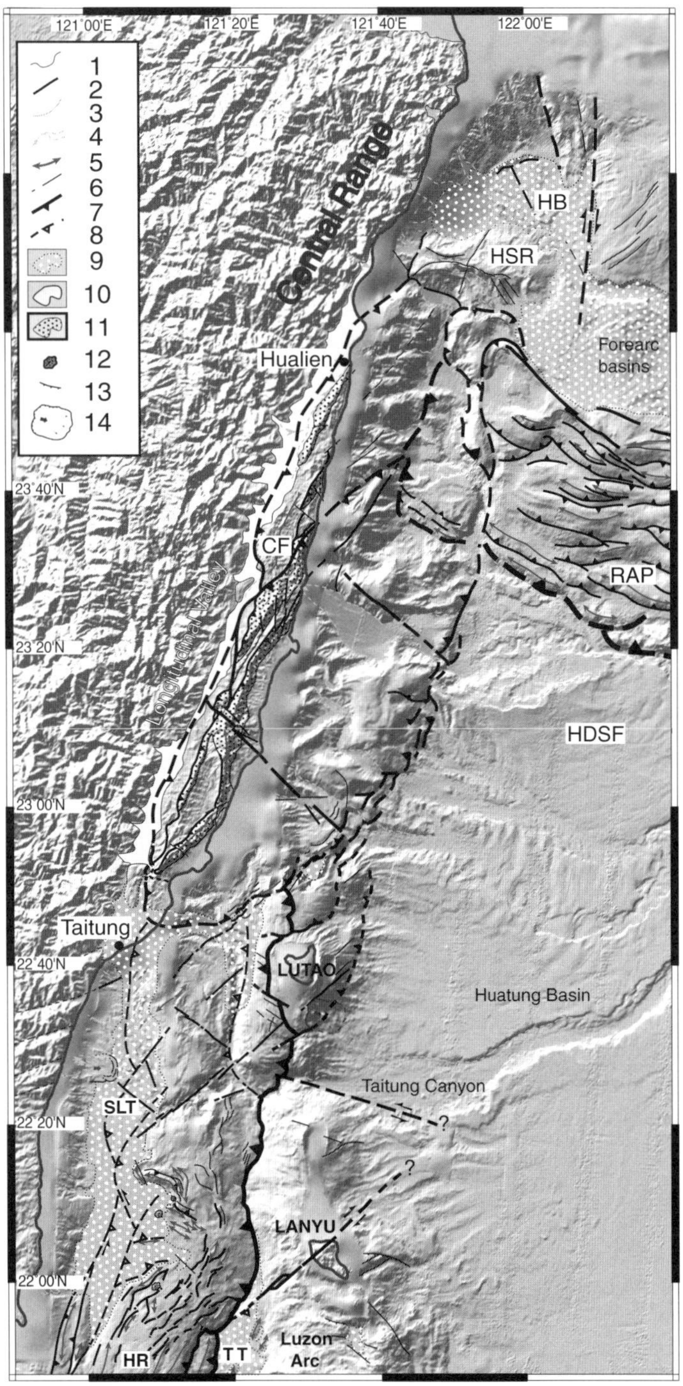

Figure 4. Structural map of the study area offshore Taiwan. Geologic features are drawn on the bathymetry (shaded view, illumination from N310°E). Symbols and patterns: 1—geologic boundary, 2—fault, 3—large-scale mud waves, 4—drainage, 5—fold axis, 6—minor fault, 7—thrust (teeth on upper plate), 8—inactive thrust (teeth on upper plate), 9—recent collisional basin, 10—Quaternary deposits of the Longitudinal Valley, 11—volcanic rocks of the Coastal Range, 12—mud volcano, 13—normal fault, 14—mass wasting. SLT—Southern Longitudinal Trough, HR—Huatung Ridge, TT—Taitung Trough, CF—Chimei fault, HSR—Hsingsen Ridge, HB—Hoping Basin, HDSF—Hualien Deep-Sea Fan, RAP—Ryukyu accretionary prism.

basin developed in a forearc position (e.g., Lundberg and Dorsey, 1988; Lundberg et al., 1997; Fuh et al., 1997). The trough is elongated in a north-south direction between 21°50′N and 22°40′N. Its shape is complex; it is about 90 km long and 15 km wide, and it narrows toward the south. The seafloor of the trough deepens from 800 m to the north to 1300 m to the south. It is bounded to the west by the 10°E-dipping structural slope of the Hengchun Peninsula and to the east by an antiformal, narrow ridge that extends westward from the Huatung Ridge near 22°05′N and that is about 600 m higher than the seafloor of the Southern Longitudinal Trough.

Today, sediments are mainly supplied from the subaerial peninsula through numerous east-trending gullies. Most of the sediments coming from the Central Range through the Longitudinal Valley are dumped into the Taitung Canyon (see north-looking perspective view in Fig. 6), but some of them are collected by the tributaries of another canyon, which begins in the northern part of the Southern Longitudinal Trough, turns sharply east, meanders across the Huatung Ridge near latitude 22°20′N, and finally flows into the Taitung Canyon position (Fig. 4) (Lundberg and Dorsey, 1988; Lundberg et al., 1997). The Central Range source for Holocene orogenic sediments was confirmed by box coring into the basins conducted during two cruises of the R/V *Ocean Researcher I* in 1988 and 1989 (Huang et al., 1992). Samples cored in the southern part of the Southern Longitudinal Trough consisted of hemipelagic mud with slate chips, and samples cored at the head of the Taitung Canyon contained angular metamorphic detritus. The Taitung Canyon crosses the Luzon arc and the Huatung Basin, reaches the Ryukyu Trench, and finally emerges at a distal fan east of the Gagua Ridge (see the detailed study by Schnürle et al., 1998). The flat floor of the Taitung Canyon is strongly reflective in backscattering images, suggesting the presence of coarse sediments or gravels (Fig. 7). The east-trending gullies on the western border of the Southern Longitudinal Trough are also reflective, whereas the braided channels on the floor of the Southern Longitudinal Trough are much less reflective. This difference could be interpreted as marking a slope break at the heads of the gullies, which provide sediments to the channels, or may be caused by the change in slope steepness.

Migrated, six-channel, seismic reflection profiles across the Southern Longitudinal Trough, acquired during the MW9006 cruise of the R/V *Moana Wave* in 1990, reveal that much of the basin is filled by growth strata, which record relative uplift of the Huatung Ridge, forming the dam of this elongated basin (Lundberg et al., 1992, 1997). Uppermost strata are nearly flat lying, but with increasing depth, the strata dip westward, documenting progressive tilting, probably resulting from uplift of the eastern part of the basin. Gently west-dipping thrusts have been imaged from east-west seismic lines (Lundberg et al., 1997).

Figures 8, 9, and 10 represent time-migrated, six-channel, seismic lines ACT 49, ACT 48, and ACT 47, respectively, recorded along the Southern Longitudinal Trough (see Fig. 2 for

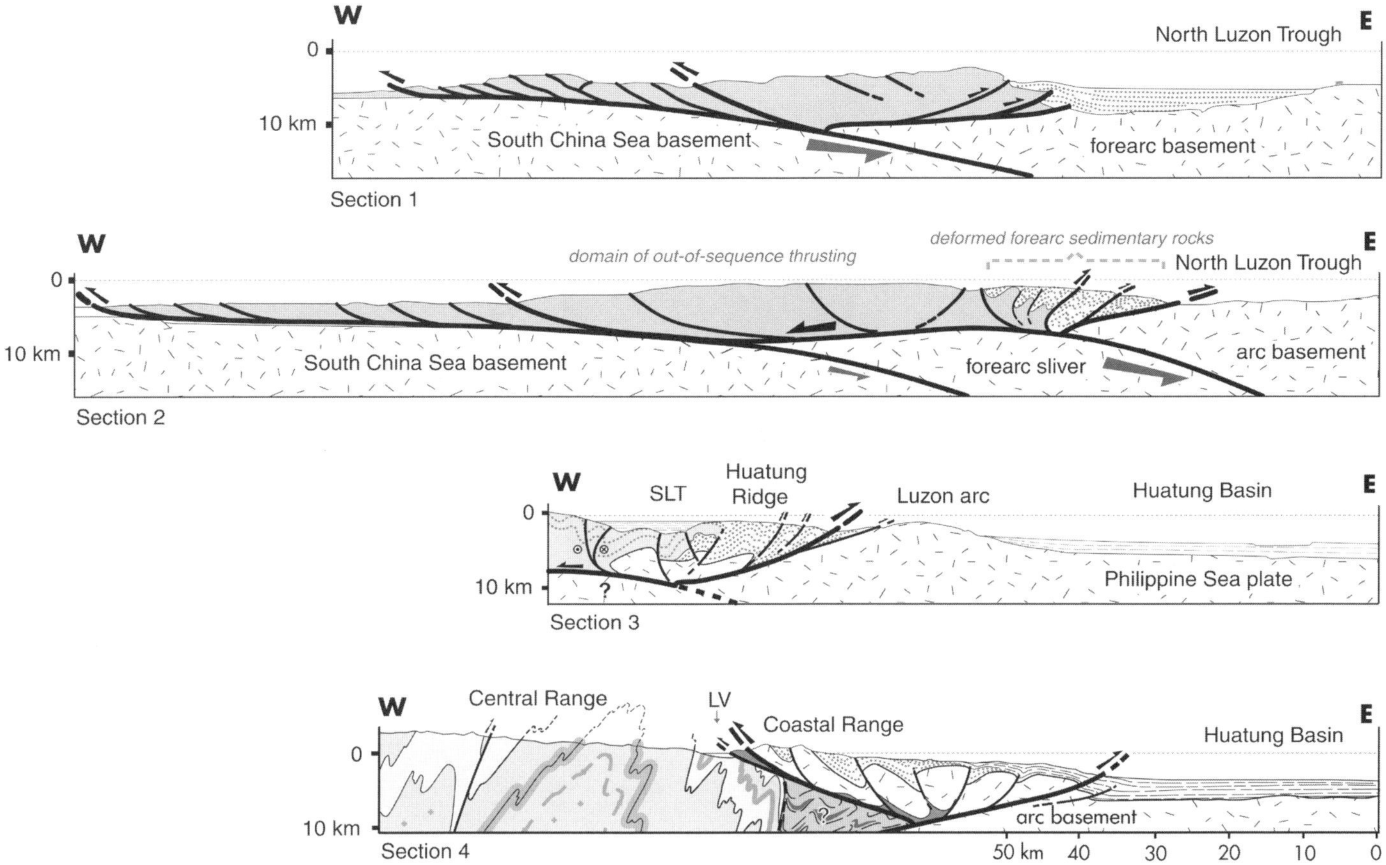

Figure 5. Geologic cross sections (all are at the same scale with no vertical exaggeration). Sections 1 and 2 are based on Reed et al. (1992). See location of sections in Figure 1. LV—Longitudinal Valley; SLT—Southern Longitudinal Trough.

location). These three parallel north-south lines illustrate the complex structure of the trough. The line ACT 49 (Fig. 8) follows the western part of the basin. The upper strata of the basin are relatively flat, onlapping the Huatung Ridge to the south, whereas at depth, the lower strata overlying an irregular basement are gently deformed. An unconformity may separate the two sedimentary sequences. On the line ACT 48 (Fig. 9), located on the median part of the trough, the thickness of the younger strata above the unconformity remains the same, but the underlying sequences are thicker and more deformed than those from the western side of the trough. The lower sequences of the basin appear to be contained in three minor troughs: a southern one, which shows folding between two basement structures, a relatively undeformed middle trough dipping north, and a northern trough, with a gentle syncline structure, separated from the previous basin by a deformed zone of uplifted basement. ACT 47 line (Fig. 10) crosses the eastern part of the Southern Longitudinal Trough and shows that the undeformed sequences in the upper part of the basin exist only in the southern trough (see also Fig. 4). These sequences lie unconformably on top of older, folded and faulted sequences of the former trough. In the northern domain, the lower sequences have undergone complex deformation involving thrusting.

Our observations suggest that the Southern Longitudinal Trough was developed in a complex tectonic setting during at least two different stages. First, the trough was represented by an early stage of basin development in an area of active shortening. The shape of the thrusts leads us to interpret them as positive flower structures, and the orientation of the folds oblique to the north trend of the trough strongly suggest that deformation occurred between two growing ridges in a domain of sinistral transpressional tectonics, probably as a result of the oblique convergence of the Philippine Sea plate. A more recent second stage occurred during which the sediments were deposited unconformably on top of the earlier-deformed strata. This second stage occurred after the main region of deformation migrated eastward toward the Huatung Ridge.

After correlation of the strata on both the east-west and north-south lines at intersecting points, we observed that the "nearly flat-lying" uppermost strata along east-west transects are gently folded along axes oblique to the north trend of the basin (Fig. 6); the strata locally dip about 10°. These units probably represent the younger generation of growth strata recording recent deformation.

The structural analysis shows that the present collisional basin is tilted to the west because of the growing Huatung Ridge

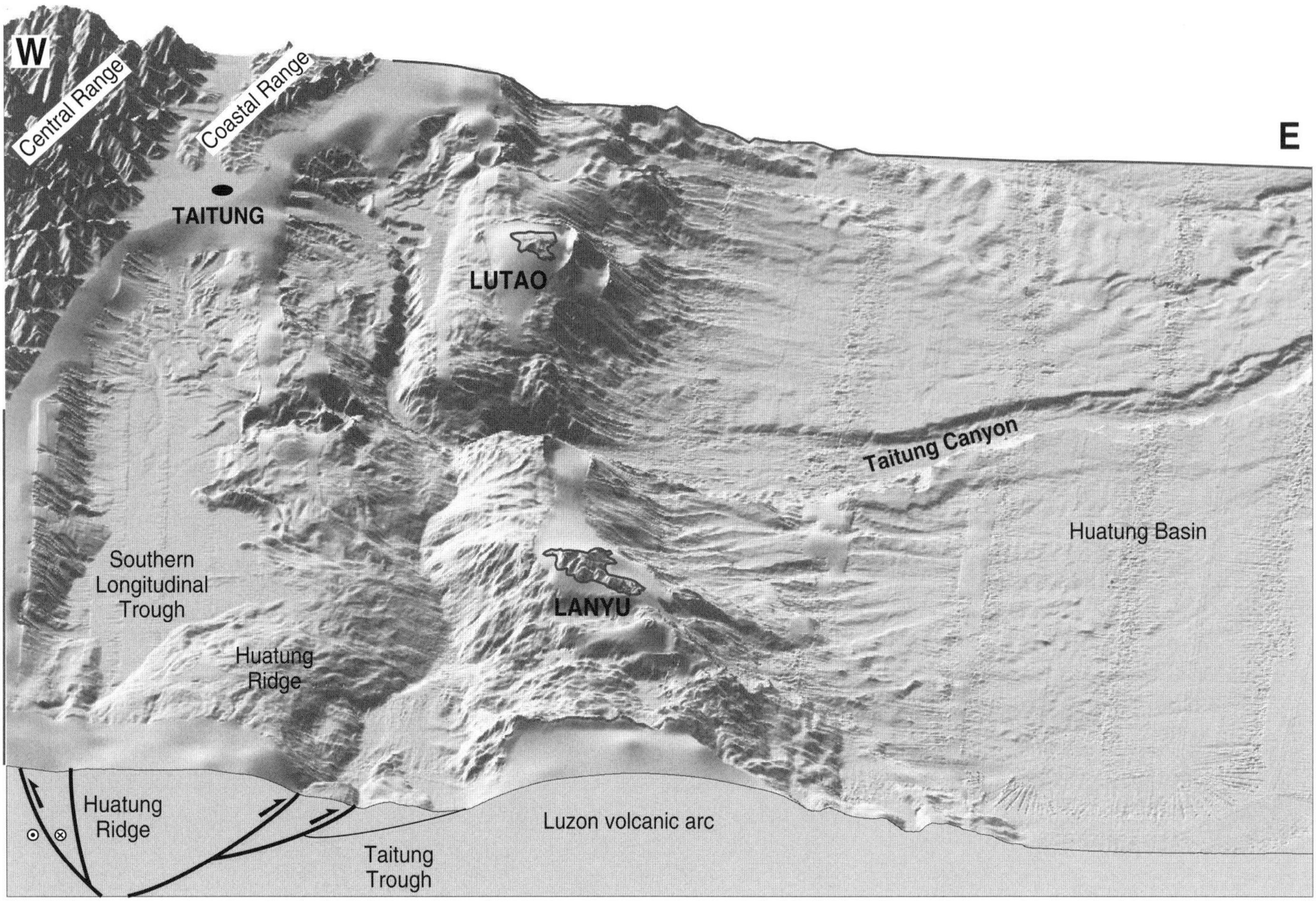

Figure 6. Shaded hill relief of the southeastern area showing front-view cross section and the main morphostructural features (perspective view looking north).

on its eastern border, but it is also deformed by transverse structures. Such deformations are likely due to the local effects of the oblique indentation of the growing Huatung Ridge by the irregular western slope of the Luzon island arc.

As the Longitudinal Valley is currently characterized by east-dipping thrusting (e.g., Angelier et al., 1997), we expected before the cruise to find the same active deformation in the Southern Longitudinal Trough. Despite the morphological continuity between the Longitudinal Valley on land and the offshore trough, the most surprising observation was the absence of any sign that could reveal the activity of east-dipping reverse faults in the Southern Longitudinal Trough (Lundberg et al., 1997). To the contrary, we observe that (1) the eastern flank of the Hengchun Peninsula is continuous from onshore to offshore (10°E-dipping structural slope), (2) the orogenic north-south-elongated sedimentary basin follows the N30°E-trending Longitudinal Valley basin but is 1000 m deeper south of Taitung, and (3) the Southern Longitudinal Trough is controlled by the east-verging antiformal structures of the Huatung Ridge, not by equivalents of the east-dipping thrusts of the Coastal Range.

The Huatung Ridge. The preceding observations imply that at least the upper layers of the Huatung Ridge consist of growing orogenic strata of the Southern Longitudinal Trough. Huang et al. (1992) noticed that, despite the morphological continuity between the Coastal Range and the Huatung Ridge, the ridge exhibits negative values of both free-air and magnetic anomalies (Liu et al., 1992). They thus concluded that it consists mainly of a sedimentary mass. Reed et al. (1992) suggested that the deformed forearc basin forms an important constituent of the Huatung Ridge. Samples cored on the Huatung Ridge revealed sandstone and mudstone subangular cemented blocks within a sheared mudstone matrix, similar to the onshore "Lichi mélange" (see the geologic map of eastern Coastal Range [Huang et al., 1993]). Part of the Huatung Ridge could be the offshore equivalent of the Lichi mélange, which is interpreted as an olistostromal syntectonic mélange formed during the arc-continent collision (e.g., Page and Suppe, 1981; Huang et al., 1992).

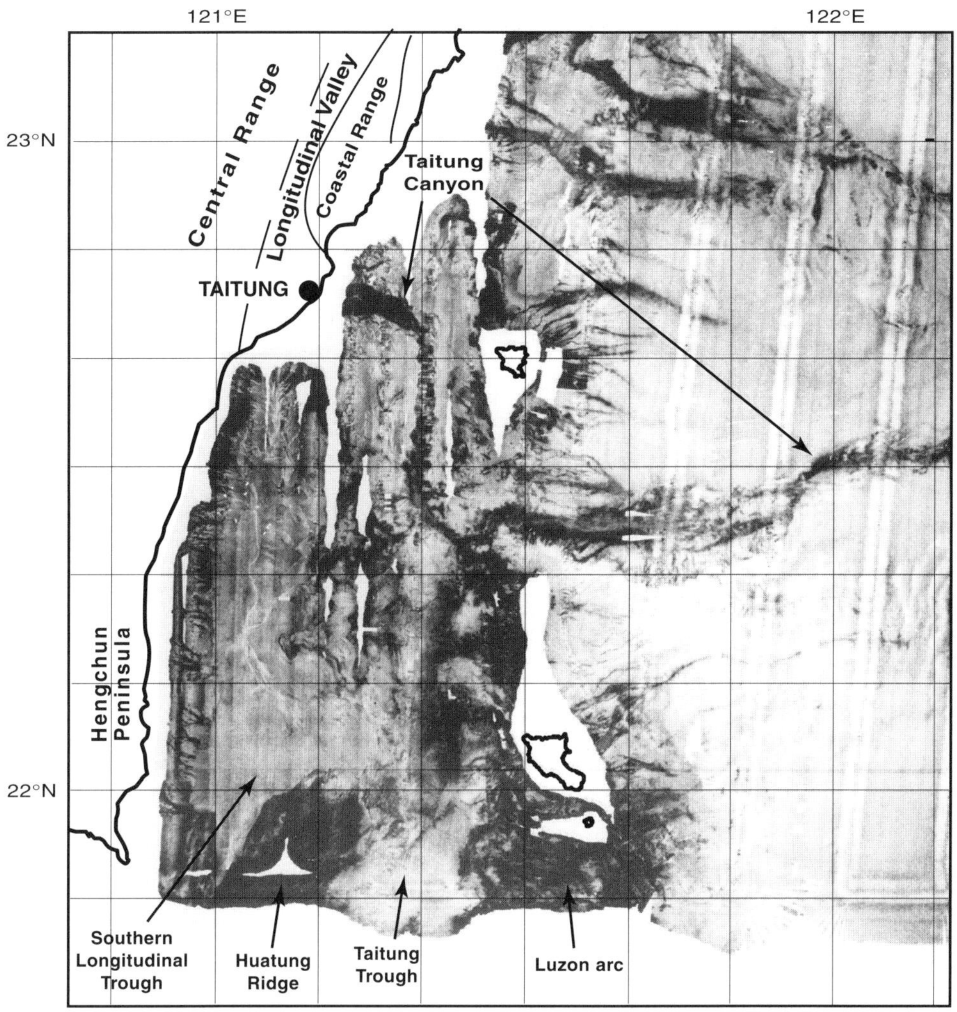

Figure 7. Side-scan sonar imagery of the study area. Note the strong reflectivity (black) of the Taitung Canyon floor, top of Huatung Ridge, and Luzon volcanic arc and the low reflectivity of the small channels of the Southern Longitudinal Trough and of the southeastern steep slope of the Huatung Ridge.

The ridge is not continuous. North of 22°02′N, it trends north, is 20 to 30 km wide, and reaches 600 m in relief with respect to the adjacent Southern Longitudinal Trough floor (Fig. 4). South of 22°02′N, it trends N20°E and is 25 km wide, and its top is 500 m higher than the adjacent basin floor. The northern part shows a complex seafloor reflectivity whereas the top of the southern part is strongly reflective (Fig. 7), suggesting the presence of outcrops of coarse sediments or well-lithified sediments.

The moderately reflective eastern flank of the Huatung Ridge is steeper and higher than the western flank. Its height reaches a water depth of 1200 m at 22°10′N and increases to 2400 m to the north and to 2100 m to the south. The average slope is 10° but reaches 20° in the northern part of the ridge (Fig. 4). South of the study area, the east-west ACT 53 seismic profile (Fig. 11) confirms that the ridge is thrusting over the Taitung Trough termination. In fact, such backthrusting was formerly observed during the POP2 cruise onboard the R/V *Jean Charcot* in 1984 (J.F. Stéphan, personal communication, 1996) (Pelletier, 1985) and later during the R/V *Moana Wave* MW9006 cruise in 1990 (Lundberg et al., 1997).

South of 21°30′N, the Huatung Ridge disappears. The Hengchun Ridge directly dams the arc sediments in the North Luzon Trough, which is the forearc basin associated with the Manila Trench subduction zone. Backthrusting at the rear of the wedge controls the development of the forearc basin (section 1 in Fig. 5) (Reed et al., 1992; see also Larroque et al., 1995, for the mechanisms of basin development). At 22°N, the Taitung Trough narrows drastically and extends northward into a V-shaped valley. The seafloor in the trough shallows symmetrically northward and southward from a depth of 3000 m toward a pass, 2000 m deep, at 22°10′N (Figs. 1 and 4). This morphology, together with the westward tilting of the overlying strata of the Southern Longitudinal Trough, shows that the Huatung Ridge is thrusting over the Luzon volcanic arc at least between 22°N and 22°25′N.

The general map of Figure 1 and the ACT 99 line acquired during the transit between eastern and western studied domains

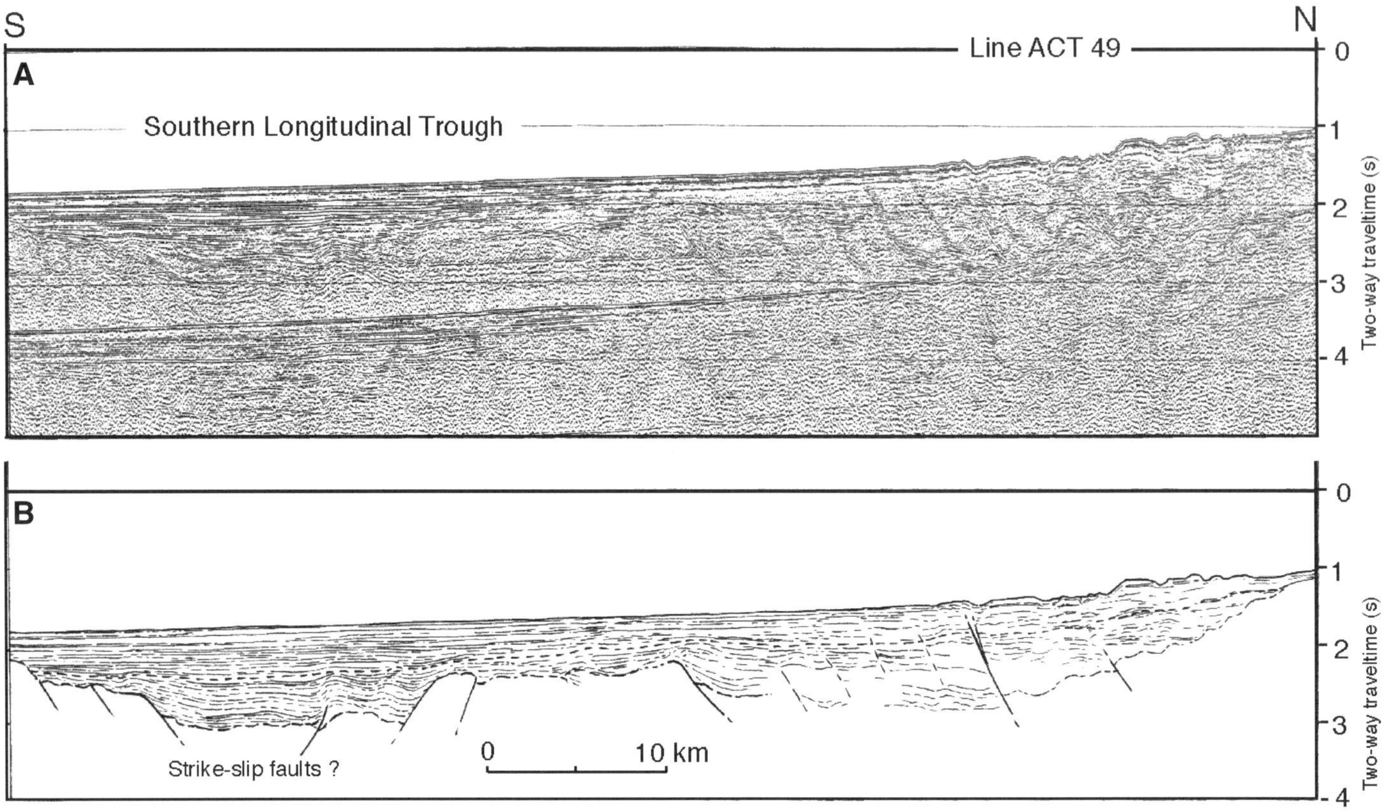

Figure 8. North-south seismic profile across the eastern part of the Southern Longitudinal Trough. (A) Migrated six-channel profile. (B) Line drawing showing the main geologic features.

shows the same backthrusting of the Huatung Ridge more to the south, down to 21°20′N (also described in Lundberg et al. [1997]). North of 22°25′N, the Huatung Ridge widens and is cut by the head of the Taitung Canyon, which extends southward to 22°25′N where it turns east and crosses the volcanic edifice to join the Huatung Basin to the east. The north-trending part of the canyon cuts into the flat-lying sediments of a recent basin, which previously filled the narrow trough developed between the Huatung Ridge and the volcanic edifice of Lutao Island. The thick sequence of sediments in this basin remain tectonically undisturbed as revealed by east-west reflection seismic lines (Huang et al., 1992) and north-south seismic profiles (this study). At about 22°22′N, a sinuous canyon incises the Huatung Ridge and joins the east-trending segment of the Taitung Canyon (Fig. 12).

These last observations suggest that early in the history of the Huatung Ridge formation, most of the sediments coming from the Central Range, via the Longitudinal Valley and the former paleo–Taitung Canyon, filled the developing Southern Longitudinal Trough and then crossed the ridge and the volcanic arc to be deposited in the Huatung Basin. It seems that because of the rapid growth of the Huatung Ridge and the increasing thrusting to the east over the arc, the paleo–Taitung Canyon was progressively abandoned or became less efficient. Its new bed was uplifted, resulting in the north-south-oriented canyon currently crossing the Taitung Basin and deeply incising the strata.

The ACT data show that the shortening between the Luzon arc (islands of Lutao and Lanyu) and the Hengchun Peninsula of Taiwan is mainly absorbed along the west-dipping thrust beneath the Huatung Ridge; however, the shortening is partially accommodated by folding and thrusting within the ridge, as previously mentioned in Lundberg et al. (1997). South of Taitung, the difference between GPS data obtained on the arc islands (Lutao and Lanyu) and on east coast of the Hengchun Peninsula confirms that about 4 cm/yr (half of the total convergence) (Yu et al., 1997) is accommodated offshore.

Deformation of the northern structural domain

Offshore deformation east of the Coastal Range. On land, two main tectonic units are characterized by active deformation. The boundary between the southern and northern units is marked by the southeast-dipping Chimei fault (Fig. 4).

Offshore, east of the Coastal Range, the slope is actively deforming. Minor northwest-trending strike-slip faults (probably left lateral), west-dipping thrusts, and north-northeast–trending folds characterize the deformation (Figs. 13 and 14). About 1.7 cm/yr should be accommodated there as indicated

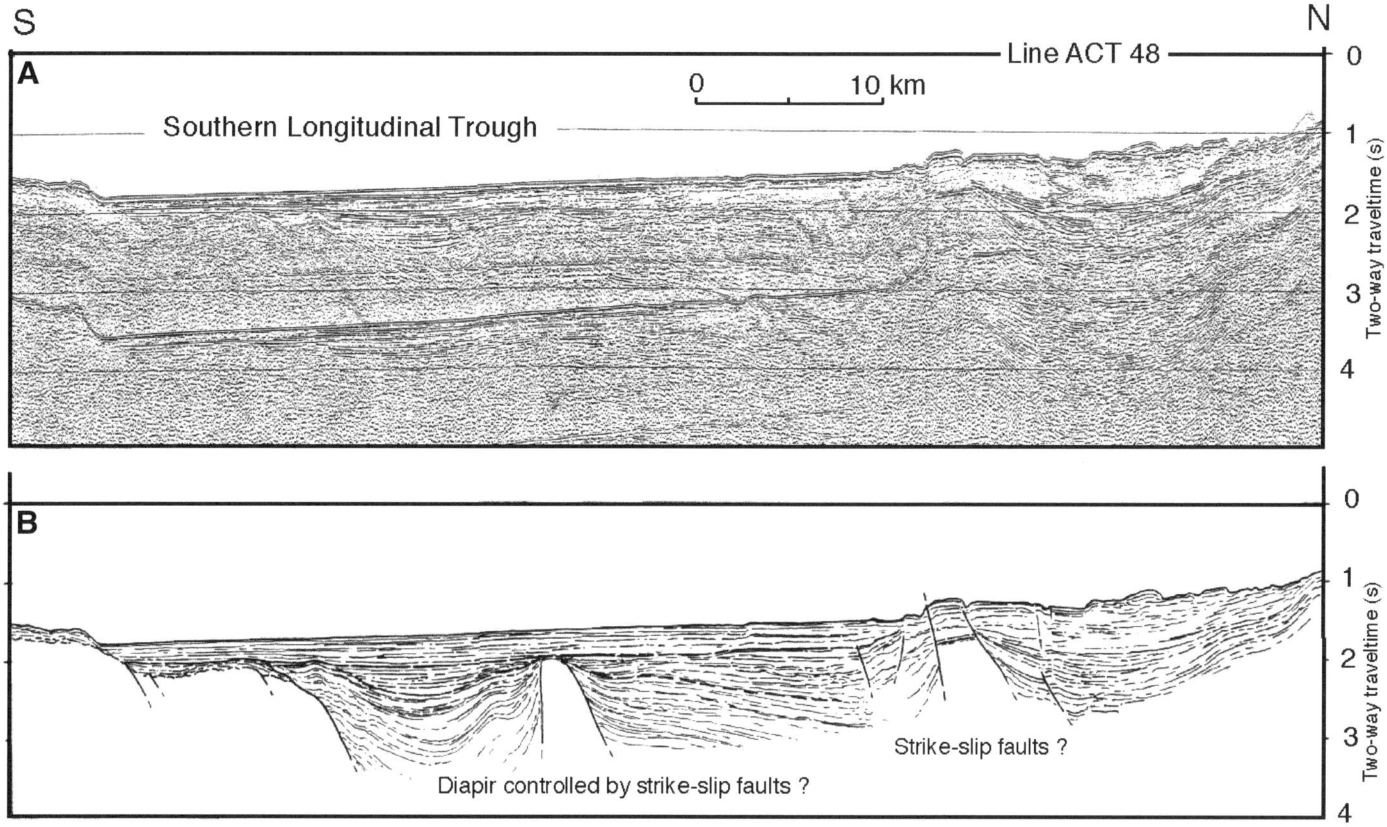

Figure 9. North-south seismic profile across the central part of the Southern Longitudinal Trough. (A) Migrated six-channel profile. (B) Line drawing showing the main geologic features.

by published geodetic data (Yu et al., 1997), probably mainly along thrust faults marked by the steep scarps cutting the range to the east. A seafloor vertical offset of about 500 m is observed downstream of the flat-floored Chimei Canyon (see the perspective shaded view Fig. 13). It has been interpreted from the seismic line EW 9509-05 (Fig. 14) (Liu et al., 1998), slightly oblique to the scarp, as caused by west-northwest–dipping reverse faults. North-south ACT seismic profiles show that most of the offshore slope is covered by sediments that could be derived from the Coastal Range and deposited in former intra-arc or forearc basins. The strata are locally folded and faulted, showing that the whole unit has undergone shortening.

The submarine Hualien Canyon flows roughly north-south through the Ryukyu accretionary wedge and follows the trench where it runs eastward joining the Taitung Canyon close to the Gagua Ridge (Fig. 2). The Chimei Canyon provides a conduit for the sediments coming from the growing Central Range through a unique valley crossing the Coastal Range. The submarine extension of the canyon extends across a large sedimentary fan in the northwest corner of the Huatung Basin and joins the Hualien Canyon at the trench. Several other minor canyons coming from the eastern flank of the Coastal Range mountains flow downslope toward the Huatung Basin (Fig. 4).

From Taitung to 23°40′N, 3 cm/yr are taken up along the east-dipping thrusts of the Longitudinal Valley, and 1.7 cm/yr of shortening occurs offshore, through west-dipping thrusts (calculated from geodetic data [Yu et al., 1997]).

From 23°30′N to 24°10′N, the area of Hualien is actively deformed, as indicated by the intense seismicity, geodetic data, and the steepness of the eastern flank of the range. Accretion of the northernmost part of the Luzon arc lithosphere may thus be achieved at these latitudes. The small northern Coastal Range tectonic unit terminates northward against the Hsincheng Ridge, an uplifting sedimentary mass, which is supposed to be a remnant part of the inner domains of the Ryukyu accretionary prism. Lallemand et al. (1999) have studied the area immediately north of the Coastal Range. They have interpreted the Hsincheng Ridge (Fig. 4) as the last possible surface indication of the presence of the Luzon arc to the north, suggesting that (1) the arc never extended northward and (2) only the oceanic crust of the Philippine Sea plate subducts beneath northern Taiwan.

North of the Longitudinal Valley, the offshore steep slope of the eastern flank of the Central Range suggests faulting between the arc domain (the trough filled by the Hoping Basin) and the Central Range (Fig. 2). This area corresponds to the place where drastic changes in kinematics are suggested by GPS data (Lallemand and Liu, 1998). Indeed, instead of west-

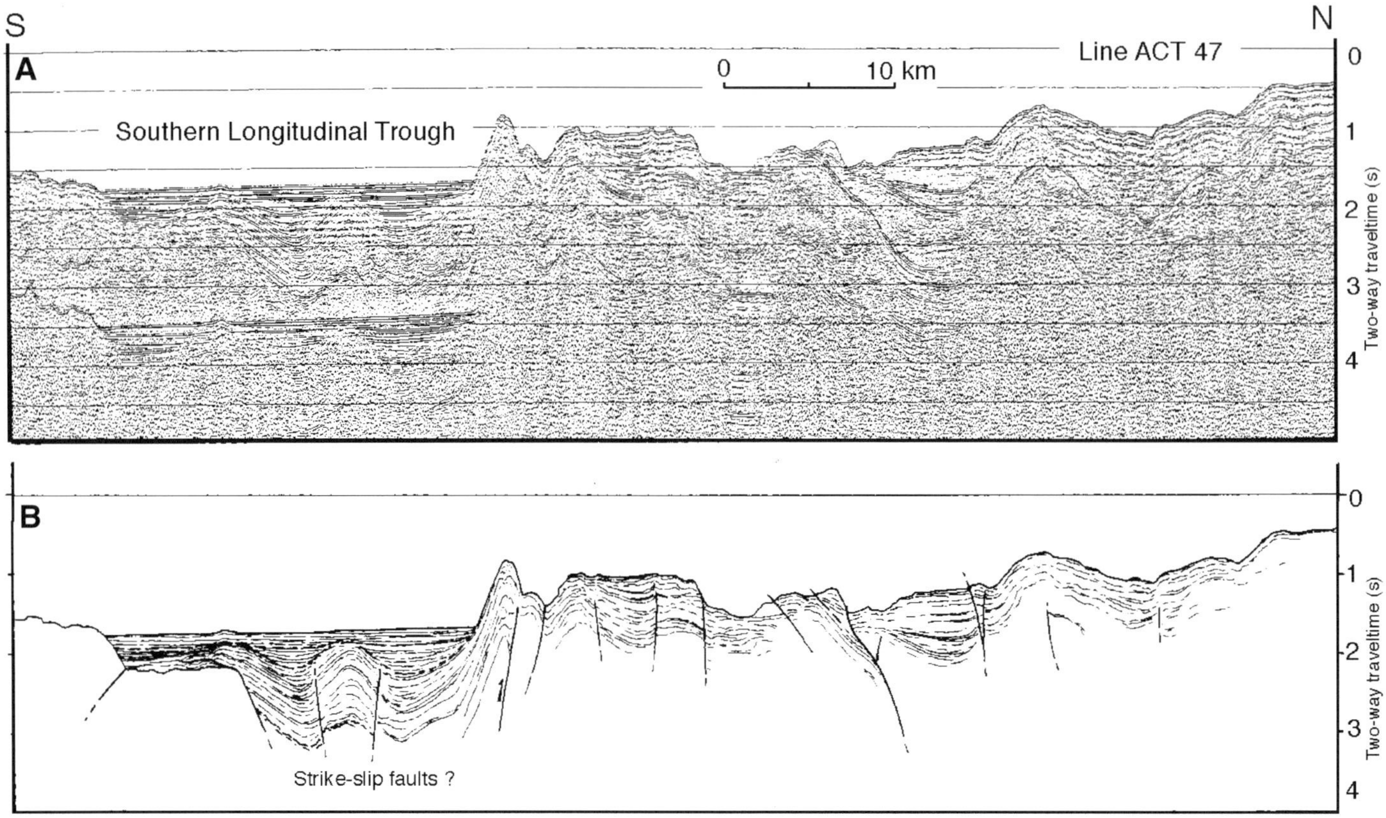

Figure 10. North-south seismic profile across the western part of the Southern Longitudinal Trough. (A) Migrated six-channel profile. (B) Line drawing showing the main geologic features.

ward convergence, the displacement vectors show that the northern Coastal Range domain is not moving relative to the Chinese continental margin. In this area, the eastern boundary of the backthrust front is not easy to recognize because the Hualien Canyon deeply incises the materials of both the forearc and the accretionary wedge. Bathymetric and morphologic evidence suggest that large volumes of sediments were previously deposited in this area and are currently being removed by erosion. The wide deep-sea fan, which was formed early at the boundary between the frontal slope of the wedge and the submarine Coastal Range, is also deeply eroded. Wide, east-trending canyons (Chimei Canyon, for example) also cut the east-dipping slope of the island, developing deep-sea fans of gravelly sediments and transporting detritus far toward the east in the Ryukyu Trench (Fig. 13). These observations suggest that the general tectonic and sedimentologic environment has changed recently.

Major tectonic changes at the latitude of Taitung. An important tectonic boundary forms the southern limit of the Coastal Range and its offshore continuation (Fig. 4). A northeast-trending, curved fault zone characterized by a series of imbricate scarps joins the southern tip of the Coastal Range to the main thrust, which marks the base of the eastern slope.

The north-south seismic profiles acquired during the ACT cruise show evidence of northwest-dipping thrusts in this area. For example, line ACT 36 (Fig. 15) shows a northwest-dipping thrust that offsets Holocene sediments of a small ponded basin lying on the back side of a submarine volcano, north of Lutao Island (Fig. 4). This zone of active faulting may be responsible for the drainage changes in the area east of Taitung. The seafloor in the upper part of a wide canyon, located near the base of the slope (Fig. 4) seems to have been offset by the fault scarp. Before thrusting and uplift in this area, this canyon may have been an active branch of the Taitung Canyon, allowing the sediments to be transported directly to the Huatung Basin.

Clockwise rotation of the Coastal Range is documented by paleomagnetic studies (Lee et al., 1991a, 1991b). Rapid clockwise rotation of about 25°–30° has occurred since the late Pliocene, with a propagation rate of the collision at about 70 ± 10 km/m.y. from north to south. This result is in good agreement with the existence of a major transfer zone of deformation at the latitude of Taitung, which coincides with the 1000 m of difference in elevation between the Longitudinal Valley and the Southern Longitudinal Trough, the clockwise rotation of the Coastal Range tectonic units, and the change in trend of the island arc (Fig. 4). Today, this transfer zone may account for the reversal of thrusting polarity, which occurs near Taitung, by allowing simultaneous activity on both

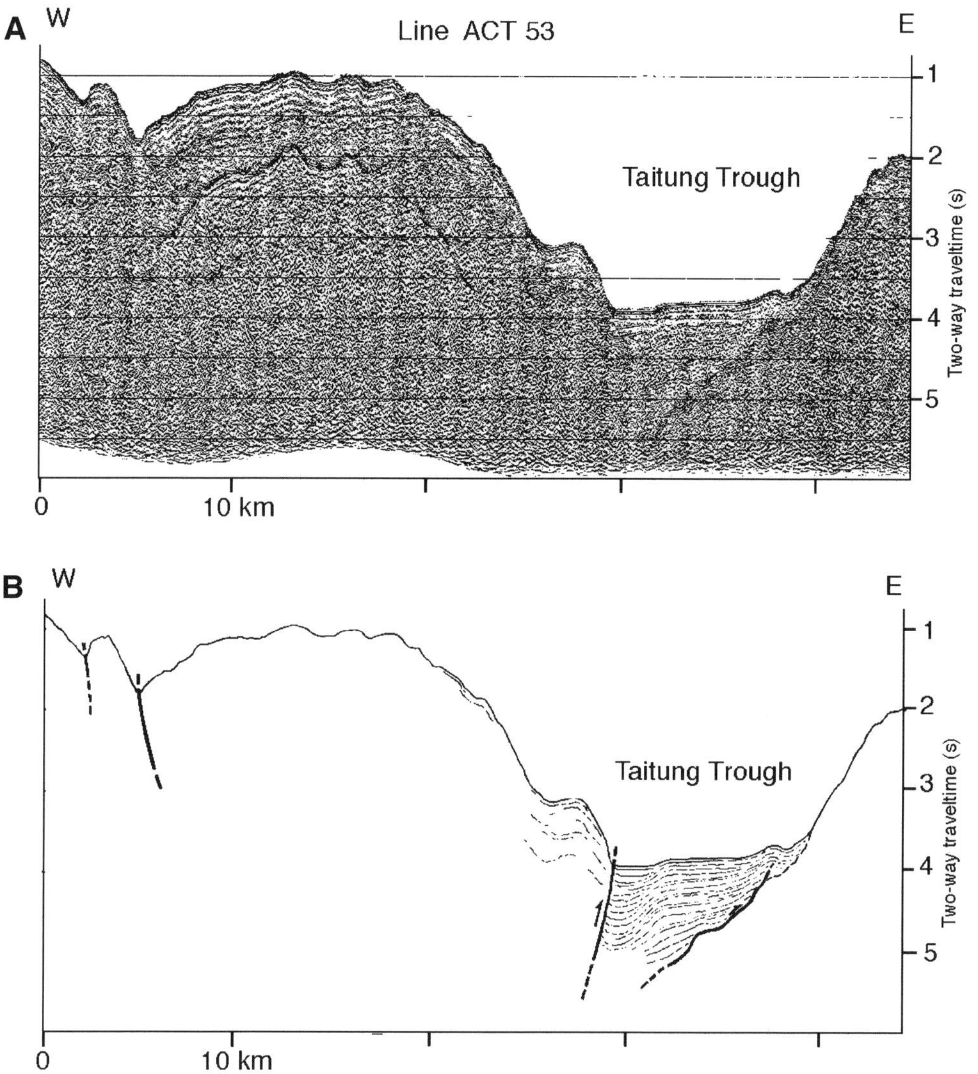

Figure 11. East-west seismic profile south of the study area. (A) Migrated six-channel profile. (B) Line drawing showing the main geologic features.

east-dipping thrust faults of the Coastal Range and west-dipping backthrust faults of the Huatung Ridge onto the arc.

South of Taitung, the Lutao volcanic block is apparently rotated about 10° clockwise with respect to the Lanyu segment, assuming that this part of the arc was linear. GPS measurements on this island indicate a clockwise change in the azimuth of convergence direction of about 6° and a slight decrease in the convergence rate (about 3 mm/yr) from Lanyu to Lutao (Fig. 3). The 60-km-long segment of the volcanic edifice in this region is bounded to the west by the growing Huatung Ridge and is probably more intensively involved in the collision process in this area than to the south. The subsequent shortening and rotation would require thrusting along the northeastern flank of the arc segment, which is suggested by the break at the base of the slope and local mass wasting south-southeast of Lutao (Fig. 4). This hypothesis is not yet confirmed by seismic data, but we can consider that the Luzon volcanic arc undergoes collision-related deformation north of 22°25′N.

DISCUSSION

Bathymetric and seismic data show that the forearc domain is very narrow along eastern Taiwan and that the arc is probably absent to the north of 24°N. Two hypotheses can account for this situation: (1) the volcanic arc was poorly developed or missing in this region of the Philippine Sea plate, or (2) the frontal part of the arc was subducted during an earlier stage of the collision. Figure 16 presents two interpretive, lithospheric-scale cross sections in the collision zone that take into account our observations. The first (Fig. 16A) is located at the latitude of the Hengchun Peninsula; the second (Fig. 16B) is across the Central Range. The structural map of Figure 3 shows the two different tectonic domains involved in the collision process and their kinematics relative to Eurasia.

In the northern domain (north of 22°40′N), simple considerations based on geometry imply that the whole forearc area of the Luzon arc has been underthrust beneath the Coastal Range

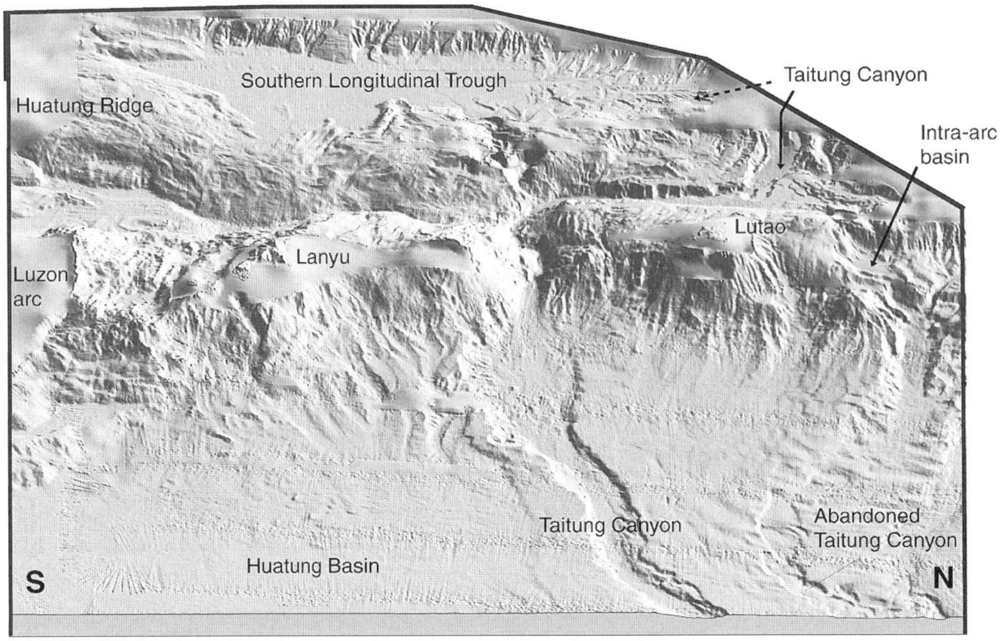

Figure 12. Shaded hill relief of the southern part of the study area (perspective view looking west).

between 22°50′N and 24°10′N (cross section 4 of Fig. 5; also Fig. 16B). The detailed mechanism remains somewhat enigmatic, because a sliver of about 90 ± 20 km width of forearc basement is suspected to have completely disappeared, and there is no direct surface expression of such intra-arc subduction.

To arrive at the present situation in which the arc remnants of the Coastal Range tectonic unit (including dismembered parts of volcanic arc edifices, associated forearc, intra-arc, and collisional basins) directly overthrust the already-exhumed metamorphic parts of the orogenic wedge, the forearc domain must be completely closed. The only way to do this is by underthrusting the basement lithosphere of the forearc beneath the arc. This process would allow enough time to exhume the internal parts of the Taiwan thrust wedge, by processes involving uplift of metamorphic rocks associated with major backthrusting against the rigid backstop formed by the basement of the Luzon forearc, tectonic underplating at depth beneath the Central Range, and erosion. The last (present) event is the westward thrusting of the Coastal Range on the Central Range.

In the southern domain (south of 22°40′N), the collision process is less evolved, and the continental margin of China is not yet subducting beneath the Philippine Sea plate oceanic lithosphere, as suggested by the geometry of the intersection between the ocean-continent boundary with the Manila Trench. The major structural elements and morphology are different from those described to the north. The basement of the Hengchun Ridge and the Hengchun Peninsula could be either the Luzon arc basement or the core of the former Manila oceanic accretionary wedge, as suggested by the absence of a magnetic signature (Liu et al., 1992). There are several ways to interpret the presence of this ridge: (1) it could result from uplift above the subducting and underplating slope sediments beneath the older accretionary wedge (e.g., Reed et al., 1991); (2) it could result from duplexing of subducted crust below the ridge (Reed et al., 1992, 1997); (3) it could also correspond to the southernmost extension of the Central Range of Taiwan; or (4) it could be caused by the underthrusting of part of the forearc domain under the arc, inducing intense shortening and uplift in the back part of the accretionary wedge. The last interpretation is shown in Figure 16A. This hypothesis is favored because south of 22°25′N, the Luzon arc is apparently slightly deformed by strike-slip faults (Lewis and Hayes, 1989), but the forearc domain is intensely shortened. Thus, the forearc basement (between the volcanic line and the accretionary wedge) should be subducting beneath the arc. The surface expression of forearc subduction is marked in the morphology and structures reflecting shortening in the orogenic wedge by (1) the major backthrusting of the Huatung Ridge and its piggyback Southern Longitudinal Trough collisional basin and (2) out-of-sequence thrusting in the wedge that is induced by the eastward jump of the velocity break from the tip of the forearc basement (previous backstop of the wedge) to the front of the basement of the arc edifice (new backstop).

The Southern Longitudinal Trough could be the offshore equivalent today of the collisional basin in which the Lichi mélange was deposited earlier, but farther to the north. The Huatung Ridge has been shown to be made in part by folded and faulted sedimentary deposits of the basin. In such a case, they represent equivalents of the on-land Lichi mélange, which is interpreted as the most recent suture zone between the Luzon

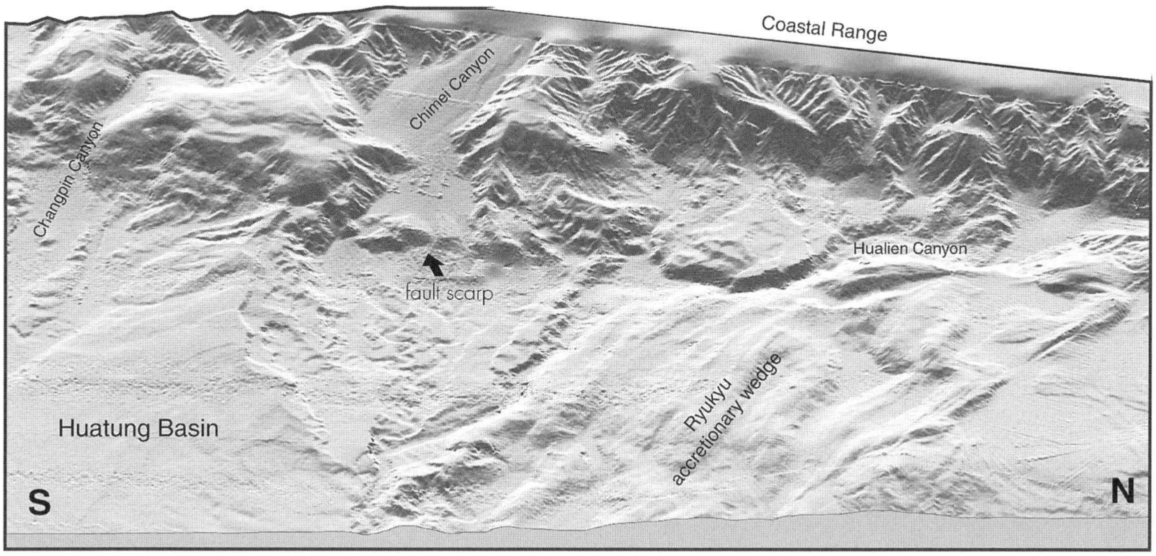

Figure 13. Shaded hill relief of the eastward-dipping slope offshore of the Coastal Range (perspective view from the east). The relief map shows the prominent north-trending fault scarp cutting the base of the slope in the middle of the figure.

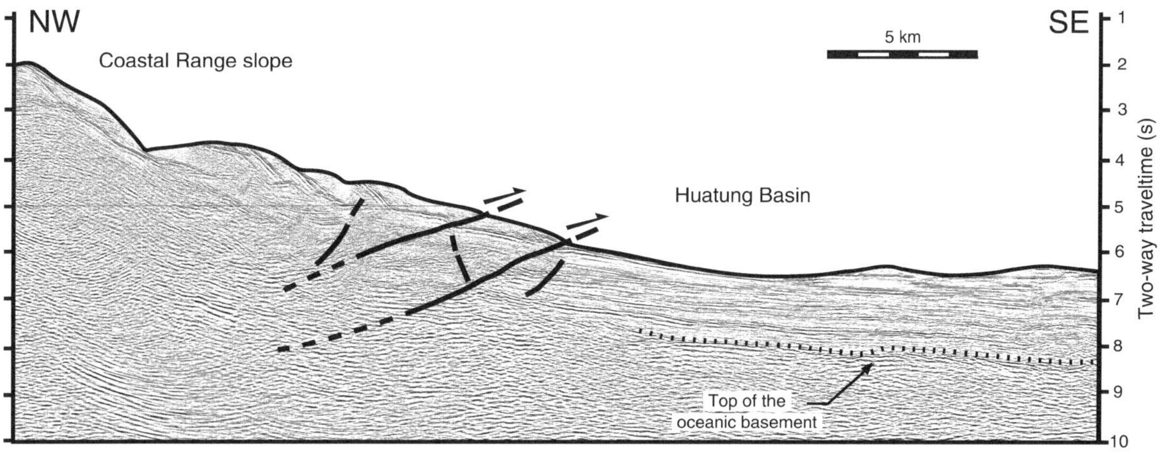

Figure 14. Northwest-southeast seismic profile showing the west-dipping thrusts cutting the base of the slope. Ewing data, 160-channel profile.

arc and the Eurasian margin in Taiwan. This second "suture" would mark the recent jump of basement underthrusting to the east in the forearc. The age of the mélange is disputed (e.g., Pelletier, 1985; Lu and Hsü, 1992; Huang et al., 1992). According to the observations made after the ACT survey, the age of the matrix must be younger toward the south, because it consists of sediments that have been recently deposited in the Southern Longitudinal Trough, south of the Longitudinal Valley. The Longitudinal Valley, whose morphology is well expressed on land, has no similar expression offshore either to the north or to the south.

Because a 50–80-km-wide segment of the forearc basement (from magnetism) (Liu et al., 1992) has disappeared from section A to B (Fig. 16), we also suggest that the frontal part of the Luzon arc basement has been thrust down the Luzon subduction system along an east-dipping thrust fault. This segment of crust could be thus now located at depth beneath the Longitudinal Valley (see cross section B in Fig. 16). This interpretation, based on the geometry of the system, implies that the western side of a paleo–Longitudinal Valley fault (currently buried under the Coastal Range) had a top-to-the-east component of motion during convergence, allowing the uplift and exhumation of the Central Range together with the underthrusting of the Luzon forearc basement. Such forearc subduction with compressional failure and exhumation along the arc has been studied by physical and numerical modeling (Chemenda et al., 1997, 2001; Tang et al., this volume) and application of the models to Taiwan. Exhumation of the Central Range, associated

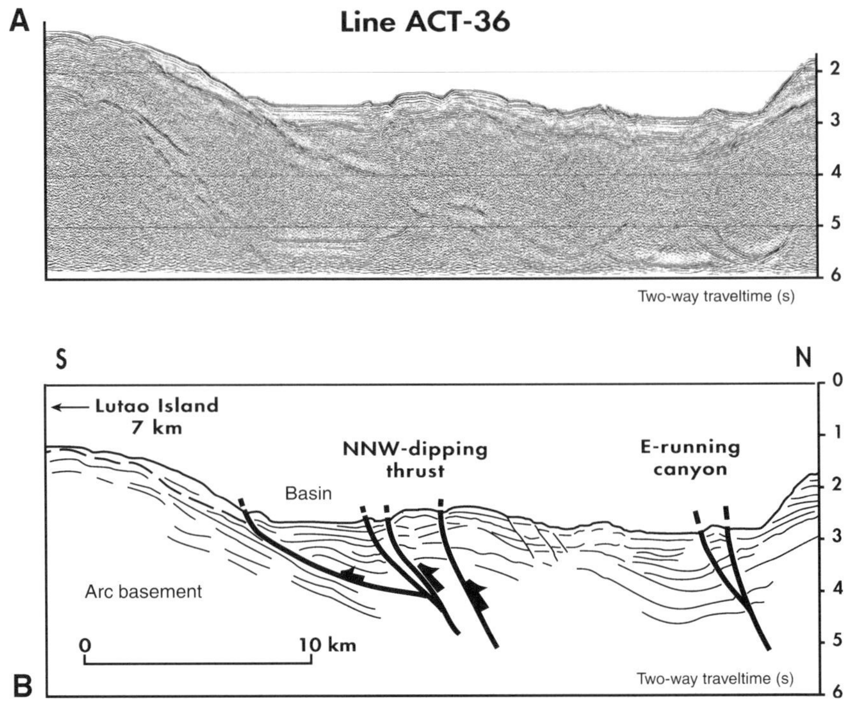

Figure 15. North-south seismic profile showing the transfer zone of deformation that develops at the southeastern tip of the Coastal Range. (A) Migrated six-channel profile. (B) Line drawing showing the main geologic features.

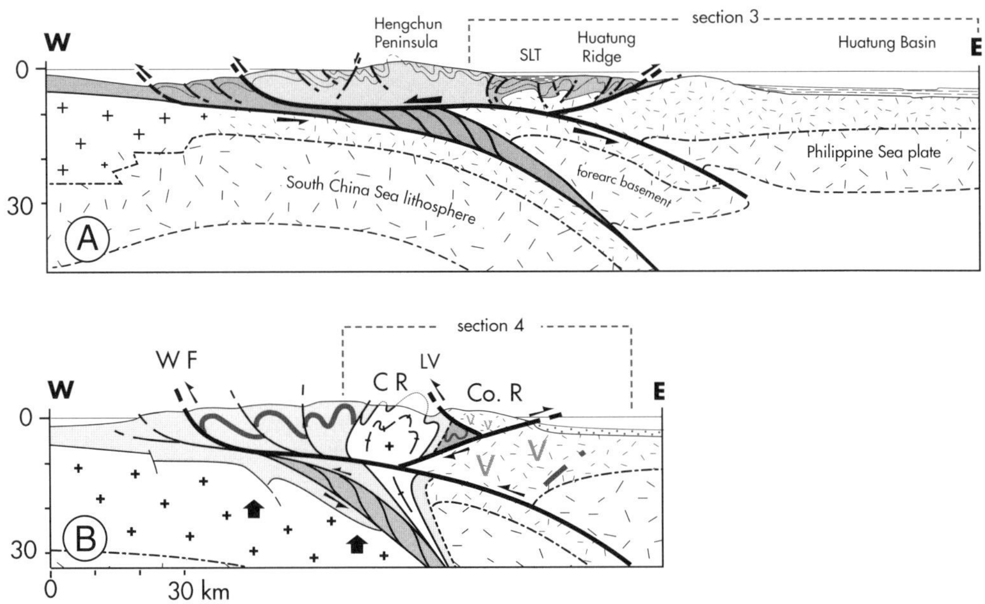

Figure 16. Interpretive geologic cross sections (A) south of the collision zone and (B) at the latitude of the Coastal Range. WF—Western Foothills, CR—Central Range, LV—Longitudinal Valley, Co. R—Coastal Range, SLT—Southern Longitudinal Trough. Locations of sections in Figure 1.

with synorogenic surficial extension and apparent normal faulting along the eastern side of the range, has already been proposed by Crespi et al. (1996) on the basis of structural data collected across the Central Range.

Several observations based on ACT cruise data suggest that the collisional process is not continuous, but rather occurs in separate stages including reversal of thrust polarity, rotation, shortening, and accretion. Indeed, it is very difficult to consider that the stage observed at 23°N results from the simple evolution of a stage at 22°N, which in turn results from the increasing compression of a stage at 21°N. Because of the obliquity of the plate convergence and the geometry of the ocean-continent transition in the Eurasian plate, the duration of the collisional stage increases toward the north (Suppe, 1981) and probably reaches a maximum at, and north of, 24°N where GPS stations show no significant northwestward motion of the Coastal Range relative to mainland China (Yu et al., 1997). North of 24°N, compressional stress is released offshore mainly along the Ryukyu Trench, but probably also in a more complex manner offshore Hualien (Lallemand et al., 1997b). The western termination of the Ryukyu accretionary wedge and the complex area that marks the junction between the collision zone and the Ryukyu Trench are exhaustively described in Lallemand et al. (1999), Font et al. (1999), and Dominguez et al. (1998).

Tectonic evolutionary model

Figure 17 depicts a proposed model for the three-dimensional tectonic evolution of the Taiwan collision since 3 Ma. To allow a better understanding of collision mechanisms, this sketch shows only the limits of the different lithosphere units. Sediments of the accretionary wedges have been omitted; thus, the growth of the collision thrust wedge is only suggested on the Eurasian plate. Figure 18 depicts the same stages of the tectonic evolution along a reference east-west cross section located at the same latitude as the present Central Range of Taiwan; the cross sections show the main tectonic units involved in the collision.

First stage. The Philippine Sea plate was obliquely subducted beneath Eurasia to the north, and the South China Sea oceanic lithosphere was subducted under the Luzon arc to the south (Fig. 17A). At that time (3 Ma), a major transform fault connected the two oppositely dipping subduction zones. The section of Figure 18A shows the beginning of continental-margin subduction beneath the Manila accretionary wedge.

Second stage. The northwestern corner of the Philippine Sea plate deformed the continental margin of China at 1.5 Ma (Fig. 17B), which began to subduct, inducing a stress increase in the forearc lithosphere of the Philippine Sea plate. Figure

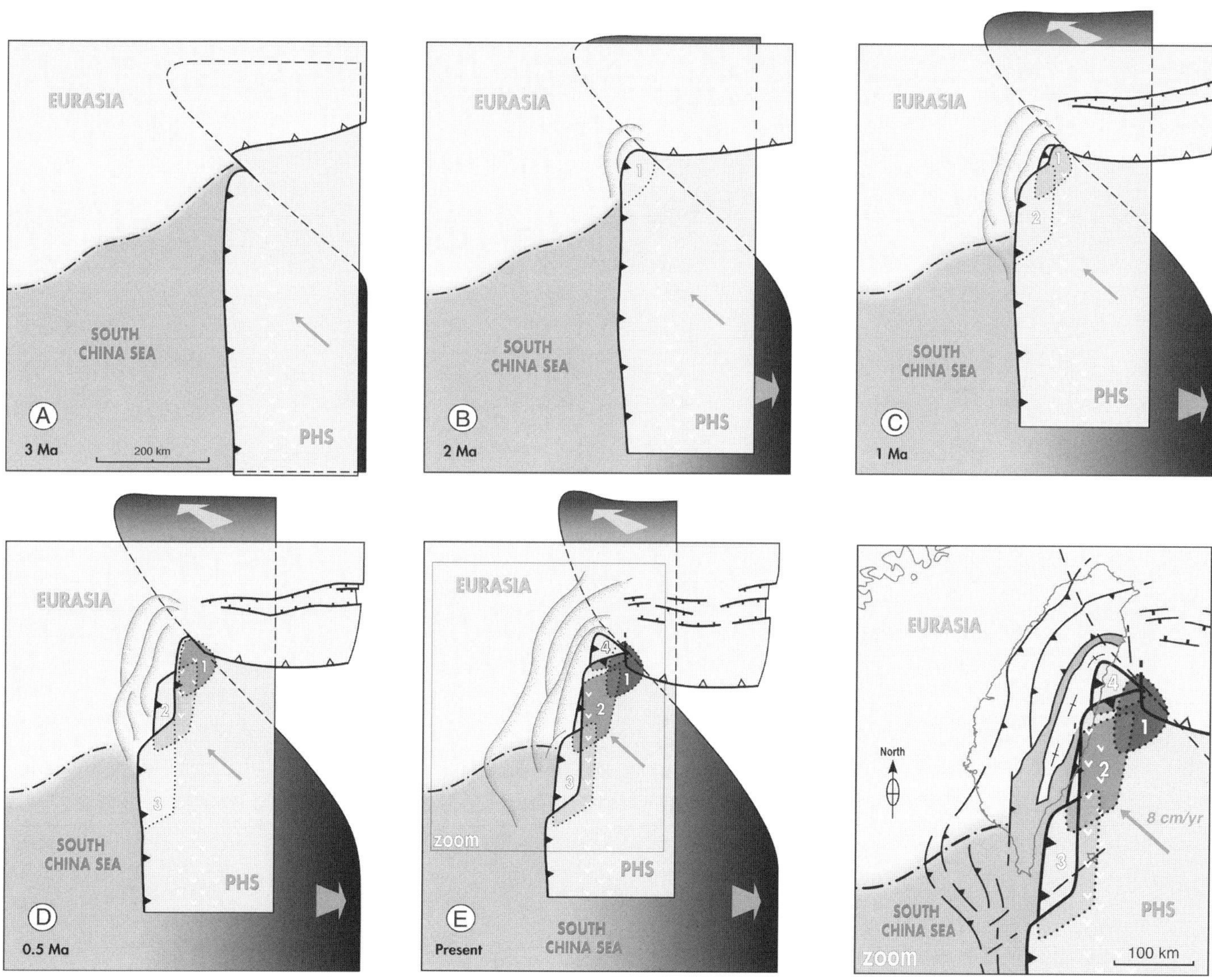

Figure 17. Three-dimensional geodynamic model of the Taiwan orogen. (A–E) The different stages since 3 Ma (see explanation in the text). Only the plate boundaries and the limits of the basement thrusts in the Philippine Sea plate (PHS) are shown; thrust-wedge sediments have been omitted.

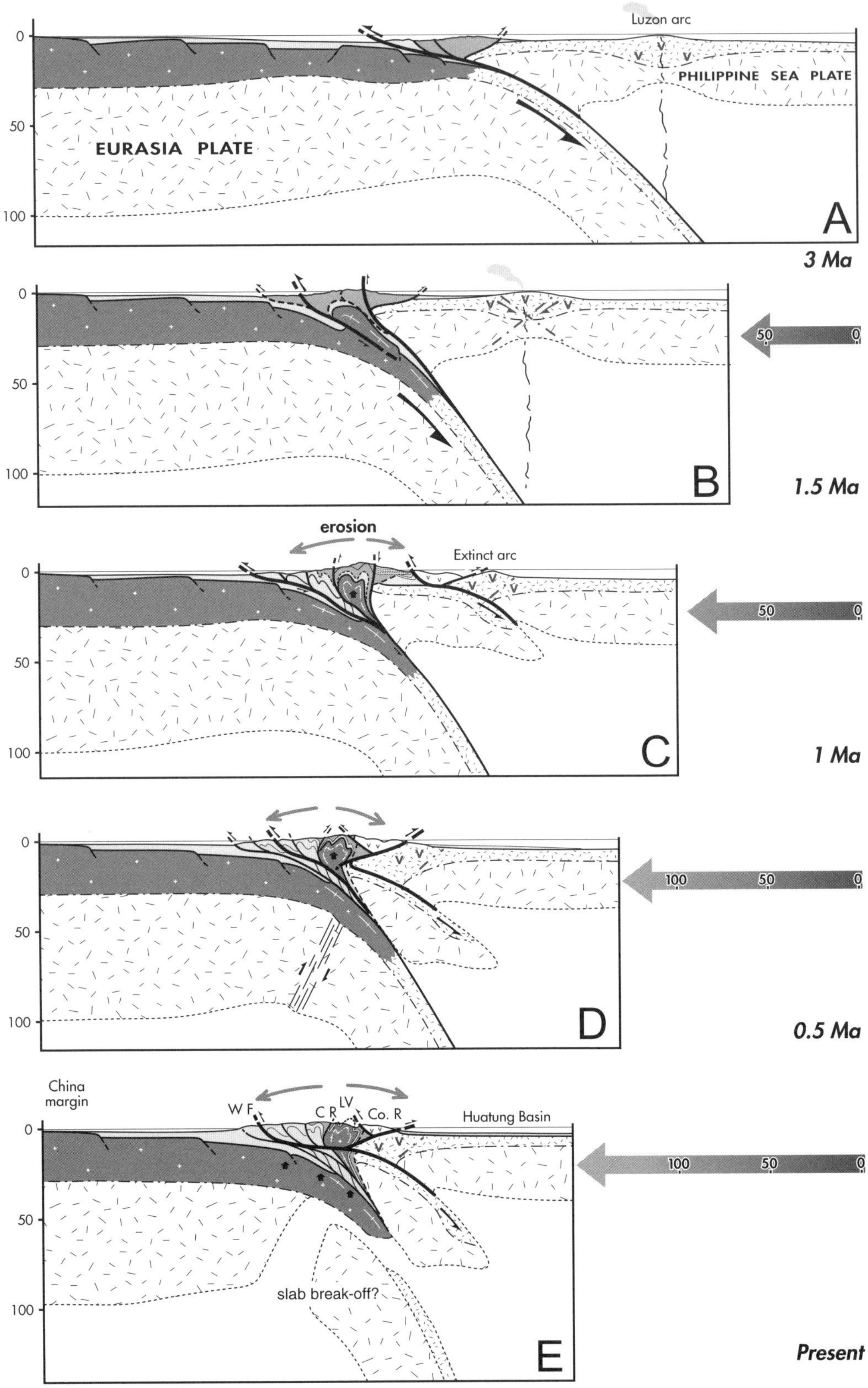

Figure 18. Tectonic evolutionary model of the Taiwan orogen since the beginning of the collision (see explanation in the text). The sections are correctly balanced. WF—Western Foothills, CR—Central Range, LV—Longitudinal Valley, Co. R—Coastal Range.

18B shows the deformation of the margin and the early stages of development of the Central Range. This tectonic unit corresponds to an offscraped slice of the upper crust in the basement of the continental margin, strongly deformed and metamorphosed during burial under the Philippine Sea plate, which forms a backstop. Increasing stress in the Philippine Sea plate induces failure of the forearc lithosphere in the weak domain of the volcanic arc (and arc subsidence). A set of conjugate thrusts develops within the arc lithosphere.

Third stage. At 1 Ma, a slice of the forearc lithosphere (unit 1 in Fig. 17C) was thrust beneath the arc. Collision-related deformation migrated to the south, involving stress increase in a new domain of the Philippine Sea plate. Figure 18C shows the different processes that occurred during this complex evolutionary stage. To the east, the volcanic activity of the arc stopped because of the subduction of the forearc sliver, which cut the magma source. West of the extinct arc, a complex collisional basin developed, mainly filled by sediments eroded from the Taiwan mountain belt. A new thrust wedge developed against the arc edifice, composed of volcaniclastic sediments of the early forearc basin, detrital sediments of the collisional basin, and offscraped volcanic rocks from the forearc basement. This feature is the northern early equivalent of the present-day Huatung Ridge. To the west, continuing subduction of the continental margin involved the growth of a large sedimentary wedge by stacking of thick sedimentary sequences of the passive margin. In the core of the growing wedge, metamorphic rocks of the Central Range were progressively exhumed and rose through the overlying rocks of the old oceanic wedge owing to the combined effects of erosion of the internal domain of the wedge and probable underplating at depth favoring uplift. The metamorphic rocks rose along the subvertical surface marking the western extent of the backstop formed by the forearc lithosphere. The present-day western boundary of the Central Range is characterized by subvertical faults with apparent normal offsets. Depending on their dip during the uplift, they could be normal faults (eastward dip) developed in a convergent setting or steep backthrusts (westward dip). As in classical accretionary wedges, the dip of such faults would be controlled by the shape and dip angle of the backstop (Malavieille et al., 1991) (in our case, the shape of the forearc lithosphere, which is poorly known). During uplift of the internal parts of the mountain belt, the rocks of the early oceanic accretionary wedge (including ophiolite blocks of former mélanges) resting on top of it, were deeply eroded and became the source for the olistostromal blocks currently included in the Lichi mélange. Most of these sediments were deposited in collisional basins during shortening of the forearc domain and were later involved in thrusting.

Fourth stage. The southward migration of deformation and the subduction of a second slice of the forearc lithosphere (unit 2 of Fig. 17D) occurred at 0.5 Ma, whereas farther south, another domain (unit 3 of Fig. 17D) of forearc lithosphere was subjected to a stress increase. Foreland deformation developed to the west in the Taiwan Strait. Offscraped remnants of the forearc domain including forearc and intra-arc sediments were tectonically mixed with sediments of former collisional basins. A new mechanism allowed exhumation to continue. The asymmetrical mechanism of wedge growth by forward thrusting (e.g., Malavieille, 1984) that had been operating in the Taiwan region up to this time was replaced by a more symmetrical evolution with significant backthrusting (Fig. 18D). Uplift therefore occurred because of the conjugate effects of east-dipping, forward out-of-sequence thrusting, and west-dipping backthrusting. The forearc domain was shortened significantly at this stage.

Present situation. Figure 17E shows a third unit of forearc lithosphere to the south that has begun subducting, whereas lithosphere of unit 2 is completely subducted. The tip of the arc basement now enters in the collision process, indenting the Central Range, which is now cut by a east-dipping, out-of-sequence thrust (Fig. 18E). The main tectonic features of present-day Taiwan are presented in the Figure 17F.

Two main hypothesis may account for the present evolution of the collisional process in this domain.

1. Increasing stress in the northernmost domain of collision has induced the propagation of a northwest-trending transform-tear fault that bounds the Ryukyu slab to the southwest. Failure in the oceanic lithosphere has resulted in the detachment of a fourth tectonic unit to the north (unit 4 of Fig. 17E). In this domain, most of the convergence occurs southeast of this unit probably on east-dipping blind thrusts merging below the easternmost part of the Ryukyu accretionary wedge and joining the basal décollement in the Ryukyu Trench system. This hypothesis is depicted in Figure 17E.

2. As proposed by Chemenda et al. (1997), after removal of the forearc crust, a major west-dipping thrust developed in response to indentation by the arc lithosphere and cut through the continental lithosphere of the China margin (see Fig. 12D, p. 267, and Fig. 13, p. 268, in Chemenda et al. [1997]). This process results in the development of a new subduction zone that propagates southward.

Both of these mechanisms could explain the lack of displacement shown today by GPS measurements in northern Taiwan.

Geophysical constraints. Numerous geophysical studies help to define the general structure of the Taiwan collisional area. Among them, several recent seismological observations may be consistent with the underthrusting of part of the forearc crust under the arc (Rau and Wu, 1995, 1998; Wu et al., 1997; Kao et al., 1998; Cheng et al., 1998 and this volume). Figure 19 shows three interpretive sections (locations in Fig. 1) of the present stage of the collision process and its evolution in space from north to south of the orogen. Seismicity data from global (Engdahl et al., 1998) and regional networks (Wu et al., 1997; Cheng et al., 1998) have been projected onto the three sections arranged from north to south.

In particular, in the section that crosses southern Taiwan (Fig. 19B), two east-dipping bands of seismicity can be distin-

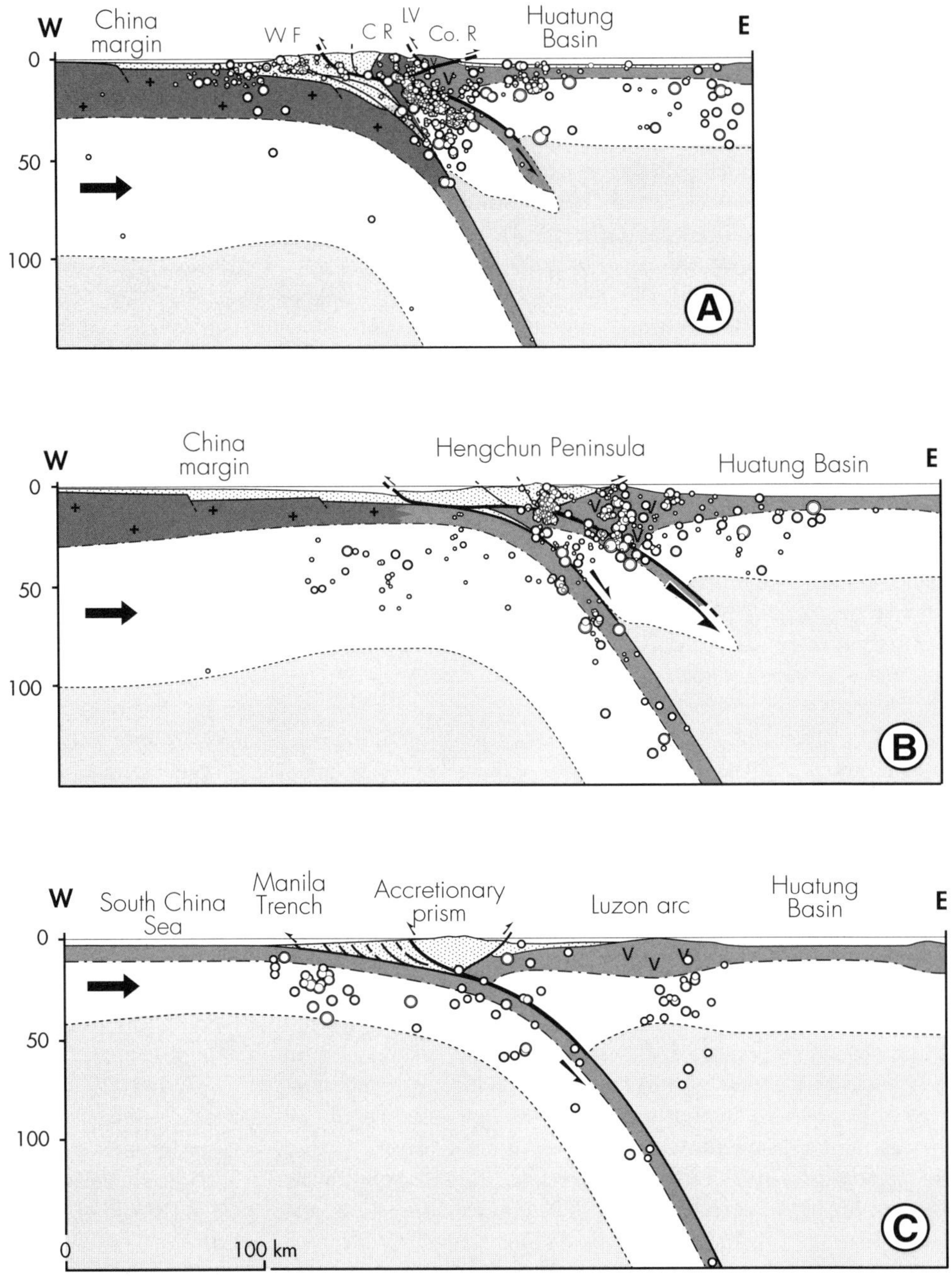

Figure 19. East-west sections of the present stage of collision showing its evolution in space from the south to the north of the orogen (for location, see Fig. 1). Seismicity data from global (Engdahl et al., 1998) and regional networks (Wu et al., 1997; Cheng et al., 1998) have been projected onto the three sections. (A) Section at 23°N, across the Taiwan mountains. (B) Section at 22°20′N, across the Hengchun Peninsula. (C) Section at 21°N. WF—Western Foothills, CR—Central Range, LV—Longitudinal Valley, Co. R—Coastal Range.

guished: a broad, steeply dipping (50°–60°) band (on the left), which represents the plate boundary and the subducting lithosphere of the Chinese margin, and a narrower, more shallowly dipping band beneath the arc at 20–40 km depth (on the right in Fig. 19B). The latter corresponds well to the 20°E-dipping underthrusting focal plane shown in recently published fault-plane solutions (Kao et al., 1998) and supports eastward-directed underthrusting of the forearc beneath the Luzon arc south of Lanyu. A high-velocity zone, at a depth range of 15–50 km, has been imaged by Rau and Wu (1995) under the Coastal Range and the Philippine Sea. Detailed seismic tomography studies by Cheng et al. (1998 and this volume) also show two zones of high seismicity and a prominent high-velocity anomaly in the middle to lower crust, which could correspond to the underthrusting of forearc crust.

SUMMARY AND CONCLUSIONS

The progressive subduction of the continental margin of China induces (1) to the south, major eastward backthrusting and shortening of the forearc domain between the former oceanic accretionary wedge and the Luzon arc volcanic edifice and

(2) to the north, accretion of parts of the arc domain to the collisional belt associated with westward thrusting and block rotation.

Our marine observations show the complex interactions of tectonic deformation, sediment dispersal, and basin development along eastern Taiwan.

In the southern domain, the Longitudinal Valley disappears offshore to the south below the Southern Longitudinal Trough, which is a collisional basin filled with Holocene orogenic sediments lying unconformably on deformed and folded strata of the former forearc domain. Transpressive deformation has played an important role during the development of the Southern Longitudinal Trough. Present deformation has jumped to the east, and it is characterized by the growth of a sedimentary ridge (the Huatung Ridge, the rear part of the former Manila oceanic accretionary wedge including forearc and intra-arc sequences), which overthrusts the basement of the island arc. Part of the island arc (Lutao area) is affected by subsequent shortening.

In the northern domain, at the base of the eastern slope of the Coastal Range, prominent fault scarps suggest active eastward thrusting of parts of the arc (volcanic edifices and intra-arc or forearc sedimentary deposits) onto the oceanic crust of the Philippine Sea plate. The northernmost part of the Coastal Range is presumably being accreted to the rest of Taiwan.

A N50°E-trending transfer zone of deformation accommodates the differential motion and rotation of the northern and southern tectonic units.

The relationship of the Coastal Range and the offshore structures along eastern Taiwan to the distribution of canyons and deformation offshore Hualien and the Ryukyu Trench fits the evolution described for the region.

Structural observations show that the forearc basement of the Luzon arc no longer exists north of the latitude of Taitung. A model is proposed for the "missing" part of forearc crust in the region of collision and the tectonic evolution of the Taiwan orogen.

ACKNOWLEDGMENTS

We thank Insitut National des Sciences de l'Univers, the French Institute in Taipei (Ministry of Foreign Affairs), and the National Science Council (Taiwan) for funding and supporting this collaborative work, IFREMER for providing R/V *L'Atalante* ship time and equipment, and GENAVIR officers, technicians and crew. We thank Marc-André Gutscher for his suggestions and help with the English. We acknowledge B. Engdahl, F. Wu, and W.-B. Cheng for kindly providing regional seismicity data. We also thank D. Fischer and D. Reed for their constructive reviews and G.-J. Jahn for his help and kindness during our Taiwan field trips. Many thanks to A. Delplanque for helping draft most of the figures in this paper. Figures 1, 2, 4, 7, 12, and 13, were produced with GMT software (Wessel and Smith, 1995).

REFERENCES CITED

Angelier, J., Chu, H.-T., and Lee, J.C., 1997, Shear concentration in a collision zone: Kinematics of the Chihshang fault as revealed by outcrop-scale quantification of active faulting, Longitudinal Valley, eastern Taiwan: Tectonophysics, v. 274, p. 117–143.

Angelier, J., Bergerat, F., Chu, H.-T., and Lee, T.Q., 1990, Tectonic analysis and the evolution of a curved collision belt: The Hsuehshan Range, northern Taiwan: Tectonophysics, v. 183, p. 77–96.

Barrier, E., and Angelier, J., 1986, Active collision in eastern Taiwan: The Coastal Range: Tectonophysics, v. 125, p. 39–72.

Biq, C., 1972, Dual-trench structure in the Taiwan-Luzon region: Proceedings of the Geological Society of China, no. 15, p. 65–75.

Briais, A., and Pautot, G., 1992, Reconstructions of the South China Sea from structural data and magnetic anomalies, *in* Xianglong, J., Kudrass, H.R., and Pautot, G., eds., Marine geology and geophysics of the South China Sea: China Ocean Press, p. 60–69.

Chai, B.H.T., 1972, Structure and tectonic evolution of Taiwan: American Journal of Science, v. 272, p. 389–422.

Chemenda, A.I., Yang, R.K., Stéphan, J.-F., Konstantinovskaia, E.A., and Ivanov, G.M., 2001, New results from physical modeling of arc-continent collision in Taiwan: Evolutionary model: Tectonophysics, v. 333, p. 159–178.

Chemenda, A.I., Yang, R.K., Hsieh, C.-H., and Groholsky, A.L., 1997, Evolutionary model for the Taiwan collision based on physical modelling: Tectonophysics, v. 274, p. 253–274.

Chen, W.P., and Juang, W.S., 1986, Seafloor physiography off southeastern Taiwan: Acta Oceanographica Taiwanica, v. 16, p. 1–7.

Cheng, W.-B., Wang, C., Shyu, C.-T., and Shin, T.-C., 1998, A three-dimensional Vp model of the southeastern Taiwan area and its tectonic implications: TAO, Terrestrial, Atmospheric and Oceanic Sciences, v. 9, p. 425–452.

Clark, M.B., Fisher, D.M., Lu, C.-Y., and Chen, C.H., 1993, Kinematic analysis of the Hsuehshan Range, Taiwan: A large-scale pop-up structure: Tectonics, v. 12, p. 205–217.

Crespi, J.M., Chan, Y.-C., and Swaim, M.S., 1996, Synorogenic extension and exhumation of the Taiwan hinterland: Geology, v. 24, p. 247–250.

Davis, D., Suppe, J., and Dahlen, F.A., 1983, Mechanics of fold and thrust belts and accretionary wedges: Journal of Geophysical Research, v. 88, p. 1153–1172.

Déramond, J., Delcaillau, B., Souquet, P., Angelier, J., Chu, H.-T., Lee, T.-Q., Lee, J.-F., Liew, P.-M., Lin, T.S., and Teng, L., 1996, Signatures de la surrection et de la subsidence dans les bassins d'avant-chaîne actifs: Les Foothills de Taiwan (de 8 Ma à l'actuel): Bulletin de la Société Géologique de France, v. 167, p. 111–123.

Dominguez, S., Lallemand, S., Malavieille, J., and Schnurle, P., 1998, Oblique subduction of the Gagua Ridge beneath the Ryukyu accretionary wedge system: Insights from marine observations and sandbox experiments: Marine Geophysical Research, v. 20, p. 383–402.

Dorsey, R.J., 1992, Collapse of the Luzon volcanic arc during onset of arc-continent collision: Evidence from a Miocene-Pliocene unconformity, eastern Taiwan: Tectonics, v. 11, p. 177–191.

Engdahl, E.R., van der Hilst, R.D., and Buland, R., 1998, Global tele-seismic earthquake relocation with improved travel times and procedures for depth relocation: Bulletin of the Seismological Society of America, v. 88, p. 722–743.

Font, Y., Lallemand S., and Angelier J., 1999, Etude de la transition entre l'orogène actif de Taiwan et la subduction des Ryukyu—Apport de la sismicité: Bulletin de la Société Géologique de France, v. 170, p. 3271–3283.

Fuh, S.C., Liu, C.S., Lundberg, N., and Reed, D., 1997, Strike-slip faults offshore southern Taiwan: Implications for the oblique arc-continent collision processes: Tectonophysics, v. 274, p. 25–39.

Hayes, D.E., and Lewis, S.D., 1984, A geophysical study of the Manila Trench,

Luzon, Philippines. 1. Crustal structure, gravity, and regional tectonic evolution: Journal of Geophysical Research, v. 89, p. 9171–9195.

Ho, C.-S., 1982, Tectonic evolution of Taiwan and explanatory text of the tectonic map of Taiwan: Taipei, Taiwan, Republic of China, Central Geological Survey, Ministry of Economic Affairs, 126 p.

Ho, C.-S., 1986, A synthesis of the geologic evolution of Taiwan: Tectonophysics, v. 125, p. 1–16.

Ho, C.-S., 1988, An introduction to the geology of Taiwan (second edition) and explanatory text of the geologic map of Taiwan: Taipei, Taiwan, Republic of China, Central Geological Survey, Ministry of Economic Affairs, 192 p.

Hsu, S.-K., Liu, C.-H., Shyu, C.-T., Liu, S.-Y., Sibuet, J.-C., Lallemand, S., Wang, C., and Reed, D., 1998, New gravity and magnetic anomaly maps in the Taiwan-Luzon region and their preliminary interpretation: TAO, Terrestrial, Atmospheric, and Oceanic Sciences, v. 9, p. 509–532.

Huang, C.-Y., Wu, W.-Y., Chang, C.P., Tsao, S., Yuan, P.B., Lin, C.W., and Kuan-Yuan, X., 1997, Tectonic evolution of accretionary prism in the arc-continent collision terrane of Taiwan: Tectonophysics, v. 281, p. 31–51.

Huang, C.-Y., Shyu, C.-T., Lin, S.B., Lee, T.-Q., and Sheu, D.D., 1992, Marine geology in the arc-continent collision zone off southeastern Taiwan: Implications for Late Neogene evolution of the Coastal Range: Marine Geology, v. 107, p. 183–212.

Huang, T, Wang, Y., and Chen, W.S., 1993, Geologic map of eastern Coastal Range: Taiwan, Central Geological Survey, Ministry of Economic Affairs, scale 1:100 000, 2 sheets.

Jahn, B.M., 1972, Reinterpretation of geologic evolution of the Coastal Range, eastern Taiwan: Geological Society of America Bulletin, v. 83, p. 241–248.

Juang, W.S., and Bellon, H., 1984, The potassium-argon dating of andesites from Taiwan: Proceedings of the Geological Society of China, no. 27, p. 86–100.

Kao, H., Jian, P.-R., Ma, K.-F., Huang, B.-S., and Liu, C.-C., 1998, Moment-tensor inversion for offshore earthquakes east of Taiwan and their implications to regional collision: Geophysical Research Letters, v. 25, p. 3619–3622.

Karig, D.E., and Ingle, J.C., Jr., et al., 1975, Initial Reports of the Deep Sea Drilling Project, Volume 31: Washington, D.C., U.S. Government Printing Office, 972 p.

Lallemand, S.E., and Tsien, H.-H., 1997, An introduction to active collision in Taiwan: Tectonophysics, v. 274, p. 1–4.

Lallemand, S.E., Liu, C.-S., and the ACT scientific crew, 1997a, Swath bathymetry mapping reveals details of the active arc-continent collision offshore Taiwan: Eos (Transactions, American Geophysical Union), v. 78, p. 173–175.

Lallemand, S.E., Liu, C.-S., and Font, Y., 1997b, A tear fault boundary between the Taiwan orogen and the Ryukyu subduction zone: Tectonophysics, v. 274, p. 171–190.

Lallemand, S., and Liu, C.-S., 1998, Geodynamic implications of present-day kinematics in the southern Ryukyus: Journal of the Geological Society of China, v. 41, p. 551–564.

Lallemand, S., Liu, C.-S., Dominguez, S., Schnürle, P., Malavieille, J., and the ACT scientific crew, 1999, Trench-parallel stretching and folding of forearc basins and lateral migration of the accretionary wedge in the southern Ryukyus: A case of strain partition caused by oblique convergence: Tectonics, v. 8, p. 231–247.

Larroque, C., Calassou, S., Malavieille, J., and Chanier, F., 1995, Experimental modeling of forearc basin development during accretionary wedge growth: Basin Research, v. 7, p. 255–268.

Lee, J.-C., Angelier, J., and Chu, H.-T., 1997, Polyphase history and kinematics of a complex major fault zone in the northern Taiwan mountain belt: The Lishan Fault: Tectonophysics, v. 274, p. 97–115.

Lee, T.-Q., Kissel, C., Barrier, E., Lai, C., and Chi, W.R., 1991, Paleomagnetic evidence for a diachronic clockwise rotation of the Coastal Range: Earth and Planetary Science Letters, v. 104, p. 245–257.

Lee, T.Q., Angelier, J., Chu, H.-T., and Bergerat, F., 1991, Rotations in the northeastern collision belt of Taiwan: Preliminary results from paleomagnetism: Tectonophysics, v. 199, p. 109–120.

Letouzey, J., and Kimura, M., 1985, Okinawa Trough genesis: Structure and evolution of a back-arc basin developed in a continent: Marine Petroleum Geology, v. 2, p. 111–130.

Letouzey, J., Sage, L., and Müller, C., 1988, Geological and structural map of eastern Asia: Introductory notes: American Association of Petroleum Geologists, 52 p.

Lewis, S.D., and Hayes, D.E., 1983, The tectonics of northward propagating subduction along eastern Luzon, Philippine Islands, *in* Hayes, D.E., ed., The tectonic and geologic evolution of the southeast Asian seas and islands, part 2: American Geophysical Union Geophysical Monograph 27, p. 57–78.

Lewis, S.D., and Hayes, D.E., 1984, A geophysical study of the Manila Trench, Luzon, Philippines. 2. Fore arc basin structural and sratigraphic evolution: Journal of Geophysical Research, v. 89, p. 9196–9214.

Lewis, S.D., and Hayes, D.E., 1989, Plate convergence and deformation, North Luzon Ridge, Philippines: Tectonophysics, v. 168, p. 221–237.

Liu, C.-S., Liu, S.-Y., Kuo, B.-Y., Lundberg, N., and Reed, D., 1992, Characteristics of the gravity and magnetic anomalies off southern Taiwan: Acta Geologica Taiwanica, v. 30, p. 123–130.

Liu, C.-S., Huang, I.L., and Teng, L.S., 1997, Structural features off southwestern Taiwan: Marine Geology, v. 137, p. 305–319.

Liu, C.-S., Schnürle, P., Lallemand, S., and Reed, D., 1998, Crustal structures of the Philippine Sea Plate near Taiwan arc-continent collision zone [abs.]: 8th International Symposium on Deep Seismic Profiling of the Continents and their Margins, Barcelona, Spain, p. 31–32.

Liu, C.-S., Liu, S.-Y., Lallemand, S., Lundberg, N., and Reed, D., 1998, Digital elevation model offshore Taiwan and its tectonic implications: TAO, Terrestrial, Atmospheric, and Oceanic Sciences, v. 9, p. 705–738.

Liu, C.-S., Reed, D.L., Lundberg, N., Moore, G.F., McIntosh, K.D., Nakamura, Y., Wang, T.K., Chen, T.H., and Lallemand, S., 1995, Deep seismic imaging of the Taiwan arc-continent collision zone: Eos (Transactions, American Geophysical Union), v. 76, p. F635.

Lu, C.-Y., and Hsü, K.-J., 1992, Tectonic evolution of the Taiwan mountain belt: Petroleum Geology of Taiwan, v. 27, p. 21–46.

Lu, C.-Y., and Malavieille, J., 1994, Oblique convergence, indentation and rotation tectonic in Taiwan Mountain belt: Insights from experimental modeling: Earth and Planetary Science Letters, v. 121, p. 477–494.

Lu, C.-Y., Jeng, F.S., Chang, K.J., and Jian, W.T., 1998, Impact of basement high on the structure and kinematics of the western Taiwan Thrust Wedge: Insights from sandbox models: TAO, Terrestrial, Atmospheric, and Oceanic Sciences, v. 9, p. 533–550.

Lundberg, N., and Dorsey, R.J., 1988, Synorogenic sedimentation and subsidence in a Plio-Pleistocene collisional basin, eastern Taiwan, *in* Kleinspehn., K.L., and Paola, C., eds., New perspectives in basin analysis: New York, Springer-Verlag, p. 265–280.

Lundberg, N., and Dorsey, R.J., 1990, Rapid Quaternary emergence, uplift and denudation of the Coastal Range, eastern Taiwan: Geology, v. 18, p. 638–641.

Lundberg, N., Reed, D.L., and Lieskes, J.H., Jr., 1992, Structural controls on orogenic sedimentation, submarine Taiwan collision: EOS (Transactions, American Geophysical Union), v. 73, p. 539.

Lundberg, N., Reed, D.L., Liu, C.-S., and Lieskes, J.H., Jr., 1997, Forearc-basin closure and arc accretion in the submarine suture zone south of Taiwan: Tectonophysics, v. 274, p. 5–24.

Malavieille, J., 1984, Modélisation expérimentale des chevauchements imbriqués: Application aux chaînes de montagnes: Bulletin de la Société Géologique de France, v. 26, p. 129–138.

Malavieille, J., Calassou, S., Larroque, C., Lallemand, S., and Stéphan, J.-F., 1991, Modelisation analogique des prismes d'accrétion océaniques: Montage d'animation vidéo de 27′, Serie Cours M037, Formation exploration, Elf SNEAP.

Page, B.M., and Suppe, J., 1981, The Pliocene Lichi melange of Taiwan: Its plate-tectonic and olistostromal origin: American Journal of Science, v. 281, p. 193–227.

Pautot, G., and Rangin, C., 1989, Subduction of the South China Sea axial ridge below Luzon (Philippines): Earth and Planetary Science Letters, v. 92, p. 57–69.

Pelletier, B., 1985, De la fosse de Manille à la Chaîne de Taiwan [Ph.D. thesis]: Brest, France, Brest University, 268 p.

Rau, R.-J., and Wu, F.T., 1995, Tomographic imaging of lithospheric structure under Taiwan: Earth and Planetary Science Letters, v. 133, p. 517–532.

Rau, R.-J., and Wu, F.T., 1998, Active tectonics of Taiwan orogeny from focal mechanisms of small-to-moderate-sized earthquakes: TAO, Terrestrial, Atmospheric and Oceanic Sciences, v. 9, p. 755–778.

Reed, D.L., Lundberg, N., Liu, C.-S., and Macintosh, K.D., 1991, Evidence of frontal thrust propagation and fluid migration in an offscrapped sedimentary basin sequence, offshore Taiwan: Taipei, Taiwan, Proceedings of the TAICRUST Workshop, Abstracts, p. 103–106.

Reed, D.L., Lungberg, N., Liu, C.H., and Kuo, B.Y., 1992, Structural relations along the margins of the offshore Taiwan accretionary wedge: Implications for accretion and crustal kinematics: Acta Geologica Taiwanica, v. 30, p. 105–122.

Reed, D.L., Chi, W.-C., Lundberg, N., Moore, G.F., Nakamura, Y., Macintosh, K. D., Nguyen, T., Liu, C.-S., Zhao, Z., and Wang, C.-Y., 1997, Geophysical cross-sections at the transition from subduction to collision: Results from the TAICRUST experiment: International Conference and Sino-American Symposium on Tectonics of East Asia, Chungli, Taiwan, Republic of China, Abstract, p. 132–133.

Richard, M., Bellon, H., Maury, R., Barrier, E., and Juang, W.-S., 1986, Miocene to recent calc-alkalic volcanism in eastern Taiwan: K-Ar ages and petrography: Tectonophysics, v. 125, p. 87–102.

Schnürle, P., Liu, C.-S., Lallemand, S.E., and Reed, D.L., 1998, Structural controls of the Taitung Canyon in the Huatung Basin east of Taiwan: TAO, Terrestrial, Atmospheric, and Oceanic Sciences, v. 9, p. 453–472.

Schnürle, P., Liu, C.-S., Lallemand, S.E., and Reed, D.L., 1998, Structural insight into the south Ryukyu margin: Effects of the subducting Gagua Ridge: Tectonophysics, v. 288, p. 237–250.

Seno, T., 1977, The instantaneous rotation vector of the Philippine Sea plate relative to the Eurasian Plate: Tectonophysics, v. 42, p. 209–225.

Seno, T., Stein, S., and Grip, A.E., 1993, A model for the motion of the Philippine Sea plate consistent with NUVEL-1 and geologic data: Journal of Geophysical Research, v. 98, p. 17941–17948.

Sibuet, J.-C., Deffontaines, B., Hsu, S.-K., Thareau, N., Le Formal, J.-P., Liu, C.-S., and the ACT party, 1999, Okinawa trough backarc basin: Early tectonic and magmatic evolution: Journal of Geophysical Research, v. 103, p. 30245–30267.

Sibuet, J.-C., and Hsu, S.-K., 1997, Geodynamics of the Taiwan arc-arc collision: Tectonophysics, v. 274, p. 221–251.

Suppe, J., 1981, Mechanics of mountain building and metamorphism in Taiwan: Memoir of the Geological Society of China, v. 4, p. 67–89.

Suppe, J., 1980, A retrodeformable cross section of northern Taiwan: Proceedings of the Geological Society of China, no. 23, p. 46–55.

Suppe, J., 1987, The active Taiwan mountain belt, *in* Schaer, J.P., and Rogers, J., eds., The anatomy of mountain ranges: Princeton, New Jersey, Princeton University Press, p. 277–293.

Suppe, J., 1988, Tectonics of arc-continent collision on both sides of the South China Sea: Taiwan and Mindoro: Acta Geologica Taiwanica, Science reports of the National Taiwan University, no. 26, p. 1–18.

Tillman, K.S., and Byrne, T.B., 1995, Kinematic analysis of the Taiwan Slate Belt: Tectonics, v. 14, p. 322–341.

Teng, L.S., Wang, Y., Tang, C.H., Huang, C.Y., Huang, T.C., Yu, M.S., and Ke, A., 1991, Tectonic aspects of the Paleogene depositional basin of northern Taiwan: Proceedings of the Geological Society of China, v. 34, p. 313–336.

Wang, T.-K., McIntosh, K., Nakamura, Y. and Liu, C.-S., 1996, OBS refraction survey and imaging offshore eastern Taiwan: EOS (Transactions, American Geophysical Union), v. 77, p. F720.

Wessel, P., and Smith, W.H.F., 1995. New version of the generic mapping tools released: EOS (Transactions, American Geophysical Union), v. 76, p. 329.

Wu, F., 1978, Recent tectonics in Taiwan: Journal of Physics of the Earth, v. 26 (suppl.), p. 265–299.

Wu, F., Rau, R.-J. and Salzberg, D., 1997, Taiwan orogeny: Thin-skinned or lithospheric collision?: Tectonophysics, v. 274, p. 191–220.

Yang, T.F., Lee, T., Chen, C.-H., Cheng, S.-N., Knittel, U., Punongbayan, R.S., and Rasdas, A.R., 1996, A double island arc between Taiwan and Luzon: Consequence of ridge subduction: Tectonophysics, v. 258, p. 85–101.

Yen, T.P., 1973, Plate tectonics in the Taiwan region: Proceedings of the Geological Society of China, no. 16, p. 7–21.

Yu, S.-B., Chen, H.-Y., and Kuo, L.-C., 1997, Velocity field of GPS stations in the Taiwan area: Tectonophysics, v. 274, p. 41–59.

Manuscript Accepted by the Society April 12, 2001

Printed in the U.S.A.